HEALTH AND ENVIRONMENT IN AQUACULTURE

Edited by **Edmir Daniel Carvalho, Gianmarco Silva David** and **Reinaldo José da Silva**

INTECHOPEN.COM

Health and Environment in Aquaculture
Edited by Edmir Daniel Carvalho, Gianmarco Silva David and Reinaldo José da Silva

CBS, Edition **2017**

Published by InTech
Janeza Trdine 9, 51000 Rijeka, Croatia

Publishing Process Manager Molly Kaliman

Technical Editor InTech DTP team

Cover Designer InTech Design Team

Additional hard copies can be obtained from orders@intechopen.com

Health and Environment in Aquaculture,
Edited by Edmir Daniel Carvalho, Gianmarco Silva David and Reinaldo José da Silva
p. cm.
ISBN 978-953-51-0497-1

INTECH

open science | open minds

Contents

Contents

Contents

Preface

Aquaculture is a modality of food production that has been experiencing continuous expansion in many countries worldwide. This expansion brings the challenge of developing reliable tools for disease control, to assure high productivity of healthy seafood. The increase of farmed fish production raises the issue of achieving a sustainable and environmental friendly aquaculture. The adoption of best management practices in the whole production chain, based on "state of the art" scientific knowledge, is the key for sustainable health management. In this book, experts from several countries bring updated information about some of the main health issues that currently affects aquaculture. Topics concerning pathogens, antibiotics, probiotics, cell biology, ecological interactions, and safety are included in the six sections of this book.

The first section is entitled as "Parasitic diseases", addressing issues and impacts of parasites upon aquaculture. The first chapter is the "Transmission biology of the myxozoa", which explains about the diseases that some myxozoans cause in marine and freshwater fish, and how they can be a problem for aquaculture and fishery industries. It also elucidates the life cycle of myxozoans, that involves invertebrates, and a vertebrate host that is typically a fish. However, there are no commercially available chemotherapeutants and vaccines to treat myxozoan infections. This review summarizes the current knowledge on the transmission biology of myxozoans, which would be useful for designing management strategies for related diseases. The second chapter is "Metazoan parasites of cultured European sea bass *Dicentrarchus labrax* (Linneaus 1758) from Corsica". It is a study relating that parasitic infections and associated diseases have emerged in aquaculture systems in many regions of Europe, resulting in significant economical losses. This study points out that wild fish are believed to be the primary reservoirs of parasite infection for fish farmed in cages, and environmental conditions in culture systems may favor disease transmission, threatening production activity. In this sense, it is considered that animals reared in sea-cages are exposed to a large number of parasitic agents. The third chapter is "Parasitic diseases in cultured marine fish in Northwest Mexico". This chapter summarizes the main parasitic diseases that affect marine fish species with aquaculture potential in the Norwest Pacific coast of Mexico, emphasising proper strategies for their control. The study shows the need to perform parasite treatment and control applying prophylactic and therapeutic measures.

In the section "Bacterial diseases", the fourth chapter "Updated information of *Aeromonas* infections and furunculosis derived from molecular methods" focuses on the bacteria *Aeromonas salmonicida,* the causal agent of furunculosis, considered a particularly important fish pathogen mainly due to its widespread distribution and ability to infect a diverse range of hosts, causing massive mortalities and economic losses. Additionally, climate change has been considered to play a role in the appearance and impact of furunculosis. The study undertakes molecular techniques in *Aeromonas* infections in fish, including significant advances in genomics and taxonomy of these microorganisms. The fifth chapter is "An overview on virulence-associated factors of Gram-negative fish pathogenic bacteria", which addresses the issue of bacterial outbreaks causing important economic losses for aquaculture. Gram-negative bacteria have long been recognized as a cause of the most prevalent fish pathologies in the aquaculture industry. The application of *in vivo* and *in vitro* molecular techniques to fish pathogenic bacteria resulted in the characterization of novel virulence determinants and allowed to increase the knowledge of bacterial pathogenic mechanisms. This review deals with representative species of gram-negative fish pathogenic bacteria in the context of the analysis of well-established virulent factors produced by these pathogens.

In the section "Antibiotics and probiotics", the sixth chapter is "Antibiotics in aquaculture: use, abuse and alternatives". This study argues that unpredictable mortalities in aquaculture production may be due to negative interactions between fish and pathogenic bacteria. To solve this problem, farmers frequently use antibiotic compounds to treat bacterial diseases. The concerns about the increase in bacterial resistance and antibiotic residues have aroused great caution in the use of antibiotics in aquaculture, which has encouraged research to obtain alternatives. The aim of this chapter is to provide information about the current knowledge in antibiotic use in aquaculture systems, including information about mechanisms of action and resistance. The seventh chapter is "The use of antibiotics in shrimp farming". This is an important study, considering that shrimp cultivation has been the most expanding aquaculture activity. Nevertheless, this industry faces major problems with viral and bacterial diseases, and large quantities of chemical and antibiotic products are frequently used to counteract this. The study demonstrates the importance of applying appropriate therapies with antibiotics, seeking greater effectiveness for the control of bacterial infections. The eighth and ninth chapters, within this section, deal with probiotics in aquaculture, which has been considered a key factor for fish health management, due to the increasing demand for environment friendly aquaculture. The eighth chapter is "Probiotics in aquaculture: benefits to the health, technological applications and safety". This study points out that, currently, a number of preparations of probiotics are commercially available and have been introduced to fish, shrimp and molluscan farming as feed additives. Thus, there is a commercial and academic interest of increasing our knowledge in effective preparation, technological applications, and safety evaluation of probiotics. The ninth chapter is "Probiotics in aquaculture of Kuwait: current state and prospect", and mentions the application of autochthonous probiotics. In this experimental study, a protocol for the isolation, screening and selection of candidate probiotic bacteria based on several selective criteria was accomplished. This study showed that the methods were suitable to certain extent to assess the antagonism ability of probiotic bacteria on pathogenic bacteria, and these findings can be applied to other cultured fish.

In the section entitled as "Applied topics of cellular and molecular biology", the tenth chapter is the "Use of microarray technology to improve DNA vaccines in fish aquaculture: the rhabdoviral model". Rhabdovirosis are one of the most important diseases affecting farmed fish worldwide, and are amongst the few fish diseases for which there is an efficacious DNA vaccine. Understanding the induced molecular events occurring after fish immunization with rhabdoviruses and their DNA vaccines might contribute to improve vaccines to other fish pathogens. This study focus on data published on the use of microarrays for the identification of rhabdoviral-induced genes, with properties that make them candidate adjuvants for the improvement of fish DNA vaccines. The eleventh chapter is "Fighting virus and parasites with fish cytotoxic cells", which is a review on the fish cell-mediated cytotoxic activity as the main cellular immune mechanism against tumors, parasites and viral infections. It also addresses the modulation of this activity by means of immunostimulants, stress, pollution, and vaccines. This research contributes to understand fish cytotoxic cells and their activity from an evolutionary point of view. Furthermore, the lack of commercial antiviral and anti-parasitic vaccines for fish makes necessary to increase the knowledge on the cell-mediated cytoxic activity of fish. The twelfth chapter is "Bacteriocins of aquatic microorganisms and their potential applications in the seafood industry". Narrow killing spectrum bacteriocins are recognized as a promising alternative to broad-spectrum antibiotics, whose efficacy has been compromised by the evolution of resistant bacteria. This study aims to provide an overview of the diversity of bacteriocins produced by marine microorganisms, their role in mediating microbial interactions in the marine environment, and their potential applications in the seafood industry. The thirteenth chapter is "Molecular characterization of pathological bone development in Atlantic salmon (*Salmo salar*)". This study argues that spinal disorders are a recurrent problem for aquaculture, and until recently, their molecular development in fish has received relatively little attention. In this review, the current knowledge on the cellular and molecular mechanisms for skeletal homeostasis and aberrant development of bone in the Atlantic salmon vertebrae is referred.

In the section "Ecological impacts of fish farming", the fourteenth chapter is "Ecological features of large Neotropical reservoirs related to health of cage reared fish". This study raises the subject of fish cage culture in hydroelectric reservoirs in Brazil. Wild native fish species and a farmed fish species, *Oreochromis niloticus*, were searched for ectoparasites, which showed that the cultured fish presented high rates of parasitic infection. This research attempted to identify interferences of fish cage farming upon water quality, wild fish assemblages and parasitic diseases in large freshwater reservoirs. The fifteenth chapter is "Spread of pathogens from marine cage aquaculture: a potential threat for wild fish assemblages under protection regimes?" focusing on the exchange of viruses between farmed and wild populations, and further, the potential impact on natural ecosystems. The study reviews the effects of a serious disease, Viral Nervous Necrosis (VNN), which affects more than 40 fish species worldwide. Likewise, betanodaviruses are the most important viral pathogens reported in marine aquaculture within the Mediterranean region.

The last section is "Work-related hazards: prevention and mitigation", with the sixteenth chapter: "Aquacultural Safety and Health" showing that occupational hazards in aquaculture are associated with different rearing technologies. Farm operators are encouraged to adopt or develop inherently safety technologies by first eliminating, then guarding against, and finally warning about the hazard. A model safety manual presents contents that can be adapted to aquaculture.

The challenge of editing this book could only be accomplished with the help of some colleagues. Therefore, we would like to thank Professor Dr. Fernanda Natália Antoneli, from Federal University of Mossoró (RN, Brazil), who has assisted us with her background on cell biology; Dra. Fabiana Garcia Scaloppi, from Sao Paulo State Agency of Agribusiness Technology (APTA at Votuporanga, SP, Brazil) who has collaborated with her expertise on parasitology; Professor Dra. Mara Renata Dega, from Marechal Rondon Faculty (at Sao Manuel, SP, Brazil) who has helped with pharmacology themes. Finally, I would like to give especial thanks to the biologist Aline Angelina Acosta, a graduate student in Zoology, who has collaborated throughout the edition process with her English skills.

Dr. Edmir Daniel Carvalho
Dr. Reinaldo José da Silva
Dr. Gianmarco Silva David
Sao Paulo State University
Brazil

Part 1

Parasitic Diseases

1

Transmission Biology of the Myxozoa

Hiroshi Yokoyama[1], Daniel Grabner[2] and Sho Shirakashi[3]
[1]*The University of Tokyo*
[2]*University of Duisburg-Essen*
[3]*Kinki University*
[1,3]*Japan*
[2]*Germany*

1. Introduction

Myxozoans are spore-forming parasites of both freshwater and marine fishes (Lom & Dyková, 1992, Kent et al., 2001; Feist & Longshaw, 2006). The Myxozoa were previously classified as protozoans, although the multicellular state and functional specialization of the cells composing spores were considered to exceed protozoan level (Lom & Dyková, 1992). Indeed, molecular studies demonstrated that myxozoans are metazoans (Smothers et al., 1994, Siddal et al., 1995). However, there were two conflicting views concerning the phylogenetic origin of myxozoans; the Bilateria (Smothers et al., 1994, Schlegel et al., 1996, Anderson et al., 1998, Okamura et al., 2002) vs. the Cnidaria (Siddal et al., 1995). More recently, the Cnidaria-hypothesis has been strongly supported by phylogenetic analyses of protein-coding genes of myxozoans (Jimenez-Guri et al., 2007, Holland et al., 2010). The phylum Myxozoa, of which more than 2100 species in 58 genera are described to date, is divided into two classes, Myxosporea and Malacosporea (Lom & Dyková, 2006). Most of myxozoans are not harmful to host fish, however, some species cause diseases in cultured and wild fish which are problems for aquaculture and fishery industries worldwide. Generally, freshwater myxosporeans appear to be specific at the family or the genus level of the host, while some marine myxosporeans have a low host-specificity. Some examples are mentioned below.

For freshwater species, myxozoans infecting salmonids have been relatively well studied. For example *Myxobolus cerebralis*, the causative agent of whirling disease, *Tetracapsuloides bryosalmonae*, the cause of proliferative kidney disease (= PKD), and *Ceratomyxa shasta*, causing ceratomyxosis, have fatal effects on farmed salmonid fish (Table 1). Salmonid ceratomyxosis is a local disease which is restricted only to North America (Bartholomew et al., 1997), while whirling disease and PKD are widely distributed in the world (Hedrick et al., 1993, 1998). *M. cerebralis* infects cartilage tissue and causes a whirling behaviour (tail-chasing swimming), a black tail, and skeletal deformities of affected fish. Whirling disease was previously known as a hatchery disease, but recently, it has been recognized as one of the causes for the decline of natural rainbow trout populations in several western states of the USA (Hedrick et al, 1998). Symptoms of PKD in salmonid fish are a swollen kidney (Fig. 1A) and anemic gills, evoked by chronic inflammation of the kidney interstitium. The

Myxozoans	Disease names or typical signs	Fish	References
Ceratomyxa shasta	Ceratomyxosis	Salmonids	Bartholomew et al. (1997)
Chloromyxum truttae	Hypertrophy of gall bllader	Salmonids	Lom & Dyková (1992)
Henneguya ictaluri	Proliferative gill disease (PGD)	*Ictalurus punctatus*	Pote et al. (2000)
Henneguya salminicola	Milky condition	Salmonids	Awakura & Kimura (1977)
Hoferellus carassii	Kidney enlargement disease (KED)	*Carassisus auratus*	Yokoyama et al. (1990)
Myxidium giardi	Systemic infection	*Anguilla* spp.	Ventura & Paperna (1984)
Myxobolus artus	Muscular myxobolosis	*Cyprinus carpio*	Yokoyama et al. (1996)
Myxobolus cerebralis	Whirling disease	Salmonids	Hedrick et al. (1998)
Myxobolus cyprini	Malignant anemia	*Cyprinus carpio*	Molnár & Kovács-Gayer (1985)
Myxobolus koi	Gill myxobolosis	*Cyprinus carpio*	Yokoyama et al. (1997a)
Myxobolus murakamii	Myxosporean sleeping disease	*Oncorhynchus masou*	Urawa et al. (2009)
Myxobolus wulii	Cysts in gill or hepatopancreas	*Carassius auratus*	Zhang et al. (2010b)
Parvicapsula pseudobranchicola	Inflammation and necrosis of filaments	*Salmo salar*	Karlsbakk et al. (2002)
Sphaerospora dykovae	Swimbladder inflammation (SBI)	Cyprinids	Dyková & Lom (1988)
Tetracapsuloides bryosalmonae	Proliferative kidney disease (PKD)	Salmonids	Hedrick et al. (1993)
Thelohanellus hovorkai	Hemorrhagic thelohanellosis	*Cyprinus carpio*	Yokoyama et al. (1998)

Table 1. Economically important freshwater myxosporeans.

causative agent of PKD has not been identified for a long time, and thus the organism was previously called PKX (Hedrick et al., 1993). It was assigned to the Myxozoa in 1999 and initially called *Tetracapsula bryosalmonae* (Canning et al., 1999). Canning et al. (2000) erected the new class Malacosporea in the Myxozoa, and later, in the course of nomenclature changes by Canning *et al.* (2002) *Tetracapsula bryosalmonae* was renamed to *Tetracapsuloides bryosalmonae* (Fig. 1B). Salmonids suffering from ceratomyxosis show abdominal distension and exophthamia, possibly caused by osmotic imbalance due to *C. shasta* infection in the internal organs (Bartholomew et al., 1997). *Henneguya salminicola* produces cysts in the musculature of anadromous salmonid fish (Fig. 1C, D). This parasite does not cause a health

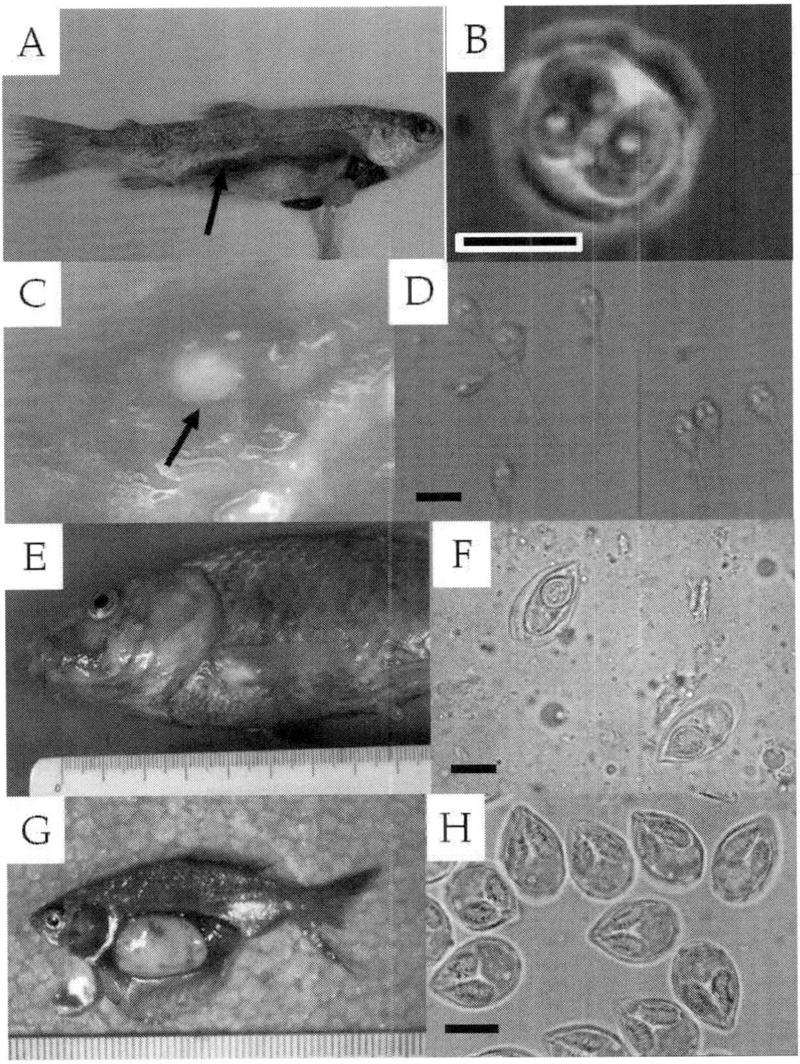

A & B: Proliferative kidney disease of rainbow trout (*Oncorhynchus mykiss*). Note the swollen kidney (arrow). Malacospore of *Tetracapsuloides bryosalmonae* from bryozoans host (B). C & D: Milky condition of pink salmon (*Oncorhynchus gorbuscha*). White exudate (arrow) filled with spores of *Henneguya salminicola* (D). Photos of courtesy by Dr. T. Awakura. E & F: Hemorrhagic thelohanellosis of common carp (*Cyprinus carpio*). Note extensive haemorrhages in mouth and abdomen caused by *Thelohanellus hovorkai* (F) in the subcutaneous tissue. G & H: Creamy appearance of enlarged hepatopancreas of goldfish (*Carassius auratus*) infected with *Myxobolus wulii* (H). Scale bars for B, D, F and H are 10μm.

Fig. 1. Myxozoan diseases of freshwater fish and the causative myxozoan parasites.

problem of the host, but renders the infected fish unmarketable due to the milky condition of the flesh (Awakura & Kimura, 1977). Myxosporean sleeping disease is caused by *Myxobolus murakamii* infecting the peripheral nerve of masu salmon (*Oncorhynchus masou*). This disease has been known only in Hiroshima Prefecture, in south-western Japan, although *M. murakamii* occurs also in Hokkaido, the northernmost area of Japan. It remains to be clarified why the sleeping disease does not occur in Hokkaido (Urawa et al., 2009). *Chloromyxum truttae* infects the gallbladder of brood stock of rainbow trout (*Oncorhynchus mykiss*), while it infects the yearlings of Atlantic salmon (*Salmo salar*). Affected fish showed loss of appetite, yellow colouration of body, and hypertrophic gall bladder (Lom & Dyková, 1992). Pseudobranch infection with *Parvicapsula pseudobranchicola* has been reported in Atlantic salmon in Norway, showing lethargy, disorganized swimming, exophthalmia and low-grade to significant mortalities (Karlsbakk et al., 2002). Affected fish exhibited eye bleeding and cataracts, possibly due to obstruction of the blood supply to the choroid bodies of the eyes.

Myxobolus koi, *Thelohanellus hovorkai*, and *Sphaerospora dykovae* (= *S. renicola*) are well-known pathogens in cultured common carp (*Cyprinus carpio*) in Europe and Asia (Dyková & Lom, 1988, Yokoyama et al., 1997a, 1998). *M. koi* infects the gills and causes a respiratory disfunction of carp juveniles. Yokoyama et al. (1997a) reported that there are two types of *M. koi* infections; the one forms large-type (pathogenic) cysts in the gill filaments, while the other forms small-type (non-pathogenic) cysts in the gill lamellae. *T. hovorkai* infecting the connective tissue is the causative agent of the hemorrhagic thelohanellosis of common carp (Yokoyama et al., 1998). Spore dispersion of *T. hovorkai* in subcutaneous connective tissue causes extensive hemorrhages and edema, finally resulting in death of affected fish (Fig. 1E, F). *S. dykovae*, the cause of swimbladder inflammation (SBI) was previously known as *S. renicola*, but has recently been renamed as *S. dykovae* in association with revised taxonomy of the genus *Leptotheca* (Gunter & Adlard, 2010). The target organ (spore forming site) for *S. dykovae* is the kidney, but the extrasporogonic stage of *S. dykovae* proliferates in the swimmbladder, which causes SBI of carp (Dyková & Lom, 1988). *Myxobolus artus* produced rice bean-like cysts in the musculature of common carp. Adult carp (over 1-year old) do not die of the disease but lose their commercial value. In contrast, juvenile carp (0-year old) heavily infected with *M. artus* exhibit hemorrhagic anemia and increased mortality rate. After degeneration of *M. artus* cysts in the musculature, spores engulfed by macrophages are transferred into gills, where numerous spores accumulate and pack within the lamellae. As a result, the gill epithelia are exfoliated, causing the hemorrhagic anemia (Yokoyama et al., 1996). *Myxobolus cyprini* infecting the skeletal muscle of common carp was also reported to cause the malignant anemia (Molnár & Kovács-Gayer, 1985), but it is unknown whether the disease mechanisms are the same as *M. artus*. *Thelohanellus kitauei* forms large cysts in the intestinal mucosa of common carp so that the intestine was occluded to emaciate the infected fish.

Hoferellus carassii infecting the kidney of goldfish (*Carassius auratus*) is the causative agent of kidney enlargement disease (KED). This parasite does not cause a high mortality of affected fish, but a low marketability as an ornamental fish (Yokoyama et al., 1990). *Myxobolus wulii* forms numerous cysts in the gills of goldfish in some cases, whereas large cysts are formed in the hepatopancreas in other cases (Fig. 1G, H). In both cases, infection of fish results in high mortality (Zhang et al., 2010b). Gill infections with *Henneguya ictaluri* and *H. exilis* are typical myxosporean diseases in catfish culture. *H. ictaluri* causes proliferative gill disease of catfish (*Ictalurus punctatus*) (Pote et al. 2000). *Myxidium giardi* infects multiple organs including gills and kidney of several eel species, *Anguilla anguilla, A. rostorata,* and *A. japonica*. Infected elvers exhibit dropsy, ascites, and swollen kidney (Ventura & Paperna, 1984).

Compared to freshwater myxosporeans, many marine species have a broad host range, such as *Kudoa thyrsites, K. yasunagai* and *Enteromyxum leei* (Table 2). *K. thyrsites* lowers the

Myxozoans	Disease names or typical signs	Fish	References
Enteromyxum leei	Enteromyxosis or myxosporean emaciation disease	*Diplodus puntazzo, Sparus aurata, Paralichthys olivaceus, Pagrus major, Takifugu rubripes*	Diamant (1997) Yasuda et al. (2002)
Enteromyxum scophthalmi	Enteromyxosis		Palenzuela et al. (2002)
Henneguya lateolabracis	Cardiac henneguyosis	*Lateolabrax* sp.	Yokoyama et al. (2003)
Henneguya pagri	Cardiac henneguyosis	*Pagrus major*	Yokoyama et al. (2005a)
Kudoa amamiensis	Kudoosis amami	*Seriola quinqueradiata*	Yokoyama et al. (2000)
Kudoa iwatai	Cysts in multiple organs	*Dicentrarchus labrax, Lateolabrax japonicus, Mugil cephalus, Sparus aurata, Pagrus major, Oplegnatus punctatus*	Diamant et al. (2005)
Kudoa lateolabracis	Post-mortem myoliquefaction	*Lateolabrax* sp., *Paralichthys olivaceus*	Yokoyama et al. (2004)
Kudoa lutjanus	Systemic infection	*Lutjanus erythropterus*	Wang et al. (2005)
Kudoa neurophila	Meningoencephalomyelitis	*Latris lineata*	Grossel et al. (2003)
Kudoa shiomitsui	Cysts in the heart	*Takifugu rubripes, Thunnus orientalis*	Zhang et al. (2010)
Kudoa thyrsites	Post-mortem myoliquefaction	*Salmo salar, Paralichtys olivaceus, Coryphaena hyppurus*	Moran et al. (1999a)
Kudoa yasunagai	Abnormal swimming	*Lateolabrax japonicus, Oplegnathus fasciatus, Seriola quinqueradiata, Takifugu rubripes, Thunnus orientalis, Plotosus lineatus*	Zhang et al. (2010a)
Myxobolus acanthogobii	Myxosporean scoliosis or skeletal deformity	*Seriola quinqueradiata, Scomber japonicus*	Yokoyama et al. (2005b)
Sphaerospora epinepheli	Disorientation, hemorrhage	*Epinephelus malabaricus*	Supamattaya et al. (1991)
Sphaerospora fugu (= *Leptotheca fugu*)	Myxosporean emaciation disease	*Takifugu rubripes*	Tin Tun et al. (2000)

Table 2. Economically important marine myxosporeans (see also Fig. 2).

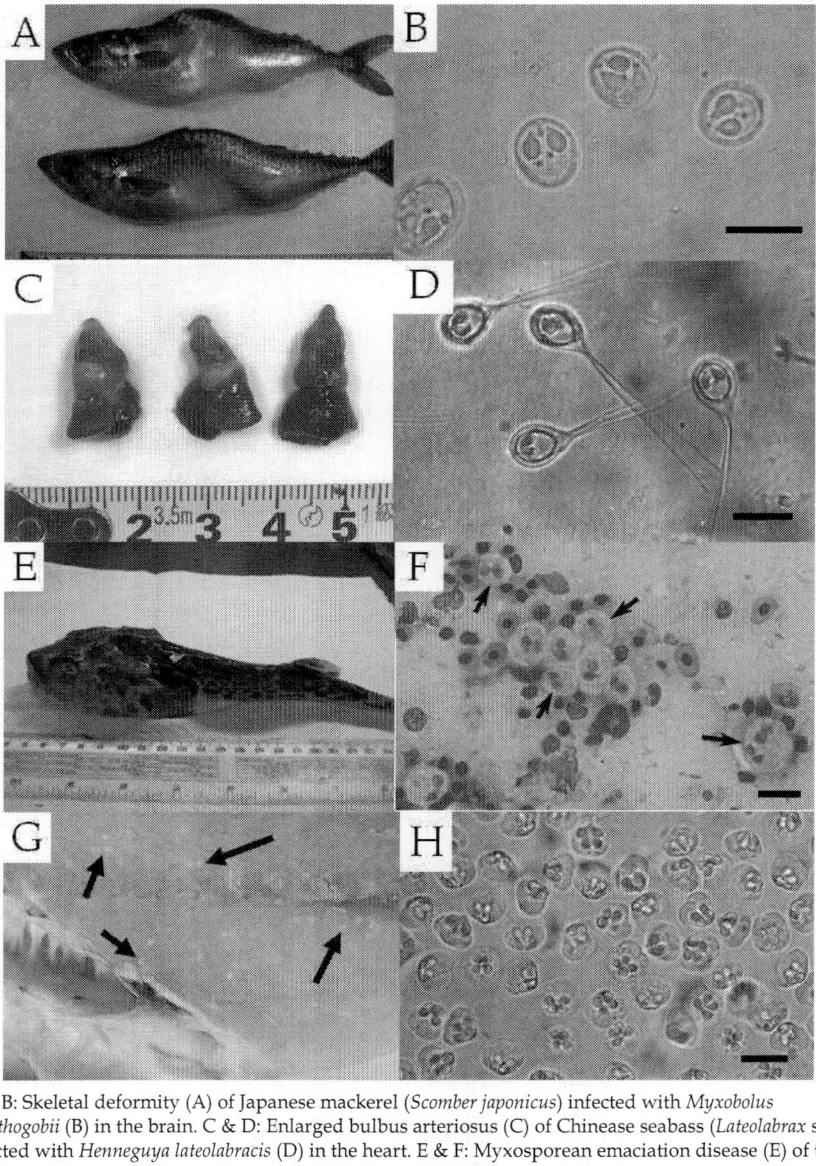

A & B: Skeletal deformity (A) of Japanese mackerel (*Scomber japonicus*) infected with *Myxobolus acanthogobii* (B) in the brain. C & D: Enlarged bulbus arteriosus (C) of Chinease seabass (*Lateolabrax* sp.) infected with *Henneguya lateolabracis* (D) in the heart. E & F: Myxosporean emaciation disease (E) of tiger puffer (*Takifugu rubripes*) infected with developmental stages (arrows) of *Enteromyxum leei* (F) in the intestine. Diff-Quik stain (F). G & H: Cysts (arrows) in the skeletal muscle (G) of red sea bream (*Pagrus major*). Cysts are packed with spores of *Kudoa iwatai* (H). Scale bars for B, D, F and H are 10 µm.

Fig. 2. Myxosporean diseases of marine fish and the causative myxozoan parasites.

commercial value of various cultured marine fish species, particularly Atlantic salmon (*Salmo salar*) in North America, by causing post-mortem myoliquefaction (Moran et al., 1999a). *K. yasunagai* forms numerous cysts in the brain, probably causing disorder of swimming performance of many fish species (Zhang et al., 2010a). Recently, enteromyxosis or myxosporean emaciation disease, caused by *E. leei*, has emerged as a new threat in various cultured marine fish, e.g. gilthead sea bream (*Sparus aurata*) in Mediterranean countries and tiger puffer (*Takifugu rubripes*) in Japan (Diamant, 1997, Yasuda et al., 2002). In contrast, *Enteromyxum scophthalmi* and *Sphaerospora fugu* (= *Leptotheca fugu*) have been found only in the intestine of turbot (*Psetta maxima*) and tiger puffer (*Takifugu rubripes*), respectively, although the signs of the disease appear to be similar to *E. leei* infection (Tin Tun et al., 2000, Palenzuela et al., 2002). Heart infections have been documented such as *Henneguya lateolabracis, H. pagri,* and *Kudoa shiomitsui*. The former two species are highly pathogenic to Chinese sea bass (*Lateolabrax* sp.) and red sea bream (*Pagrus major*), respectively (Yokoyama et al., 2003, 2005a), whereas the pathogenic effects of *K. shiomitsui* are not clear (Zhang et al., 2010a). Many *Kudoa* infections in skeletal muscle may render the infected fish unmarketable by producing cysts (e.g., *K. amamiensis* and *K. iwatai*) or causing myoliquefaction (e.g., *K. lateolabracis* and *K. neothunni*). *K. neurophila* has become an impediment to the juvenile production of striped trumpeter (*Latris lineata*) in Tasmania, due to meningoencephalomyelitis of hatched larvae (Grossel et al., 2003). *Myxobolus acanthogobii* infects the brain and causes the myxosporean scoliosis in yellowtail (*Seriola quinqueradiata*), while infected Japanese mackerel (*Scomber japonicus*) exhibits the lordosis (dorso-ventral deformity) and infected goby (*Acanthogobius flavimanus*) is subclinical (Yokoyama et al., 2005b). *Sphaerospora epinepheli* infects the kidney of *Epinephelus malabaricus*, which shows disorientation of the body and hemorrhages (Supamattaya et al., 1991).

2. Myxosporeans

The class Myxosporea is comprised of the two orders, Bivalvulida and Multivalvulida. Bivalvulids include 52 genera with more than 2100 species described from freshwater and marine fishes, while multivalvulids contain 5 genera with more than 60 species predominantly from marine fish (Lom & Dyková, 2006). Morphology, life cycle, phylogeny, and biology of myxosporeans are summarized below.

2.1 Morphology of myxosporean

Myxosporean spores are composed of shell valves, sporoplasms, and polar capsules containing coiled polar filaments (Fig. 3). Number of valves and polar capsules, arrangement of the polar capsules, and ornamentation of spores allow the genus-level diagnosis of myxosporeans. Identification at the species-level is based on spore dimensions. Species description of myxospores should follow the guidelines of Lom & Arthur (1989). For bivalvulids, spore length and spore width in frontal view, spore thickness in side view, length and width of polar capsules are measured (Fig. 3). If ornamentations such as the caudal appendages for *Henneguya* are present, the length is also measured. For multivalvulids, spore length (including the apical projections, if present) in side view, spore width and spore thickness in top view, length and width of polar capsules are determined. Care must be taken to avoid confusion of thickness and width of spores, because multivalvulids are radially symmetrical.

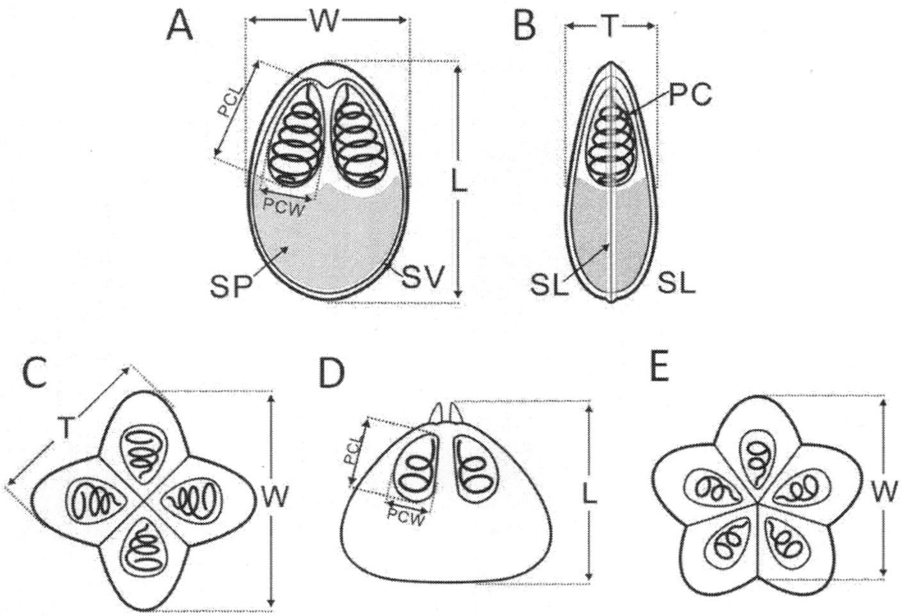

PC: polar capsule, SP: sporoplasm, SV: shell valve, SL: sutural line, L: spore length, W: spore width, T: spore thickness, PCL: polar capsule length, PCW: polar capsule width.

Fig. 3. Diagrams of bivalvulid (A: frontal view, B: side view) and multivalvulid (C & E, top view, D: side view) myxosporean spores.

2.2 Life cycle of myxosporeans

The first myxozoan life cycle was discovered for *M. cerebralis* by Wolf & Markiw in 1984 and was later confirmed by many other researchers, who reported similar life cycles for more than 30 myxosporean species. These life cycles involve an annelid invertebrate (mainly oligochaetes for freshwater species and polychaetes for marine species) and a vertebrate host which is typically a fish (Fig. 4). In the latter, myxosporean spore stages (= myxospores) develop. Myxospores are ingested by annelids, in which the polar filaments extrude to anchor the spore to the gut epithelium. Opening of the shell valves allows the sporoplasms to penetrate into the epithelium. Subsequently, the parasite undergoes reproduction and development in the gut tissue, and finally produces usually eight actinosporean spore stages (= actinospores) within a pansporocyst. After mature actinospores are released from their hosts they float in the water column (El-Matbouli & Hoffmann, 1998). Upon contact with skin or gills of fish, sporoplasms penetrate through the epithelium, followed by development of the myxosporean stage. Myxosporean trophozoites are characterized by cell-in-cell state, where the daughter (secondary) cells develop in the mother (primary) cells. The presporogonic stages multiply, migrate via nervous or circulatory systems, and develop into sporogonic stages. At the final site of infection, they produce mature spores within mono- or disporic pseudoplasmodia, or polysporic plasmodia (El-Matobouli & Hoffmann, 1995).

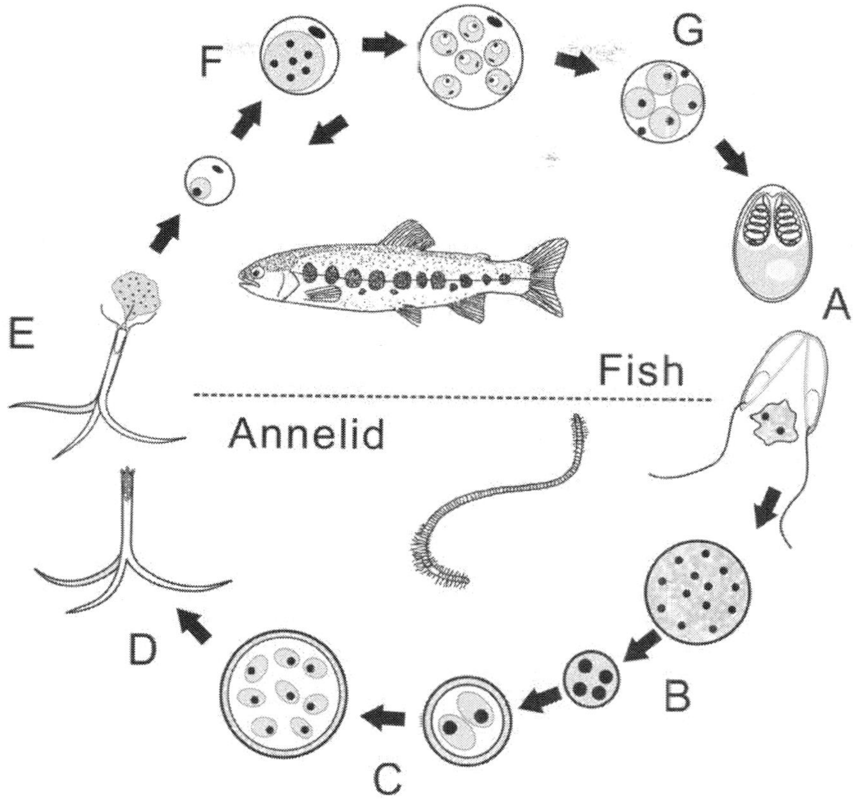

A: The polar filaments are extruded to anchor the spore to the gut epithelium, followed by opening of shell valves of myxospore. B: Gametogony. C: Sporogony of actinosporean phase. D: Mature actinospore stages develop in a pansporocyst, and actinospores are released into the water. E: Upon contact of actinospores with the skin or gills of the fish host, polar filaments extrude to anchor the spore to the skin or gills, facilitating invasion of the sporoplasms into the fish. F: Presporogonic multiplication in a cell-in-cell state. G: Sporogony of myxosporean phase.

Fig. 4. Diagram of the life cycle of myxosporean alternating fish and annelid hosts.

2.3 Morphology of actinospores

Actinospores that are formed in the invertebrate hosts have a triradiate form with exclusively 3 polar capsules and mostly 3 caudal processes (Figs. 5 & 6). To characterize actinosporean stages, researchers should follow the guidelines of Lom et al. (1997); shape of the caudal processes (straight, curved or branched), presence of the style (small stalk below the spore body) and formation of spore nets (pattern of connection between several spores), number of daughter cells in the spore body, and measurements of the spore body, style, polar capsules and processes (Fig. 6).

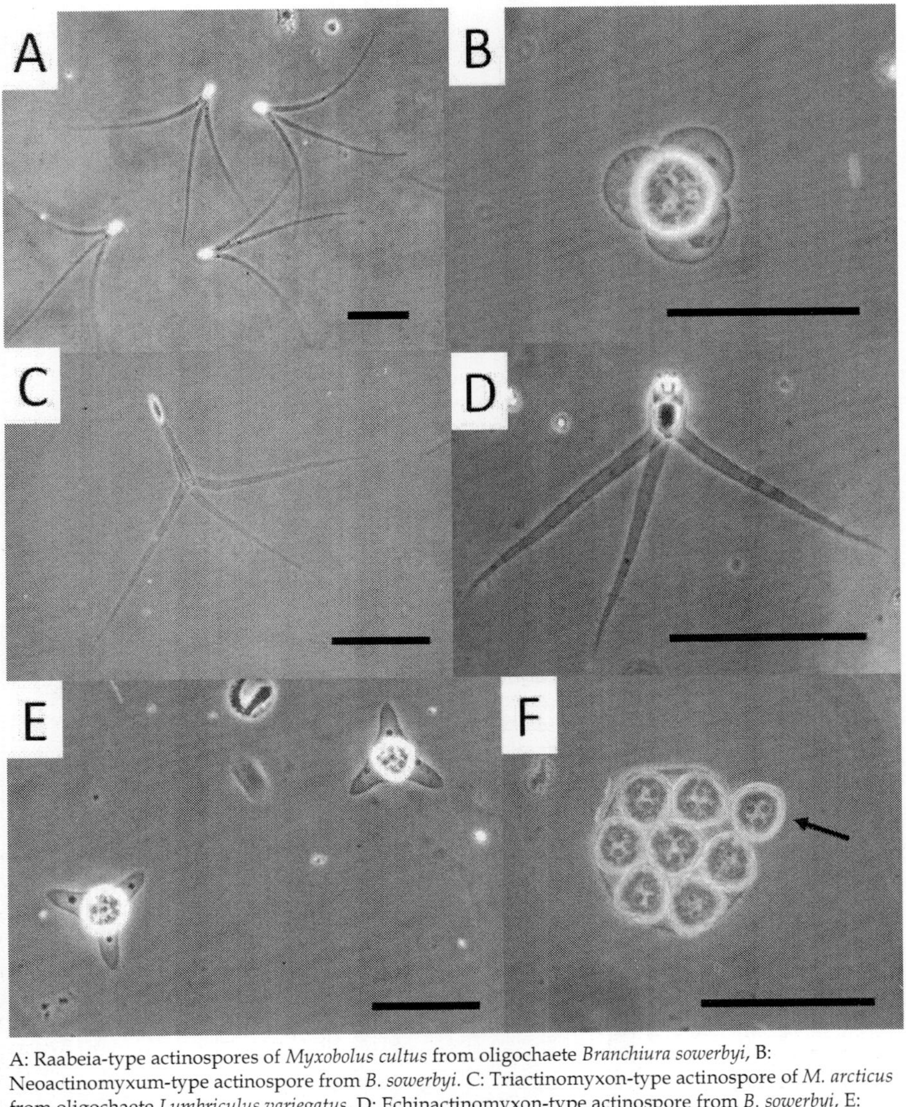

A: Raabeia-type actinospores of *Myxobolus cultus* from oligochaete *Branchiura sowerbyi*, B: Neoactinomyxum-type actinospore from *B. sowerbyi*. C: Triactinomyxon-type actinospore of *M. arcticus* from oligochaete *Lumbriculus variegatus*, D: Echinactinomyxon-type actinospore from *B. sowerbyi*, E: Aurantiactinomyxon-type actinospore of *Thelohanellus hovorkai* from *B. sowerbyi*, F: Sphaeractinomyxon-type actinospores from unidentified marine oligochaete, which was collected in May 1990, on the coast of Mie Prefecture, the middle part of Japan. Arrow shows an actinospore released from a pansporocyst which develops 8 actinospores. Scale bars for A, C and D are 100 µm, and those for B, E and F are 50 µm.

Fig. 5. Several morphotypes of actinosporean spores.

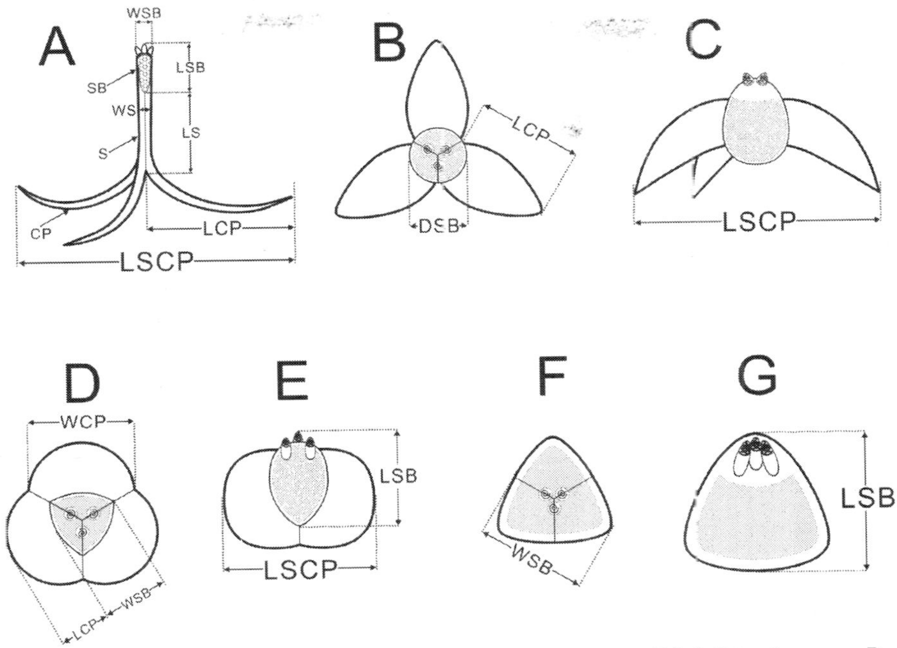

A: Triactinomyxon, B & C: Aurantiactinomyxon, D & E: Neoactinomyxum, F & G: Tetractinomyxon.B, D & F: top views , C, E & G: side view. SB: spore body, LSB: length of spore body; WSB: width of spore body; S: style; LS: length of style; WS: width of style; CP: caudal process; LCP: length of caudal process (regardless of curvature). LSCP: largest span of between the tips of the caudal processes; PC: polar capsule; DSB: diameter of spherical spore body

Fig. 6. Diagram of actinosporean spores.

There have been 18 collective groups described thus far (Lom & Dyková, 2006, Rangel et al., 2011). Based on the total length of spore (or interconnected spore mass), they are distinctly divided into two morphotypes; the small-type ranges from 15 to 40 µm, e.g., Endocapsa, Sphaeractinomyxon, Tetraspora, Tetractinomyxon, Aurantiactinomyxon, Neoactinomyxum and Guyenotia, while the large-type ranges from approximately 100 to 400 µm, e.g., Echinactinomyxon, Raabeia, Triactinomyxon, Pseudotriactinomyxon, Hexactinomyxon, Ormieractinomyxon, Siedleckiella, Synactinomyxon, Antoactinomyxon, Hungactinomyxon and Unicapsulactinomyxon. From the practical point of view, the large-type actinospores are more likely to be removed by filtration systems than the small-type actinospores. Thus it is important to determine the type of the corresponding actinospore, not only for parasitology, but also for disease management in aquaculture.

Practical key for determination of actinospore-types:

1. a. Processes are absent..2
   b. Processes are present...4

2. a. Spores are tetrahedral with a single binucleate sporoplasm ....**Tetractinomyxon**
   b. Spores are subspherical with polar capsules embedded beneath the spore surface...3
3. a. Eight spores, rounded in side view, are developed within a pansporocyst...**Sphaeractinomyxon**
   b. Four spores, flattened in side view, are developed within a pansporocyst...**Tetraspora**
4. a. Spores do not connect each other...5
   b. Spores connect each other at the end of the processes, forming a net structure........7
5. a. Processes are reduced to bulge-like swellings...6
   b. Spores with curved leaf-like processes resemble an orange with partly opened peel ....**Aurantiactinomyxon**
   c. Spores have a subspherical spore body with 3 finger-like processes..........**Guyenotia**
   d. Spores have an ovoid spore body with 3 straight spine-like processes...**Echinactinomyxon**
   e. Spores have an elipsoidal spore body with 3 curved, and sharp-tipped processes...**Raabeia**
   f. Spores have an elongated spore body with a style and 3 anchor-like processes...**Triactinomyxon**
   g. Spores are similar to triactinomyxon, but the processes have longitudinal sutures, which remain fused over all their length...**Pseudotriactinomyxon**
   h. Spores have an elongated spore body with a style and 3 diverged (in total 6) processes...**Hexactinomyxon**
   i. Spores have a single and large polar capsule in an elliptical spore body...**Unicapsulactinomyxon**
6. a. Flattened in side view, and polar capsules are embedded below the spore surface ....**Endocapsa**
   b. Rounded triangular in top view, and polar capsules protrude at the spore apex..**Neoactinomyxum**
7. a. Spore units are echinactinomyxon whose 4 processes of different spores form the junction..**Antoactinomyxon**
   b. Spores units are triactinomyxon whose 3 processes of different spores form the junction...**Siedleckiella**
   c. Spore units are echinactinomyxon whose 8 processes have anchor-like hooks at the end, adhering together..**Ormieractinomyxon**
   d. Spores have two wing-like and one short, conical process, forming a star-like structure..**Synactinomyxon**
   e. Four spores form a cube-like net interlaced with another cube made of 4 spores...**Hungactinomyxon**

2.4 Phylogeny of myxosporeans and actinospore-types

As far as we know, the corresponding actinospore stages have been identified for 39 myxosporean species. Among them, 18S rDNA sequences of 33 species were registered in GenBank for either myxospore or actinospore stages, or both (Table 3). Cladistic analysis of myxosporean and actinospore-types revealed a lack of taxonomic congruity between the

Myxosporean species	GenBank No.	Actinospore type	GenBank No.
Ceratomyxa shasta	AF001579	Tetractinomyxon	nr
Ceratomyxa auerbachi	EU616730	Tetractinomyxon	EU616733
Chloromyxum schurovi	AJ581917	Neoactinomyxum	AJ582007
Chloromyxum truttae	AJ581916	Aurantiactinomyxon	AJ582006
Ellipsomyxa gobii	GQ229235	Tetractinomyxon	AY505127
Ellipsomyxa mugilis	AF411336	Tetractinomyxon	EU867770
Gadimyxa atlantica	EU163416	Tetractinomyxon	EU163412
Henneguya exilis	AF021881	Aurantiactinomyxon	nr
Henneguya ictaluri	AF195510	Aurantiactinomyxon	nr
Henneguya nuesslini	AY669810	Triactinomyxon	nr
Myxidium giardi	AJ582213	Aurantiactinomyxon	nr
Myxidium truttae	AJ582061	Raabeia	AJ5820009
Myxobilatus gasterostei	EU861210	Triactinomyxon	EU861209
Myxobolus arcticus	AB353130	Triactinomyxon	AB353128
Myxobolus bramae	AF507968	Triactinomyxon	nr
Myxobolus cerebralis	EF370478	Triactinomyxon	MCU96492
Myxobolus cultus	HQ613409	Raabeia	AB121146
Myxobolus dispar	AF507972	Raabeia	nr
Myxobolus hungaricus	AF448444	Triactinomyxon	nr
Myxobolus intimus	AY325285	Triactinomyxon	nr
Myxobolus lentisuturalis	AY278563	Raabeia	nr
Myxobolus macrocapsularis	AF507969	Triactinomyxon	nr
Myxobolus parviformis	AY836151	Triactinomyxon	AY495704
Myxobolus pavlovskii	HM991164	Echinactinomyxon	nr
Myxobolus portucalensis	AF085182	Triactinomyxon	nr
Myxobolus pseudodispar	AF380144	Triactinomyxon	EF466088
Myxobolus rotundus	EU710583	Triactinomyxon	FJ851447
Parvicapsula minibicornis	HQ624972	Tetractinomyxon	DQ231038
Sphaerospora dykovae (= *S. renicola*)	AY735410	Neoactinomyxum	nr
Sphaerospora truttae	AJ581915	Echinactinomyxon (?)	nr
Thelohanellus hovorkai	DQ231155	Aurantiactinomyxon	DQ231155
Thelohanellus nikolskii	DQ231156	Aurantiactinomyxon	nr
Zschokkella nova	DQ377690	Siedleckiella	nr

Table 3. List of myxosporean species and the corresponding actinosporean types registered in GenBank. nr: not registered.

two stages (Xiao & Desser, 2000a). Different phenotypes may be subject to environmental factors. Since the first study of molecular relationship between myxosporean and actinospore-types based on the 18S rDNA by Holzer et al. (2004), some more life cycles of marine myxosporeans have been discovered. Thus we update the phylogenetic analysis of species, where both life stages are described using the data available in GenBank. It is widely accepted that freshwater and marine myxosporeans are separated into two major branches (Kent et al., 2001, Fiala, 2006), and the phylogenetic tree in the present study also supports this (Fig. 7). Further, the close relationship between the marine clade myxosporeans and the

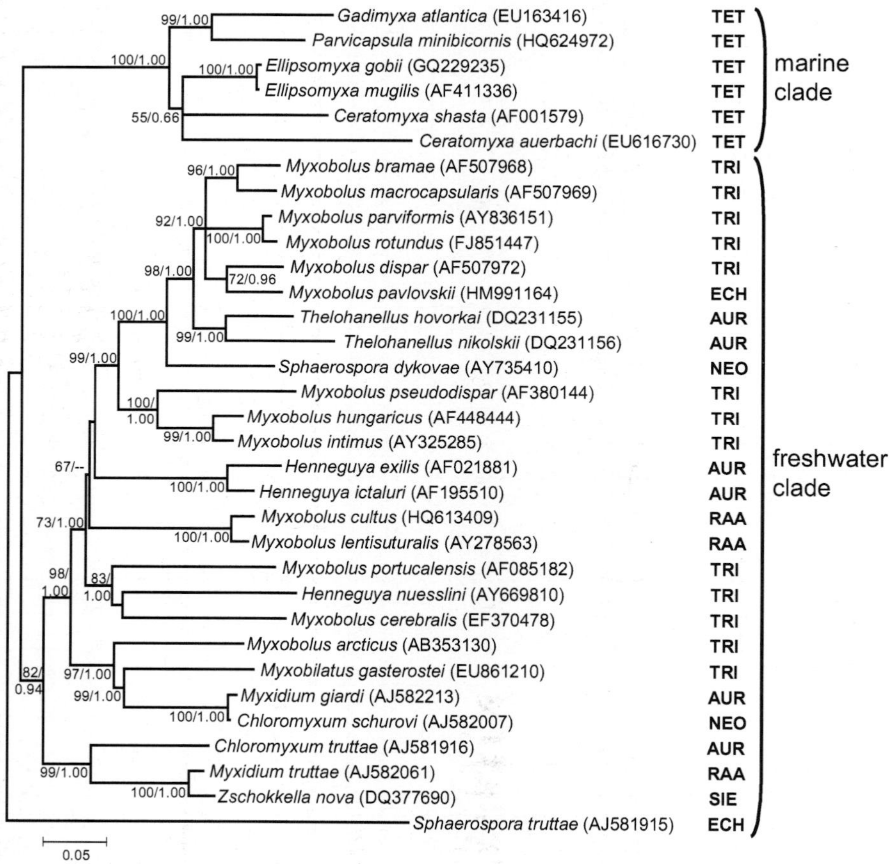

TET: tetractinomyxon, TRI: triactinomyxon, ECH: echinactinomyxon, AUR: aurantiactinomyxon, NEO: neoactinomyxum, RAA: raabeia, SIE: siedleckiella.

Fig. 7. Phylogram of myxosporeans based on 18S rDNA. Bayesian and maximum likelihood analyses. Myxosporean species names were followed by GenBank accession numbers in parenthesis and the corresponding actinospore-types.

tetractinomyxon-type actinospores has been strongly supported, whereas no obvious pattern was observed for actinospore morphology for the freshwater clade myxozoans.

To date, the information on myxozoan life cycles is still limited. Therefore, the species used for this phylogenetic analysis only covers a small portion of the wide myxozoan diversity. This leads to instability in some parts of the tree. The marine "tetractinomyxon-clade" was well defined, only the position of *C. auerbachi* and the rather aberrant *C. shasta* was not resolved properly. Considering that the hosts for *C. shasta* are anadromous salmonids and a polychaete which is typically marine, *C. shasta* may be an originally marine parasite, which migrated secondarily to the freshwater environment. The inclusion of further *Ceratomyxa*-species would probably help to stabilize this placement, but unfortunately *C. shasta* and *C. auerbachi* are the only species of the genus where both actinospore and myxospore are known. The exact positions of *M. cerebralis, H. exilis + H. ictaluri, H. nuesslini, M. portucalenis* and *M. cultus + M. lentisuturalis* at the base of the "*Myxobolus*-clade" were not clarified. Again, the inclusion of more species might stabilize their branches. The placement of *M. rotundus* with *M. parviformis* is quite different compared to the analysis of Fiala (2006), but the sequence of the actinospore was used in the present study, because it exhibits a higher quality at the 3′ end compared to the available myxospore-sequence of this species.

According to the hypothesis of Fiala & Bartošová (2010), who stated that the common ancestor of myxozoans was a freshwater species, the congruence of the marine clade actinospore-type (tetractinomyxon) might reflect the divergence of freshwater and marine myxozoans. When colonizing polychaetes as hosts, the tetractinomyxon type of spore developed or was already present in the freshwater ancestor of marine myxozoans. This actinospore-type persisted at least in most myxozoans parasitizing marine polychaetes that we know to date. Knowledge of more life cycles of marine myxozoans is necessary to provide information on marine actinosporean diversity. At present, there are 13 marine actinosporeans for which the myxosporean stage of the life cycle is still unknown (Table 4); 2 types of endocapsa from oligochaetes, 3 types of sphaeractinomyxon from polychaetes, 4 tetractinomyxon from polychaetes and sipunculids, 2 tetraspora from oligochaetes, 1 triactinomyxon from oligochaete, and 1 unicapsulactinomyxon from polychaete. Among them, most of the oligochaetes are benthic living in beach sediments whereas most of the polychaetes are sedentary tube worms (fan worm) attaching on the rocks or shells in coastal areas. The sipunculid (peanut worm) lives in shallow waters, either in burrows or in discarded shells.

2.5 Biology of actinosporeans

Since the discovery of the life cycle of *M. cerebralis* by Wolf & Markiw (1984), many scientists have focused on biological studies of actinosporeans, such as emergence from annelid hosts, waterborne stage, invasion mechanisms, and the portals of entry into fish host. Invasion process has been also investigated in relation to the mechanisms in the host specificity of the parasites. The current knowledge on the aforementioned points is summarized below.

2.5.1 Methodology for actinosporean biology

To obtain materials for research on actinosporeans, it is desirable to maintain the life cycle of the model-myxosporean in the laboratory. Released actinospores can be harvested by

Actinospore type	Invertebrate host	Corresponding myxosporean
Endocapsa rosulata (Hallett et al., 1999)	*Heterodrilus* cf. *keenani* (Oligochaeta)	nd
Endocapsa stepheni (Hallett et al., 1999)	*Heterodrilus* cf. *keenani* (Oligochaeta)	nd
Sphaeractinomyxon stolci (Caullery & Mesnil, 1904)	*Clitellio, Peloscolex, Tubifex* (Oligochaeta)	nd
Sphaeractinomyxon ersei (Hallett et al., 1998)	*Doliodrilus diverticulatus* (Oligochaeta)	nd
Sphaeractinomyxon leptocapsula (Hallett et al., 1999)	*Heronidrilus* sp. (Oligochaeta)	nd
Tetractinomyxon (Køie et al., 2008)	*Chone infunduliformis* (Sabellidae: Polychaeta)	*Ceratomyxa auberbachi*
Tetractinomyxon (Køie et al., 2004)	*Nereis diversicolor* and *N. succinea* (Nereididae: Polychaeta)	*Ellipsomyxa gobii*
Tetractinomyxon (Køie et al., 2007)	*Spirorbis* sp.(Spirorbidae: Polychaeta)	*Gadimyxa atlantica*
Tetractinomyxon (Køie, 2002)	*Hydroides norvegica* (Polychaeta)	nd
Tetractinomyxon (Køie, 2005)	Unidentified spionid (Polychaeta)	nd
Tetractinomyxon intermedium (Ikeda, 1912)	*Nephasoma minuta* (Sipunculidae: Sipuncula)	nd
Tetractinomyxon irregulare (Ikeda, 1912)	*Nephasoma minuta* (Sipunculidae: Sipuncula)	nd
Tetraspora discoidea(Hallett & Lester, 1999)	*Doliodrilus diverticulatus* (Oligochaeta)	nd
Tetraspora rotundum (Hallett & Lester, 1999)	Tibificidae spp. (Oligochaeta)	nd
Triactinomyxon (Roubal et al., 1997)	*Duridrilus* sp. (Oligochaeta)	nd
Unicapsulactinomyxn (Rangel et al., 2011)	*Diopatra neapolitana* (Polychaeta)	nd

Table 4. Marine actinosporeans from annelids or sipunculids. nd: not determined.

filtering of the aquarium water through mesh screens (El-Matbouli et al., 1995). If a laboratory system is not available, study materials are obtained from naturally infected wild invertebrate worms. Yokoyama et al. (1991) developed a multi-well plate method to collect actinospores of a single myxozoan species. Oligochaetes are placed individually in wells filled with dechlorinated tapwater. One of the advantages of this method is that even small-size actinospores which are hard to trap by filtration can be collected easily from wells. However, it may be difficult to apply this method to fragile or large-size worms. Also, if actinospores are released after host death, the well plate method will be inapplicable

(Rangel et al., 2009). In that case, worms may be crushed on a glass slide with gentle pressure. However, Rangel et al. (2011) successfully obtained marine actinosporeans of *Zschokkela mugilis* from the coelomic fluid of the polychaete host with a hypodermic needle and syringe. To determine the viability of actinospores, presence or absence of the sporoplasms in the spore body has been used as an indicator (Yokoyama et al., 1993, Xiao & Desser, 2000b), because aged actinospores spontaneously release sporoplasms so that spores become empty. Alternatively, a vital staining technique with fluorescein diacetate (FDA) and propidium iodide (PI) can be applied (Markiw, 1992, Yokoyama et al., 1997b, Wagner et al., 2003, Kallert et al., 2005).

2.5.2 Emergence pattern

Most of freshwater actinosporeans infect the intestinal epithelium of oligochaetes and emerge into the environment by defecation (Fig. 8A, B), whereas *C. shasta* actinosporeans

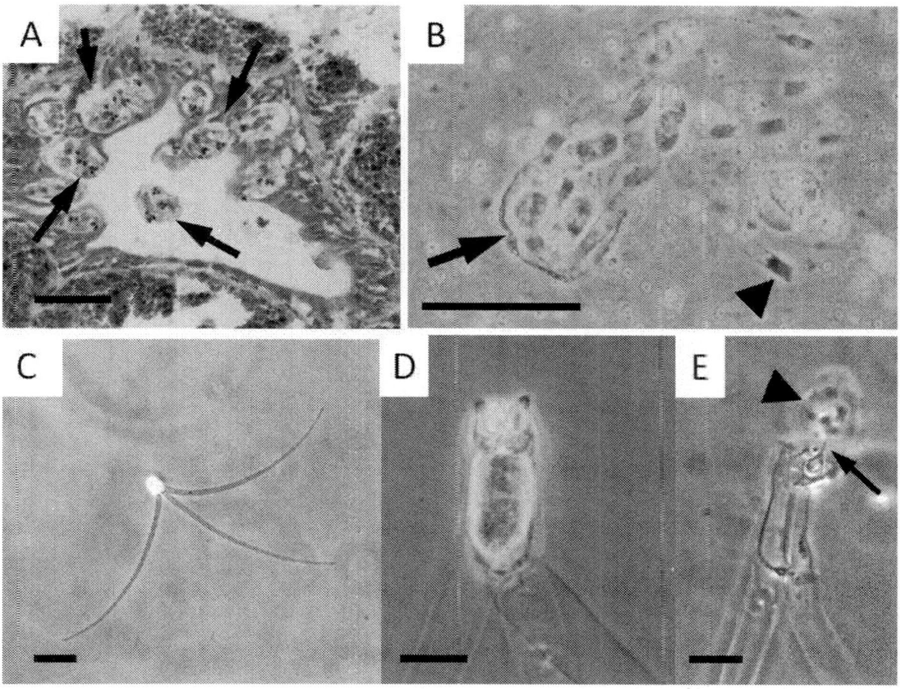

A: Pansporocysts (arrows) develop in the intestinal epithelium of *Branchiura sowerbyi*. B: Pansporocyst is excreted from *B. sowerbyi*. As the pansporocyst membrane (arrow) is ruptured, actinospores are released. Tip of the caudal process is still folded (arrowhead). C: Free actinospore. Note completely unfolded processes. D & E: Chemical response of actinospore to fish mucus. D: Intact spore. E: Empty spore releasing sporoplasm (arrowhead) immediately after contact with mucus. Polar filaments (arrow) are discharged. Scale bars for A, B and C are 50 µm, and those for D and E are 10 µm.

Fig. 8. Process of emergence, floating and invasion of *Myxobolus cultus* actinospores.

develop in the epidermis of the polychaete *Manayunkia speciosa* and actinospores are released directly from the epidermis into the water column (Meaders & Hendrickson, 2009). In many myxosporean species, actinospores are shed from the annelid hosts between spring and summer (Yokoyama et al., 1993, El-Mansy et al., 1998a, b, Özer & Wootten, 2002), which may be an adaptation to synchronize with hatching and growing seasons of larval fish. However, in some species, actinospores are released throughout the year. Prevalence of infection in the invertebrate hosts has been reported to be relatively low, 0.1-4% (Yokoyama et al., 1993, Özer & Wootten, 2002), but in some cases, it reached extremely high value of over 90% (El-Mansy et al., 1998a). Actinospore release may persist for the natural life-span of oligochaete hosts, at least for 2 years in case of *Tubifex tubifex* infected with *M. cerebralis* (Gilbert & Granath, 2001). Actinospore emergence follows a circadian rhythm with a significant peak in the middle of the night or early morning (Yokoyama et al., 1993, Özer & Wootten, 2001). It is unclear if this daily pattern in spore release is due to the rhythm of the oligochaete itself or of the actinosporean, and the ecological significance of this phenomenon for transmission to the next host remains to be investigated. Alteration of the photoperiods affected the release pattern of actinospores (Yokoyama et al., 1993), and thus artificial control of lighting condition may have some effects on myxosporean transmissions in the field.

2.5.3 Waterborne stage

Actinosporeans with long processes are buoyant (Fig. 8C) and can remain suspended in the water column for more than 24 hours (Kerans & Zale, 2002). Longevity of actinospores in the water ranges from 4 to 25 days, depending on temperature and species (Markiw, 1992, Yokoyama et al., 1993, Xiao & Desser, 2000b). Life-span decreases with increasing temperature (Yokoyama et al., 1993, Özer & Wootten, 2002). At ambient temperature (20 °C), viability of raabeia actinospores persisted for 10 days, while echinactinomyxon spores survived for 21 days (Yokoyama et al., 1993). In contrast, Özer & Wootten (2002) reported that raabeia and synactinomyxon spores remain viable only for 2-3 days at 22 °C. Markiw (1992) showed that the infectivity of actinospores of *M. cerebralis* persisted for 3-4 days at 12.5 °C, whereas El-Matbouli et al. (1999a) indicated that *M. cerebralis* actinospores survived and maintained their infectivity for 15 days at 15 °C. Using morphological characteristics and vital staining technique, Kallert & El-Matbouli (2008) showed that actinospores of the myxosporean species survive longer at lower temperature (4 °C vs. 12 °C). *M. cerebralis* actinospores were most sensitive and showed a significant decrease of viability already after 1 d at 12 °C, while *M. pseudodispar* and *Henneguya nuesslini* survived longer, even at 12 °C. Water flow has been recognized as an environmental factor which have some effects on myxsporean infections (Hallett & Bartholomew, 2008, Bjork & Bartholomew, 2009). Higher water velocity resulted in lower infection prevalence of *C. shasta* in polychaete and decreased infection severity in fish (Bjork & Bartholomew, 2009). During the planktonic phase of actinospores, high flow velocity may cause mechanical damages and dilution effects on actinospores. Also, high flow rates may limit the time for actinospores to encounter and attach to the fish host (Hallett & Bartholomew, 2008).

2.5.4 Invasion mechanisms

Polar filament discharge and sporoplasm release of actinospores are induced by chemical responses to fish mucus (Fig. 8D & E), suggesting the role of chemoreception in the host attachment of actinospores (Yokoyama et al., 1993, 1995, Uspenskaya, 1995, McGeorge et al., 1997, Xiao & Desser, 2000b). However, the percentage of actinospores reacting to the mucus varied among fish and parasite species (Yokoyama et al., 1993, Özer & Wootten, 2002). Thus, it is not clearly understood whether the chemical stimulation with fish mucus reflects the host specificity of myxosporeans. Actinospores of *M. cultus* reacted not only to the skin mucus from natural host but also to the mucus from abnormal host (Yokoyama et al., 1993) and even to mucin from bovine submaxillary gland (Yokoyama et al., 1995). Further, purification of the reactants from fish mucus by gel filtration and ultrafiltration revealed that they were low-molecular-weight (<6000 MW) substances (Yokoyama et al., 1995). Yokoyama et al. (2006) indicated that *M. arcticus* actinospores reacted to the mucus of the susceptible host, masu salmon (*Oncorhynchus masou*) as well as non-susceptible hosts, sockeye salmon (*O. nerka*) and goldfish (*Carassius auratus*), whereas *T. hovorkai* actinospores reacted only to the susceptible host, common carp (*Cyprinus carpio*). In contrast, actinospores of *Myxobolus cerebralis* did not react to fish mucus alone (El-Matbouli et al., 1999b) and required both mechanical and chemical stimuli (Kallert et al., 2005). Nevertheless, *M. cerebralis* actinospores were unable to specifically detect susceptible fish (salmonids), but also penetrated gills of carp at the same rate as gills of trout (Kallert et al., 2009). Further, Kallert et al (2007) revealed the process of host invasion of *M. cerebralis* actinospores in detail; immediately after filament discharge of actinospores, contraction of the filaments brings the actinospore apex to contact with the host surface. Then, opening of the apical valves is followed by penetration of the sporoplasms through the epithelium. The active fraction inducing the polar filament discharge of *M. cerebralis* actinospores was small molecular, amphiphilic to slightly hydrophobic organic substances (Kallert et al., 2010). More recently, several nucleosides derived from surface mucus of fish, inosine, 2′-deoxyinosine and guanosine have been determined by HPLC method as 'chemical cues' triggering host recognition for *M. cerebralis* actinospores (Kallert et al., 2011).

2.5.5 Portals of entry into fish

Entry of myxozoans into the fish host via the skin, fins and buccal cavity was first demonstrated in rainbow trout experimentally exposed to actinospores of *Myxobolus cerebralis* by Markiw (1989). Within 5-10 min of exposure, aggregates of sporoplasms were observed in the epithelia of exposed fish (Markiw, 1989, El-Matbouli et al., 1995). Further, El-Matbouli et al. (1999b) revealed by scanning electron microscopy that *M. cerebralis* actinospores penetrate into the secretory openings of the mucous cells of the epidermis. Belem & Pote (2001) showed by indirect fluorescent antibody test that *Henneguya ictaluri* has the multiple entry sites; the gut mucosa, skin and buccal cavity of the channel catfish (*Ictalurus punctatus*). Some actinospores may be able to enter the fish through different portals of entry. *Sphaerospora truttae* and *Ceratomyxa shasta* utilize predominantly the gills as entry site (Holzer et al., 2003, Bjork & Bartholomew, 2010). Yokoyama & Urawa (1997) suggested that small actinospore (aurantiactinomyxon) invade the fish through the gills, whereas large actinospores (triactinomyxon and raabeia) penetrate mainly through the fin and skin.

2.5.6 Other biological characteristics

Effects of physical and chemical treatments on viability of actinosporeans were investigated, although the information is available only for *Myxobolus cerebralis* and *Myxobolus cultus*. For

M. cerebralis, drying at room temperature for 15 min, freezing at -20°C for 1 hour, temperatures above 75 °C for 5 min and sonication (47 kHz, 130 W) for 10-13 min were effective in killing actinospores, but pressure of 6.2 x 10^7 Pa (9000 psi) was not (Wagner et al., 2003). To inactivate actinospores of *M. cerebralis* chemically, chlorine of 13 ppm for 10 min, hydrogen peroxide of 10% for 10 min, and povidone-iodine of 50% solution (5000 ppm active iodine) for 60 min were effective (Wagner et al., 2003). Electricity with a pulse length of 99 μsec at 3 kV induced polar filament discharge of *M. cerebralis* actinospores, suggesting a potential use of direct current as a means of disinfection (Wagner et al., 2002). For *M. cultus*, drying at 5 °C for 1 day and ultraviolet irradiation at 600 mW s cm^{-2} were highly effective in killing actinospores, whereas sodium chloride of 0.5% had a moderate effect (Yokoyama et al., 1997b). However, even high concentrations of malachite green (10 ppm), metrophonate (5 ppm) and formalin (1000 ppm) did not affect the treated spores (Yokoyama et al., 1997b).

Actinosporean infections are also influenced by various biological and ecological factors, such as host (annelid) susceptibility, water temperature, and sediment type. Susceptibility to *M. cerebralis* varied among different genetic strains of *T. tubifex* (Beauchamp et al., 2002). Development and release of *M. cerebralis* actinospores from *T. tubifex* were temperature-dependent; High temperatures above 20 °C were lethal for the parasite, whereas low temperatures between 5 and 10 °C delayed development, and moderate temperatures between 15 and 20 °C accelerated development, and increased the number of spores released (El-Matbouli et al., 1999a). Blazer et al. (2003) also reported a similar pattern of temperature effects on development of *M. cerebralis* actinospores in *T. tubifex*. Environmental factors like substratum and water quality may influence the actinosporean production. Blazer et al. (2003) indicated that the mud substrate produced the highest total number of *M. cerebralis* actinospores in *T. tubifex*, whereas the leaf litter was the least productive substratum in number of actinospores released. Aquatic oligochaetes have habitat preferences which are closely associated with some environmental parameters, such as substrate type, texture, nutritional potentials, and anaerobic conditions (Koprivnikar et al., 2002, Liyanage et al., 2003). Actinospore production of *M. cerebralis* is also affected by environmental pollutants (Shirakashi & El-Matbouli, 2010).

3. Fish-to-fish transmission of marine myxosporeans

Enteromyxum leei develops within the gut epithelium of marine fish, and the developmental stages are excreted to the water (Fig. 9). Released stages are orally ingested by other fish, resulting in establishment of horizontal infection (Diamant, 1997, Yasuda et al., 2002, 2005, Sitja-Bobadilla et al., 2007). This route of transmission may occur only in intensive culture systems, where it facilitates rapid spread of the parasite. Broad host range of *E. leei* also appears to assist the parasite's dispersion (Diamant et al., 2006). Indeed, an episode of enteromyxosis in 25 different fish species in an exhibition aquarium was reported (Padrós et al., 2001). *E. scophthalmi* and *E. fugu* also transmit from fish to fish directly, but their host ranges are narrow.

3.1 Infective developmental stages of *Enteromyxum* spp. in water column

Although actinosporean stages for *Enteromyxum* spp. have not been discovered, some biological characteristics of infective developmental stages have been investigated. Viability

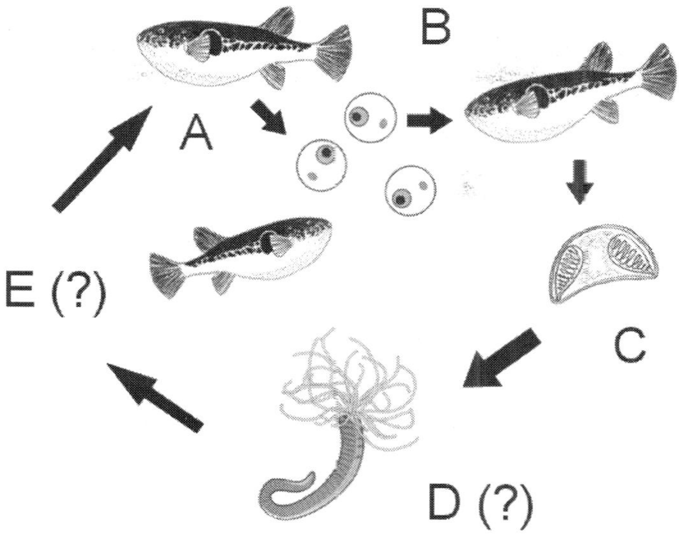

A: *E. leei* develops in the intestine, followed by excretion through the vent. B: Developmental stages are horizontally transmitted to other fish by oral ingestion. C: Mature myxospores are released into water column. D: Myxospore possibly infects marine annelids. E: Actinospore is possibly released from annelids, followed by infection to fish.

Fig. 9. Diagram of fish-to-fish transmission and putative life cycle of *Enteromyxum leei*

of *Enteromyxum* spp. stages was determined *in vitro* by dye-exclusion assays (Redondo et al., 2003, Yokoyama & Shirakashi, 2007), tetrazolium-based cell-proliferation assay (Redondo et al., 2003), and vital staining with fluorescent dyes, Hoechst 33342 and propidium iodide (Yokoyama et al., 2009). Longevities of *E. leei* and *E. scophthalmi* were estimated to be at most 1 day in seawater (Redondo et al., 2003, Yokoyama et al., 2009). However, intestinal mucosal remnants covering the parasites may protect them from osmotic shock, resulting in retaining their viability in seawater (Redondo et al., 2002, Yokoyama et al., 2009). Survivability of developmental stages of *E. leei* decreased significantly in low salinity of less than 8‰ (Yokoyama & Shirakashi, 2007). Also, fish size and parasite dose likely affect the success of fish-to-fish transmission (Sitja-Bobadilla et al., 2007, Yokoyama & Shirakashi, 2007).

3.2 Invasion and development of *Enteromyxum* spp. in fish host

Following ingestion of the infective stages, the first barrier is the intestinal mucosa of the fish. A role of lectin/carbohydrate interaction in the turbot-*E. scophtahlmi* relationship was suggested (Redondo & Alvarez-Pellitero, 2009). Further, attachment and invasion of *E. scophthalmi* to the turbot intestinal epithelium were inhibited by pre-treatments of parasites by some lectins, Con A and SBA, suggesting the involvement of N-acetyl-galactosamine and galactose residues and also of mannose/glucose residues (Redondo & Alvarez-Pellitero, 2010). After penetration of the developmental stages into the intestinal epithelium, several factors are involved in the progression of the disease (Quiroga et al., 2006). One of the most important factor is water temperature.Yanagida et al. (2006) showed that temperatures below 15 °C suppressed the development of *E. leei* and onset of the disease, but a temperature increase to 20 °C promoted *E. leei* development. Similarly, infection with *E. scophthalmi* was established earlier at higher temperature (Redondo et al., 2002).

4. Malacosporeans

The class Malacosporea is a recently discovered group, and only three species belonging to two genera have been described to date (Canning & Okamura, 2004; Canning et al., 2007). All of them are known to be parasites of freshwater bryozoans, but the life cycle is described only for *T. bryosalmonae*. Besides the bryozoan stage, it involves the infection of salmonid fish (Saulnier et al., 1999) where the parasite causes the Proliferative Kidney Disease (PKD). According to recent findings, species of the second malacosporean genus *Buddenbrockia* might also require a fish host in their life cycles (Grabner & El-Matbouli, 2010a).

4.1 Morphology of malacosporeans in bryozoan hosts (bryozoa-spores)

Malacosporean spores developing in the bryozoan host (bryozoa-spores) are small (15 - 20 µm), approximately spherical without appendices (Fig. 10). They consist of two haploid sporoplasms including one secondary sporoplasm cell each, four capsulogenic cells and eight valve cells (Canning et al., 2000, McGurk et al., 2005). Only minimal morphological differences have been recorded between spores of different malacosporean species. Morris et al. (2002) documented ornamented spores with a mean diameter of 19.0 µm in the bryozoan *Plumatella repens* infected with worm-like malacosporean stages. The spores observed by McGurk et al. (2006a), also released by a worm-shaped malacosporean in *P. repens*, were spherical and 17.7 µm in diameter.

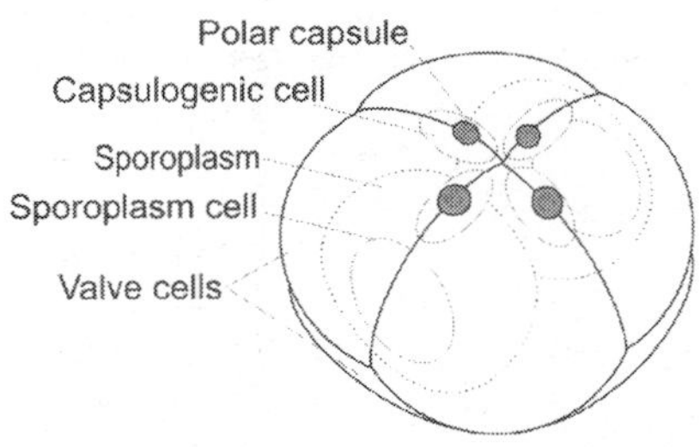

Fig. 10. Diagram of malacosporeans in bryozoans (bryozoa-spores).

4.2 Morphology of malacosporean in fish hosts (fishmalacospores)

To date, fresh and mature fishmalacospores were only described for *T. bryosalmonae*. They are about 12 × 7 µm in size and bear two polar capsules with 4 to 6 turns of their polar filament, one sporoplasm and four valve cells (Kent et al., 2000, Hedrick et al., 2004, Morris & Adams, 2008). Apparently, only few fishmalacospores are released at a time by *T. bryosalmonae*-infected fish, because only small numbers of spores were found in urine samples from infected fish over a prolonged period of time (Hedrick et al., 2004).

4.3 Life cycle of malacosporeans

Life cycles of malacosporeans still remain mysterious, but recent studies have revealed most parts of the development and transmission of *T. bryosalmonae* (Canning et al., 1999, Kent et al., 2000, Morris & Adams, 2006a). This parasite develops as sac like stages in the body cavity of freshwater bryozoans, followed by release of malacospores to the surrounding water. When the spores come in contact to the skin or gills of a fish host (salmonid), the sporoplasm penetrates the epithelium and is transported by the blood into the kidney interstitium, causing PKD. Sporogony commences after migration to the kidney tubules and mature spores are released with the urine to the water, where they are infective for bryozoans (Fig. 11). The whole development, beginning with the penetration into the fish, to the presence of mature spores in the kidney tubules takes about 9 weeks in brown trout (Morris & Adams 2006a).

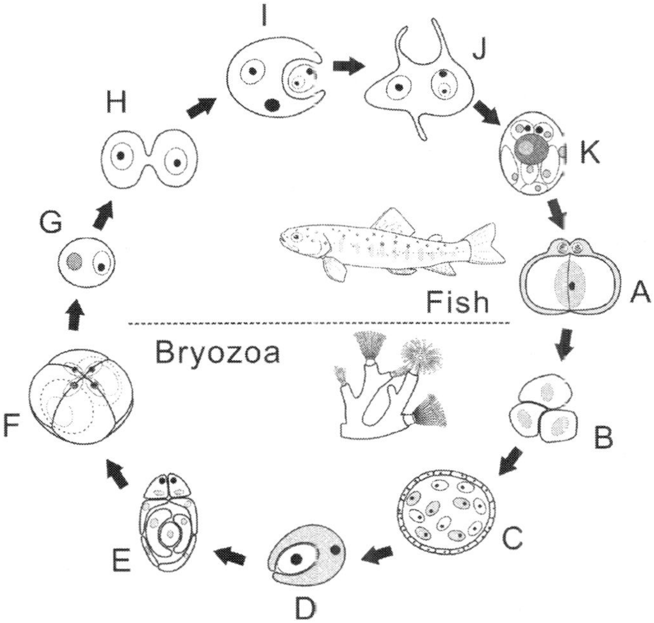

A: Fishmalacospore infects freshwater bryozoans. B: Presaccular cell aggregates in coelomic cavity of bryzoans. C: Early spore sac floating in bryozoan coelomic fluid. It contains stellate and sporogenic cells. D: Sporogenic cell becomes enclosed by stellate cells. E: Maturing spore with casulogenic cells, valve cell and forming sporoplasms. F: Mature bryozoa-spore infects fish. G: Proliferative stage (cell doublet with primary cell and secondary cell inside) in kidney interstitium. These stages are in close contact to host phagocytes (not shown). H: Division of cell doublet resulting in 2 cell doublets. I: Engulfment of one cell doublet by another resulting in a S-T-doublet (primary cell enclosing one secondary and one secondary with tertiary cell). J: S-T-doublet in kidney tubule. Note that contact to the host phagocyte is lost during migration through the tubule epithelium. K: Sporogony inside of primary cell (pseudoplasmodium).

Fig. 11. Diagram of the life cycle of *T. bryosalmonae* (Malacosporea), alternating between fish and bryozoan host.

T. bryosalmonae can infect a wide variety of salmonid fish. Most affected are species of the genera *Salmo* and *Oncorhynchus*, but also *Salvelinus* species (Hedrick et al., 1993, El-Matbouli & Hoffmann, 1994). Severe outbreaks of the disease were also noted in grayling (*Thymallus thymallus*) (Hoffmann & Dangschat, 1981). Northern Pike (*Esox lucius*) is the only non-salmonid fish species, in which extrasporogonic stages similar to those of *T. bryosalmonae* were found (Seagrave et al., 1981, Morris et al., 2000a). It was observed that fish become resistant against reinfection with *T. bryosalmonae* after surviving the disease (Ferguson, 1981, Foott & Hedrick, 1987). However, in some fish species sporogonic stages seem to persist after clinical infection and possibly continue to form spores chronically (Kent et al., 1998, Kent et al., 2000). Recently it was shown by transmission experiments conducted with European parasite lineages that brown trout (Morris & Adams, 2006a) and brook trout (Grabner & El-Matbouli, 2008) can transmit the parasite to bryozoans. In contrast, rainbow trout and grayling became infected, but no infection appeared in bryozoans cohabitated with these fish. But as mature fishmalacospores were reported from rainbow trout infected with *T. bryosalmonae* in North America, it seems likely that there is a regional difference in host specificity (Morris & Adams, 2006a). Additionally, infection experiments have shown that common carp (*Cyprinus carpio*) and minnow (*Phoxinus phoxinus*) can become infected by *Buddenbrockia* species, but the proof for the completion of the life cycle is still missing (Grabner & El-Matbouli, 2010a). Additionally, intra-bryozoan cycles without involvement of a fish host might be possible for some malacosporeans (Hill & Okamura 2007).

4.4 Biology of malacosporeans

Knowledge on the biology of malacosporeans is still limited. Most information exists for *T. bryosalmonae*, while the understanding of life cycles of other malacosporeans is still in its infancy. The information concerning the transmission of malacosporeans will be summarized below.

4.4.1 Emergence pattern

Occurrence of PKD is seasonal and occurs from spring till autumn. This can be explained by the higher abundance of the bryozoan host in warmer months and therefore higher spore load in the water, but also by increase in severity of infection in fish at higher temperatures (Foot & Hedrick, 1987, Hedrick et al., 1993). It has to be noted, that in most cases infections with *T. bryosalmonae* become apparent only in trout farms. Mortalities or diseased fish in the wild are not found in most cases. Therefore, the dynamics of natural life cycles are difficult to investigate (Okamura et al., 2011).

4.4.2 Waterborne stage

Malacosporean bryozoa-spores do not possess hard valves for protection against external damage. Therefore, they are very short-lived and lose their infectivity after about 24h (de Kinkelin et al., 2002). Hedrick et al. (2004) described that fishmalacospores degrade already within minutes on a microscope slide. The floating characteristics of malacosporean spores are not investigated, but the lack of processes that might prevent sinking down in the water column and short life-span suggest that contact to the host must occur soon after release of spores.

4.4.3 Portals of entry into fish

The *T. bryosalmonae* spores released from parasitized Bryozoa most likely enter the fish through the gills (Morris et al., 2000b, Holzer et al., 2006, Grabner & El-Matbouli, 2010b) or the mucus cells of the skin (Longshaw et al., 2002), while the blood stream was considered to be the most probable route to the target organs (Morris et al., 2000b, Holzer et al., 2006). The infection seems to be very effective that one single spore is sufficient to infect a fish and to cause clinical symptoms of PKD (McGurk et al., 2006b).

4.4.4 Other biological characteristics

Transfer of malacosporeans to new habitats can also occur by fragmentation and reattachment of bryozoan colonies, which was found to be common for *Fredericella sultana*-colonies (Morris & Adams, 2006b). Another way for propagation of malacosporeans without a fish host might be the infection of durable stages (statoblasts) of bryozoans (Hill & Okamura, 2007). Water quality, especially increase of organic material, seems to influence disease outbreaks, most likely by fostering growth of bryozoan colonies and thereby increasing numbers of infective stages in the water (El-Matbouli & Hoffmann, 2002).

5. Control strategies of myxozoans

To date, there are no commercially available chemotherapeutants and vaccines to treat myxozoan infections. Thus, the current disease control strategies can only be based on the biology of myxozoans. Compared to myxospores, waterborne actinospores are generally short-lived and highly susceptible to several treatments. The actinospore stage can be considered as 'weak point' in the life cycle of myxozoans and should be targeted for the control strategy. This paragraph deals with possible control strategies of myxozoan diseases with emphasis on prevention of transmission to fish hosts.

5.1 Eradication of invertebrate hosts

In case of most myxozoans with indirect life cycle, transmission success largely depends on the size of population of invertebrate hosts. The most effective way for prevention of myxozoan transmission is to eradicate the invertebrate hosts in the aquaculture environment. Habitat manipulation may be an effective means to remove oligochaetes for example by dredging mud from the pond bottom or by conversion of earthen ponds to concrete raceways. Replacing the muddy substrate with coarse sand reduced the number of *Branchiura sowerbyi* which is the alternate oligochaete host for *Thelohanellus hovorkai*, mitigating the hemorrhagic thelohanellosis of carp (Liyanage et al., 2003). This was explained by a delicate body surface of *B. sowerbyi* was damaged by rugged-edged sand particles. Removing the vegetation upstream of the water inlet to a fish farm with PKD problems is considered as a possibility for prevention of PKD-outbreaks because it reduces habitats for bryozoan and spore load in the water, but this measure is not be feasible in most cases (de Kinkelin et al., 2002). Besides the substrate amendment for eliminating the habitat of invertebrate hosts, use of a benthos-eating fishes as a biological control of oligochaete abundance is worth considering in fish farms (Yokoyama et al., 2002).

5.2 Avoidance of infective period

If actinospore emergence occurs only in a certain time of the year, rearing fish outside the infective period may be useful. To reduce PKD-related losses, it is recommended to delay the transfer of young fish to the endemic water until autumn when water temperature is decreasing. As bryozoan populations decline under low temperature conditions in autumn, the number of *T. bryosalmonae* spores in the water becomes significantly low at this time. Additionally, low water temperature prevents the clinical outbreak of the disease and the fish usually become resistant against this pathogen in the subsequent years (Foott & Hedrick, 1987).

5.3 Removal of actinospores

Sand filtration is highly effective in removing actinospores of *M. cerebralis* (Arndt & Wagner, 2003, Nehring et al., 2003). However, it may be applied only for large-type actinospores. The sand filtration of water supply was also suggested to prevent enteromyxosis caused by *E. scophthalmi* in turbot farms, though the actinospores of this species have not been determined (Quiroga et al., 2006). In contrast, the infective stage of *K. thyrsites* was not removed by filtration of seawater (Moran et al., 1999b). Ozone and ultraviolet treatments of water supply are also effective for disinfection of *M. cerebralis* and *C. shasta* (Sanders et al., 1972, Tipping, 1988, Hedrick et al., 2000). Increase of water velocity may dilute the density of actinospores and reduce infection severity in culture facilities or rivers where water release is managed e.g. by dams (Hallett & Barthomew, 2008, Bjork & Bartholomew, 2009). Chemical treatments with toxic compounds may be effective but not be environmentally acceptable. Biological filtration of floating actinospores using a planktonic copepod (*Cyclops* spp.) may be practical in fish farms (Rácz et al., 2006). Murakami (1983) reported based on his empirical observations that rearing of rainbow trout, which is non-susceptible to *M. murakamii,* upstream of the masu salmon farm reduced the myxosporean sleeping disease, suggesting that rainbow trout plays a role of biological filter of the waterborne infective stage of *M. murakamii*. This may be explained by the nonspecific response of actinospore to fish mucus. The same was shown experimentally for *M. cerebralis* by preincubation of carp with actinospores of the parasite. Thereby, infection rate in susceptible rainow trout was reduced significantly (Kallert et al., 2009).

5.4 Interception of fish-to-fish transmission of *Enteromyxum* spp.

Fish-to-fish transmission of *Enteromyxum* spp. should be considered as an exceptional case. Because of their direct life cycle, epidemics of *Enteromyxum* frequently occurs in closed aquaculture systems. Effective control may be achieved with the integrated management strategies. The foremost strategy is to prevent infected fish from entering the culture system. Early diagnosis using a highly sensitive PCR assay greatly reduces the potential for dispersal of *E. leei* via infected juveniles (Yanagida et al., 2005). Fish farmers should minimize their risk of pathogen introduction by cultured fish from uncertified sources. Fish cages with different age classes should not be set up in a close proximity, because infection rate may increase with fish age, and older fish could spread the pathogen to younger fish. Infected fish must be removed as soon as the disease becomes apparent. Fallowing is also effective in intercepting the transmission cycle of *Enteromyxum* spp., if all fish farmers in the area cooperate in this practice at the same time. However, wild fish living in the farming area may act as carrier or reservoir of the parasite. In this case, the fallowing practice becomes useless, though *Enteromyxum* infection has never been detected from wild fish so far. Increase of water flow possibly lowers the chance of ingestion of infective parasites by dilution effect. Evacuation of sea cages to offshore area with faster water flow may reduce further transmission of *E. leei*. In case of land-based culture farms, increasing water exchange rates of rearing water would help to flush the waterborne parasite out of fish tank (Yokoyama et al., 2009). Development of *E. leei* is strongly influenced by water temperature. Rearing of Malabar grouper (*Epinephelus malabaricus*) at 30 °C had both for preventive and curative effects on *E. leei* infection, although a similar treatment was not effective in tiger puffer. Hyposalinity treatment below 1/4 seawater (8‰) was effective in killing developmental stages of *E. leei* in *in vitro* (Yokoyama & Shirakashi, 2007), but *in vivo* trials where tiger puffer were reared in low salinity seawater were unsuccessful for prevention of the disease. Further studies are required to clarify these inconsistent results.

6. Acknowledgment

This review is based on an invited lecture given in "*Kudoa thyrsites* workshop, emphasizing the current knowledge of Myxozoa biology, host-pathogen interaction, diagnostic methods, and impact on fish farm and fisheries industry", which was held at Campbell River, B.C., Canada, on 7-8 December, 2010. We greatly thank Dr. Luis Afonso of BC Centre for Aquatic Health Sciences, and Dr. Simon Jones of Pacific Biological Station, for their kind invitation to this workshop.

7. References

Anderson, C.L., Canning, E.U. & Okamura, B. (1998). A triploblast origin for Myxozoa? *Nature*, Vol. 392, No. 6674, March 1998, p. 346, ISSN 0028-0836

Arndt, R.E. & Wagner, E.J. (2003). Filtering *Myxobolus cerebralis* triactinomyxons from contaminated water using rapid sand filtration. *Aquacultural Engineering*, Vol. 29, No. 3-4, Dec. 2003, pp. 77-91, ISSN 0144-8609

Awakura, T. & Kimura, T. (1977). On the milky condition in smoked coho salmon (*Oncorhynchus kisutch*) caused by myxosporidian parasite. *Fish Pathology*, Vol. 12, No. 3, Dec. 1977, pp. 179-184, ISSN 0388-788X

Bartholomew, J.L., Whipple, M.J., Stevens, D.G. & Fryer, J.L. (1997). The life cycle of *Ceratomyxa shasta*, a myxosporean parasite of salmonids, requires a freshwater polychaete as an alternate host. *Journal of Parasitology*, Vol. 83, No. 5, pp. 859-868, ISSN 0022-3395

Beauchamp, K.A., Gay, M., Kelley, G.O., El-Matbouli, M., Kathman, R.D., Nehring, R.B. & Hedrick, R.P. (2002). Prevalence and susceptibility of infection to *Myxobolus cerebralis*, and genetic differences among populations of *Tubifex tubifex*. *Diseases of Aquatic Organisms*, Vol. 51, No. 2, Aug. 2002, pp. 113-121, ISSN 0177-5103

Belem, A.M.G. & Pote, L.M. (2001). Portals of entry and systemic localization of proliferative gill disease organisms in channel catfish Ictalurus punctatus. *Diseases of Aquatic Organisms*, Vol. 48, No. 1, Dec. 2000, pp. 37-42, ISSN 0177-5103

Bjork, S.J. & Bartholomew, J.L. (2009). The effects of water velocity on the *Ceratomyxa shasta* infection cycle. *Journal of Fish Diseases*, Vol. 32, No. 2, Feb. 2009, pp. 131-142, ISSN 0140-7775

Bjork, S.J. & Bartholomew, J.L. (2010). Invasion of *Ceratomyxa shasta* (Myxozoa) and comparison of migration to the intestine between susceptible and resistant fish hosts. *International Journal for Parasitology*, Vol. 40, No. 9, Aug. 2010, pp. 1087-1095, ISSN 0020-7519

Blazer, V.S., Waldrop, T.B., Schill, W.B., Densmore, C.L. & Smith, D. (2003). Effects of water temperature and substrate type on spore production and release in eastern *Tubifex tubifex* worms infected with *Myxobolus cerebralis*. *Journal of Parasitology*, Vol. 89, No. 1, Feb. 2003, pp. 21-26, ISSN 0022-3395

Canning, E.U., Curry, A., Feist, S.W., Longshaw, M. & Okamura, B. (1999). *Tetracapsula bryosalmonae* n. sp. for PKX organism, the cause of PKD in salmonid fish. *Bulletin of the European Association of Fish Pathologists*, Vol. 19, No. 1, Feb. 2000, pp. 203-206, ISSN 0022-3395

Canning, E.U., Curry, A., Feist, S.W., Longshaw, M. & Okamura, B. (2000). A new class and order of myxozoans to accommodate parasites of bryozoans with ultrastructural observations on *Tetracapsula bryosalmonae* (PKX organism). *Journal of Eukaryotic Microbiology*, Vol. 47, No. 5, Sep.-Oct. 2000, pp. 456-468, ISSN 1066-5234

Canning, E.U., Tops, S., Curry, A., Wood, T.S. & Okamura, B. (2002). Ecology, development and pathogenicity of *Buddenbrockia plumatellae* Schröder, 1910 (Myxozoa, Malacosporea) (syn. *Tetracapsula bryozoides*) and establishment of *Tetracapsuloides* n. gen. for *Tetracapsula bryosalmonae*. *Journal of Eukaryotic Microbiology*, Vol. 49, No. 4, Jul.-Aug. 2002, pp. 280-295, ISSN 1066-5234

Canning, E.U. & Okamura, B. (2004). Biodiversity and Evolution of the Myxozoa, In: *Advances in Parasitology*, Baker, J. R., Muller, R. & Rollinson, D., Vol. 56, pp. 43–131, Academic Press, ISBN 0-12-031751-6, London, England.

Canning, E.U., Curry, A., Hill, S.L.L. & Okamura, B. (2007). Ultrastructure of *Buddenbrockia allmani* n. sp. (Myxozoa, Malacosporea), a parasite of *Lophopus crystallinus* (Bryozoa, Phylactolaemata). *Journal of Eukaryotic Microbiology*, Vol. 54, No. 3, May-Jun. 2007, pp. 247-262, ISSN 1066-5234

de Kinkelin, P., Gay, M. & Forman, S. (2002). The persistence of infectivity of *Tetracapsula bryosalmonae*-infected water for rainbow trout, *Oncorhynchus mykiss* (Walbaum). *Journal of Fish Diseases*, Vol. 25, No. 1, Aug. 2002, pp. 477-482, ISSN 0140-7775

Diamant, A. (1997). Fish-to-fish transmission of a marine myxosporean. *Diseases of Aquatic Organisms*, Vol. 30, No. 2, Aug. 1997, pp. 99-105. ISSN 0177-5103

Diamant, A., Ucko, M., Paperna, I., Colomi, A. & Lipshitz, A. (2005). *Kudoa iwatai* (Myxosporea: Multivalvulida) in wild and cultured fish in the Red Sea: Redescription and molecular phylogeny. *Journal of Parasitology* Vol. 91, No. 6, Dec. 2005, pp. 1175-1189, ISSN 0022-3395

Diamant, A., Ram, S. & Paperna, I. (2006). Experimental transmission of *Enteromyxum leei* to freshwater fish. *Diseases of Aquatic Organisms*, Vol. 72, No. 2, Oct. 2006, pp. 171-178, ISSN 0177-5103

Dyková, I. & Lom, J. (1988). Review of pathogenic myxosporeans in intensive culture of carp (*Cyprinus carpio*) in Europe. *Folia Parasitologica*, Vol. 35, No. 4, pp. 289-307, ISSN 0015-5683

El-Mansy, A., Székely, C. & Molnár, K. (1998a). Studies on the occurrence of actinosporean stages of fish myxosporeans in a fish farm of Hungary, with description of Triactinomyxon, Raabeia, Aurantiactinomyxon and Neoactinomyxon types. *Acta Veterinaria Hungarica*, Vol. 46, No. 2, pp. 259-284, ISSN 0236-6290

El-Mansy, A., Székely, C. & Molnár, K. (1998b). Studies on the occurrence of actinosporean stages of fish myxosporeans in Lake Balaton, Hungary, with description of Triactinomyxon, Raabeia and Aurantiactinomyxon types. *Acta Veterinaria Hungarica*, Vol. 46, No. 4, pp. 437-450, ISSN 0236-6290

El-Matbouli, M. & Hoffman, R.W. (1994). Proliferative kidney disease (PKD) as an important myxosporean infection in salmonid fish, In: *Parasitic Diseases of Fish*, Pike, A.W. & Lewis, J.W., pp. 3-15, Samara Publishing Ltd, ISBN 1873692021, Samara House, Tresaith, Dyfed, UK.

El-Matbouli, M., Hoffmann, R.W. & Mandok, C. (1995). Light and electron microscopic observations on the route of the triactinomyxon-sporoplasm of *Myxobolus cerebralis* from epidermis into rainbow trout cartilage. *Journal of Fish Biology*, Vol. 46, No. 6, pp. 919-935, ISSN 1095-8649

El-Matbouli, M. & Hoffmann, R.W. (1998) Light and electron microscopic studies on the chronological development of *Myxobolus cerebralis* to the actinosporean stage in *Tubifex tubifex*. *International Journal for Parasitology*, Vol. 28, No. 1, Jan. 1998, pp. 195-217, ISSN 0020-7519

El-Matbouli, M., McDowell, T.S., Antonio, D.B., Andree, K.B. & Hedrick, R.P. (1999a). Effect of water temperature on the development, release and survival of the triactinomyxon stage of *Myxobolus cerebralis* in its oligochaete host. *International Journal for Parasitology*, Vol. 29, No. 4, Apr. 1999, pp. 627-641, ISSN 0020-7519

El-Matbouli, M., Hoffmann, R.W., Shoel, H., McDowell, T.S. & Hedrick, R.P. (1999b). Whirling disease: host specificity and interaction between the actinosporean stage of *Myxobolus cerebralis* and rainbow trout *Oncorhynchus mykiss*, *Diseases of Aquatic Organisms*, Vol. 35, Vol. 1, Jan. 1999, pp. 1-12, ISSN 0177-5103

El-Matbouli, M. & Hoffmann, R.W. (2002). Influence of water quality on the outbreak of proliferative kidney disease - field studies and exposure experiments. *Journal of Fish Diseases*, Vol. 25, No. 8, Aug. 2002, pp. 459-467, ISSN 0140-7775

Feist, S.W. & Longshaw, M. (2006). Phylum Myxozoa, In: *Fish Diseases and Disorders, Protozoan and Metazoan Infections,* Vol. 1, P.T.K. Woo, (Ed.), 230-296, CABI, ISBN 0851990150, Wallingford, U.K.

Ferguson, H.W. (1981). The effects of water temperature on the development of proliferative kidney disease in rainbow trout, *Salmo gairdneri* Richardson. *Journal of Fish Diseases*, Vol. 4, No. 2, Mar. 1981, pp. 175-177, ISSN 1365-2761

Fiala, I. (2006). The phylogeny of Myxosporea (Myxozoa) based on small subunit ribosomal RNA gene analysis. *International Journal for Parasitology*, Vol. 36, No. 14, Dec. 2006, pp. 1521-1534, ISSN 0020-7519

Fiala, I., & Bartosová, P. (2010). History of myxozoan character evolution on the basis of rDNA and EF-2 data. *BMC evolutionary biology*, Vol. 10, 228. doi: 10.1186/1471-2148-10-228, ISSN 1471-2148, http://www.biomedcentral.com/1471-2148/10/228.

Foott, J.S., Hedrick, R.P. (1987). Seasonal occurrence of the infectious stage of proliferative kidney disease (PKD) and resistance of rainbow trout, *Salmo gairdneri* Richardson, to reinfection. *Journal of Fish Biology*, Vol. 30, No. 4, Apr. 1987, pp. 477-483, ISSN 1095-8649

Gilbert, M.A. & Granath, Jr., W.O. (2001). Persistent infection of *Myxobolus cerebralis*, the causative agent of salmonid whirling disease, in *Tubifex tubifex*. *Journal of Parasitology*, Vol. 87, No. 1, Feb. 2001, pp. 101-107, ISSN 0022-3395

Grabner, D.S. & El-Matbouli, M. (2008). Transmission of *Tetracapsuloides bryosalmonae* (Myxozoa: Malacosporea) to *Fredericella sultana* (Bryozoa: Phylactolaemata) by various fish species. *Diseases of Aquatic Organisms*, Vol. 79, No. 2, Apr. 2008, pp. 133-139, ISSN 0177-5103

Grabner, D.S. & El-Matbouli, M. (2010a). Experimental transmission of malacosporean parasites from bryozoans to common carp (*Cyprinus carpio*) and minnow (*Phoxinus phoxinus*). *Parasitology*, Vol. 137, No. 4, Apr. 2010, pp. 629-639, ISSN 0031-1820

Grabner, D.S. & El-Matbouli, M. (2010b). Investigations on the portal of entry of *Tetracapsuloides bryosalmonae* (Myxozoa; Malacosporea) into the fish. *Diseases of Aquatic Organisms*, Vol. 90, No. 3, July 2010, pp. 199-208, ISSN 0177-5103

Grossel, G.W., Dyková, I., Handlinger, J. & Munday, B.L. (2003). *Pentacapsula neurophila* n. sp. (Multivalvulida) from the central nervous system of striped trumpeter, *Latris lineate* (Forster). *Journal of Fish Diseases*, Vol. 26, No. 6, Jun. 2003, pp. 315-320, ISSN 0140-7775

Gunter, N. & Adlard, R. (2010). The demise of *Leptotheca* Thélohan, 1895 (Myxozoa: Myxosporea: Ceratomyxidae) and assignment of its species to *Ceratomyxa* Thélohan, 1892 (Myxosporea: Ceratomyxidae), *Ellipsomyxa* Køie, 2003 (Myxosporea: Ceratomyxidae), *Myxobolus* Bütschli, 1882 and *Sphaerospora* Thélohan, 1892 (Myxosporea: Sphaerosporidae). *Systematic Parasitology*, Vol. 75, No. 2, Feb. 2010, pp. 81-104, ISSN 0165-5752

Hallett, S.L., O'donoghue, P.J. & Lester, R.J.G. (1998). Structure and development of a marine actinosporean, *Sphaeractinomyxon ersei* n. sp. (Myxozoa). *Journal of Eukaryotic Microbiology*, Vol. 45, No. 1, Jan. 1998, pp. 142-150, ISSN 1066-5234

Hallett, S.L. & Lester, R.J.G. (1999). Actinosporeans (Myxozoa) with four developing spores within a pansporocyst: *Tetraspora discoidea* n. g. n. sp. and *Tetraspora rotundum* n. sp. *International Journal for Parasitology*, Vol. 29, No. 3, March 1999, pp. 419-427, ISSN 0020-7519

Hallett, S.L., Erséus, C. & Lester, R.J.G. (1999). Actinosporeans (Myxozoa) from marine oligochaetes of the Great Barrier Reef. *Systematic Parasitology*, Vol. 44, No. 1, Sep. 1999, pp. 49-57, ISSN 0165-5752

Hallett, S.L. & Bartholomew, J.L. (2008). Effects of water flow on the infection dynamics of *Myxobolus cerebralis*. *Parasitology*, Vol. 135, No. 3, March 2008, pp. 371-384, ISSN 0031-1820

Hedrick, R.P., MacConnell, E. & De Kinkelin, P. (1993). Proliferative kidney disease of salmonid fish. *Annual Review of Fish Diseases*, Vol. 3, pp. 277-290, ISSN 0044-8486

Hedrick, R.P., El-Matbouli, M., Adkison, M.A. & MacConnell, E. (1998). Whirling disease: re-emergence among wild trout. *Immunological Reviews*, Vol. 166, Vol. 1, Dec. 1998, pp. 365–376, ISSN 0105-2896

Hedrick,R.P., McDowell, T.S., Marty, G.D., Mukkatira, K., Antonio, D.B., Andree, K.B., Bukhari, Z. & Clancy, T. (2000). Ultraviolet irradiation inactivates the waterborne infective stages of *Myxobolus cerebralis*: a treatment for hatchery water supplies. *Diseases of Aquatic Organisms*, Vol. 42. No. 1, Aug. 2000, pp. 53-59, ISSN 0177-5103

Hedrick, R.P., Baxa, D.V., De Kinkelin, P. & Okamura, B. (2004). Malacosporean-like spores in urine of rainbow trout react with antibody and DNA probes to *Tetracapsuloides bryosalmonae*. *Parasitology Research*, Vol. 92, No. 2, Nov. 2003, pp. 81-88, ISSN 0932-0113

Hill, S.L.L. & Okamura, B. (2007). Endoparasitism in colonial hosts: pattern and processes. *Parasitology*, Vol. 134, No. Pt. 6, Feb. 2007, pp. 841-852, ISSN 0031-1820

Hoffmann, R. & Dangschat, H. (1981). A note on the occurrence of proliferative kidney disease in Germany. *Bulletin of the European Association of Fish Pathologists*, Vol. 1, No. 2, p. 33, ISSN 0108-0288

Holland, J.W., Okamura, B., Hartikainen, H. & Secombes, C.J. (2010). A novel minicollagen gene links cnidarians and myxozoans. *Proceedings of the Royal Society B., Biological Sciences*, Vol. 278, No. 1705, Sep. 2010, pp. 546-553, ISSN 1471-2954

Holzer, A.S., Sommerville, C., & Wootten, R. (2003). Tracing the route of *Sphaerospora truttae* from the entry locus to the target organ of the host, *Salmo salar* L., using an optimized and specific in situ hybridization technique. *Journal of Fish Diseases*, Vol. 26, No. 11-12, Nov. 2003, pp. 647 -655, ISSN 0140-7775

Holzer, A.S., Sommerville, C. & Wootten, R. (2004). Molecular relationships and phylogeny in a community of myxosporeans and actinosporeans based on their 18S rDNA sequences. *International Journal for Parasitology*, Vol. 34, No. 10, Sep. 2004, pp. 1099-1111, ISSN 0020-7519

Holzer, A.S., Sommerville, C. & Wootten, R. (2006). Molecular studies on the seasonal occurrence and development of five myxozoans in farmed *Salmo trutta* L. *Parasitology*, Vol. 132, No. Pt. 2, Oct. 2005, pp. 193-205, ISSN 0031-1820

Ikeda, J. (1912). Studies on some sporozoan parasites of sipunculoids. I. The life history of a new Actonomyxidian, *Tetractinomyxon intermedium* g. et. sp. nov. *Archiv für Protistenkunde*, Vol. 25, pp. 240-242, ISSN 0003-9365

Jimenez-Guri, E., Philippe, H, Okamura, B. & Holland, P.W.H. (2007). *Buddenbrockia* is a cnidarian worm. *Science*, Vol. 317, No. 5834, Jul. 2007, pp. 116-118, ISSN 0036-8075

Kallert, D.M., El-Matbouli, M. & Haas, W. (2005). Polar filament discharge of *Myxobolus cerebralis* actinospores is triggered by combined non-specific mechanical and chemical cues. *Parasitology*, Vol. 131. No. 5, Nov. 2005, pp. 609-616, ISSN 0031-1820

Kallert, D.M., Ponader, S., Eszterbauer, E., El-Matbouli, M. & Haas, W. (2007). Myxozoan transmission via actinospores: new insights into mechanisms and adaptations for host invasion. *Parasitology*, Vol. 134, No. 12, Nov. 2007, pp. 1741-1750, ISSN 0031-1820

Kallert, D. M., & El-Matbouli, M. (2008). Differences in viability and reactivity of actinospores of three myxozoan species upon ageing. *Folia Parasitologica,* Vol. 55, No. 2, pp. 105-110, ISSN 0015-5683

Kallert, D., Eszterbauer, E., Grabner, D., & El-Matbouli, M. (2009). *In vivo* exposure of susceptible and non-susceptible fish species to *Myxobolus cerebralis* actinospores reveals non-specific invasion behaviour. *Diseases of Aquatic Organisms,* Vol. 84, No. 2, April 2009, pp. 123-130, ISSN 0177-5103

Kallert, D., Ponader, S., Adelt, S., Kaese, P., Geyer, R, Haas, W. & El-Matbouli, M. (2010). Analysis of rainbow trout *Oncorhynchus mykiss* epidermal mucus and evaluation of semiochemical activity for polar filament discharge in *Myxobolus cerebralis* actinospores. *Journal of Fish Biology,* Vol. 77, No. 7, Nov. 2010, pp. 1579-1598, ISSN 0022-1112

Kallert, D., Bauer, W., Haas, W. & El-Matbouli, M. (2011). No shot in the dark: Myxozoans chemically detect fresh fish. *International Journal for Parasitology,* Vol. 41, No. 3-4, Mar. 2011, pp. 271-276, ISSN 0020-7519

Karlsbakk, E., Sæther, B.A., Høstlund, C., Fjellsøy, K.R. & Nylund, A. (2002). *Parvicapsula pseudobranchicola* n. sp. (Myxozoa), a myxoxporidian infecting the pseudobranch of cultured atlantic salmon (*Salmo salar*) in Norway. *Bulletin of the European Association of Fish Pathologists,* , Vol. 22, Vol. 6, pp. 381-387, ISSN 0022-3395

Kent, M.L., Khattra, J., Hervio, D.M.L. & Devlin, R.H. (1998). Ribosomal DNA sequence analysis of isolates of the PKX myxosporean and their relationship to members of the genus *Sphaerospora. Journal of Aquatic Animal Health,* Vol. 10, No. 1, pp. 12-21, ISSN 0899-7659

Kent, M.L., Khattra, J., Hedrick, R.P. & Devlin, R.H. (2000). *Tetracapsula renicola* n. sp. (Myxozoa: Saccosporidae); the PKX myxozoan – the cause of proliferative kidney disease of salmonid fishes. *Journal of Parasitology,* Vol. 86, No. 1, Feb. 2000, pp. 103-111, ISSN 0022-3395

Kent, M.L., Andree, K.B., Bartholomew, J.L., El-Matbouli, M., Desser, S.S., Devlin, R.H., Feist, S.W., Hedrick, R.P., Hoffmann, R.W., Khattra, J., Hallett, S.L., Lester, R.J., Longshaw, M., Palenzuela, O., Siddal, M.E. & Xiao, C. (2001). Recent advances in our knowledge of the Myxozoa. *Journal of Eukaryotic Microbiology,* Vol. 48, No. 4, Jul. 2001, pp. 395-413, ISSN 1066-5234

Kerans, B.L. & Zale, A.V. (2002). The ecology of *Myxobolus cerebralis.* In: *Whirling Disease: Reviews and Current Topics,* J.L. Barthomew & J.C. Wilson, (Eds.), American Fisheries Society Symposium 29, pp. 61-76, ISBN 1-888569-37-9

Køie, M. (2002). Spirorbid and sepulid polychaetes are candidates as invertebrate hosts for Myxozoa. *Folia Parasitologica,* Vol. 49, No. 2, pp. 160-162. ISSN 0015-5683

Køie, M. (2005). The Spionidae (Polychaeta) act as invertebrate hosts for marine Myxozoa. *Bulletin of the European Association of Fish Pathologists,* 25, No. 4, pp. 179-181, ISSN 0022-3395

Køie, M., Whipps, C.M. & Kent, M.L. (2004) *Ellipsomyxa gobii* (Myxozoa: Ceratomyxidae) in the common goby *Pomatoschistus microps* (Annelida: Polychaeta) uses *Nereis* spp. (Annelida: Polychaeta) as invertebrate hosts. *Folia Parasitologica,* Vol. 51, No. 1, pp. 14-18. ISSN 0015-5683

Køie, M., Karlsbakk, E. & Nylund, A. (2007) A new genus *Gadimyxa* with three new species (Myxozoa, Parvicapsulidae) parasitic in marine fish (Gadidae) and the two-host life cycle of *Gadimyxa atlantica* n. sp. *Journal of Parasitology*, Vol. 93, No. 6, Dec. 2007, pp. 1459-1467, ISSN 0022-3395

Køie, M., Karlsbakk, E., & Nylund A. (2008) The marine herring myxozoan *Ceratomyxa auerbachi* (Myxozoa: Ceratomyxidae) uses *Chone infundibuliformis* (Annelida: Polychaeta: Sabellidae) as invertebrate host. *Folia Parasitologica*, Vol. 55, No. 2, pp. 100-104, ISSN 0015-5683

Koprivnikar, J., Koehler, A., Rodd, F.H. & Desser, S.S. (2002). Environmental factors affecting the distribution and abundance of cyst-forming *Myxobolus* spp. and their cyprinid hosts in 3 lakes in Algonquin Park, Ontario, *Journal of Parasitology*, Vol. 88, No. 3, Jun. 2002, pp. 467-473, ISSN 0022-3395

Liyanage, Y.S., Yokoyama, H. & Wakabayashi, H. (2003) Evaluation of vector-control strategy of haemorrhagic thelohanellosis in carp, caused by *Thelohanellus hovorkai* (Myxozoa). *Diseases of Aquatic Organisms*, Vol. 55, No. 1, Jun. 2003, pp. 31-35, ISSN 0177-5103

Lom, J. & Arthur, J.R. (1989). A guideline for the preparation of species descriptions in Myxosporea. *Journal of Fish Diseases*, Vol. 12, No. 2, March, 1989, pp. 151-156, ISSN 1365-2761

Lom, J. & Dyková, I. (1992). *Protozoan Parasites of Fishes. Developments in Aquaculture and Fisheries Science*, Vol. 26, Elsevier, ISBN 0-444-89434-9, Amsterdam, The Netherlands

Lom, J. & Dyková, I. (2006). Myxozoan genera: definition and notes on taxonomy, life-cycle terminology and pathogenic species *Folia Parasitologica*, Vol.53, No. 1, pp. 1-36, ISSN 0015-5683

Lom, J., McGeorge, J., Feist, S.W., Morris, D. & Adams, A. (1997): Guidelines for the uniform characterisation of the actinosporean stages of parasites of the phylum Myxozoa. *Diseases of Aquatic Organisms*, Vol. 30, No. 1, Jul. 1997, pp. 1-9, ISSN 0177-5103

Longshaw, M., le Deuff, R.-M., Harris, A.F. & Feist, S.W. (2002). Development of proliferative kidney disease in rainbow tout, *Oncorhynchus mykiss* (Walbaum), following short-term exposure to *Tetracapsula bryosalmonae* infected bryozoans. *Journal of Fish Diseases*, Vol. 25, No. 8, Aug. 2002, pp. 443-449, ISSN 1365-2761

Markiw, M.E. (1989). Portals of entry for salmonid whirling disease in rainbow trout. *Diseases of Aquatic Organisms*, Vol. 6, pp. 7-10. ISSN 0177-5103

Markiw, M. E. (1992) Experimentally induced whirling disease. II Determination of longevity of the infective triactinomyxon stage of *Myxobolus cerebralis* by vital staining. *Journal of Aquatic Animal Health*, Vol. 4, No. 1, pp. 44-47. ISSN 0899-7659

McGeorge, J., Sommerville, C. & Wootten, R. (1997). Studies on actinosporean myxozoan stages parasitic in oligochaetes from the sediments of a hatchery where Atlantic salmon harbour *Sphaerospora truttae* infection. *Diseases of Aqautic Organisms*, Vol. 30, No. 2, Aug. 1997, pp. 107-119, ISSN 0177-5103

McGurk, C., Morris, D.J., Bron, J.E. & Adams, A. (2005). The morphology of *Tetracapsuloides bryosalmonae* (Myxozoa: Malacosporea) spores released from *Fredericella sultana* (Bryozoa: Phylactolaemata). *Journal of Fish Diseases*, Vol. 28, No. 5, May 2005, pp. 307-312, ISSN 1365-2761

McGurk, C., Morris, D.J. & Adams, A. (2006a). Sequential development of *Buddenbrockia plumatellae* (Myxozoa: Malacosporea) within *Plumatella repens* (Bryozoa: Phylactolaemata). *Diseases of Aquatic Organisms*, Vol. 73, No. 2, Dec. 2006, pp. 159-169, ISSN 0177-5103

McGurk, C., Morris, D.J., Auchinachie, N.A. & Adams, A. (2006b). Development of *Tetracapsuloides bryosalmonae* (Myxozoa: Malacosporea) in bryozoan hosts (as examined by light microscopy) and quantitation of infective dose to rainbow trout *(Oncorhynchus mykiss)*. *Veterinary Parasitology*, Vol. 135, No. 3-4, Feb. 2006, pp. 249-257, ISSN 0304-4017

Meaders, M.D. & Hendrickson, G.L. (2009). Chronological development of *Ceratomyxa shasta* in the polychaete host, *Manayunkia speciosa*. *Journal of Parasitology*, Vol. 95, No. 6, Dec. 2009, pp. 1397-1407, ISSN 0022-3395

Molnár, K. & Kovács-Gayer, E. (1985). The pathogenicity and development within the host fish of *Myxobolus cyprini* Doflein, 1898. *Parasitology*, Vol. 90, No. 3, Jun. 1985, pp. 549-555, ISSN 0031-1820

Moran, J.D.W., Whitaker, D.J. & Kent, M.L. (1999a). A review of the myxosporean genus *Kudoa* Meglitsch, 1947, and its impact on the international aquaculture industry and commercial fisheries. *Aquaculture*, Vol. 172, No. 1-2, March 1999, pp. 163-196, ISSN 0044-8486

Moran, J.D.W., Whitaker, D.J. & Kent, M.L. (1999b). Natural and laboratory transmission of the marine myxozoan parasite *Kudoa thyrsites* to Atlantic salmon. *Journal of Aquatic Animal Health*, Vol. 11, No. 2, pp. 110-115, ISSN 0899-7659

Morris, D.J., Adams, A., Feist, S.W., McGeorge, J. & Richards, R.H. (2000a). Immunohistochemical and PCR studies of wild fish for *Tetracapsula bryosalmonae* (PKX), the causative organism of proliferative kidney disease. *Journal of Fish Diseases*, Vol. 23, No. 2, Mar. 2000, pp. 129-135, ISSN 1365-2761

Morris, D.J., Adams, A. & Richards, R.H. (2000b). In situ hybridisation identifies the gill as a portal of entry for PKX (Phylum Myxozoa), the causative agent of proliferative kidney disease in salmonids. *Parasitology Research*, Vol. 86, No. 12, Dec. 2000, pp. 950-956, ISSN 0932-0113

Morris, D.J., Morris, D.C. & Adams, A. (2002). Development and release of a malacosporean (Myxozoa) from *Plumatella repens* (Bryozoa: Phylactolaemata). *Folia Parasitologica (Praha)*, Vol. 49, No. 1, pp. 25-34, ISSN 0015-5683

Morris, D.J. & Adams, A. (2006a). Transmission of *Tetracapsuloides bryosalmonae* (Myxozoa: Malacosporea), the causative organism of salmonid proliferative kidney disease, to the freshwater bryozoan *Fredericella sultana*. *Parasitology*, Vol. 133, No. 6, Dec. 2006, pp. 701-709, ISSN 0031-1820

Morris, D.J. & Adams, A. (2006b). Transmission of freshwater myxozoans during the asexual propagation of invertebrate hosts. *International Journal for Parasitology*, Vol. 36, No. 3, Mar. 2006, pp. 371-377, ISSN 0020-7519

Morris, D.J. & Adams, A. (2008). Sporogony of *Tetracapsuloides bryosalmonae* in the brown trout *Salmo trutta* and the role of the tertiary cell during the vertebrate phase of myxozoan life cycles. *Parasitology*, Vo. 135, No. 9, Aug. 2008, pp. 1075-1092, ISSN 0031-1820

Murakami, Y. (1983). Studies on the control of fish diseases. (1) Sleeping disease of yamame and amago. *Annual Report of Freshwater Fisheries Experimental Station of Hiroshima Prefecture, Fiscal 1981*, pp. 14-15. (In Japanese.)

Nehring, R.B., Thompson, K.G., Taurman, K., & Atkinson, W. (2003). Efficacy of passive sand filtration in reducing exposure of salmonids to the actinospore of *Myxobolus cerebralis. Diseases of Aquatic Organisms*, Vol. 57, No. 1-2, Dec. 2003, pp. 77-83. ISSN 0177-5103

Okamura, B., Curry, A., Wood, T.S. & Canning, E.U. (2002). Ultrastructure of *Buddenbrockia* identifies it as a myxozoan and verifies the bilaterian origin of the Myxozoa. *Parasitology*, Vo. 124, No. 2, Feb. 2002, pp. 215-223, ISSN 0031-1820

Okamura, B., Hartikainen, H., Schmidt-Posthaus, H., & Wahli, T. (2011). Life cycle complexity, environmental change and the emerging status of salmonid proliferative kidney disease. *Freshwater Biology*, Vol. 56, No. 4, Apr. 2011, pp. 735-753, ISSN 1365-2427

Özer, A. & Wootten, R. (2001). Release of actinosporean and myxosporean spores from their hosts, with special reference to both stages of *Sphaerospora truttae* (Myxozoa, Myxosporea). *Acta Parasitologica*, Vol. 46, No. 2, Apr. 2001, pp. 103-112, ISSN 1230-2821

Özer, A. & Wootten, R. (2002). Biological characteristic of some actinosporeans. *Journal of Natural History*, Vol. 36, No. 18, pp. 2199-2209,ISSN 0022-2933

Padrós, F., Palenzuela, O., Hispano, C., Tosas, O., Zaraza,, C., Crespo, S. & Alvarez-Pellitero, P. (2001). *Myxidium leei* (Myxozoa) infections in aquarium-reared Mediterranean fish species. *Diseases of Aquatic Organisms*, Vol. 47, No. 1, Oct. 2001, pp. 57-62, ISSN 0177-5103

Palenzuela, O., Redondo, M.J. & Alvarez-Pellitero, P. (2002). Description of *Enteromyxum scophtahlmi* gen. nov., sp. nov. (Myxozoa), an intestinal parasite of turbot (*Scophtahlmus maximus* L.) using morphological and ribosomal RNA sequence data. *Parasitology*, Vol. 124, No. 4, Apr. 2002, pp. 369-379, ISSN 0031-1820

Pote, L.M., Hanson, L.A. & Shivaji, R. (2000). Small subunit ribosomal RNA sequences link the cause of proliferative gill disease in channel catfish to *Henneguya* n. sp. (Myxozoa: Myxosporea). *Journal of Aquatic Animal Health*, Vol. 12, No. 3, pp. 230-240, ISSN 0899-7659

Quiroga, M.I., Redondo, M.J., Sitjà-Bobadilla, A., Palenzuela, O., Riaza, A., Macías, A., Vázquez, S., Perez, A., Nieto, J.M. & Alvarez-Pellitero, P (2006). Risk factors associated with *Enteromyxum scophthalmi* (Myxozoa) infection in cultured turbot, *Scophthalmus maximus* (L.). *Parasitology*, Vol. 133, No. 4, Jun 2006, pp. 433-442, ISSN 0031-1820

Rácz, O., Székely, Cs. & Molnár, K. (2006). The role of copepods (*Cyclops* spp.) in eliminating the actinospore stages of fish-parasitic myxozoans. *Acta Veterinaria Hungarica*, Vol. 54, No. 1, Jan. 2006, pp. 61-70, ISSN 0236-6290

Rangel, L.F., Santos, M.J., Cech, G. & Székely, Cs. (2009). Morphology, molecular data, and development of *Zschokkella mugilis* (Myxosporea, Bivalvulida) in a polychaete host, *Nereis diversicolor. Jounal of Parasitology*, Vol. 95, No. 3, Jun. 2009, pp. 561-569, ISSN 0022-3395

Rangel, L.F., Cech, G., Székely, Cs. & Santos, M.J. (2011). A new actinospore type Unicapsulactinomyxon (Myxozoa), infecting the marine polychaete, *Diopatra neapolitana* (Polychaeta: Onuphidae) in the Aveiro Estuary (Portugal). *Parasitology*, Vol. 138, No. 6, May 2011, pp. 698-712, ISSN 0031-1820

Redondo, M.J., Palenzuela, O., Riaza, A., Macias, A & Alvarez-Pelitero, P. (2002). Experimental transmission of *Enteromyxum scophthalmi* (Myxozoa), an enteric parasite of turbot *Scophthalmus maximus*. *Journal of Parasitology*, Vol. 88, No. 3, Jun. 2002, pp. 482-488, ISSN 0022-3395

Redondo, M.J., Palenzuela, O. & Alvarez-Pellitero, P. (2003). *In vitro* studies on viability and proliferation of *Enteromyxum scophthalmi* (Myxozoa), an enteric parasite of cultured turbot *Scophthalmus maximus*. *Diseases of Aquatic Organisms*, Vol. 55, No. 2, Jul. 2003, pp. 133-144, ISSN 0177-5103

Redondo, M.J. & Alvarez-Pellitero, P. (2009). Lectin histochemical detection of terminal carbohydrate residues in the enteric myxozoan *Enteromyxum leei* parasitizing gilthead seabream *Sparus aurata* (Pisces: Teleostei) : a stdy using light and transmission electron microscopy. *Folia Parasitologica*, Vo. 56, Vol. 4, Dec. 2009, pp. 259-267, ISSN 0015-5683

Redondo, M.J. & Alvarez-Pellitero, P. (2010). The effect of lectins on the attachment and invasion of *Enteromyxum scophthalmi* (Myxozoa) in turbot (*Psetta maxima* L.) intestinal epithelium *in vitro*. *Experimental Parasitlogy*, Vol. 126, No. 4, Dec. 2010, pp. 577-581, ISSN 0014-4894

Roubal, F.R., Hallett, S.L. & Lester, R.J.G. (1997). First record of triactinomyxon actinosporean in marine oligochaete. *Bulletin of the European Association of Fish Pathologists*, Vol. 17, No. 3-4, pp. 83-85, ISSN 0022-3395

Sanders, J.E., Fryer, J.L.D., Leith, A. & Moore, K.D. (1972) Control of the infectious protozoan *Ceratomyxa shasta* by treating hatchery water supplies. *Progressive Fish-Culturist*, Vol. 34, No. 1, pp. 13–17, ISSN 0033-0779

Saulnier, D., Philippe, H. & De Kinkelin, P. (1999). Molecular evidence that the proliferative kidney disease organism unknown (PKX) is a myxosporean. *Diseases of Aquatic Organisms*, Vol. 36, No. 3, May 1999, pp. 209-212, ISSN 0177-5103

Schlegel, M., Lom, J., Stechmann, A., Bernhard, D., Leipe, D., Dykova, I. & Sogin, M.L. (1996). Phylogenic analysis of complete small subunit ribosomal RNA coding region of *Myxidium lieberkuehni*: evidence that Myxozoa are Metazoa and related to the Bilateria. *Archiv für Protistenkunde*, Vol. 147, No. 1, May 1996, pp. 1-9, ISSN 0003-9365

Seagrave, C.P., Bucke, D., Hudson, E.B. & McGregor, D. (1981). A survey of the prevalence and distribution of proliferative kidney disease (PKD) in England and Wales. *Journal of Fish Diseases*, Vol. 4, No. 5, Sept. 1981, pp. 437-439, ISSN 1365-2761

Shirakashi, S. & El-Matbouli (2010). Effect of cadmium on the susceptibility of *Tubifex tubifex* to *Myxobolus cerebralis* (Myxozoa), the causative agent of whirling disease. *Diseases of Aquatic Organisms*, Vol. 89, No. 1, Feb. 2010, pp. 63-70, ISSN 0177-5103

Siddall, M.E., Martin, D.S., Bridge, D., Desser, S.S. & Cone, D.K. (1995). The demise of a phylum of protists: phylogeny of the Myxozoan and other parasitic cnidaria. *Journal of Parasitology*, Vol. 81, No. 6, pp. 961-967, ISSN 0022-3395

Sitjà-Bobadilla, A., Diamant, A., Palenzuela, O. & Alvarez-Pellitero, P. (2007). Effect of host factors and experimental conditions on the horizontal transmission of *Enteromyxum leei* (Myxozoa) to gilthead sea bream, *Sparus aurata* L., and European sea bass, *Dicentrarchus labrax* (L.). *Journal of Fish Diseases*, Vol. 30, No. 4, Apr. 2007, pp. 243-250., ISSN 0140-7775

Smothers, J.F., von Dohlen, C.D., Smith, Jr., L.H. & Spall, R.D. (1994). Molecular evidence that the myxozoan protists are metazoans. *Science*, Vol. 265, No. 5179, Sep. 1994, pp. 1719-1721, ISSN 0036-8075

Supamattaya, K., Fischer-Scherl, T., Hoffmann, R.W. & Boonyaratpalin, S. (1991). *Sphaerospora epinepheli* n. sp. (Myxosporea, Sphaerosporidae) observed in grouper (*Epinephelus malabaricus*). *Journal of Protozoology*, Vol. 38, Vol. 5, Sep. 1991, pp. 448-454, ISSN 1066-5234

Tin Tun, Yokoyama, H., Ogawa, K. & Wakabayashi, H. (2000). Myxosporeans and their hyperparasitic microsporeans in the intestine of emaciated tiger puffer. *Fish Pathology*, Vol. 35, No. 3, Sep. 2000, pp. 145-156, ISSN 0388-788X

Tipping, J.M. (1988). Ozone control of ceratomyxosis: survival and growth benefits to steelhead and cutthroat trout. *Progressive Fish-Culturist*, Vol. 50, No. 4, pp. 202-210 ISSN 0033-0779

Urawa, S., Iida, Y., Freeman, M.A., Yanagida, T., Karlsbakk, E. & Yokoyama, H. (2009). Morphological and molecular comparisons of *Myxobolus* spp. in the nerve tissues of salmonid fishes with the description of *Myxobolus murakamii* n. sp., the causative agent of myxosporean sleeping disease. *Fish Pathology*, Vol. 44, No. 2, Jun. 2009, pp. 72-80, ISSN 0388-788X

Uspenskaya, A.V. (1995). Alternation of actinosporean and myxosporean phases in the life cycle of *Zschokkella nova* (Myxozoa). *Journal of Eukaryotic Microbiology*, Vo. 42, No. 6, Nov. 1995, pp. 665-668, ISSN 1066-5234

Ventura, M.T., & Paperna, I. (1984). Histopathology of *Myxidium giardi* Cépède 1906 infection in European eels, *Anguilla anguilla* L., in Portugal, *Aquaculture*, Vol. 43, No. 4, Dec. 1984, pp. 357–368, ISSN 0044-8486

Wagner, E.J., Cannon, Q., Smith, M., Hillyard, R. & Arndt, R. (2002). Extrusion of polar filaments of the *Myxobolus cerebralis* triactinomyxon by salts, electricity, and other agents. In: *Whirling Disease: Reviews and Current Topics*, J.L. Barthomew & J.C. Wilson, (Eds.), American Fisheries Society Symposium 29, pp. 61-76, ISBN 1-888569-37-9

Wagner, E.J., Smith, M., Arndt, R. & Roberts, D.W. (2003). Physical and chemical effects on viability of the *Myxobolus cerebralis* triactinomyxon. *Diseases of Aquatic Organisms*, Vol. 53, No. 2, Feb. 2003, pp. 133-142, ISSN 0177-5103

Wang, P.C., Huang, J.P., Tsai, M.A., Cheng, S.Y., Tsai, S.S., Chen, S.D., Chen, S.P., Chiu, S.H., Liaw, L.L., Chang, L.T. & Chen, S.C. (2005). Systemic infection of *Kudoa lutjanus* n. sp. (Myxozoa: Myxosporea) in red snapper *Lutjanus erythropterus* from Taiwan. *Diseases of Aquatic Organisms* Vol. 67, No. 1-2, Nov. 2005, pp. 115-124, ISSN 0177-5103

Wolf, K. & Markiw, M.E. (1984). Biology contravenes taxonomy in the Myxozoa: New discoveries show alternation of invertebrate and vertebrate hosts. *Science*, Vol. 225, No. 4669, Sep. 1984, pp. 1449-1452, ISSN 0036-8075

Xiao, C. & Desser, S.S. (2000a). Cladistic analysis of myxozoan species with known alternating life-cycles. *Systematic Parasitology*, Vol. 46, No. 2, Jun. 2000, pp. 81-91, ISSN 0165-5752

Xiao, C. & Desser, S.S. (2000b) The longevity of actinosporean spores from oligochaetes of Lake Sasajewun, Algonquin Park, Ontario, and their reaction to fish mucus. *Journal of Parasitology*, Vol. 86, No. 1, Feb. 2000, pp. 193-195, ISSN 0022-3395

Yanagida, T., Freeman, M.A., Nomura, Y., Takami, I., Sugihara, Y., Yokoyama, H. & Ogawa, K. (2005). Development of a PCR-based method for the detection of enteric myxozoans causing the emaciation disease of cultured tiger puffer. *Fish Pathology*, Vol. 40, No. 1, March 2005, pp. 23-28, ISSN 0388-788X

Yanagida, T., Sameshima, M., Nasu, H., Yokoyama, H. & Ogawa, K. (2006). Temperature effects on the development of *Enteromyxum* spp. (Myxozoa) in experimentally infected tiger puffer, *Takifugu rubripes* (Temminck & Schlegel). *Journal of Fish Diseases*, Vol. 29, No. 9, Sep. 2006, pp. 561-567, ISSN 0140-7775

Yasuda, H., Ooyama, T., Iwata, K., Tin Tun, Yokoyama, H. & Ogawa, K. (2002). Fish-to-fish transmission of *Myxidium* spp. (Myxozoa) in cultured tiger puffer suffering from emaciation disease. *Fish Pathology*, Vol. 37, No. 1, March 2002, pp. 29-33, ISSN 0388-788X

Yasuda, H., Ooyama, T., Nakamura, A., Iwata, K., Palenzuela, O. & Yokoyama, H. (2005). Occurrence of the myxosporean emaciation disease caused by *Enteromyxum leei* in cultured Japanese flounder *Paralichthys olivaceus*. *Fish Pathology*, Vol. 40, No. 4, Dec. 2005, pp. 175-180, ISSN 0388-788X

Yokoyama, H., Ogawa, K. & Wakabayashi, H. (1990). Light and electron microscopic studies on the development of *Hoferellus carassi* (Myxosporea), the causative organism of kidney enlargement disease of goldfish. *Fish Pathology*, Vol. 25, No. 3, Sep. 1990, pp. 149-156, ISSN 0388-788X

Yokoyama, H., Ogawa, K., & Wakabayashi, H. (1991). A new collection method of actinosporeans - a probable infective stage of myxosporeans to fishes - from tubificids and experimental infection of goldfish with the actinosporean, *Raabeia* sp. *Fish Pathology*, Vol. 26, No. 3, Sep. 1991, pp. 133-138, ISSN 0388-788X

Yokoyama, H., Ogawa, K. & Wakabayashi, H. (1993). Some biological characteristics of actinosporeans from the oligochaete *Branchiura sowerbyi*. *Diseases of Aquatic Organisms*, Vol. 17, pp. 223-228. ISSN 0177-5103

Yokoyama, H., Ogawa, K. & Wakabayashi, H. (1995). Chemoresponse of actinosporean spores of *Myxobolus cultus* to skin mucus of goldfish *Carassius auratus*. *Diseases of Aquatic Organisms*, Vol. 21, No. 1, Feb. 1993, pp. 7-11. ISSN 0177-5103

Yokoyama, H., Danjo, T., Ogawa, K., Arima, T. & Wakabayashi, H. (1996). Hemorrhagic anemia of carp associated with spore discharge of *Myxobolus artus* (Myxozoa: Myxosporea). *Fish Pathology*, Vol. 31, No. 1, March 1996, pp. 19-23, ISSN 0388-788X

Yokoyama, H. & Urawa, S. (1997). Fluorescent labelling of actinospores for determining the portals of entry into fish. *Diseases of Aquatic Organisms*, Vol. 30, No. 3, Sep. 1997, pp. 165-169. ISSN 0177-5103

Yokoyama, H., Inoue, D., Kumamaru, A. & Wakabayashi, H. (1997a). *Myxobolus koi* (Myxozoa: Myxosporea) forms large-and small-type "cysts" in the gills of common carp. *Fish Pathology*, Vol. 32, No. 4, Dec. 1997, pp. 211-217, ISSN 0388-788X

Yokoyama, H., Danjo, T., Ogawa, K. & Wakabayashi, H. (1997b). A vital staining technique with fluorescein diacetate (FDA) and propidium iodide (PI) for the determination of viability of myxosporean and actinosporean spores. *Journal of Fish Diseases*, Vol. 20, No. 4, Jul. 1997, pp. 281-286, ISSN 0140-7775

Yokoyama, H., Liyanage, Y.S., Sugai, A. & Wakabayashi, H. (1998). Hemorrhagic thelohanellosis of color carp caused by *Thelohanellus hovorkai* (Myxozoa: Myxosporea). *Fish Pathology*, Vol. 33, No. 2, Jun. 1998, pp. 85-89, ISSN 0388-788X

Yokoyama, H., Inoue, D., Sugiyama, A. & Wakabayashi, H. (2000). Polymerase chain reation and indirect fluorescent antibody technique for the detection of *Kudoa amamiensis* (Multivalvulida: Myxozoa) in yellowtail *Seriola quinqueradiata*. *Fish Pathology*, Vol. 35, No. 3, Sep 2000, pp. 157-162, ISSN 0388-788X

Yokoyama, H., Liyanage, Y. & Wakabayashi, H. (2002). A new strategy against myxosporean diseases: vector control in hemorrhagic thelohanellosis of cultured carp. *Fisheries Science*, Vol. 68, Suppl I., Nov. 2002, pp. 825-828, ISSN 0919-9268

Yokoyama, H., Kawakami, H., Yasuda, H. & Tanaka, S. (2003). *Henneguya lateolabracis* sp.n. (Myxozoa: Myxosporea), the causative agent of cardiac henneguyois in Chinese sea bass *Lateolabrax sp*. *Fisheries Science*, Vol. 69, No. 6, Dec. 2003, pp. 1116-1120, ISSN 0919-9268

Yokoyama, H., Whipps, C.M., Kent, M.L., Mizuno K. & Kawakami. H. (2004). *Kudoa thyrsites* from Japanese flounder and *Kudoa lateolabracis* n. sp. from Chinese sea bass: causative agent of post-mortem myoliquefaction. *Fish Pathology*, Vol. 39, No. 2, Jun. 2004, pp. 79-85, ISSN 0388-788X

Yokoyama, H., Itoh, N. & Tanaka, S. (2005a). *Henneguya pagri n.* sp. (Myxozoa: Myxosporea) causing cardiac henneguyosis in red sea bream, *Pagrus major* (Temminck & Schlegel). *Journal of Fish Diseases*, Vol. 28, No. 8, Aug. 2005, pp. 479-487, ISSN 0140-7775

Yokoyama, H., Freeman, M.A., Itoh, N. & Fukuda, Y. (2005b). Spinal curvature of cultured Japanese mackerel *Scomber japonicus* associated with a brain myxosporean, *Myxobolus acanthogobii*. *Diseases of Aquatic Organisms*, Vol. 66, No. 1, Aug. 2005, pp. 1-7, ISSN 0177-5103

Yokoyama, H., Kim, J.-H. & Urawa, S. (2006). Differences in host selection of actinospores of two myxosporeans, *Myxobolus arcticus* and *Thelohanellus hovorkai*. *Journal of Parasitology*, Vol. 92, No. 4, Aug. 2006, pp. 725-729, ISSN: 0022-3395.

Yokoyama, H. & Shirakashi, S. (2007). Evaluation of hyposalinity treatment on infection with *Enteromyxum leei* (Myxozoa) using anemonefish *Amphiprion* spp. as experimental host. *Bulletin of the European Association of Fish Pathologists*, Vol. 27, No. 2, pp. 74-78, ISSN 0022-3395

Yokoyama, H., Kageyama, M., Yanagida, T. & Ogawa, K. (2009). Seawater survival of *Enteromyxum leei* (Myxozoa) evaluated by *in vitro* viability and *in vivo* infectivity assays. *Fish Pathology*, Vo. 44, No. 4, Dec. 2009, ISSN 0388-788X

Zhang, J.Y., Meng, F., Yokoyama, H., Miyahara. J., Takami, I. & Ogawa K. (2010a). Myxosporean and microsporidian infections in cultured Pacific bluefin tuna *Thunnus orientalis* in Japan. *Fisheries Science*, Vol. 76, No. 6, Nov. 2010, pp. 981-990 ISSN 0919-9268

Zhang, J.Y., Yokoyama, H., Wang, J.G., Li, A.H., Gong, X.N., Ryu-Hasegawa, A., Iwashita, M. & Ogawa, K. (2010b). Utilization of tissue habitats by *Myxobolus wulii* Landsberg & Lom, 1991 in different carp hosts and disease resistance in allogynogenetic gibel carp: redescription of *M. wulii* from China and Japan. *Journal of Fish Diseases*, Vol. 33, No. 1, Jan. 2010, pp. 57-68, ISSN 0140-7775

2

Metazoan Parasites of the European Sea Bass *Dicentrarchus labrax* (Linnaeus 1758) (Pisces: Teleostei) from Corsica

Laetitia Antonelli and Bernard Marchand
Laboratory "Parasites et Ecosystèmes Méditerranéens", University Pasquale Paoli, Corsica France

1. Introduction

As in many parts of the world, aquaculture production in the Mediterranean has been expanding rapidly over recent years (Basurco & Lovatelli, 2005). The continuous decline of ocean fisheries stocks has provided impetus for rapid growth in aquaculture. Global production of farm fish has more than doubled in the past 20 years with the growing demands of an expanding population, and it seems that grow is set to continue. It has been the fastest growing food production sector since 1970, with an increase at a compounded rate about 9.2% per year (Guo & Woo, 2009; Naylor et al., 2000). Alongside, and perhaps partly due to this rapid expansion, the welfare of farmed fish has received increasing attention. Fish welfare is an important issue for the industry, not just for public perception, marketing and product acceptance, but also often in terms of production efficiency, quality and quantity (Ashley, 2006). Many factors influenced the rapid increase in the production of cultured fish species in the last two decades (FAO, 2010). One of the most important factors is the great development and diffusion of sea-cage culture. The success of the sea-cage farming is essentially correlated with the possibility to reduce production cost. On the other hand, aquaculture of fish is not without problems. Cages constitute an open system, which allow free exchange between wild and caged organisms, leading to the emergence of characteristic diseases (Ghittino et al., 2003; Merella et al., 2006). The maintenance of a good health is a major problem in animal populations. Rapid development of fish culture in marine cages has been associated with an emergence of parasitic diseases (Fioravanti et al., 2006; Nowak, 2007). The difficulty of managing disease within sea-cages is regarded as a major drawback (Mladineo, 2006). Fish parasites are a major component of aquatic biodiversity, and their monitoring is considered an essential element of the management of the health for animals. Parasites that have low rates of infestation and minor pathological effects on their hosts under natural conditions can easily spread in populations confined to rearing systems, causing serious outbreaks and epizootics diseases resulting in significant economical losses (Combes, 1990; Ghittino et al., 2003; Johnson et al., 2004; Naylor et al., 2000). The environmental conditions in culture systems, in particular increased density of fish, repeated introduction of hosts, homogeneous host populations, fast growth and a potential decrease in genetic diversity have an important effect on commercial production and could prevent the expansion of the industry (FAO, 2010; Fernandez-Jover et al., 2010; Nowak 2007). Numerous studies on the parasites of marine fish were carried out, specifically on species with a great economical interest such as the European sea bass *Dicentrarchus labrax* or the Gilthead sea bream *Sparus aurata* (Antonelli et al., 2010a; Euzet & Combes, 1998; Focardi et al., 2005; Reversat et al., 1992). They have long been recognized to have the potential to affect the survival of their hosts (Johnson et al., 2004). Investigations revealed that infections through the attachment of parasites and active feeding on mucus and epithelial cells of host fish by large populations can cause severe damages such as necroses, haemorrhages, inflammation, and mucus hyperproduction (Manera & Dezfuli, 2003; Noga, 2000; Paperna & Baudin Laurencin, 1979).

The European sea bass *Dicentrarchus labrax,* is one of the main marine fish species extensively reared in the Mediterranean area. It represents an important economical source, especially in Corsica. Fish production increases from 1 000 tons in 1994 to 2 250 tons today (Antonelli et al., 2009; Antonelli et al., 2010b).

They are many publications on the parasitofauna of cultured sea bass in various Mediterranean areas, where fish farming represents a significant economic activity (Cecchini et al., 1994; Johnson et al., 2004; Sitja-Bobadilla et al., 2006). Despite, the historical and commercial activities related to *D. labrax* in Corsica, few studies have been reported on their parasites. Investigations have revealed that many parasites species are globally frequent on this fish species, but large scales of mortality have not yet been registered (Antonelli et al., 2009; Antonelli, 2010).

The objective of the present study is to relate the occurrence of parasites to climatic conditions. Our survey aims to establish seasonal changes and spatial dynamics of parasites, and their importance for animal health for an improvement in the monitoring and management of populations of *D. labrax* of Corsica. The variation in the patterns of parasites communities was examined by taking into account environmental factors such as temperature, and physiological parameters related to host. Several studies have shown an increase of parasite richness with host age and that age and/or size related recruitment influenced parasite aggregation (Cable & Van Oosterhout, 2007; Poulin & Rohde, 1997; Silan, 1984). All animals are not equally infested according to their age. Our work will allow us to identify the most sensitive animals. Prevalence and abundances of the infections in different culture systems, fish stocks, and sampling seasons are reported.

2. Materials and methods

2.1 Sampling and examinations

A total of 470 European sea bass *D. labrax* from several culture systems on the South coasts of Corsica were sampled during the period ranging from January 2007 to December 2008. Fish were caught for parasitological examinations from six fish farms. Their names are not revealed for reasons of confidentiality: each farm is designated by a number (Figure 1). Seasonal samples varied from 10 to 15 specimens for each farm.

The production cycle of *D. labrax* cultured in Corsica is known. On average, larger pre-fattened sea bass (10 grams) reach first commercial size in about 1 year, while smaller juveniles (5 grams) reach the same size in about 16 months. Generally the fish are marketed

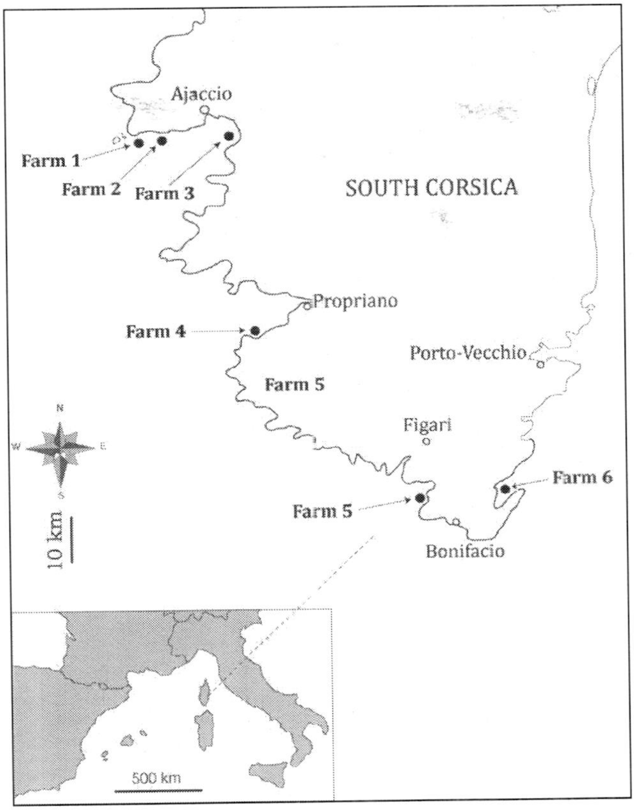

Fig. 1. Localisation of fish farms studied.

at age 2 years. In order to have consistent and significant results for a population we collected our samples still in the same cage for each farm throughout the study.

The fish were randomly collected directly from cages and immediately transported to the laboratory for analysis. Biometrical measures, sex and maturity stage of each specimen of *D. labrax* were recorded. Body surface (skin, fins) and gills of the fish were examined. Gill arches were carefully removed and studied in a fresh condition. The eight arches were separated (four right, four left) and immersed in Petri dishes containing seawater to replicate the best initial environmental conditions, and individually examined. Parasites were collected alive and immediately fixed in absolute ethanol. Parasites were identified using a light microscope according to the description done in previous studies (Cabral et al., 1984; Ozel et al., 2004; Toksen et al., 2008). Some specimens were fixed in glutaraldehyde to be studied in scanning electronic microscopy.

In order to asses the relationship of the infection with the host size and age, we considered two classes to discriminate younger from older fish. The first class (I) included fish under 30 cm total length, and the second class (II) fish over 30 cm. These classes were chosen taking into account the size ranged on sampled specimens. According to that study, classes would mainly include specimens with age 1+ (I) and 2+ (II).

At each sampling, surface water temperature (°C) and salinity (‰) in cages was measured.

2.2 Statistical analyses

Relations between rates of infestation were investigated using statistical indices frequently used in ecology. Prevalence (proportion of the population infected), abundance (mean number of parasites of both infected and uninfected fish) and mean intensity (mean number of parasites of infected hosts) were calculated and applied according to Bush et al. (1997).

Statistical tests were conducted to determine the influence of several factors on the distribution of *Lernanthropus kroyeri*. Possible differences in infection parameters between seasons were evaluated using one-way analysis of variance. If data failed to meet parametric testing requirements, comparisons of the mean number of parasites found on *D. labrax* were performed using the Mann-Whitney *U* test (two groups) and Kruskal-Wallis test (more than two groups) according to Dagnelie (1975) and Sprent & Ley (1992). Significance for all the statistical analyses was established with 95% confidence intervals. Calculations were performed using the statistical software Systat 12 and SigmaStat 3.5. In order to be sure of our values we will then use the Bonferroni test which consists of correcting the mistake on the risk of alpha error when using the same test several times (Bland & Altman, 1995).

3. Results

3.1 Distribution of parasites in host population

Five species of metazoan parasites included in four groups were identified: one monogenean *Diplectanum aequans*, two copepods, *Lernanthropus kroyeri* and *Caligus minimus*, one isopod *Cerotothoa oestroides* and the myxosporidian *Ceratomyxa labracis*.

428 (91.1%) of the 470 *D. labrax* examined were infected. The number of parasites per host ranged from 1 to 169. Externally, all fish used for the study appeared to be in good health, in body condition and in coloration. A total of 7 376 parasites were collected. Water temperature varied from 13°C in winter to 24°C in summer (Fig. 2). Rates of salinity (‰) are reported in table I.

Farms	Seasons								
	W2007	S2007	Su2007	A2007	W2008	S2008	Su2008	A2008	Mean values
1	38.2	37.6	37.8	37.6	38.0	37.8	37.9	37.9	37.9
2	38.2	37.6	37.8	37.6	38.0	37.8	37.9	37.9	37.9
3	36.0	37.0	37.1	36.5	36.0	36.1	37.0	36.9	36.6
4	37.8	37.3	38.0	37.8	37.6	37.8	37.6	38.1	37.8
5	37.6	37.6	38.2	37.8	37.6	37.5	37.6	38.2	37.8
6	38.0	38.4	38.6	38.2	38.0	38.2	38.3	38.3	38.3

Table 1. Seasonal rates of salinity (‰) in all fish farms studied (W: winter; S: spring; Su: summer; A: autumn; 2007: first year of sampling; 2008: second year of sampling).

3.2 Presence of parasites in fish farms

3.2.1 Monogenea

D. aequans was collected in all farms studied. 408 fish (86.8%) were infected.

Farm 1: *D. aequans* occurred throughout the sampling period. Prevalence values varied from 50% to 90% (Figure 2). However, there is no statistically significant difference between seasons (Kruskal-Wallis, $p < 0.5$). The highest mean abundance value was recorded in winter. Values ranged from 4.2 to 10.4 during the study (Figure 3). The mean intensity shows the same trend as abundance with highest values recorded in autumn and winter (Figure 4). The infestation shows a clear seasonal pattern statistically verified with Kruskal-Wallis test ($p < 0.001$).

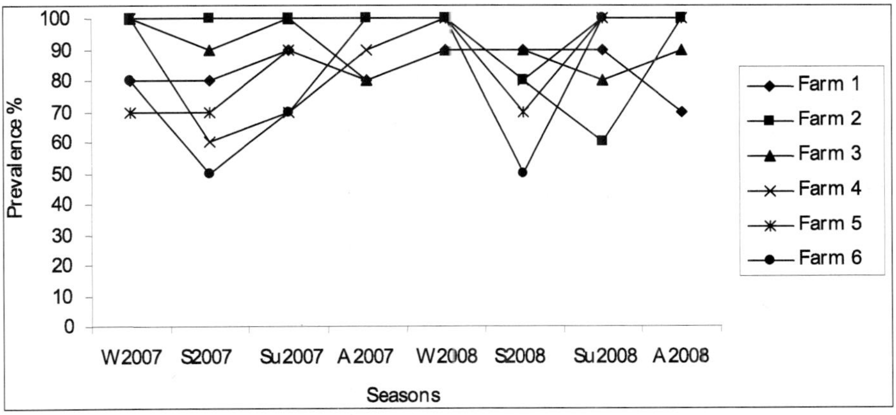

Fig. 2. Prevalence of *Diplectanum aequans* in fish farm studied according to season (W: winter; S: spring; Su: summer; A: autumn; 2007: first year of sampling; 2008: second year of sampling).

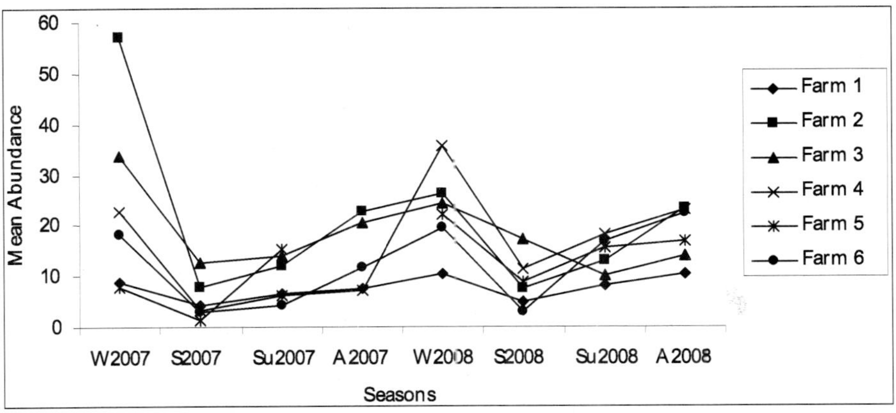

Fig. 3. Abundance of *Diplectanum aequans* in fish farm studied according to season (W: winter; S: spring; Su: summer; A: autumn; 2007: first year of sampling; 2008: second year of sampling).

Farm 2: *D. aequans* were identified on all fish examined during the first year of study. After winter 2008, values sharply decreased to summer 2008 and increased gradually to autumn 2008 (Figure 2). The highest mean abundance was recorded in winter. Values ranged from 7.5 to 57.1 and the Kruskal-Wallis test shows a significant difference between seasons ($p < 0.001$) (Figure 3). Mean highest intensity was recorded in winter (Figure 4) and statistically significant difference was revealed between spring and winter (Kruskal-Wallis, $p < 0.001$).

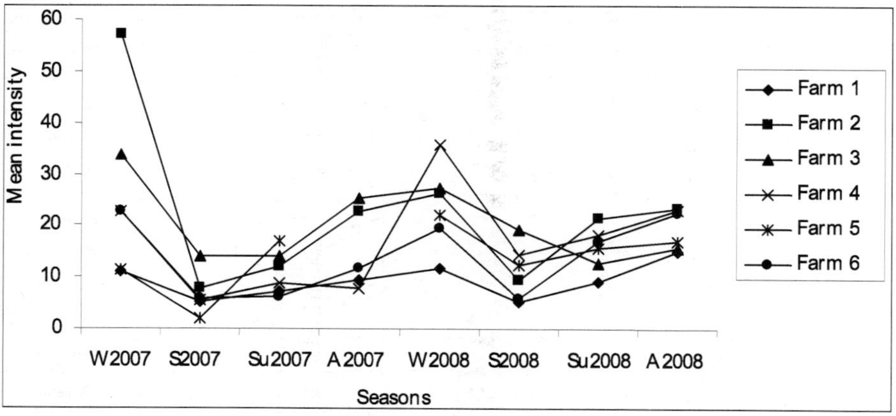

Fig. 4. Mean intensity of *Diplectanum aequans* in fish farm studied according to season (W: winter; S: spring; Su: summer; A: autumn; 2007: first year of sampling; 2008: second year of sampling).

Farm 3: the number of infected fish ranged from 80% to 100%, with highest prevalence values recorded in winter (Figure 2). Mean abundance values ranged from 10 to 33.8 (Figure 3). Mann-Whitney *U* test revealed significant difference between winter and spring ($p < 0.001$). The highest mean intensity values were recorded in winter and the lowest in summer (Figure 4). A statistically significant difference was revealed between winter and summer (Mann-Whitney *U* test, $p < 0.001$).

Farm 4: prevalence values ranged from 60% to 100% (Figure 2). Abundances and intensity values are highest in winter and lowest in spring (Figure 3, 4). Statistically, significant difference was highlighted between spring and winter (Mann-Whitney *U* test, $p < 0.005$).

Farm 5: the number of infected ranged from 70% to 100% (Figure 2). The mean abundance and mean intensity show the same as prevalence with highest values recorded in winter (Figures 3, 4). However, no significant difference were found between seasons (Kruskal-Wallis, $p = 0.25$). In the case of farm 5, it is difficult to analyze correctly data obtained during the sampling period because we had an interruption in our samples in autumn 2007.

Farm 6: prevalence values varied from 50% to 100%, with highest values registered in autumn and winter and lowest in spring. After winter 2007, values decrease to spring 2007 and increase to autumn 2008; it decreases again in spring 2008 (Figure 2). The mean abundance was highest in autumn, with values ranging to 3 to 22.4 (Figure 3). The infestation shows a clear seasonal pattern verified with Kruskal-Wallis test (p <0.001). The mean intensity show a similar trend to abundance with highest values recorded in autumn and winter (Figure 4), with a significant difference of seasonal infestation (Kruskal-Wallis, p < 0.001).

3.2.2 Copepoda

Specimens of *L. kroyeri* were collected only in the farm 6. The investigations revealed the presence of parasites in 65% of *D. labrax* examined in this farm. A total of 283 specimens of *L. kroyeri* were collected (mean abundance: 3.7, mean intensity: 5.4). *L. kroyeri* occurred throughout the sampling period. The number of infected fish varied from 30% to 90% with highest prevalence values recorded in summer and lowest in winter (Figure 5).

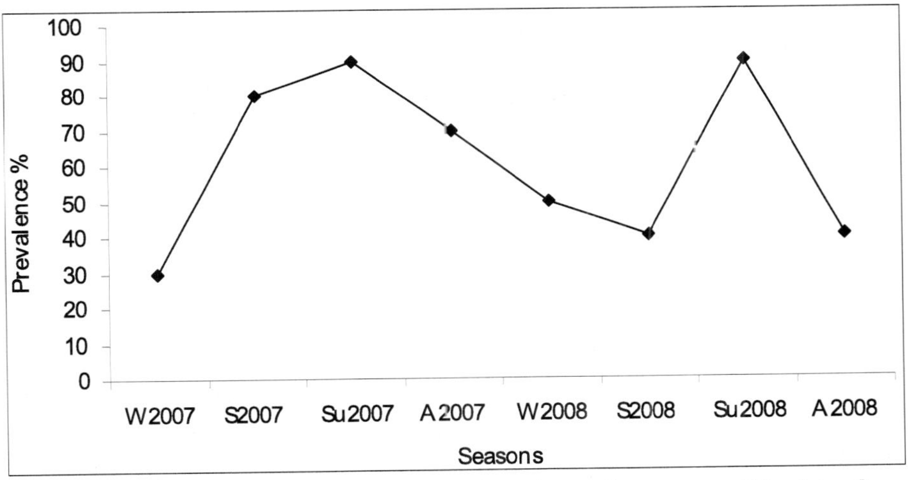

Fig. 5. Prevalence of *Lernanthropus kroyeri* in fish farm 6 according to season (W: winter; S: spring; Su: summer; A: autumn; 2007: first year of sampling; 2008: second year of sampling).

Mean abundance values ranged from 0.7 to 9.3, and the highest values were recorded in spring 2007 separated by intervals of decreasing values, with a significant decline during winter 2008. After winter 2008, values sharply increased to summer 2008 (Figure 6).

The mean intensity shows the same trend as the abundance for the second year of study, with highest values recorded in summer (Figure 7). Statistically, the occurrence of infection showed a clear seasonal pattern verified with Kruskal-Wallis test ($p < 0.001$), with significant differences between winter and summer.

As *L. kroyeri*, *C. minimus* is identified only in the farm 6. Rates of infestations are lowest than those of *L. kroyeri*. Prevalence values ranged from 0 to 60% with highest values recorded in spring (Figure 8).

Abundance values shows that rates of infestation with *C. minimus* are relatively stable throughout the seasons, with an infestation a little more important for spring (Figure 9).

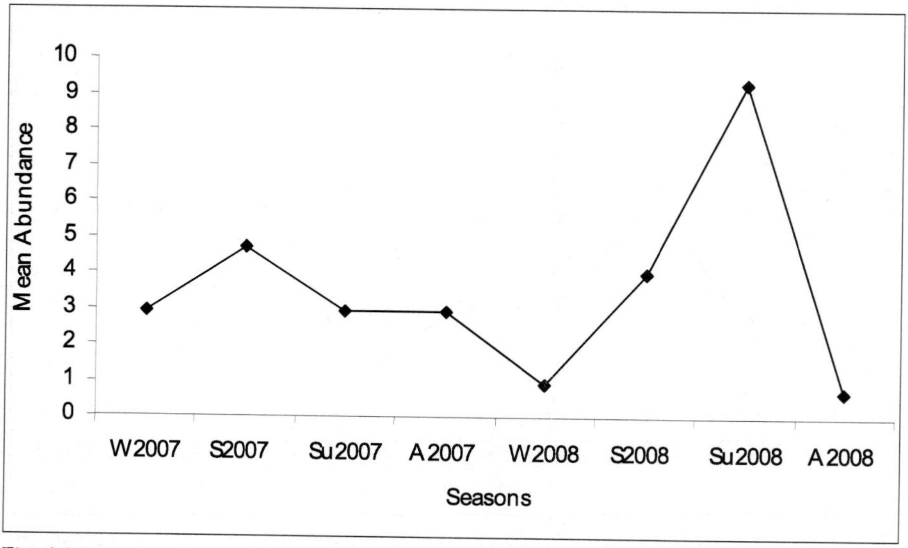

Fig. 6. Mean abundance of *Lernanthropus kroyeri* in fish farm 6 according to season (W: winter; S: spring; Su: summer; A: autumn; 2007: first year of sampling; 2008: second year of sampling).

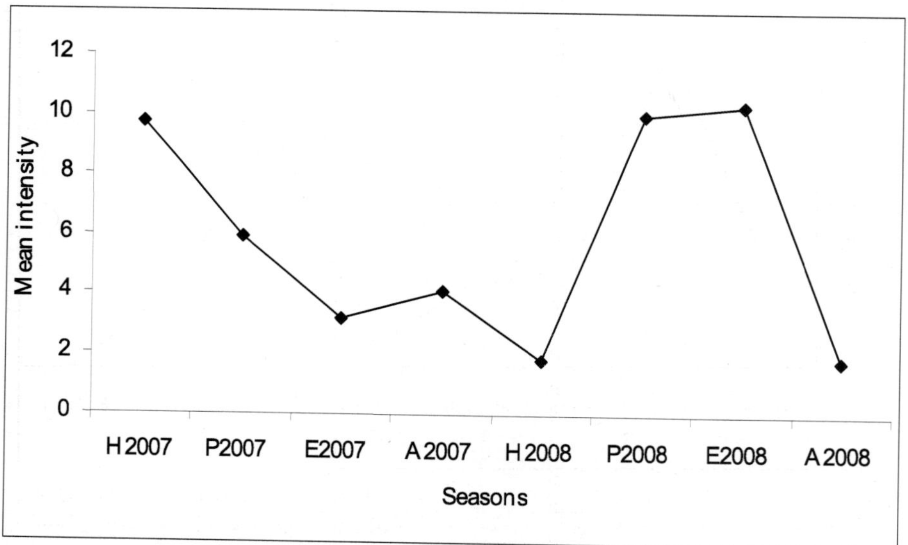

Fig. 7. Mean intensity of *Lernanthropus kroyeri* in fish farm 6 according to season (W: winter; S: spring; Su: summer; A: autumn; 2007: first year of sampling; 2008: second year of sampling).

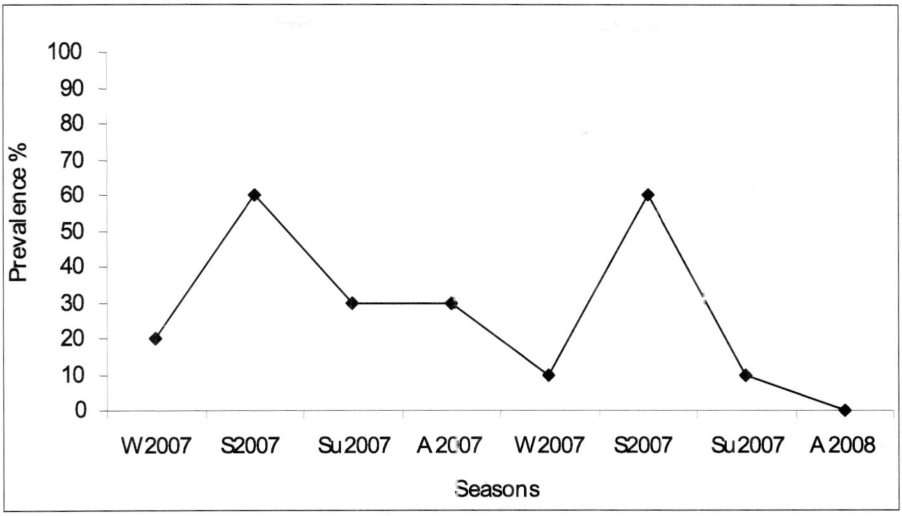

Fig. 8. Prevalence of *Caligus minimus* in fish farm 6 according to season (W: winter; S: spring; Su: summer; A: autumn; 2007: first year of sampling; 2008: second year of sampling).

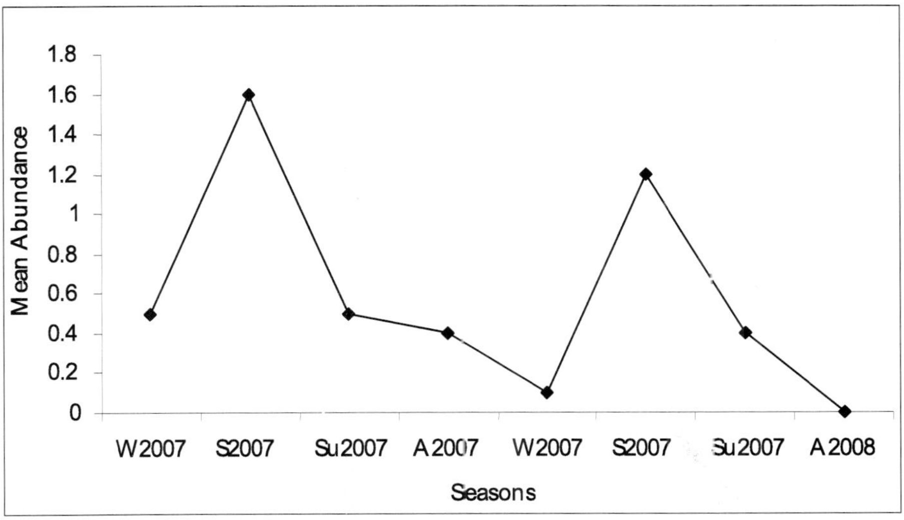

Fig. 9. Mean abundance of *Caligus minimus* in fish farm 6 according to season (W: winter; S: spring; Su: summer; A: autumn; 2007: first year of sampling; 2008: second year of sampling).

In 2008, the mean intensity increases with the elevation of water temperature. Highest values were recorded in spring and summer (Figure 10). However, no significant difference was revealed between season (Kruskal-Wallis, $p = 0.39$) because values are lower.

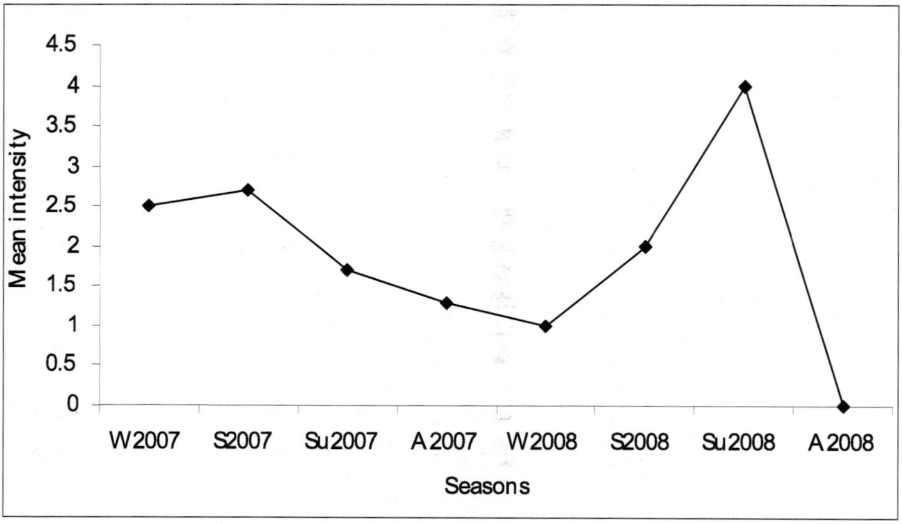

Fig. 10. Mean intensity of *Caligus minimus* in fish farm 6 according to season (W: winter; S: spring; Su: summer; A: autumn; 2007: first year of sampling; 2008: second year of sampling).

3.2.3 Isopoda

All specimens were found at the end of spring in 22 fish of the same farm. However, rates of infestation are to lower to have a significant result (prevalence: 4.7%, mean abundance: 0.1).

3.2.4 Myxosporea

All specimens were found at the end of spring. As *C. oestroides* the number of parasites collected is to lower to have a significant result (prevalence: 0.43%, mean abundance: 0.2). However, the study of mean intensity shows that in spite of the small number of infested fish *C. labracis* possesses highest value of mean intensity (69.04).

3.3 Influence of the host size on parasite population

3.3.1 Monogenea

The highest prevalence of *D. aequans* was registered for class II with 92%. Class I possesses a lowest prevalence with 81%. Our results showed an increase of mean abundances and mean intensities in the courses of generations. Values are higher for class II with 19.4 and 21.1 respectively. Statistical test show significant difference in abundances according to age classes (Kruskall-Wallis, $p < 0{,}001$). Older fish are more infested (Table 2).

3.3.2 Copepoda

Considering all samples as a whole, it was found that rates of infestation varied with host size. Results (prevalence, abundance and mean intensity) are presented in table 2.

Concerning *L. kroyeri*, the lowest prevalence values were observed for class I (55%) and the highest values for class II. The mean abundance and mean intensity have shown an irregular and significant pattern with size class (Kruskal-Wallis, $p = 0.01$): values were higher in the largest specimens. Each size class corresponds to an age class. Class I represents younger fish while class II corresponds to the oldest fish. Rates of infestation were relatively similar but values of mean abundance and mean intensity showed that larger and older fish tend to be more parasitized.

Prevalence values of *C. minimus* increase between class I and class II. Older fish are more infected. However, there is no significant difference between age classes (Kruskal-Wallis, p = 0.237). Mean abundance and mean intensity show values relatively close between generations.

Parasites species	Age class	N Fish examined	Prevalence %	Mean abundance	Mean Intensity
D. aequans	1+	210	81	9	11
	2+	260	92	19.4	21.1
L. kroyeri	1+	33	55	1.7	3.1
	2+	47	72.3	4.8	6.7
C. minimus	1+	33	21.2	0.3	1.3
	2+	47	32	0.8	2.5

Table 2. Rates of infestation of *D. labrax* between age classes.

3.3.3 Isopoda

Four parasites were collected. The number of parasites is too low to establish a relation between infestation and host age.

3.3.4 Myxosporea

Colonies of *C. labracis* were collected in the stomach of two fish. A total of 120 colonies were counted (prévalence: 1 %, abundance: 0.5) but our data are not sufficient to establish a hypothesis between host age and the number of parasites.

4. Discussion

4.1 Distribution of parasites in host population

The relationship of the number of parasites to severity of disease can be dependent of some factors. A variety of environmental and biological factors and management practices that may influence rates of infestation of parasites have been identified. Parameters including oceanographic factors (temperature or water circulation) associated with physiological factors (immunological alterations and hormonal changes) are the most frequent causes suggest to explained seasonal fluctuations of parasitic infections. They are highly variable and may vary seasonally or annually (Gonzalez-Lanza et al., 1991, Oliver, 1982; Robertsen et al., 2008; Winger et al., 2008).

4.2 Presence of parasites in fish farms

Seasonal fluctuations in prevalence and abundance are common in many parasites species infecting marine fish (Oliver, 1982). Our results indicated pronounced changes in the cycle of infection, with variation in parasite distribution. The fluctuation in rates of infestations showed that water temperature, which generates a definite seasonal cycle, has a great influence on the population dynamics of parasites. Many studies conducted on the behaviour of parasites in culture conditions have shown that water temperature is the major abiotic factor affecting reproduction and population growth (Oliver, 1987). Generally, high water temperature levels promote faster hatching of eggs and parasites propagate rapidly (Gannicott & Tinsley, 1998; Kim et al., 2001; Silan & Maillard, 1989). Cyclic peaks of many parasites species have been mainly registered in sea cages during summer months (Andersen and Buchmann, 1998; Cecchini et al., 1998; Gelnar, 1987).

As monogeneans, in contrast to other helminths, which need intermediate hosts to complete their life cycle, have a direct life cycle, temperature appears to be the most significant in explaining the seasonal periodicities. Generally, monogeneans propagate rapidly on the host fish in warm water seasons. Our results differ in some extent from those reports in the Mediterranean. This study demonstrates that transmission of *D. aequans* in fish farms is seasonal, with invasion maximized during the colder part of the year and almost negligible infection over summer. Influence of temperature causing a rapid development on the life cycle of *D. aequans* was reported in *D. labrax*. Cecchini (1994) mentioned that between 20°C and 30°C, hatching of *D. aequans* takes place a few days after laying, whereas at 15°C and 10°C it occurs respectively between the 11th and the 19th day. However, the present study revealed that *D. aequans* exhibit clear and significant seasonal changes with highest rates of infestation recorded in winter and the lowest during summer, following a contrary temperature pattern. These variations tend to confirm that *D. aequans* display spatial and temporal stability. Similar data has already been reported for some monogeneans, including *D. aequans*, and may be explained by changes in parasites behaviour (Antonelli, 2010; Robertsen et al., 2008; Silan, 1984). These results may indicate that *D. aequans* is adapted to cold temperature. Lambert & Maillard (1974) and Gonzalez-Lanza et al., (1991) demonstrated that seasonal changes in composition of parasites population of *D. labrax* could be the sign of a parasitic adaptation, and suggest continuous recruitment and their persistence throughout the year.

Temperature is also assumed to influence the immune defence in fishes, and may affect intensities of parasites indirectly (Johnson et al., 2004; Oliver, 1982). Variations in rates of infestation through the seasons can also be explained by changes in the fish behaviour. Abundance peak recorded in winter for *D. aequans* can coincide with an immunodeficiency of fish linked to brutal changes in temperature during the transition from autumn to winter (Faliex et al., 2008). These periods of decreasing of water temperature result in a stress response of the European sea bass (Hadj Kacem et al., 1987). This may reflect a weakened immune system of fish, and therefore a greater vulnerability, leaving the opportunity for parasites to increase their populations rapidly (Oliver, 1982).

Results found for copepod species *L. kroyeri* agree with data previously reported in the Mediterranean (Bahri et al., 2002; Manera & Dezfuli, 2003). The development of *L. kroyeri* appears to be directly linked to water temperature; highest rates of infestation were recorded in spring and summer, coinciding with an increase of water temperature. Seasonal changes in prevalence and abundance suggest that *L. kroyeri* have an annual life cycle, with one period of heavy infestation. Like most copepods, *L. kroyeri* have a direct life cycle and environmental factors such as temperature appears to be the most significant in explaining the seasonal periodicities (Noga, 2000). Host tissue reactions were observed on gill filaments in the immediate vicinity of the copepod attachment sites. This hyperplasia or inflammation of gill tissue is interpreted as a host tissue immune response to parasite aggression.

C. minimus, C. oestroides and *C. labracis* were identified in our study but rates of infestation appear very low. Two hypotheses are considered to explain these results:

- The fish is adapted to the presence of these parasites species and rates of infestation decreased as it was already demonstrated by Mladineo & Marsic-Lucic (2007) for the Gilthead sea bream *Sparus aurata* ;
- This lowest infestation explains by a recent arrival of the parasites in the environment.

Only a long-term surveillance would allow us to verify these hypotheses.

Specimens of *C. oestroides* and *C. labracis* have been collected in the same fish farm, but our data values are too low to comment statistically our results. However, the presence of these isolated individuals could be the result of a parasitic transfer. Numerous wild fish revolved all around cages looking for food. There is an international speculation that parasite transmission from farmed to wild fish leads to increase the incidence of parasitism in wild fish. However, our previous study tend to confirm that some diseases are transmitted directly to fish by cohabitation with infected fish or by horizontal transmission from wild marine fish. The analyse of the parasitic fauna of wild fish living around cages highlighted important rates of infestation with *C. oestroides* (Antonelli et al., 2009; Antonelli 2010). Wild fish are believed to be the primary reservoirs of parasites infection for fish farmed in sea-cages, and can negatively impact upon the health of farmed fish (Bragoni et al., 1984; Hutson et al., 2007). In our case, *D. labrax* can constitute a new host to colonize for new parasites species. Physiological parameters of some parasites species, e.g. female *C. oestroides* constantly mature, could entrain a rapid parasitic development, causing highest mortalities (Bragoni et al., 1984).

A few studies highlighted infestation with myxosporean in *D. labrax*. Alvarez-Pellitero & Sitjà-Bobadilla (1993) mentioned that infestation due to *C. labracis* is more important in cultured fish than in wild fish. The infestation in fish farm 5 is relatively low but it must be controlled. *C. labracis* is highly pathogen and can provoke serious outbreaks on fish, such as inflammatory reaction of epithelial cells.

Our observations were also consistent with the idea that local conditions can influence rates of infestation. The high densities living conditions in sea-cages can favour parasites infestation, and fish became targets for infection at high levels. Parasites having low rates of infestation in the wild can easily spread in populations confined to rearing systems. This phenomenon is partly due to the direct biological cycle of parasites without intermediate host (Buchmann and Lindenstrøm, 2002; Fioravanti et al., 2006; Hayward et al., 2007). They are likely to establish and proliferate in aquaculture because they may reproduce rapidly (Hutson et al., 2007). The presence of parasites throughout the year may be due to close and prolonged contact between fish in the same cage (Mladineo, 2006).

In addition to the season, the presence of copepods in a single farm suggests that several factors may operate in the distribution of this parasite, but also the location of the farm. The fact that the farm 6 was isolated on the East coast of Corsica may not have allowed the parasite to infect other farms. The presence of copepods on a single farm, could explain according to salinity. The farm 6 is situated in a zone of shallow depth. In littoral waters (some tens of meters) the contributions of continental fresh water generate vertical stratifications which can modify the parameters of temperature and salinity (Aminot & Kerouel, on 2004). This difference could also be due to the circulating sea currents around the island, in particular the Liguro-Provençal current. This one goes raises of Sicily towards French coasts, by the canal to Corsica, situated on the eastern coast, with waters of surface charged in minerals. In the Gulf of Lions, these waters are then going to mix with waters of the western Mediterranean Sea, much less salted.

Copepods require concentrations in salt less important than monogeneans (Devreker et al., 2009; Mc Allen & Brennan, 2009). We can suppose that monogeneans are thus less present in the fish farm 6 because the environment is less salty than on the western coast. It also coincides with the results obtained for the monogenea *D. aequans* in the fish farm 3 and suggests that the salinity could be a factor conditioning the presence or the absence of certain parasites of *D. labrax* in Corsica.

Heavy rates of infestation could result also of breeding conditions. Highest densities and thus contact between fish in cages could favor the apparition and development of pathogens (Combes, 1990; Johnson et al., 2004; Nowak, 2007; Sterud, 2002). While there are species-specific differences in optimum stocking density intensively cultured fish are usually kept at greater densities than occur in the wild. This increases host proximity and therefore improves the ability of a parasite to locate a host. High densities in cages can also promote a parasite infestation. For example, sea bass showed higher stress levels at high densities resulting on chronic stress situation and innate immune response favouring parasites development (Vazzana et al., 2002).

The introduction of a new species in an ecosystem without sanitary control could constitute a serious risk of parasites transfer responsible for the break of for the break of natural host-parasite interactions (Euzet & Pariselle, 1996). In mariculture, naive fish are introduced from hatcheries to sea cages and can increase the frequency of infections. Intensive aquaculture of fish is not without its problems, and these include disease outbreaks and consequences of introducing parasites to new host and/or new localities with the transportation of live fish (Combes, 2001). Severe epizootic periods have occurred because of fastest accumulation of pathogens in fish populations.

4.3 Influence of the host size on parasite population

We found significant relationships between the parasite burden and host-related factors. Dominant parasites species showed a positive relationship between host age/size and parasite load. Rates of infestation (prevalence, abundance and intensity) showed significant changes with host age/size (body length is closely related to age in fish). Older fish are more parasitized. This positive correlation between age (and therefore the size) of fish abundance and parasite occurs relatively frequently for different fish parasites including monogeneans (Cable & Van Oosterhout, 2007; Poulin & Rohde, 1997). A higher parasitisation level in older fish has also been observed by Silan (1984) for *D. aequans*. Our results for copepods differ from those reported by Manera & Dezfuli (2003), which reported that small fish are more parasitised than large, but prevalence of copepods increased with host size up to fish of intermediate length before decreased.

The influence of host age and/or size on parasites populations has been demonstrated by several authors, and two hypotheses have been proposed to explain these relationships.

- The first hypothesis followed the island size hypothesis, which is also linked to the "theory of island biogeography" proposed by Mac Arthur & Wilson (1967). Dogiel (1964) compared the host to an island and concluded that the larger the host, the more parasites can settle on it.
- The second hypothesis refers directly to host age. Host accumulates parasites as they get older because time increases the likelihood of acquiring parasites.

Previous studies showed that the variation of the parasitism depending on the size and age of fish can be depending on several factors. The availability of space may be one of these factors involved in the infection of European sea bass. The abundance of gill-infecting parasites has been correlated to host size, where an increase in body length is accompanied by modification in gill surface. An increase in the branchial surface can provide a larger area of infestation, and thus more potential sites for attachment of these parasites (El Madhi & Belghyti, 2006). Changes in size of gills filaments were associated with an increase of the volume of water passing through the gill cavity, may affect attachment capability of parasites and enhancing the possibility of being detached by respiratory currents (Timi & Lanfranchi, 2006). Parasites were attached horizontally, facing into a seawater flow supplied by a circulating water system. This may indicate that only the most resistant parasites can be securely attached to a fish of a given size, and may influence transmission, increasing or decreasing the chances of invasion by the larvae of parasites (Buchmann & Lindenstrøm, 2002). Also, in their natural environment, larger individuals have higher physical (ventilation volume) or chemical (mucus) stimuli which increase their attractiveness to parasites (Kearn, 1967). showed that copepods were able to respond to visual and mechanical cues by increasing their swimming activities. All of these factors should logically increase the probability of encountering free infective stages.

5. Conclusion

Even though the overall prevalence was higher, parasites were never abundant in our study. Mortalities related to parasites were not reported and all fish examined appear to be in good health. Our results differ in some extent from those reported by Manera & Dezfuli (2003) who reported clinical signs such as occlusion of branchial circulation, destruction of gills and massive mucus cell proliferation observed on fish. Fish sampled during our work do not display the symptoms described in previous studies (Manera & Dezfuli, 2003; Toksen et al., 2008). Parasites were present at too low abundances and intensities to affect the health of cultured sea bass. However, we must remain vigilant and we have to monitor the evolution of the parasite. We expect that the parasites infection scenarios reported here illustrate basic characteristics of the host–parasite interaction and would correspond with infection regimes affecting fish farms at different times of the year. The fact that transmission is seasonal will allow the introduction of measures to decrease parasite recruitment.

6. References

Alvarez-Pellitero, P. & Sitjà-Bobadilla , A. (1993). *Ceratomyxa spp.* (Protozoa, Myxosporea) infections in wild and cultured se bass, *Dicentrachus labrax*, from the Spanish Mediterranean area. *Journal of fish biology*, Vol. 42, No. 6 (June 1993), pp. (889-901), ISSN 1095-8649.

Aminot, A. & Kerouel, R. (2004). Hydrologie des écosystèmes marins: paramètres et analyses. Collection Méthodes d'analyse en milieu marin, IFREMER, pp. 336.

Andersen, P.S & Buchmann, K. (1998). Temperature dependent population of *Gyrodactylus derjavini* on rainbow trout *Oncorhynchus mykiss*. *Journal of Helminthology*, Vol. 72, No. 1 (March 1998), pp. (9-14), ISSN 0022-149X.

Antonelli, L. (2010). Impact du parasitisme sur la pisciculture en Corse. Suivi des parasitoses et étude des transferts de parasites depuis la faune sauvage vers les poisons élevés en mer ouverte. Thèse, pp. 254.

Antonelli, L., Ternengo, S., Agostini, S. & Marchand, B. (2009). Suivi parasitaire des fermes aquacoles de Corse (campagnes 2007-2008). Action collective, Contrat Syndicat des Aquaculteurs Corses / Université de Corse. *Université de Corse*, pp. 47.

Antonelli, L., Quilichini, Y. & Marchand, B. (2010a). *Sparicotyle chrysophrii* (Van Beneden and Hesse 1863) (Monogenea: Polyopisthocotylea) parasite of cultured Gilthead sea bream *Sparus aurata* (Linnaeus 1758) (Pisces: Teleostei) from Corsica: ecological and morphological study. *Parasitology Research*, Vol. 107, No. 2 (April 2010), pp. (389-398), ISSN: 0932-0113.

Antonelli L., Quilichini Y. & Marchand B. (2010b). Biological study of *Furnestinia echeneis* Euzet and Audoin 1959 (Monogenea: Diplectanidae), parasite of cultured Gilthead sea bream *Sparus aurata* (Linnaeus 1758) (Pisces: Teleostei) from Corsica. *Aquaculture*, Vol. 307, No. 3-4 (September 2010), pp. (179-186), ISSN 0044-8486.

Ashley, P.J. (2006). Fish welfare: Current issues in aquaculture. *Applied Animal Behaviour Science*, Vol. 104, No. 3-4 (May 2007), pp. (199-235) ISSN 0168-1591.

Bahri, L., Ben Hamida, J. & Ben Hassine, O.K. (2002). Use of parasitic copepod, *Lernanthropus kroyeri* (Van Beneden, 1851) (Lernanthropidae) as bio-indicator of two fish populations, *Dicentrarchus labrax* (Linnaeus, 1758) and *Dicentrarchus punctatus* (Bloch, 1792) (Moronidae) in Tunisian inshore areas. *Crustaceana*, Vol. 75, No. 3-4 (March-April 2001), pp. (253-267), ISSN 0011-216X.

Basurco, B. & Lovatelli, A. (2005). *The aquaculture situation in the Mediterraneans sea: predictions for the future*, Retrieved from http://www.iasonnet.gr./abstract/Basurco.pdf.

Bland, J.M. & Altman, D.G. (1995). Multiple significance tests: the Bonferroni method. *British Medical Journal*, Vol. 310, No. 6973 (January 1995), pp. 170, ISSN 0959-8146.

Bragoni, G., Romestand, B. & Trilles, J.P. (1984). Parasitoses à Cymothoadiens chez le loup *Dicentrarchus labrax* (Linneaus, 1758) en élevage. 1. Ecologie parasitaire dans le cas de l'étang de Diana (Haute-Corse) (Isopoda, Cymothoidae). *Crustaceana*, Vol. 47, No. 1 (January 1984), pp. (44-51), ISSN 0011-216X.

Buchmann, K. & Lindenstrøm, T. (2002). Interactions between monogenean parasites and their fish hosts. *International Journal for Parasitology*, Vol. 32, No. 3 (March 2002), pp. (309-319), ISSN 0020-7519.

Bush, A.O., Lafferty, K.D. & Lotz, J.M. (1997). Parasitology meets ecology on its own terms: Margolis et al. revisited. *Journal of Parasitology*, Vol. 83, No. 4 (August 1997), pp. (575-583), ISSN 0022-3395.

Cable, J. & Van Oosterhout, C. (2007). The impact of parasites on the life history evolution of guppies (*Poecilia reticulate*): the effects of host size on parasite virulence. *International Journal for Parasitology*. Vol. 37, No. 13 (November 2007), pp. (1449-1458), ISSN: 0020-7519.

Cabral, P., Coste, F. & Raibaut, A. (1984). Cycle évolutif de *Lernanthropus kroyeri* (Van Beneden, 1851) Copépode branchial du loup *Dicentrarchus labarx* (Linné, 1758) dans des populations naturelles et en élevage. *Annales de Parasitologie Humaine et Comparée*, Vol. 59, No. 2, pp. (189-207), ISSN 0003-4150.

Cecchini, S. (1994). Influence of temperature on the hatching of eggs of *Diplectanum aequans*, a parasite of sea bass. *Aquaculture International*, Vol. 2, No. 4 (November 1994), pp. (249-253), ISSN 0967-6120.

Cecchini, S., Saroglia, M., Berni, P. & Cognetti-Varriale, A.M. (1998). Influence of temperature on the life cycle of *Diplectanum aequans* (Monogenea, Diplectanidae), parasitic on sea bass, *Dicentrarchus labrax* (L.). *Journal of Fish Diseases*, Vol. 21, No. 1 (January 1998), pp. (73-75), ISSN 0140-7775.

Combes, C. (1990). When is a parasite pathogen? *Metazoan parasitic diseases: concluding remarks*. In: P*athology in marine science*, Academic Press Inc. Publisher, pp. (331-340).

Combes, C. (1995). *Interactions durables. Ecologie et évolution du parasitisme*. Ed. Masson, Paris, collection écologie. 523 pp., ISBN: 2-10-005753-7.

Dagnelie, P. (1975). *Analyses statistiques à plusieurs variables*, Presses agronomiques de Gembloux, Bruxelles. 362 pp., ISBN 978-2-87016-019-0.

Devreker, D., Souissi, S., Winkler, G., Forget-Leray, J. & Leboulenger, F. (2009). Effects of salinity, temperature and individual variability on the reproduction of Eurytemora affinis (Copepoda; Calanoida) from the Seine estuary: A laboratory study. *Journal of Experimental Marine Biology and Ecology*, Vol. 368, No.2 (January 2009), pp. (113-123) ISSN 0022-0981.

Dogiel, V.A. (1964). General parasitology. Translation 1.2, Kabata, Oliver and Boyd, Edinburgh and London. 516 pp.

El Madhi, Y., Belghyti, D. (2006). Distribution de deux Monogènes dans les individus hôtes de *Trachinotus ovatus* (L, 1758) de la côte de Mehdia. *Biologie et Santé*, Vol. 6, No.2, pp. (65-76).

Euzet, L. & Combes, C. (1998). The selection of habitats among the monogenea. *International Journal for Parasitology*, Vol. 28, No 10 (October 1998), pp. (1645-1652), ISSN 0020-7519.

Euzet, L. & Pariselle, A. (1996). Le parasitisme des poissons Siluroidei: un danger pour l'aquaculture? *Aquatic Living Resources*, Vol. 9, supplement S1 (Special Edition: The biology and culture of catfishes) (November 1996), pp. (145-151), ISSN 0990-7440.

Faliex, E., Da Silva, C., Simon, G. & Sasal, P. (2008). Dynamic expression of immune response genes in the sea bass *Dicentrarchus labrax* experimentally infected with the monogenean *Diplectanum aequans*. *Fish and Shellfish Immunology*, Vol. 24, No. 6 (June 2008), pp. (759-767) ISSN 1050-4648.

FAO. (2010). *Food and Agriculture Organization of the United Nations: State of world aquaculture*. 197 pp., ISSN 1020-5489.

Fernandez-Jover, D., Faliex, E., Sanchez-Jerez, P., Sasal, P. & Bayle-Sempere, J.T. (2010). Coastal fish farming does not affect the total parasite communities of wild fish in SW Mediterranean. *Aquaculture*, Vol. 300,No 1-4 (February 2010), pp. (10-16), ISSN 0044-8486.

Fioravanti, M.L., Caffara, M., Florio, D., Gustinelli, A., Marcer, F. & Quaglio, F. (2006). Parasitic diseases of marine fish: epidemiological and sanitary considerations. *Parassitologia*, Vol. 48, No. 1-2 (June 2006), pp. (15-18), ISSN 0048-2951.

Focardi, S., Corsi, I. & Franchi, E. (2005). Safety issues and sustainable development of European aquaculture: new tools for environmentally sound aquaculture. *Aquaculture International*, Vol. 13, No. 1-2 (January 2005), pp. (3-17), ISSN 0967-6120.

Gannicott, A.M. & Tinsley, R.C. (1998). Larval survival characteristics and behaviour of the gills monogenean *Discocotyle sagittata*. *Parasitology*, Vol. 177, No. 5 (November 1998), pp. (491-498), ISSN 0031-1820.

Ghittino, C., Latini, M., Agnetti, F., Panzieri, C., Lauro, L., Ciapelloni, R. & Petracca, A. (2003). Emerging pathologies in aquaculture: effects on production and food safety. *Veterinary Research Communication*, Vol. 27, supplement 1 , pp. (471-479) ISSN 0165-7380.

Gonzalez-Lanza, C., Alvarez-Pellitero, P. & Sitjà-Bobadilla, A. (1991). Diplectanidae (Monogenea) infestations of sea bass, *Dicentrarchus labrax* (L.), from the Spanish area. Histopathology and population dynamics under culture conditions. *Parasitology Research*, Vol. 77, No. 4 (April 1991), pp. (307-314), ISSN 0932-0113.

Gelnar, M. (1987). Experimental verification of the effect of water temperature on micropopulation growth of *Gyrodactylus katharineri* Malmberg, 1964 (Monogenea) paraziting carp fry (*Cyprinus carpio* L.). *Folia Parasitologica*, Vol. 34, No. 1 (January-February 1987), pp. (19-23), ISSN 0015-5683.

Guo, F.C. & Woo, P.T.K. (2009). Selected parasitosis in cultured and wild fish. *Veterinary Parasitology*, Vol. 163, No 3 (August 2007), pp. (207-216), ISSN 0304-4017.

Hadj Kacem, N.H., Aldrin, J.F. & Romestand, B. (1987). Effets immediats d'une augmentation rapide de temperature sur certains parameters sanguins du loup, *Dicentrarchus labrax*. *Aquaculture*, Vol. 64, No. 4 (August 1987), pp. (325-331), ISSN 0044-8486.

Hayward, C.J., Aiken, H.M. & Nowak, B.F. (2007). Metazoan parasites on gills of Southern Bluefin Tuna (*Thunnus maccoyii*) do not rapidly proliferate after transfer to sea cages. *Aquaculture*, Vol. 262, No. 1 (February 2007), pp. (10-16), ISSN 0044-8486.

Hutson, K.S., Ernst, I. & Whittington, I.D. (2007). Risk assessment for metazoan parasites of yellowtail kingfish *Seriola lalandi* (Perciformes: Carangidae) in South Australian sea-cage aquaculture. *Aquaculture*, Vol. 271, No. 1-4 (October 2007), pp. (85-99), ISSN 0044-8486.

Kearn, G.C. (1967). Experiments on host-findings and host-specificity in the monogenean skin parasite *Entobdella soleae*. *Parasitology*, Vol. 57, No. 3 (August 1967), pp. (585-605), ISSN 0031-1820.

Kim, K.H., Ahn, K.J. & Kim, C.S. (2001). Seasonal abundances of *Prosomicrocotyla gotoi* (Monogenea) and *Opecoelus sphaericus* (Digenea) from greenlings *Hexagrammos otakii* in a southern coastal area in Korea. *Aquaculture*, Vol. 192, No. 2-4, (January 2001), pp. (147-153), ISSN 0044-8486.

Johnson, S.C., Treasurer, J.W., Bravo, S., Nagasawa, K. & Kabata, Z. (2004). A review of the impact of parasitic Copepods on marine aquaculture. *Zoological Studies*, Vol. 43, No. 2 (April 2004), pp. (229-243), ISSN 1021-5506.

Lambert, A. & Maillard, C. (1974) (Monogenea, Monopisthocotylea) parasites simultanés de *Dicentrarchus labrax* (Téléostéen). *Annales de Parasitologie Humaine et Comparée*, Vol. 50, pp. (691-699), ISSN 0003-4150.

MacAllen, R. &, Brennan, E. (2009). The effect of environmental variation on the reproductive development time and output of the high-shore rockpool copepod *Tigriopus brevicornis*. *Journal of Experimental Marine Biology and Ecology*, Vol. 368, No. 1 (January 2009), pp (75-80), ISSN 0022-0981.

MacArthur, R.H. & Wilson, E.O. (2001). *The theory of island biogeography*. Princeton, New Jersey.: Princeton University Press.

Manera, M. & Dezfuli, B.S. (2003). *Lernanthropus kroyeri* infections in farmed sea bass *Dicentrarchus labrax*: pathological features. *Diseases of Aquatic Organisms*, Vol. 57, No. 1-2 (December 2003), pp. (177-180), ISSN 0177-5103.

Merella, P., Garippa, G. & Salati, F. (2006). Parasites of cage cultured European sea bass *Dicentrarchus labrax* and gilthead sea bream *Sparus aurata* from Sardinia (western Mediterranean): first results. *Parassitologia*, Vol. 48, No. 1-2 (June 2006), pp. 290, ISSN 0048-2951.

Mladineo, I. (2006). Parasites of Adriatic cage reared fish. *Acta Adriatica*, Vol. 47, No. 1 (June 2006), pp. (23-26), ISSN 1846-0453.

Mladineo, I. & Marsic-Lucic, J. (2007). Host switch of *Lamellodiscus elegans* (Monogenea: Monopisthocotylea) and *Sparicotyle chrysophrii* (Monogenea: Polyopisthocotylea) between Cage-reared Sparids. *Veterinary Research Communication*, Vol. 31, No. 2 (February 2007), pp. (153-160), ISSN 0165-7380.

Naylor, R.L., Goldburg, R.J., Primavera, J.H., Kautsky, N., Beveridge, M.C.M., Clay, J., Folkes, C., Lubchenco, J., Mooney, H. & Troells, M. (2000). Effects of aquaculture on world fish supplies. *Nature*, Vol. 405, No. 6790 (June 2000), pp. (1017-1024), ISSN 0028-0836.

Noga, E.J. (2000). *Fish disease. Diagnosis and treatment*, Wiley-Blackwell. 367 pp., ISBN 0813806976.

Nowak, B.F. (2007). Parasitic diseases in marine cage culture. An example of experimental evolution of parasites? *International Journal for Parasitology*, Vol. 37, No. 6 (May 2007), pp. (581-588), ISSN 0020-7519.

Oliver, G. (1982). *Microcotyle chrysophrii* Van Beneden et Hesse, 1863 (Monogenea, Polyopisthocotylea, Microcotylidae) parasite de *Sparus aurata* Linneaus, 1758 (Teleostei, Sparidae) dans les étangs littoraux du Languedoc-Roussillon (France). *Parasitologie*, Vol. 20, pp. (113-118).

Oliver, G. (1987). Les Diplectanidae Bychowsky, 1957 (Monogenea, Monopisthocotylea, Dactylogyridae). Systématique. Biologie. Ontogénie. Écologie. Essai de phylogenèse. Thèse d'État, Mention Sciences, *Université des Sciences et Techniques du Languedoc, Montpellier, France*, pp 433.

Ozel, I., Oktener, A. & Aker, V. (2004). A morphological study (SEM) on a parasitic copepod: *Lernenthropus kroyeri* (Van Beneden, 1851). *Ege University Journal of Fisheries and Aquatic Sciences*, Vol. 21, No. 3-4, pp. (335-337), ISSN 1300-1590.

Paperna, I. & Baudin Laurencin, F. (1979). Parasitic infections of sea bass, *Dicentrarchus labrax*, and gilthead sea bream, *Sparus aurata*, in mariculture facilities in France. *Aquaculture*, Vol. 16, No. 2 (February 1979), pp. (173-175), ISSN 044-8486.

Poulin, R. & Rohde, K. (1997). Comparing the richness of metazoan ectoparasite communities of marine fishes: controlling for host phylogeny. *Oecologia*, Vol. 110, No. 2, pp. (278–283), ISSN 0029-8549.

Reversat, J., Silan, P. & Maillard, C. (1992). Structure of monogenean populations, ectoparasites of the gilthead sea bream *Sparus aurata*. *Marine Biology*, Vol. 112, No. 1 (January 1992), pp. (43-47), ISSN 0025-3162.

Robertsen, G., Olstad, K., Plaisance, L., Bachmann, L. & Bakke, T.A. (2008). *Gyrodactylus salaris* (Monogenea, Gyrodactylidae) infections on resident Artic charr (*Salvelinus alpinus*) in southern Norway. *Environmental Biology of Fishes*, Vol. 83, No. 1 (September 2008), pp. (99-105), ISSN 0378-1909.

Silan, P. (1984). Biologie compare du développement de *Diplectanum aequans* et *Diplectanum laubieri*, monogènes branchiaux de *Dicentrarchus labrax*. Thèse de doctorat 3ème cycle. *Académie de Montpellier. Université des Sciences et Techniques du Languedoc*, pp. 275.

Silan, P. & Maillard, C. (1989). Biologie comparée du développement et discrimination des Diplectanidae ectoparasites du Bar (Teleostei). *Annales de Sciences Naturelles, Zoologie*, Vol. 10, pp. (31-45).

Sitjà-Bobadilla, A., Conde de Felipe, M. & Alvarez-Pellitero, P. (2006). In vivo and in vitro treatments against *Sparicotyle chrysophrii* (Monogenea: Microcotylidae) parasitizing the gills of gilthead sea bream (*Sparus aurata* L.). *Aquaculture*, Vol. 261, No. 3 (December 2006) , pp. (856–864), ISSN 044-8486.

Sprent, P. & Ley, J.P. (1992). *Pratique des statistiques non paramétriques*, Institut National de la Recherche Agronomique, Paris. 312 pp., ISBN 2-7380-0414-8.

Sterud, E. (2002). Parasites of wild sea bass *Dicentrarchus labrax* from Norway. *Diseases of Aquatic Organisms*, Vol. 48, No. 3 (April 2002), pp. (209-212), ISSN 0177-5103.

Timi, J.T. & Lanfranchi, A.L. (2006). A new species of *Cucullanus* (Nematoda: Cucullanidae) parasitizing *Conger orbignianus* (Pisces: Congridae) from Argentinean waters. *Journal of Parasitology*, Vol. 92, No. 1 (February 2006), pp. (151–154), ISSN 0022-3395.

Toksen, E., Nemli, E. & Degirmenci, U. (2008). The Morphology of *Lernanthropus kroyeri* van Beneden, 1851 (Copepoda: Lernanthropidae) Parasitic on Sea Bass, *Dicentrarchus labrax* (L., 1758), from the Aegean Sea, Turkey. *Türkiye Parazitologii Dergisi*, Vol. 32, No. 4, pp (386 – 389), ISSN 1300-6320.

Vazzana, M., Cammarata, M., Cooper, E.L. & Parrinello, N. (2002). Confinement stress in sea bass (*Dicentrarchus labrax*) depresses peritoneal leukocyte cytotoxicity. *Aquaculture*, Vol. 210, No 1-4 (July 2002), pp. (231–243), ISSN 044-8486.

Winger, A.C., Kanck, M., Kristoffersen, R. & Knudsen, R. (2008). Seasonal dynamics and persistence of *Gyrodactylus salaris* in two riverine anadromous Artic charr populations. *Environmental Biology of Fishes*, Vol. 83, No. 1 (September 2008), pp. (117-123), ISSN 0378-1909.

3

Parasitic Diseases in Cultured Marine Fish in Northwest Mexico

Emma J. Fajer-Ávila, Oscar B. Del Río-Zaragoza and Miguel Betancourt-Lozano
Research Center for Food and Development, Mazatlan Unit, Parasitoogy Lab México

1. Introduction

Parasite treatment and control, as well as proper health management of cultured marine finfishes are needed to establish mariculture as an economically viable activity. The Pacific northwest coast of Mexico harbours important commercial fish biodiversity, where some species are currently being considered for marine culture. Nowadays there are important advances in the domestication of several species of snapper, such as *Lutjanus guttatus* (spotted rose snapper), *L. peru* (Pacific red snapper), *L. argentiventris* (yellow snapper), *L. aratus* (mullet snapper), and *L. colorado* (colorado snapper). In particular, the spotted rose snapper is an important commercial fish species that has been in experimental culture for several years in Central America. In Mexico, snapper wholesale prices are among the highest, explaining the increasing interest in its cultivation that had led to several grow-out trials carried out by more than 17 coastal Pacific fishery cooperatives. Other species with potential for mariculture are the puffer-fish (*Sphoeroides annulatus*), the California halibut (*Paralichthys californicus*), the leopard groupers (*Mycteroperca rosacea*) and the yellowtail amberjack (*Seriola lalandi*), with potential for USA and Asian markets. Similarly, the Pacific coast of Mexico supports large tuna fisheries that have the highest economic value after shrimp. The capture-based tuna (*Thunnus orientalis*) aquaculture in the northern Baja California, Mexico, represents 3% of world production and is among the fastest growing forms of aquaculture in the world. It is estimated that in the future 80% of tuna will come from aquaculture (Zertuche-González et al., 2008). Snapper aquaculture in Mexico is carried out in grow-out cages (Fig. 1) maintained by coastal fishermen grouped in cooperatives in nine states of the Mexican Pacific: Baja California Sur, Sonora, Sinaloa, Nayarit, Jalisco, Colima, Michoacán, Guerrero and Oaxaca (Avilés-Quevedo, 2011), with an estimated demand of over 150,000 thousand juveniles for each production cycle. This activity nowadays depends on the capture of wild juveniles. Currently, the Mexican government (SAGARPA, CONAPESCA, INAPESCA), academia (CIAD, CICESE, CIBNOR, CRIP, UABC, UAS, others) and farmers are joining forces in the expansion and diversification of mariculture. In these sense a multispecies, 2668 m^2 pilot-scale tropical marine finfish hatchery was designed to fulfill the requirements of finfish juvenile research and development at the Research Center for Food and Development, in Mazatlan, Mexico (Alvares-Lajonchere et al., 2007). The main goals of the facility are (1) scale-up and study experimental results at a pre-commercial-scale; (2) assess technical and financial feasibility

Fig. 1. Floating cage of the spotted rose snapper in Mazatlán Sinaloa. Photograph courtesy of C. Hernández

and improve these technologies before transfer to commercial-scale; (3) adapt technology to other fish species. Other snapper hatcheries are located in Bahía Concepción, B.C.S. and in Boca de Apiza, Michoacán (CRIP), while facilities for California halibut and yellowtail amberjack are in Ensenada B.C.N. (CICECE).

The occurrence of parasites is one of the most important problems in aquaculture development. Fish farms require knowledge of parasites that could become potential pathogens. In this context, 19 species have been reported as bullseye puffer fish parasites, 38 for the spotted rose snapper, 18 for the Pacific red snapper, 5 for the Pacific cubera snapper, 10 for the yellow snapper, 2 for the mullet snapper, 36 for the Pacific bluefin tuna, 11 for the yellowtail amberjack, 4 for the leopard groupers, and 24 for the California halibut. Among these, ectoparasites represent the major threat in farming systems, where infections can be predominantly caused by protozoans, monogeneans and crustaceans. To overcome the problems posed by parasites, the use of proper prophylactic and therapeutic measures is frequently indicated, as well as immunostimulants to improve fish welfare, health and production. Therefore, the objective of this chapter is summarizing the main parasites and parasitic diseases that affect the marine fish species with aquaculture potential in the Norwest Pacific coast of Mexico, emphasising on the research approaches aimed at developing strategies for their control.

2. Diseases caused by protistants

2.1 Ciliates

Among the protozoan, the group of ciliates has more than 8000 species, with sizes ranging from about 10 μm to 4.5 mm (Lom, 2005). Ciliates can be found virtually everywhere from small ponds or streams to open oceans. Their locomotion is performed by cilia and although most are free-living, many occur as commensals, and some are parasites of invertebrates and vertebrates. The ciliates exhibit nuclear dimorphism, with a large asexual nucleus (macronucleus) responsible for the entire life activity, and the small sexual nucleus (micronucleus) for reproduction. Although in most cases the ciliates do not inflict any harm, there are some species severely pathogenic to fish.

Cryptocaryon irritans is perhaps one of the most virulent parasites of snappers causing a disease called cryptocaryonosis (Fig. 2a). The main characteristic of this holotrich ciliate is the presence of a tetra lobulated macronucleus and several smaller micronuclei (Fig. 2b) (Colorni & Diamant, 1993). Its infective phase is carried out by tomites that become pyriform free swimming theronts, which can then penetrate skin and gills. Mature trophonts leave the fish and produce cysts that eventually develop as daughter tomites (Dickerson, 2006). In the spotted red snapper, the time required to complete its life cycle is approximately 6 days at 21-24°C (unpublished results). This parasite invades the epithelial tissue of snappers cultured causing serious lesions. In our facilities, outbreaks of *C. irritans* were common in snapper juveniles (*L. guttatus*, *L. peru* and *L. argentiventris*) reared in tanks, with characteristic white pinheads on the skin, fins, eyes and gills, where each spot represented a developing trophont within an epithelial capsule or vesicle (Fig. 2). The observed infection levels were heavy (≥ 200 trophonts gill $arch^{-1}$) and caused serious damage in gill tissue such as hyperplasia, inflammation and necrosis, while mortalities rose up to 70% within the affected stock.

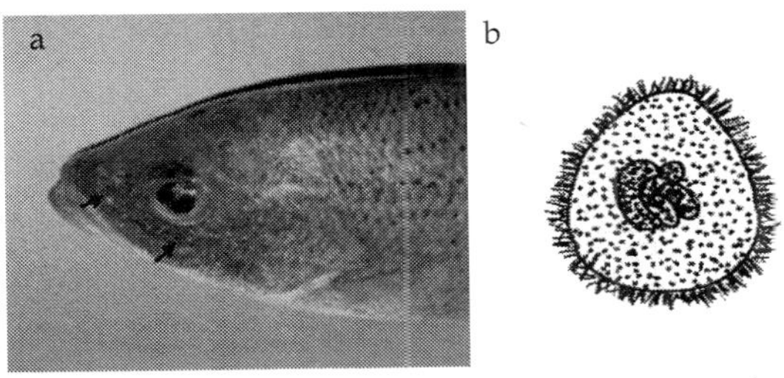

Fig. 2. Cryptocaryonosis in the spotted rose snapper (a) (from: Fajer-Ávila et al., 2011). Arrows indicate white spots. A drawing of *Cryptocaryon irritans* (b) (from: EcuRed 2012).

Although there are reports of *C. irritans* in tiger puffer fish (*Takifugu rubripes*) (Ogawa & Inouye, 1997), this ciliate has not been found on bullseye puffer fish cultured in Mexico. Ogawa & Inouye (1997) occasionally reported this parasite from tiger puffer fish cultured which seasonality on the occurrence was not clearly demonstrated. However, the occasional reports of this parasite in the tiger puffer fish, (Ogawa & Inouye, 1997) and its association to the emaciation disease in fish cultured in net cages, with clinical signs of severe emaciation, sunken eyes, bony ridges on the head and a tapered body (Tun et al., 2000), indicates that *C. irritans* could eventually become a risk for the bullseye puffer fish culture.

Another parasite that is known to affect snappers in captivity is the phylopharyngiid ciliate, *Brooklynella hostilis*. This parasite feeds on epidermis tissue of skin and gills, causing hemorrhages as also ingest blood cells with and extensible cytopharyngeal armature (Lom, 2005). In captive fish it may multiply massively causing brooklynellosis, when it can be found gliding over the gills using its ventral rows of cilia. Heavy infestations of *B. hostilis* have been associated to stress during the transportation of wild spotted rose snapper juveniles to our facilities, causing respiratory difficulties and up to 70% of fish mortality (pers. obs.). High infection levels associated to mortalities have also been observed on gills of the Pacific red snapper captivity in aquaculture facilities in La Paz, B.C.S. Mexico (Perez-Urbiola et al., 2008a).

Trichodina spp. can be considered among the parasites that could be of importance in the bullseye puffer fish culture. So far, two trichodinid species have been identified, *Trichodina jadranica* and *Trichodina* sp. (Fajer-Ávila et al., 2011), normally found on the skin and gills of wild adults from Sinaloa (Fajer-Ávila et al., 2004), where *Trichodina* sp. ocurred in 30% of the fish, and was associated to histopathological and degenerative alterations, such as hyperplasia, increase of mucous cells and inflammatory reactions (Chávez-Sánchez, pers. comm.). These degenerative changes have been previously described in other species and linked to high mortalities in wild and cultured fish (Bason & Van As, 2006; Colorni & Diamant, 2005). The high prevalence and abundance of trichodinids in wild marine fish has also been proposed as indicator organic pollution (Aksit et al., 2008).

2.2 Dinoflagelates

Amyloodiniosis is caused by the dinoflagelate *Amyloodinium ocellatum*, considered the parasite that inflicts the greatest damage to commercial marine fish cultivated in tropical waters (Fig. 3). This organism has dark brown sac-like trophozoites (trophonts; 56 to 73 µm diameter) bounded by a cell wall, internal globules and a prominent holdfasts stalk that anchors the parasite to the host.

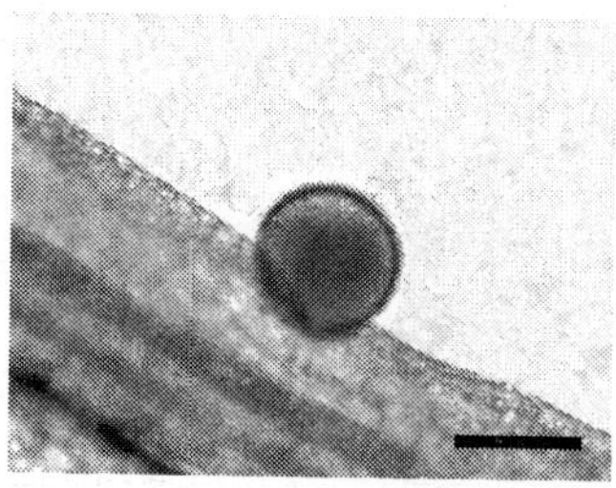

Fig. 3. *Amyloodinium ocellatum* attached to a gill filament on the spotted rose snapper. Scale bar = 100 µm (400 x).

The life cycle of *A. ocellatum* infecting captive spotted rose snapper takes 96 hours at 28°C and 34 ups of salinity (Ontiveros-García, 2008). Its growing and feeding phase occurs on the skin and gills of fish, followed by a detachment of the trophont (with stalk withdrawal), which then produces a reproductive cyst (tomont) that settles in the tank walls. The cyst performs several asexual divisions, releasing motile forms (dinospores of 12 to 15 µm) after the last division, which is then capable of infecting new hosts. Outbreaks of *A. ocellatum* have been frequently observed on bullseye puffer fish, spotted rose snapper, Pacific red snapper, Pacific cubera snapper, yellow snapper and mullet snapper causing gill lesions, impaired oxygen exchange and even 100% of fish mortality (Fajer-Ávila et al., 2011; Pérez-Urbiola et al., 2008a). The rearing conditions of juvenile leopard groupers were favorable for the proliferation of this dinoflagellate, resulting in a cumulative mortality rate of 89.5% at 7 days post-infection, where the highest virulence was observed at 48 h (Reyes-Becerril et al.,

2008). The infection is characterized by initial signs of appetite loss, followed by opacity of the skin with whitish areas and erosion of the caudal fin. Fish start swimming sideways and rubbing against the tank's bottom. Severe infection cases have been observed on the spotted rose snapper (Ontiveros-García, 2008), with hundreds of parasites per fish. A notorious fish response was an increment on erythrocyte size, which could be considered as respiratory exchange compensation due to the effect of parasites on the gills. A positive correspondence between the level of infections of *A. ocellatum* and the total leucocyte number was also observed, indicating that this is a protective response to parasitic stress (Das et al., 2006). High infection levels of *A. ocellatum* were further associated to a severe proliferative epithelial response on gills, high number of mucous cells, infiltrating inflammatory cells and lamellar fusion (Ontiveros-García, 2008).

3. Infections by monogeneans

Monogeneans are flatworm ectoparasites with a direct life cycle, frequently found in mariculture systems. The main characteristic of the monogenean group is presence of an opisthaptor, an adhesive apparatus equipped by sclerotized structures located in the posterior region of the worm. Most of monogeneans infecting marine fish are ovoviviparous (Buchmann & Bresciani, 2006). They are considered important pathogens due to their velocity of propagation among fish in culture systems (Thoney & Hargis, 1991). Mass infections by monogeneans are result of the innate susceptibility that stressful environmental conditions could elicit on fish (Buchman & Bresciani, 2006).

3.1 Capsalids

The capsalid family has a characteristic flattened, leaf-like body. The haptor is also remarkably conservative. The basic arrangement comprises a saucer-shaped attachment organ armed with three pairs of median sclerites that are usually large, 14 small hooklets at the periphery of the haptor proper and a thin, membranous marginal valve around the edge (Whittington, 2004). The capsalid monogeneans are considered important pathogens in fish culture facilities (Ogawa et al., 1995; Whittington et al., 2001a). They normally inhabit on the fish skin and under the scales, although some can be found on the gills and nostrils (Buchmann & Bresciani, 2006). These parasites feed on epithelial cells and mucus and are relatively large and flat, with prominent muscular disc haptors at the posterior end (Leong & Colorni, 2002).

Some species have been linked to epizootics of wild fishes (Paperna & Overstreet, 1981), while other species have shown low host specificity in marine fish cultured in floating net cages, such as the ones of the genera *Benedenia*, *Neobenedenia* and *Megalocotyloides* (Leong & Colorni, 2002). *Benedenia* spp., *B. epinepheli*, *B. lutjani*, and *Neobenedia* sp. were reported to the snappers *L. johni* and *L. argentimaculatus*, while *N. girellae* was reported to *L. johni* (Leong & Colorni, 2002) and *T. rubripes* (Ogawa et al., 1995). *B. epinepheli* was reported in 25 fish species, where tetraodontid fish was considered the most susceptible for severe infections (Leong & Colorni, 2002). Five *Neobenedenia* species were found in the Pacific coast and Cortez Sea, Mexico: *N. girellae*, morphologically synonymized as *N. melleni* by Whittington & Horton (1996) molecularly confirmed by Li et al. (2005), and *N. adenea*, on the groupers *M. pardalis* and *M. rosacea* in Baja California Sur, Mexico (Kohn et al., 2006); and *N. isabellae* on gills of *M. olfax* in Nayarit; *N. longiprostata* on the serranid family in Baja California Norte and *N. pacifica* on the flathead mullet (*Mugil cephalus*) in Baja California Norte (Kohn et al., 2006). Wild bullseye puffer fish were found infested by *N. melleni* at low levels (1-3 parasite per fish) (Fajer-Ávila et al., 2004), while in tanks the infection levels reached values of 28-60 parasites per fish, causing skin lesions, anorexia and mortality on juveniles (Fajer-Ávila et al., 2008). A potential new species of *Neobenedenia* (Fig. 4a) was found in spotted rose snapper causing hemorrhage of the caudal fin (Fig. 4b), emaciation and mortality in juveniles reared in tanks (unpublished results). Additional reports of *Neobenedenia* spp. on adults of *L. peru, L. aratus, L. novemfasciatus* and *L. argentiventris* reared in tanks indicates that fish show opaque eyes, mucus proliferation, coloration changes and anorexia, while this parasite was not found in juveniles of the same fish species (Pérez-Urbiola et al., 2008a). As pathogen, *Neobenedenia* is particularly important as there are reports of the introduction of this parasite into fish stocks due to unregulated fish transference across international borders (Ogawa et al., 1995).

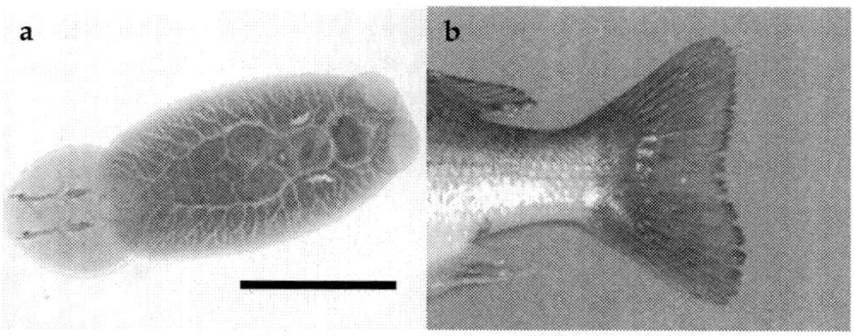

Fig. 4. *Neobenedenia* sp. (Scale bar = 1mm at 40 X; Staining Van Cleaves´ hematoxilin) (a) causing hemorrhage on the caudal fin in spotted rose snapper (b).

Benedenia seriolae is the most common capsalid species in the intensive culture of yellowtail (*Seriola quinqueradiata*), where approximately 20% of the total production costs is dedicated to control this parasite (Whittington et al., 2001b). In Bahia Magdalena, BCS, Mexico, 5% mortality of *S. lalandi* cultured in floating cages has been attributed to mismanagement and the presence of opportunistic parasites, such as *Benedenia* sp. (Avilés-Quevedo & Castelló-Orvay, 2004). *Benedenia* sp. is commonly found in wild fish such as croakers, flounders, and mackerels (Avilés-Quevedo & Castelló-Orvay, 2004). In Mexico most of the capture-based tuna aquaculture is located along the Pacific Coast of Baja California. The farms normally rear the pacific bluefin tuna a species recognized as resistant to diseases, so although several parasites have been identified, no serious problems have been reported. The capsalids reported in pacific bluefin tuna are *B. seriolae, Capsala paucispinosa* and *Capsala* sp. (Munday et al., 2003). No reports of monogeneans were found for the culture of the California halibut a common fish along the Pacific coast of North America. However, there are reports of capsalids parasitizing this fish species such as *Entobdella squamula* and *E. hippoglossi* (Castillo-Sánchez et al., 1998; Kalman, 2006).

3.2 Dactylogyrids

Dactylogyrids are common gill parasites from teleost fish distributed throughout warm seas (Wu et al., 2006). The monogeneans of the Dactylogyridae family presents two pairs of large hooks or hamuli that oppose each other and allow the parasite to attach to the secondary lamellae of the gill filament. The hamuli break the gill tissue and cause epithelial hyperplasia, oedema and haemorrhages (Whittington, 2005). Heavily affected fish may die due to asphyxia as a result of gill pathology and interference with the exchange of respiratory gases and ions (Stephens et al., 2003). Dactylogyrids could have an impact on cultured fish as they can easily multiply and disperse as result of their direct life cycle, sometimes reaching very high densities.

Some *Haliotrema* species are pathogens of marine fish. *H. johni* and *H. abaddon* were reported infecting gills of snapper *L. johni*, in floating cages in Penang, Malaysia (Leong & Wong, 1987), and the cultured Australian dhufish (*Glaucosoma hebraicum*) (Pironet & Jones, 2000). Dactylogyrid monogeneans (Fig. 5) such as *Haliotrematoides* spp. and *Euryhaliotrema* sp. are commonly found on the gills of wild and cultured spotted rose snapper (Fajer-Ávila et al., 2007).

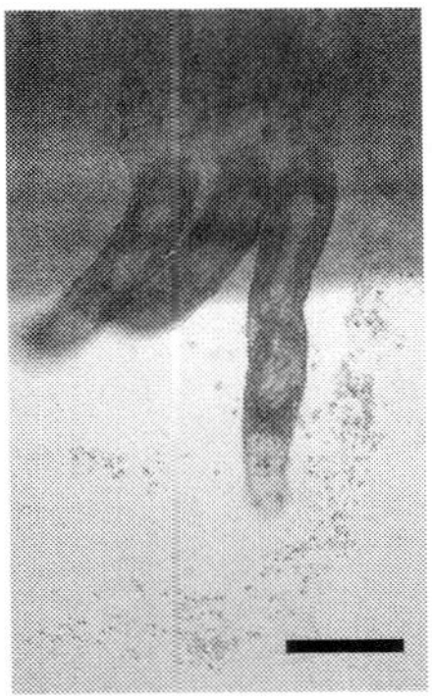

Fig. 5. Dactylogyrid monogeneans on a gill filament of spotted rose snapper.
Scale bar = 50 μm (40 x).

Three dactylogyrid genera, *Euryhaliotrema*, *Haliotrematoides* and *Tetrancistrum*, include species that have been found on the gills of snappers (Kritsky et al., 2009). Fifteen species of *Euryhaliotrema* are currently known from snappers worldwide (Fuentes-Zambrano & Silva Rojas, 2006; García-Vargas et al., 2008; Kritsky & Boeger, 2002, Li, 2005, 2006; Li et al., 2005; Pan & Zhang, 2006) and 22 of *Haliotrematoides* (Kritsky et al., 2009). In Mexico four species of *Haliotrematoides* have been reported from lutjanids from the Atlantic (Kritsky et al., 2009; Zhukov, 1976). *E. perezponcei* and *H. guttati* (García-Vargas et al., 2008), *H. spinatus*, *H. plectridium* and *Euryhaliotrematoides* sp. were found on the spotted rose snapper in northwestern coast of Mexico (Soler-Jiménez & Fajer-Ávila, in press) Perez-Urbiola et al. (2008a) reported high levels of dactylogyrids (*Haliotrema* spp.) associated to mortalities of snappers (*L. aratus*, *L. argentiventris* and *L. novemfasciatus*) confined at high densities in little ponds with low water exchange.

Experimental studies with *L. guttatus* subjected to different levels of dactylogyrid infections showed that the attachment of dactylogyrids increased the erythrocyte sedimentation rate in fish with a low level of infection (mean intensity of 22 parasites per fish). Glucose was high for all infection levels. An increase in the infection load was positively correlated with the total leucocyte count (68,000 to 152,000 leucocytes), representing an increase of more than

200 percent in fish with the infection level higher respect to control group. The infection load was also positively correlated with the thrombocyte percentage ratio and the percentage of granulocytes, whereas increased infection level was negatively correlated with the leucocyte percentage ratio and the number of lymphocytes. A high infection level (100 parasites per fish) led to severe proliferative epithelial response in gills, as well as high number of mucous cells, a moderate increase in chloride cells, infiltrating inflammatory cells and lamellar fusion (Del Río-Zaragoza et al., 2010).

3.3 Diclidophorids

Diclidophoridae monogeneans have an elongated body, which usually have an opisthohaptor with several suckers or clamps in two mostly symmetrical rows. The number of clamps never exceeds four on one side and they have also a cirrus and or genital atrium armed or unarmed (Yamaguti, 1961). The genus *Heterobothrium* is a very common diclidophorid parasite of tetraodontid fish, with four pair of sessile clamps and a ring of claw-shaped spines at the terminal end of the male copulatory organ. *Heterobothrium* has a direct cycle, invading fish hosts through free-swimming larvae, oncomiracidia. Outbreaks of diclidophorids have been reported in several Asian countries, such as *H. okamotoi* infecting gills and the branchial cavity of the tiger puffer fish, while *Neoheterobothrium hirame* infects the gills and the buccal cavity of the Japanese flounder *(P. olivaceus),* the latter parasite being considered the causative agent of anaemia among Japanese fisheries since the late 1990s (Ogawa, 2002).

In the Pacific coast of Mexico, *H. ecuadori* has been reported in the bullseye puffer fish from Oaxaca (Lamothe-Argumedo, 1967) and Sinaloa (Fajer-Ávila et al., 2004). In the Gulf of Mexico, *H. lamothei* from the gills of checkered puffer (*S. testudineus*) was described. Reports regarding the occurrence of the platyhelminth parasites on wild bullseye puffers fish from two localities in Sinaloa revealed that *H. ecuadori* was the most prevalent species, likely to be present at water temperature of 23-24.5°C (Fajer-Avila et al., 2003). The higher prevalence of *H. ecuadori* observed in some areas could be associated to the behaviour of the bullseye puffer fish, which congregates in large numbers in lagoons, such Teacapan, to spawn, facilitating the transmission of parasites with direct life cycles.

In general, the biology of diclidophorid monogeneans is poorly known (Frankland, 1955; Yasuzaki et al., 2004). Grano-Maldonado et al. (2011) described the development and life span of the oncomiracidia of *H. ecuadori* in bullseye puffer fish under experimental condition. The authors found that hatching time fluctuated between 7 and 10 days at 23±1° C and 35 ups salinity. Oncomiracidia longevity was 4 to 7 days at 21±1°C, indicating infectivity potential for a long period of time, therefore making this parasite very dangerous for fish culture. Other monogenean larvae from the same family, such as the oncomiracidia of *H. okamotoi,* were also able to survive up to 7.3 days at 20°C (Ogawa, 1998). As result, severe parasitic infestations of *H. ecuadori* can result in significant mortality in cultured bullseye puffer fish (Fajer-Avila et al., 2003). Histological findings showed that the parasites were attached to secondary gill lamellae, causing moderate damage to cells, such as inflammation, gill fusion, mucus cell number increase, hyperplasia, epithelial sloughing and necrosis (Chávez-Sánchez pers. comm.). In general, diclidophorids do not elicit great tissue damage of their delicate ways to attach and drawn blood from gill lamellaes (Cone, 1999). Infections levels higher than 100 *H. ecuadori* per gill arch on bullseye puffer fish juveniles reared in tanks were associated to pale gills and 25% mortality (personal observations), showing similarities to the effects reported by Ogawa (2005) with *H. okamotoi* on *T. rubripes*, indicating that anemia was the main pathological alteration.

3.4 Microcotylids

The monogeneans of the microcotylidae family are gill parasites that possess a haptor with numerous clamps and a mouth adapted for blood sucking (Leong & Colorni, 2002). Although reports of infections by these monogeneans are restricted to relatively few fish species, not always showing clinical signs (Leong & Colorni, 2002), it is known that microcotylids are able to parasitize a wide range of hosts in the wild (Yamaguti, 1963). However, in recent years microcotylids have caused severe losses in Mediterranean aquaculture. For instance, *Sparicotyle* (syn *Microcotyle*) *chrysophrii* has been identified as one of the most threatening ectoparasite for gilthead seabream (*Sparus aurata*) culture (Alvarez-Pellitero, 2004; Sitja-Bobadilla & Alvarez-Pellitero, 2009). *Heteraxine heterocerca* and *Zeuxapta seriolae* infecting *S. quinqueradiata* and *S. dumerili* respectively, produced similar pathological effects and were associated to mortalities. Anemia and death are normally associated to the severe gill damage and blood sucking of these monogenean parasites (Montero et al., 2004). In Korea *Microcotyle sebastis* is considered a major parasitic disease agent as it has been related to high cumulative mortalities of net-pen farmed rockfish (*Sebastes schlegeli*) (Kim & Cho, 2000). In northwest Mexico, several microcotylid species have been reported from snappers from Jalisco, Nayarit, Sinaloa and Baja California Sur (Table 1).

Microcotylids species	Fish Host	Localities	Reference
Neobivagina aniversaria	*L. guttatus*	CB	Mendoza-Garfias & Pérez-Ponce de León (1998)
Polymicrocotyle manteri	*L. guttatus*		
P. manteri	*L. jordani*		
P. manteri	*L. peru*	LP	Torres-Ayala (2008); Pérez-Urbiola et al. (2008b)
P. manteri	*L. colorado*	BB	Perez-Ponce de León et al. (1999)
Microcotyloides incisa	*L. argentiventris*	CB, LP	Mendoza-Garfias & Pérez-Ponce de León (1998); Perez-Ponce de León et al. (1999)
M. incisa	*L. jordani*	CB	
M. incisa	*L. guttatus*	BB, MB,MTB	García-Vargas (2008)
M. incisa	*L. colorado*	BB	Perez-Ponce de León et al. (1999)

Table 1. Microcotylids species found on wild snappers from Chamela Bay, Jalisco (CB), Bandera Bay, Cruz de Huanacaxtle, Nayarit (BB), Matanchen Bay, Nayarit (MTB), Mazatlan Bay, Sinaloa (MB), and La Paz, Baja California (LP) Mexico.

To our knowledge no microcotylids species has been reported in snapper fish culture. However, even if microcotylids in wild fish generally have relatively little pathological consequence on the host (Leong & Colorni, 2002), as the projected plans to rear fish in sea cages develop, the high densities in these systems could easily favor microcotylids transmission. For instance, a single-host cycle of *S. chrysophrii* could be experimentally completed in only 6 weeks post exposure in the gilthead seabream (Sitja-Bobadilla et al., 2009).

4. Infestations by copepods

The adaptations of parasite copepods are, in many cases, of such extent that there are few resemblances to the typical free-living forms (Grabda, 1991). These parasites are small dioecious crustaceans, with female carrying eggs in egg sacs attached to the genital segment. The egg hatches into nauplius that develops into copepodids and eventually adults in a rather shortened life cycle (Lester & Hayward, 2006). Copepods can be considered among the most important disease-causing parasites in both wild and aquaculture fish populations (Boxshall & Halsey, 2004). Parasites such as the sea lice (*Lepeophtheirus salmonis*), as well as cymothoid isopods, have acquired notoriety with the development of rearing fish in sea cages (Lester & Hayward, 2006).

4.1 Caligids

Caligids are dorso-ventrally flattened parasites that typically hold to their hosts by a combination of claws and suction. Their bodies are divided into an anterior cephalothorax and a post-cephalothoracic genital trunk (Boxshall, 2005). Two genus are very common parasitizing fishes: *Caligus* spp. that have frontal lunules, and *Lepeophtheirus* spp. which does not have lunules (Wootten et al., 1982). The Caligidae family is generically called sea lice, and mostly occurs in tropical and warm temperate waters. They are considered among the most important parasites in marine fish aquaculture. Sea lice infestations are particularly well known due to the problems caused to the salmonid farming industry, where the cost of treatments could represent up to 6% of the fish production (Costello, 2009a). The presence of sea lice on non-salmonid production systems has also been associated with mortality and diseases, such as the reports of *L. pectoralis* infecting the European flounder (*Platichthys flesus*) (Cavaleiro & Santos, 2007). Although the presence of caligids in the California halibut culture have not been reported, *L. bifurcatus* and *L. spatha* were reported in wild fish with a prevalence of 3% and 6%, respectively (Kalman, 2006).

L. simplex has been found in the bullseye puffer fish in Sinaloa, Mexico. This parasite showed a prevalence of 63% and a mean intensity of 2.6 copepods per fish in the Teacapan area, Sinaloa (Ho et al., 2001). In fact, a seasonality pattern of *L. simplex* was observed on bullseye puffer fish from Santa Maria La Reforma lagoon, Sinaloa, with prevalences ranging from 16 to 93% and mean intensities from 2 to 5.1 (Morales-Serna, 2010). The highest levels occurred during the warmest months (August and October), although host size was also linked to prevalence. In culture conditions, however, mean infection levels, up to 30 copepods per fish and 100% of prevalence have been observed (Fajer-Ávila & Chávez-Sánchez, 1999). *L. simplex* mobile stages (pre-adult and adult) cause the greatest damage due to their grazing habits on the host's skin, affecting the epidermal tissues due to the mouthparts sawing movements. As consequence of the infestation, bullseye puffer fish develop whitish spots and hemorrhagic points across lateral and ventral areas of its body, sloughing epithelial skin and ulceration exposing the muscular tissue, which then can led to bacterial infections and ultimately fish death.

Sometimes *L. simplex* infesting the bullseye puffer fish has been found hyperparasitised with the monogenean, *Udonella* sp., causing more damage than the copepods alone (pers. obs.). This is similar to reports of *Pseudocaligus fugu* on the *T. rubripes* hyperparasitised by *Udonella* sp. (Ogawa, 2005; Ogawa & Yokoyama, 1998).

The importance of monitoring and controlling sea lice as part of escalation of the bullseye puffer fish culture derives from the infectivity potential of this parasite (Fig. 6). In experimental conditions, sea lice nauplii and free swimming copepodids of *L. simplex* successfully infested bullseye puffer fish juveniles, taking only 12 to 14 days to complete the life cycle (Rivas-Salas, 2003). The genus *Caligus* sp. has also been reported by Fajer-Ávila et al. (2011) on the skin of bullseye puffer fish collected from Teacapan and Mazatlan areas, Sinaloa. *C. serratus*, considered as pathogenic for wild and fish culture (Costello, 2009b), was additionally found in bullseye puffer fish from Santa Maria La Reforma lagoon, Sinaloa by Morales-Serna (2010), with prevalence ranging from 14 to 70% and mean intensities from 1 to 1.5. The prevalence of *C. serratus* was positively correlated with fish size as it was found with *L. simplex* (Morales-Serna, 2010). Similarly, on the gills of wild and cage cultured spotted rose snapper adults of *Caligus* sp. were found at very low prevalence and abundance (Fajer-Ávila et al., 2011; Quispe, 2005). Pérez-Urbiola et al. (2008a) reports different *Caligus* species parasitizing wild Pacific red and yellow snappers, although indicates that the diseases caused by these parasites were developed during their captivity.

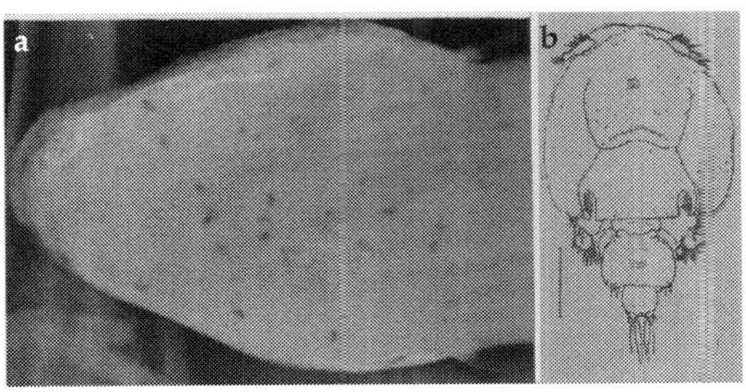

Fig. 6. Bullseye puffer fish heavily infected by *Lepeophtheirus simplex* under culture condition (a) (from: Fajer-Ávila et al., 2011). Diagram of dorsal side of *L. simplex* female. Scale bar= 500 μm (b).

4.2 Lernanthropids

The Lernanthropidae family, consisting of approximately 140 species, most from warm waters, is characterized by presenting a body compounded by cephalotorax (with well developed dorsal shield curved ventrally on each side in female, flat in male), trunk, a genital complex, and abdomen with 1 or 2 segments (Boxshall & Halsey, 2004). They attach to the gill filaments of the hosts by the antennae and maxillipeds.

Lernanthropus sp. was found on the gills of wild spotted rose snapper adults from Nayarit and Sinaloa, and juveniles in cage culture (Quispe, 2005). *Lernanthropus* can be identified by the presence of egg sacs linear, the third pair of legs not fused, somite bearing leg 4 with single dorsal plate and endopod of leg 4 as long as, or longer than, exopod (Boxshall & Halsey, 2004). In wild spotted rose snappers presented a monthly prevalence higher than

20% during sampling from May to October, with a maximum of 36% in June and a mean intensity of 0.27 parasites per fish (Quispe, 2005). However, despite their pathogenic potential there are no reports about their incidence in fish culture (Fig. 7).

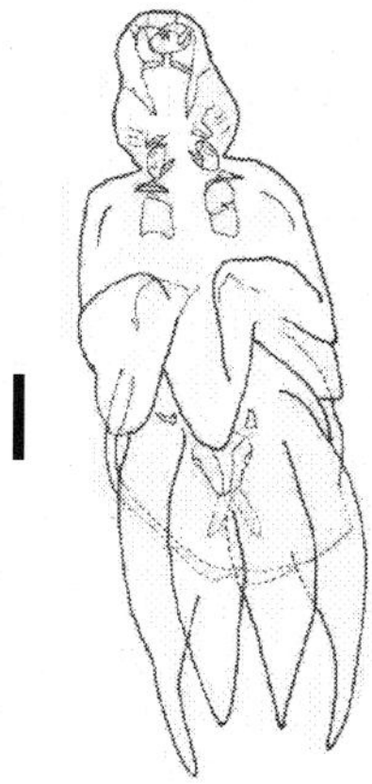

Fig. 7. Diagram of ventral side of *Lernanthropus* sp. Scale bar = 0.5 mm.

4.3 Chondracanthids

There are about 160 known species of chondracanthids that occur predominantly in the oral and branchial cavities of marine demersal fishes (Boxshall & Halsey, 2004). The body of chondracanthids is sexually dimorphic, with large transformed female typically carrying a small male. Female body is divisible into head, neck, trunk and genitoabdomen. They normally attach to their hosts using a strong, uncinate antenna (Boxshall & Halsey, 2004). Nauplii and first copepodids settle on the host and molt into a sexually differentiated second copepodid (Ho, 1970). The male copepodid then adheres to the juvenile immature female and remains attached throughout the rest of its life

The first report of the chondracanthids in fish culture was *Chondracanthus goldsmidi* on the striped trumpeter (*Latris lineate*), cultured in Australia. This parasite was reported as causing inflammation in the site of adhesion and high mortalities (Tang et al., 2007). In wild California halibut, the copepod *Acanthochondria hoi* was reported with a prevalence of 44 (Kalman, 2006). A representant of the Chondacanthidae family was found on the gills and branchial cavity of wild bullseye puffer fish from Sinaloa (Fajer-Avila et al., 2004). Additionally, *Pseudochondracanthus diceraus* was identified by Morales-Serna (2010) on bullseye puffer fish from Santa Maria La Reforma lagoon, Sinaloa (Fig. 8).

An evaluation of the pathological response of *P. dicerarus* on bullseye puffer fish showed potential to cause a disease under experimental conditions, with a lower percentage of leucocytes in those fish with the highest number of chondracanthids per fish (7 to 17) (Fajer-Ávila et al, 2011). This suggests a connection between parasitic infestations and immunological responses. The attachment area in the gill showed a characteristic response of hyperplasia and secondary lamellae fusion, atrophy of lamellar epithelium and higher

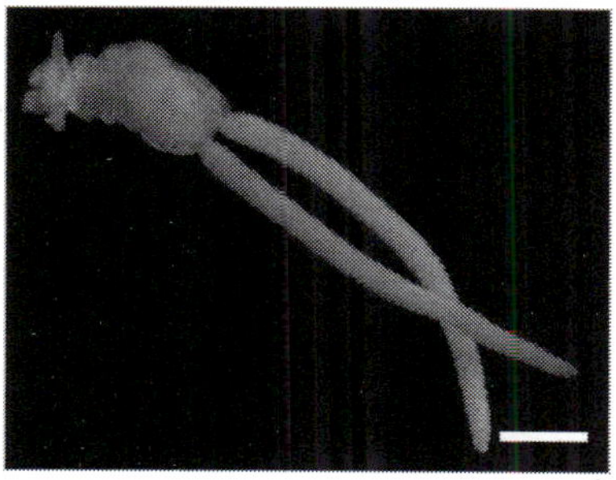

Fig. 8. Female of the *Pseudochondracanthus diceraus* from the bullseye puffer fish at 100 x. Scale bar = 200 µm.

number of mucus cells, likely produced by the claw-shaped antennae and grazing of *P. diceraus*. A description of the life cycle of *P. diceraus* under experimental conditions indicated a development from nauplius to the first adult stage (settled on juveniles of bullseye puffer fish) in only 20 days at 22°C (Guzmán-Beltrán & Zárate-Rodríguez, 2008). The results agree with the severe infection produced in only three weeks by *C. goldsmidi* in a striped trumpeter, *Latris lineata,* reared at the Tasmanian Aquaculture and Fisheries Institute, Australia (Tang et al., 2007), indicating that culture conditions can be favorable for the quick reproduction and transmission of this parasite species. Based on the preliminary results, this parasite species deserved special attention as it has the potential to become a health risk for the bullseye puffer fish culture.

5. Infection by isopods

The isopods are parasitic crustaceans, typically marine, normally inhabiting warm waters (Lester & Hayward, 2006). The body of isopods is dorso-ventrally flattened and lacking a carapace with a thorax composes of 7 free segments with 7 pairs of uniramous swimming thoracic legs (Grabda, 1991). The ventral side of female possesses a marsupium that hold the eggs. Most isopods are cymothoids ectoparasitizing marine teleosts, with adult females attaching to the gills and buccal cavity or even burrow into the fish developing a pouch. They are protandrous hermaphrodites and the first male to parasitize a fish changes into female (Lester, 2005).

Some species of cymothoids harm the fish in different ways. Fry and fingerlings are readily killed by the tissue damage produced by mancae stage juveniles, while adults could retard or inhibit both growth and reproduction (Lester & Hayward, 2006). The parasites that settle on the gill chamber are usually associated with anemia and loss of gill filaments as result of their movements and feeding habits (Lester & Hayward, 2006). In the mouth they can affect the development of oral structures and could eventually replace the tongue (Lester, 2005).

Cymothoa exigua is the only species of the genus distributed in the tropical east of the Pacific, ranging throughout the Gulf of California to Ecuador (Fig. 9). This isopod is very abundant throughout the Gulf of California, and it has been reported by Brusca (1981) and Brusca & Gilligan (1983) in *L. peru* and *L. gutattus*. Ruiz & Madrid (1992) reported *C. exigua* on Pacific red snapper from commercial fisheries in Michoacan, with isopods on the mouth and gill

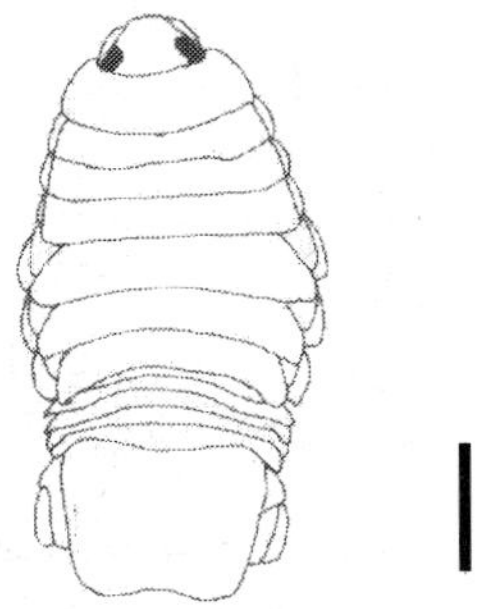

Fig. 9. Diagram of *Cymothoa exigua* dorsal side of *a*. Scale bar = 6.0 mm.

chamber of 165 from 691 snappers collected. The authors partially described the life cycle of *C. exigua* with notes about relative fecundity, marsupial development and pre-hatch mortality. This isopod was also found in mouth and gills of wild and cage cultured spotted rose snapper from Nayarit, with 12 and 8% of prevalence and a mean intensity of 0.14 and 0.6 parasites per fish, respectively, although no apparent affectation on the host was indicated (Quispe, 2005; Fajer-Ávila et al., 2011). Infections with this parasite have been reported on wild adults of Pacific red snapper from Nayarit (Torres-Ayala, 2008) and Pacific dog snapper from Baja California Sur (Pérez-Urbiola et al., 2008a).

The genus *Nerocila* has been found in groupers, seabass and snappers in Singapore and Malaysia (Leong & Colorni, 2002). Isopods are considered an emerging problem in Mediterranean sea cages (Lester & Hayward, 2006). Deficient fish growth and mortality has been attributed to infections of *N. orbignyi* on the gills of sea-caged bass (*Dicentrarchus labrax*), and gilthead sea bream (Lester, 2005). *Ceratothoa oestroides* reduced the growth rate of cultured sea bass, especially of large fingerlings (Horton & Okamura, 2001), while larvae of *C. parallela* caused great losses on young gilthead sea bream, upon introduction in an intensive cage farm in Greece (Papapanagiotou & Trilles, 2001). The transference of these parasitic isopods from wild populations to cultured ones could therefore be a risk issue for the development of this activity.

6. Treatments against ectoparasites

Studies on the effectiveness of treatments for control of specific parasites in marine fish cultures in northwest in México involve the evaluation of toxicological, metabolic and routes of exposure to ensure food safety issues, particularly if fish are used for human consumption (Fajer et al., 2011). The methods commonly used for drug administration consist on dipping into the products, from seconds to prolonged periods, although formulated diets are now widely used (Sapkota et al., 2008). The U.S. Food and Drug Administration (FDA) have several products approved for aquaculture use, such as formalin that can be used on immersion treatments to control external parasites. For instance, formalin-based products are approved against external freshwater protozoa (*Chilodonella, Costia, Epistylis, Ichthyophthirius, Scyphidia,* and *Trichodina* spp.) and the monogenetic trematodes (*Cleidodiscus, Dactylogyrus,* and *Gyrodactylus* spp.) on salmon, trout and in general all finfish (Food and Drug Administration [FDA], 2011a).

6.1 Formalin

Although FDA approves formalin, it should be administrated with caution as it is a toxic product and fish have different tolerances to chemicals depending on the species and treatment conditions (Hoffman & Meyer, 1974; Kabata, 1985). Therefore if formalin is not properly used it can potentially have undesirable effects in either the aquatic ecosystem or even in the fish subjected to treatment. Thus, it is desirable to evaluate the toxicity to the host before using it as therapeutic agent (Fajer-Ávila et al., 2003). Although information on acute toxicity of formalin to different species of freshwater fish is available (Bills et al., 1977; McKim et al., 1976; Wellens, 1982), data for marine fish are generally scarce. The toxicity to formalin was evaluated in juveniles of bullseye puffer fish. The median lethal concentration (LC_{50}) was estimated at 1095 mg L^{-1} to 30 min, 972 mg L^{-1} at 60 min and 79 mg L^{-1} at 72 hours (Fajer-Ávila et al., 2003). The *in vitro* median effective concentration (EC50) of formalin was also estimated against *H. ecuadori* with values of 225 mg L^{-1} at 30 min and 87 mg L^{-1} at 60 min. A calculation of the therapeutic index (LC_{50} / EC_{50}) was of 5 for 30 min and 11 for 60 min. This led to *in vivo* assessments of formalin on bullseye puffer fish juveniles infested with *A. ocellatum*, finding that a concentration of 51 mg L^{-1} caused a significant reduction in the number of trophozoites on the skin (97%) and gills (84%) after one hour of exposure, while 4 mg L^{-1} added to the tanks significantly reduced the number of parasites on the skin (66%) and gills (84%) after seven hours of treatment (Fajer-Ávila et al., 2003). Furthermore, the *in vitro* effect of formalin (51 mg L^{-1}) was also able to reduce the number of adults dactylogyrids (*Haliotrematoides* spp and *Euryhaliotrema* sp.) attached to gill filaments of the spotted rose snapper in exposure periods up to 60 min to achieve 72% of effectiveness. *In vivo* treatments on infected snappers showed similar effects (Fajer-Ávila et al., 2007; Velázquez-Medina, 2004).

6.2 Praziquantel

Several studies indicate that the anthelmintic compounds praziquantel can be employed as alternative for the control of monogenean ectoparasites of marine fish (Fajer-Ávila et al., 2006; Hirazawa et al., 2000; Kim & Cho, 2000; Sharp et al., 2004; Stephens et al., 2003). Praziquantel causes spastic paralysis to the parasite and induces a calcium influx across the worm tegument, while also inhibits glucose uptake, forcing the parasite to consume its glycogen stores (DEF, 2005). Five minutes of exposure to Praziquantel produces a degeneration of the worms' integument, an action that apparently is also exerted upon eggs and the encysted larvae of helminthes (DEF, 2005). The experimental *in vitro* results with Drontal™ Plus (50 mg praziquantel + 150 mg embonato de pyrantel + 150 febantel) on the spotted rose snapper showed that 15 mg L^{-1} for 240 min resulted in 100% mortality of dactylogyrids (Arambul-Muñoz, 2007). Parasites showed severe contractions prior to death. The effect of Drontal™ Plus and Vermiplex ™ (2 mg Ivermectin + 50 mg Praziquantel, Holland, Mexico) on spotted red snapper infested with dactylogyrids was also evaluated. Tested concentrations of Vermiplex ™ were 4.5 and 3.5 mg L^{-1} for 14 and 24 hours respectively, while Drontal™ Plus was tested at of 4.5 mg L^{-1} for 14 hours. All treatments had 100% effectively (Fajer-Ávila et al., 2007). Furthermore, experiments with Drontal™ Plus (5, 10, 15 and 20 mg L^{-1}) for five days showed a significant reduction in the number of dactylogyrids in snapper of the control group, but not between the concentrations tested. The effectiveness rate was increased to 82% (unpublished results).

6.3 Freshwater

The increasing concern about the possible harmful effects of chemicals and pesticides on fish and human health have led to the imposition of stringent regulations that limit the use of antibiotics and a number of chemicals harmful to the aquatic ecosystem as well as consumers. This has generated interest in seeking environment friendly alternatives, either natural or synthetic, for parasite control. The use of freshwater is possibly the most recommended treatment to control monogeneans and copepods from marine fish as it is cheap, non-toxic and readily available in aquaculture facilities. The exposure time differs between the group of parasites and host species, but there are reports of its application to control ectoparasites in the pacific bluefin tuna, yellowtail amberjack, bullseye puffer, spotted rose snapper, Pacific red snapper, Pacific cubera snapper, yellow snapper and mullet snapper, among others (Fajer-Ávila et al., 2011). In Pacific bluefin tuna cage culture, freshwater treatments could be useful to eliminate ectoparasites, while repeated baths with yellowtail amberjack can actually break the life cycle fast growing monogeneans (Avilés-Quevedo & Castelló-Orvay, 2004). Freshwater baths of 10 minutes were reported useful as prophylaxis by Pérez–Urbiola et al. (2008b) for the control of ectoparasites in Pacific red snapper, Pacific cubera snapper, yellow snapper and mullet snapper. However, after the freshwater treatments, the Pacific red snapper showed darkened body and some lied sideways on the bottom, while the yellow snapper only showed obscured body. The authors indicated that the Pacific red snapper is apparently more susceptible to change salinities than others snappers, so they recommended gradual changes of salinity for longer periods of time.

Fajer-Ávila et al. (2008) evaluated *in vitro* tolerances of nauplii and copepodids of *L. simplex* at periods from 0.5 to 16 min. The results indicated that freshwater highly effective with exposures of 1.5 to 2.5 min for nauplii and 12 and 15 min for copepodids. Furthermore, *in vivo* freshwater treatments (23 ± 0.5°C) on bullseye puffer fish infested with mobile stages of *L. simplex* and mature an immature of *Neobenedenia* sp. were able to reduce 99% of pre-adults (PA) and adults (A) of *L. simplex* at 60 min of exposure, but was unable to affect the chalimus (CH) stage. In this sense, considering that mobile stages of *L. simplex* elicit the greatest damage to the host, their rapid reduction through treatment is important. In laboratory conditions, the time that *L simplex* requires to change from CH to PA was approximately 100 h at 22°C (Rivas-Salas & Fajer-Ávila, pers. obs.), suggesting that baths should be repeated at least every 96 h to break the life cycle in order to effectively control the parasitic infection. Regarding *Neobenedenia* sp., an increase of freshwater effectiveness (24°C) was directly related to the time of exposure, where 60 min were able to remove up to 99% of immature and adult monogeneans without causing skin lesions or stress in the treated fish (Fig. 10).

The effectiveness of freshwater for the control of eggs and adults dactylogyrids infecting the gills of the spotted rose snapper was also evaluated (Fajer-Ávila et al., 2007; Velázquez-Medina, 2004). *In vitro* tests showed that after 3 h of freshwater exposure, egg viability was limited to 10 %, while freshwater *in vivo* baths for 30 min removed 100% of adult dactylogyrids, although it was observed that the treatments caused some stress in the fish. To overcome this, 5 ups salinity increments every 10 min up to 35 ups for three consecutive days may help to achieve 100% of effectiveness against dactylogyrids with minimum osmotic stress to the fish (Del Rio-Zaragoza, pers. obs.).

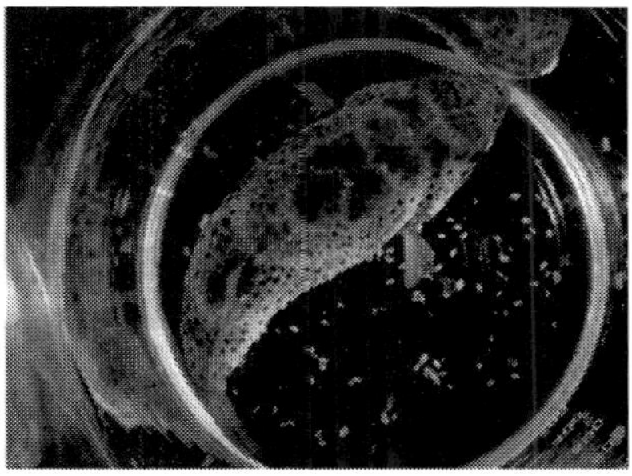

Fig. 10. Freshwater treatment on the bullseye puffer fish against *Neobenedenia* sp.

6.4 Herbal medicine

The herbal medicine has increasingly become an important alternative to treat parasite-related diseases due to the side effects attributed to synthetic drugs. A medicinal plant may contain one or more chemical compounds with potential to be used therapeutically. The utilization of herbs against fish diseases in Mexico dates from 1987, with reports of successful use in freshwater fish using lilac (*Syringa vulgaris*), pine (*Pinus teocote*), butterfly bush (*Buddleja americana*), the wax-leaf privet (*Ligustrum japonicum*), garlic (*Allium sativum*), onion (*Allium cepa*), chestnut (*Castanea sativa*) and epazote (*Chenopodium ambrosioides*) in the treatment against ectoparasites as protozoa and nematodes (Auró de Ocampo & Jiménez, 1993). In contrast, there are few studies performed about the use of medicinal plants in marine fishes.

6.4.1 Peppermint

The essence of peppermint has vermifuge properties that can be used to expel intestinal parasites. The *in vitro* effect of peppermint essence (obtained by distillation) was tested against *H. ecuadori* infecting the gills of the bullseye puffer fish (Contreras, 2001). Concentration of 1 mL L^{-1} caused mortality of the 100% of the parasites after 45 min of exposure. Concentrations of 0.6 and 0.4 mL L^{-1} caused 71 and 34% of mortalities, respectively at 75 min. It was observed that the essence of peppermint caused an immediate contraction of the parasite after being added.

6.4.2 Garlic

Garlic is one of the best studied medicinal plants, and it has been used in a variety of forms. The antiseptic properties attributed to garlic have been pointed out as one of the best remedies to kill parasites (Rojas-Alba, 2004) Garlic contains a sulfur compound called allicin, which can be extracted just by crushing fresh garlic. It is reported that garlic's parasiticidal effects could be potentiated by adding commercial dehydrated garlic sulfur (Ocampo & Auró, 2000). *In vitro* experiments using garlic powder (GNC ®) with sulfur (150 mg L^{-1} for 330 seconds) indicated that a 100% mortality of adult gill dactylogyrids from the spotted rose snapper could be obtained (Arambul-Muñoz, 2007). Spotted red snapper fed with garlic-containing diets (150 mg kg-1) for 15 days showed a reduction of 48% in the number of monogeneans compared to control treatments. It was observed that the dactylogyrids killed as consequence of the garlic treatment showed swollen and deformed bodies, and felling off the gill filaments after immobilization (Aguilar-Ibarra, com. pers).

7. Perspectives of future research and development

The need to increase productivity and develop marine fish aquaculture with new species is based on the growing worldwide demand for food and the declining of most important marine fisheries. Fish culture relies on high quality production, able to produce healthy animals, sometimes in high densities. Emerging diseases are therefore a constant threat as farming systems normally break the natural pathogen-host balance, propitiating a constant need to find new and safe approaches to control fish pathogens. Here we present a perspective of the different approaches that could help to successfully develop the aquaculture industry in northwest Mexico.

7.1 Chemical control and environmental impact

A major challenge in animal production is to implement effective strategies for pathogen control. The chemical approach faces the problem that simple compounds are frequently toxic to the infested fish, while certain commercial products, such as formalin, could be unsatisfactory due to potential undesirable side effects and low specificity. Treatments recently accepted by the FDA (FDA, 2011b), such as hydrogen peroxide (35%) have proved useful against ectoparasites such as *Neobenedenia* spp., *A. ocellatum*, *C. irritans*, and *Uronema* spp. in marine fish species. The FDA program Investigational new animal drug (INAD), that provides guidance for aquaculture drugs, also approved the product SLICE® (emamectin benzoate as active ingredient), which is orally administrated to control copepod parasites in fish species (Bowker et al., 2011).

The use of medicated feed is suggested as the most viable method for cage culture and fish breeding stations, where high stocking densities are present and where dipping treatments are impractical expensive. For example, Hirazawa et al. (2000) indicates that routine treatments for *N. girellae* infestations in off-shore net pens involved bath treatments, although this normally increase the labor cost and could be stressful to the fish (Kim & Choi, 1998). For this reason, several drugs have been proposed against parasites in marine fish aquaculture (Table 2), such as the antihelmintic praziquantel, orally administrated against skin (Okabe, 2000) and gill (Hirazawa et al., 2004; Kim et al., 2001) monogeneans.

A promissory area for pathogen control is based on the use of natural plant-based treatments. Some active components contained in traditional plant extracts have been proved effective against parasitic infections, such as phenols, polyphenols, alkaloids, quinones, terpenoids, lectines (Harikrishnan et al., 2011). In this context, investigations about these alternatives are in general incipient. For instance, the only recommendations by the FDA are the use of garlic for helminth and sea lice in marine salmonids at all life stages and onion as treatment for external crustacean parasites. There are, however, some promissory alternatives mentioned in the literature, like the reports of crude extracts of *P. elliottii* (Tóro et al., 2003), *Carica papaya* (Ekanem et al., 2004), green tea (Suzuki et al., 2006) and ginkgolic acids from *Ginkgo biloba* apparently useful to control some parasitic worms in fish (Wang et al., 2009).

Nowadays, aquaculture practices have to seriously consider the potential effects of the accumulation of chemicals either in the products (posing risk to the consumers) or the

Parasites	Hosts	Application /Compound	Reference
Protozoans			
Trichodina jadranica	*Anguilla anguilla*	Dip / Bithionol (0.1 mg L^{-1} for 1 h), Detarox AP® (45 ppm for 1 h), Virkon PF® vet. (mg L^{-1} for 3 h)	Madsen et al. (2000)
Monogeneans			
Pseudodactylogyrus anguillae	*A. anguilla*	Dip / Toltrazuril (10 µg mL^{-1} for 4 h)	Schmahal & Mehlborn (1988)
		Dip / Ginkgolic acids: C13:0 (2.5 mg L^{-1} for 48 h) and C15:1 (6.0 mg L^{-1} for 48 h)	Wang et al. (2009)
Microcotyle sebastis	*Sebastes schelegeli*	Oral / Mebendazole or bithionol (0.25 g 200 g^{-1} basal diet for 20 days)	Kim & Choi (1998)
Heterobothrium okamotoi	*Takifugu rubripes*	Oral / Praziquantel (4 g kg^{-1} basal diet for 20 days) and Caprylic acid (100 mg kg^{-1} body weigth by day for 60 days)	Hirazawa et al. (2000) Hirazawa et al. (2001)
Neobenedenia girellae	*Verasper variegatus*	Oral / Praziquantel (40 mg kg^{-1} body weight by day for 11 days)	Hirazawa et al. (2004)
Benedenia seriolae	*Seriola lalandi*	Dip/ Praziquantel (2.5 mg L^{-1} for either 24 or 48 h by two days)	Sharp et al. (2004)
Zeuxapta seriolae		Dip / Praziquantel (2.5 mg L^{-1} for either 24 or 48 h) and formalin (400 µL L^{-1} for 1 h + 5 min fresh water dip by two days)	
Sparicotyle chrysophrii	*Sparus aurata*	*In vitro*/ distilled water (5 min), limoseptic ® (1 mL L^{-1} for 5 min), formalin (300 µL L^{-1} for 30 min), or hydrogen peroxide (200 µL L^{-1} for 30 min) Oral / Praziquantel (200 mg kg^{-1} body weight by day for 6 days)	Sitjà-Bobadilla et al. (2006)
Neobenedenia girellae	*Verasper variegates*	Dip/buffer solutions (lacking Calcium and magnesium ion-free buffer for 10 min)	Ohashi et al. (2007)
Caligids			
Lepeophtheirus salmonis, *Caligus elongates*	*Salmo salar*	Oral / Emamectin benzoate (50 µg kg^{-1} biomass by day for 7 days)	Stone et al. (2000)
Isopods			
Ceratothoa gaudichaudii	*Salmo salar*	Dip/Trichlorfon (300 mg L^{-1} of Neguvon® for 60 min) and dichlorvos (3 mg L^{-1} of Nuvan 1000® for 60 min)	Sievers et al. (1995)

Table 2. Drug treatments tested against some parasites in fish aquaculture.

environment; something that has led (mainly in developed countries) to rigorous regulations that restrict the use of chemicals (Harikrishnan et al., 2011). An example of substitution of chemicals by more eco-friendly alternatives is the increasingly use of hydrogen peroxide in Japan, substituting formalin as treatment to prevent ectoparasitization in tigger puffer cage systems (Hirazawa et al., 2000).

However, although the environmental impact of hydrogen peroxide is thought to be relatively small, the viability of the fish could be affected in the summer season, while these bath treatments have shown limited effects against mature parasites embedded in the branchial cavity wall (Hirazawa et al., 2000). Comparative studies regarding the effectiveness of new and safer alternative treatments are necessary to identify the possible applications. For example, caprylic acid have shown efficacy against *H. okamotoi* infections in the tiger puffer, and is likely to be effective against other fish monogeneans, while it could be relatively innocuous to the environment (Hirazawa et al., 2000; 2001). However, further research is needed as the assessment of caprylic acid (334 mg L^{-1}; Sigma, USA) against dactylogyrids on the spotted red snapper indicated that it was not effective in reducing parasitic load, despite applying a concentration 4 times higher than that used by Hirazawa et al. (2000) to control *H. ecuadori* (Fajer-Ávila et al., 2007).

7.2 Immunostimulation

The use of inmunostimulants has become an attractive alternative to reduce the dependence of chemical treatments for parasite control. The negative impacts of the former include side effects such as immunosuppression and the appearance of drug resistant pathogens (Galina et al., 2009). Immunostimulants are particularly important during the larval and juvenile stages as they can strength the activity of nonspecific defense mechanisms of the immune system and confer protection against disease. So far there are few studies on the use of immunostimulants as a prophylactic method against parasitic diseases affecting marine fish species cultured in Mexico (Reyes-Becerril et al., 2008; Del Rio-Zaragoza et al., 2011). In this regard, little is known about the effect that immunoestimulants have on host resistance against ectoparasites. Nevertheless, immunostimulants are widely applied in culture conditions and there are many examples of the successful use of immunostimulants to improve fish welfare, health and production, such as the use of glucans in salmon diets to reduce sea lice settlement (Burrels et al., 2001; Bricknell & Dalmo, 2005). Thus, the use of in-diet immunomodulators has become widely accepted in both salmonid and non-salmonid aquaculture practices with commercially available diets. These diets were considered to be effective in managing disease outbreaks after stressful events (Bricknell & Dalmo, 2005). For example, it has been observed that oral administration of glucans is able to stimulate innate immune responses in fish, as well as increases in lysozyme and complement activity, phagocytosis, nitric oxide and respiratory burst (Bridle, 2005), increasing host resistance against bacterial pathogens in several cultured fish species (Ortuño et al., 1999; Couso et al., 2003). The development of aquaculture in northwest Mexico, particularly the one with species of economic importance needs to be complemented with novel approaches. In this sense, the implementation of immunostimulation procedures is already suggested in the scientific literature (Table 3), so more research should be encouraged to optimize and increase the profitability of this important activity.

7.3 Final considerations: Good aquaculture practices

Good aquaculture practices, particularly those with an ecological perspective, where the activity is considered as an additional component that interacts with the natural processes, are possibly the final step to achieve the sustainable development of fish production.

SOURCE	IMMUNOESTIMULANTS
Biological substances	
Bacterial derivatives	Beta glucans, lipopolysaccharides, peptidoglycans, streptococcal extracts, cells of *Vibrio anguillarum*, and *Achromobacter stenohalis*
Plant extracts	Laminaran, kelp powder, *Spirulina* (algae); carotenoids (plants and microalgae)
Animals extracts	EF-203 (Chiquen), Ete (Tunicates), Hde (abalone)
Hormones	Growth and prolactin hormones
Proteins	Cytokines (Interferon, Interleukin-2, TNF), lactoferrin
Carbohydrates	Polysaccharides (Chitin, Chitosan, glucans, lentinian, oligosaccharides, sclerotium, schizophylan)
Nutritional compounds	Vitamins (C, E y A), trace elements (zinc, iron, copper, selenium), nucleotides
Light oils	
Synthetic chemicals	Avridine, bestatin, DW-2929, FK-156, FK-565, fluoro-Chindone, Freund coadjuvant, Isoprinosin, levamisol and muramyl dipeptide.

Table 3. Immunostimulants that have shown increased non-specific and specific defense mechanisms in fish against diseases (from: Del Rio-Zaragoza & Fajer-Ávila, 2008)

Thus, aquaculture should integrate diverse aspects such as animal welfare, environmental health and consumer safety. Careful use of regulated products should ensure, to the greatest extent possible, its effectiveness, reducing both overuse and unnecessary expense. Aquaculturists that comply with the state and federal laws should be able to maintain public trust and consumer confidence in cultured aquatic animals and seafood products. The development of vaccines (not covered in this chapter) could be another useful tool where research is needed. Altogether, an integrated approach with well-established programs of quarantine procedures, disease management practices, good husbandry techniques, and chemotherapy should be aimed at preventing parasite-associated diseases (Sharp et al., 2004) with minimal environmental impacts as the best way of adding value to aquaculture products.

8. Conclusion

The ciliate *Cryptocaryon irritans*, the dinoflagelate, *Amyloodinium ocellatum*, the monogenean *Neobenedenia* spp. and the copepods *Lepeophtheirus* spp. and *Caligus* spp. represent a potential threat to the health of fish marine culture in the northwest Mexico considering their low host specificity and wide distribution. Moreover, an unregulated introduction of fish and the ability of parasites to cause mortality due to their direct life cycles and the feasibility to produce disease outbreaks in wild populations also could be matter of concern. In general wild fish could act as reservoirs of infecting agents, highlighting the need of parasitological monitoring programs as the industry moves to intensive aquaculture. Thus, the development of an integrated management program orientated to the prevention, with prompt disease diagnostics, immunization and eco-friendly treatments, will help to minimize the economic impact of parasitic diseases in mariculture systems.

9. Acknowledgment

The authors gratefully acknowledge to R.M. Medina-Guerrero and S. Abad-Rosales, and numerous students for their contribution during the studies. Most researches were funded with research grants awarded to EJFA by the National Council for Science and Technology (CONACYT: 31621B) and Fondo Sectorial de Investigación en Materia Agrícola, Pecuaria, Acuicultura, Agrobiotecnológica y Recursos Filogenéticos (SAGARPA-2002-C01378; 2003-73- 02073).

10. References

Aksit, D.; Falakali-Mutaf, B.; Göçmen, B. & Gürelli, G. (2008). Preliminary observations on *Trichodina* sp. (Ciliophora: Peritricha) on the Gills of Limpets (*Patella* spp.) in Antalya (Turkey). *North-Western Journal of Zoology,* Vol.4, No.2, (December 2008), pp. 295-299, ISSN 1843-5629

Alvarez-Lajonchere, L.; Reina-Canñez, M.A.; Camacho-Hernández, M.A. & Kraul, S. (2007) Design of a pilot-scale tropical marine hatchery for a research center at Mazatlan, Mexico. *Aquacultural Engineering,* Vol.36, No.2, (March 2007), pp.81-96, ISSN: 0144-8609

Alvarez-Pellitero, P. (2004). Report about fish parasitic diseases, In: *Mediterranean Aquaculture Diagnostic Laboratories Zaragoza (Spain),* P. Alvarez-Pellitero, JL. Barja, B. Basurco, F. Berthe & AE. Toranzo, (Eds), pp.103-129, CIHEAM/FAO. Série B: Etudes et Recherches, No. 49, Options Méditerranéennes, ISBN 2-85352-301-2, Spain

Arambul-Muñoz, E. (2007). *Evaluación in vitro e in vivo de la efectividad de agentes parasiticidas químicos y naturales sobre infestaciones mixtas de monogeneos ancirocefalinos (Haliotrema spp. y Euryhaliotrema sp.) y el dinoflagelado Amyloodinium ocellatum que parasitan las branquias del pargo lunarejo Lutjanus guttatus,* Bachelor Degree Thesis, UAS, Facimar, Mazatlán, Sinaloa, México

Auró de Ocampo, A. & Jiménez, E.M. (1993). Herbal medicine in the treatment of fish diseases in Mexico. *Veterinaria Mexicana,* Vol.24, No.4, (October-December 1993), pp. 291-295

Avilés Quevedo, A. (2011). Desarrollo de biotecnología para el cultivo de pargos. Boletin INAPESCA. Available from http://www.inapesca.gob.mx/portal/sala-de-prensa/boletines/1-desarrollo-de-biotecnologia-para-el-cultivo-de-pargos

Avilés-Quevedo, A. & Castelló-Orvay, F. (2004). *Manual para el cultivo de Seriola lalandi (Pisces: Carangidae) en Baja California Sur, México,* Instituto Nacional de Pesca, ISBN 968-800-580-0, Mexico, D.F

Bason, L. & Van As, J. (2006). Trichodinidae and other ciliophorans (Phylum Ciliophora), In: *Fish Diseases and Disorders, Volume 1: Protozoan and Metazoan Infections,* PTK. Woo, (Ed.), pp. 154-186, CABI, ISBN 0851990150, UK

Bills, T.D.; Parking, L.L. & Chandler, J.H. (1977). *Investigation in fish control formalin: Its toxicity to non target aquatic organisms, persistence and counteraction,* Vol.73. US Department of the Interior, Fish and Wildlife Service, Washington, DC, USA

Bowker, J.; Trushenski, J.; Tuttle-Lau, M.; Straus, D.; Gaikowski, M.; Goodwin, A.; Sprague, L. & Bowman, M. (February 2011). Guide to Using Drugs, Biologics, and Other Chemicals in Aquaculture, American Fisheries Society. Fish Culture Section, Available from http://www.fishculturesection.org/DrugGuide/Files/GUIDE_FEB_2011.pdf

Boxshall, G.A. & Halsey, S.H. (2004). An introduction to copepod diversity. Part I, The Ray Society, London

Boxshall, G.A. (2005). Copepoda (copepods), In: *Marine Parasitology*. K. Rohde, (Ed.), pp. 123-138, CABI, ISBN 0-643-09025-8, UK

Bricknell, I. & Dalmo, R.A. (2005). The use of immunostimulants in fish larval aquaculture. *Fish and Shellfish Immunology*, Vol.19, No.5, (November 2005), pp. 457-472, ISSN 1050-4648

Bridle, A.R; Carter, C.G.; Morrison, R.N. & Nowak, B.F. (2005). The effect of β-glucan administration on macrophage respiratory burst activity and Atlantic salmon, *Salmo salar* L., challenged with amoebic gill disease-evidence of inherent resistance. *Journal of Fish Diseases*, Vol.28, No.6, (June 2005), pp. 347-356, ISSN 1365-2761

Brusca, R.C. & Gilligan, M.R. (1983). Tongue replacement in a marine fish (*Lutjanus guttatus*) by a parasitic isopod (Crustacea: Isopods). *Copeia*, Vol.3, pp 813-816, ISSN 1938-5110

Brusca, R.C. (1981). A monograph on the Isopoda Cymothoidae (Crustacea) of the eastern Pacific. *Zoological Journal of the Linnean Society*, Vol.73, No.2, (October 1981), pp. 117-199, ISSN 1096-3642

Buchmann, K. & Bresciani, J. (2006). Monogenea (Phylum Platyhelminthes), In: *Fish Diseases and Disorders, Volume 1: Protozoan and Metazoan Infections*, PTK. Woo, (Ed.), pp. 297-344, CABI, ISBN 0851990150, UK

Burrells, C.; Williams, P.D. & Forno, P.F. (2001). Dietary nucleotides: a novel supplement in fish feeds. 1. Effects on resistance to disease in salmonids. *Aquaculture*, Vol.199, No.1-2, (July 2001), pp. 159-169, ISSN 0044-8486

Castillo-Sánchez, E.; Rosales-Casián, J.A. & Pérez-Ponce de León, G. (1998). Helminth parasites of *Paralichthys californicus* (OSTEICHTHYES: PARALICHTHYDAE) in estero de Punta Banda, Todos Santos bay and San Quintín bay, Baja California, México. *Ciencias Marinas*, Vol.24, No.4, (December 1998), pp. 443-462, ISSN 0185-3880

Cavaleiro, F.I. & Santos, M.J. (2007). Survey of the metazoan ectoparasites of the European flounder *Platichthys flesus* (Linnaeus, 1758) along the North-Central Portuguese Coast. *Journal of Parasitology*, Vol. 93, No.5, (October 2007), pp. 1218-1222, ISSN 1937-2345

Colorni, A. & Diamant, A. (2005). Hyperparasitism of trichodinid ciliates on monogenean gill flukes of two marine fish. *Diseases of Aquatic Organisms*. Vol.65, No.2, (June 2005), pp. 177-180, ISSN 0177-5103

Cone, D.K. (1999). Monogenea (Phylum Platyhelminthes). In: *Fish Diseases and Disorders, Volume 1: Protozoan and Metazoan Infections*, PTK. Woo, (Ed.), Second edition, pp. 289-327, CABI, Canada

Contreras. R. (2001). *Monogeneos del botete diana Sphoeroides annulatus (Jenyns, 1842) y evaluación "in vitro" del empleo de hierbabuena y formalina para su control*, Bachelor Degree Thesis, UAS, Facimar, Mazatlán, Sinaloa, México

Costello, M.J. (2009a). How sea lice from salmon farms may cause wild salmonids declines in Europe and North America and be a threat to fishes elsewhere. *Proceedings of the Royal Society Section B*, Vol.276, pp. 3385-3384, ISSN 0080-455X

Costello, M.J. (2009b). The global economic cost of sea lice to the salmonid farming industry. *Journal of Fish Diseases*, Vol.32, No.1, (January 2009), pp. 115-118, ISSN 1365-2761

Couso, N.; Castro, R.; Magariños, B; Obach, A. & Lamas, J. (2003). Effect of oral administration of glucans on the resistance of gilthead seabream to pasteurellosis. *Aquaculture*, Vol.219, No.1-4, (April 2003), pp. 99-109, ISSN 0044-8486

Das, P.C.; Ayyappan, S. & Jena, J.K., 2006. Haematological changes in the three Indian major carps, *Catla catla* (Hamilton), *Labeo rohita* (Hamilton) and *Cirrhinus mrigala* (Hamilton) exposed to acidic and alkaline water pH. *Aquaculture*, Vol.256, No.1-4, (June 2006), pp. 80-87, ISSN 0044-8486

DEF. (2005). *Diccionario de especialidades farmacéuticas*, ED CIENTIFICA PLM, ISBN 968-588-733-0, México, DF

Del Rio-Zaragoza, O.B. & Fajer-Ávila, E.J. (2008). El uso de inmunoestimulantes en México. *Panorama Acuícola Magazine*, (Marzo-Abril, 2008), pp. 10-20

Del Rio-Zaragoza, O.B.; Fajer-Ávila, E.J. & Almazan-Rueda, P. (2010). Haematological and gill responses to an experimental infection of dactylogyrid monogeneans on the spotted rose snapper *Lutjanus guttatus* (Steindachner, 1869). *Aquaculture Research*, Vol.41, No.11, (October 2010), pp. 1592–1601, ISSN 1365-2109

Del Rio-Zaragoza, O.B.; Fajer-Ávila, E.J. & Almazan-Rueda, P. (2011). Influence of β-glucan on innate immunity and resistance of *Lutjanus guttatus* to an experimental infection of dactylogyrid monogeneans. *Parasite Immunology*, Vol.33, No.9, (September 2001), pp. 483–494, ISSN 1365-3024

Dickerson, H.W. (2006). *Ichthyophthirius multifiliis* and *Cryptocaryon irritans* (Phylum Ciliophora), In: *Fish Diseases and Disorders Volume 1: Protozoan and Metazoan Infections*, PTK. Woo, (Ed.), pp. 116-153, CABI, ISBN 0851990150, UK

EcuRed. (2012). El *Criptocarion irritans*. Available from: http://www.ecured.cu/index.php/Criptocarion_irritans

Ekanem, A.P.; Obiekezie, A.; Kloas, W. & Knopf, K. (2004). Effects of crude extracts of *Mucuna pruriens* (Fabaceae) and *Carica papaya* (Caricaceae) against the protozoan fish parasite *Ichthyophthirius multifiliis*. *Parasitology Research*, Vol.92, No.5, (March 2004), pp. 361–366, ISSN1432-1955

Fajer-Ávila, E.J. & Chávez, M.C. (1999). Parasites and their effect on the wild bullseye puffer fish (*Sphoeroides annulatus* Jenyns, 1843). *Proceedings of Fifth International Symposium of Fish Parasites, Institute of Parasitology, Academy of Sciences*, Czech Republic August 1999

Fajer-Ávila, E.J.; Abad-Rosales, S.M; Medina-Guerrero, R.M. & Betancourt-Lozano, M. (2011). Avances sobre las investigaciones parasitológicas en el cultivo de pargo flamenco y botete diana Sinaloa, México, In: *Avances en Acuicultura y Manejo Ambiental*, A. Ruiz-Luna, CA. Berlanga-Robles & M. Betancourt-Lozano, (Eds), pp. 71-92, Trillas. ISBN 978-607-17-0754-3, México

Fajer-Ávila, E.J.; Abdo-de la Parra, M.I.; Aguilar-Zarate, G.; Contreras-Arce, R.; Zaldívar-Ramírez, R.J. & Betancourt-Lozano, M. (2003). Toxicity of formalin to bullseye puffer fish (*Sphoeroides annulatus* Jenyns, 1843) and its effectiveness to control ectoparasites. *Aquaculture*, Vol.223, No.1-4, (June 2003), pp. 41-50, ISSN 0044-8486

Fajer-Ávila, E.J.; Guzman-Beltran, L., Zárate-Rodríguez, W.C.; Del Río-Zaragoza, O.B. & Almazán-Rueda, P. (2011). Pathology caused by adult of *Pseudochondracanthus diceraus* (Copepoda: Chondracanthidae), a parasite of bullseye puffer fish *Sphoeroides annulatus*. *Revista de Biología Marina y Oceanografía*, Vol. 46, No.3, pp 293-302, ISSN: 07173326

Fajer-Ávila, E.J.; Martínez, I.; Abdo-de la Parra, M.I.; Lajonchère, L.A & Betancourt, M. (2008) Effectiveness of freshwater treatment against *Lepeophtheirus simplex* (Copepoda: Caligidae) and *Neobenedenia* sp. (Monogenea:Capsalidae), skin parasites of bullseye puffer fish, *Sphoeroides annulatus* reared in tanks. *Aquaculture*, Vol.284, No.1-4, (November 2008), pp. 277-280, ISSN 0044-8486

Fajer-Ávila, E.J.; Medina-Guerrero, R.M. & García-Vargas, F. (2006). Enfermedades parasitarias de impacto potencial en el cultivo de los pargos flamencos. *Panorama Acuícola*, Vol.11, No.4, pp. 64-71

Fajer-Ávila, E.J.; Roque, A.; Aguilar, G. & Duncan, N. (2004). Patterns of occurrence of the platyhelminth parasites of the wild bullseye puffer (*Sphoeroides annulatus*) in Sinaloa, Mexico. *Journal of Parasitology*, Vol.90, No.2, (April 2004), pp. 415-418, ISSN 1937-2345

Fajer-Ávila, E.J; Velásquez-Medina, S. & Betancourt-Lozano, M. (2007) Effectiveness of treatments against eggs, and adults of *Haliotrema* sp. and *Euryhaliotrema* sp. (Monogenea: Ancyrocephalinae) infecting red snapper, *Lutjanus guttatus*. *Aquaculture*, Vol.264, No.1-4, (April 2007), pp. 66-72, ISSN 0044-8486

FDAa. (June 2011) Reminder To Aquaculture Producers About The Use Of Formaldehyde, In: FDA, Center for Veterinary Medicine, Available from http://www.fda.gov/AnimalVeterinary/NewsEvents/CVMUpdates/ucm048274.htm

FDAb. (June 2011) Environmental Considerations and Limitations for Use of Hydrogen Peroxide in Aquaculture, In: FDA, Center for Veterinary Medicine, Available from http://www.fda.gov/AnimalVeterinary/NewsEvents/CVMUpdates/ucm045576.htm

Frankland, H. (1955). *The life history and bionomics of Diclidophora denticulate (Trematoda: Monogenea)*, PhD Thesis, University College of North Staffordshire

Fuentes-Zambrano, J.L. & Silva-Rojas, C.L. (2006) A new species of *Euryhaliotrema* Kritsky and Boeger, 2002 (Monogenea) found on the gills of *Lutjanus griseus* (Perciformes: Lutjanidae) from the Isla de Margarita, Venezuela. *Boletín del Instituto Oceanográfico de Venezuela de la Universidad de Oriente*, Vol.45, No.1, (June 2006), pp. 3-7, ISSN 0020-417X

Galina, J.; Yin, G.; Ardo, L. & Jeney, Z. (2009). The use of immunostimulating herbs in fish. An overview of research. *Fish Physiology and Biochemintry*, Vol.35, No.4, (November 2009), pp. 669-676, ISSN 1573-5168

García-Vargas, F. (2008). *Helmintos parásitos del pargo lunarejo, Lutjanus guttatus Steindachner, 1869 (Pisces: Lutjanidae) en dos localidades del Pacífico Mexicano y estructura de la comunidad de endohelmintos*, PhD Thesis, CIAD, A.C., Mazatlán, Sinaloa, México

García-Vargas, F.; Fajer-Ávila, E.J. & Lamothe-Argumedo, M.R. (2008). Two new species of Dactylogyridae (Monogenoidea) on rose spotted snapper, *Lutjanus guttatus* (Osteichthyes: Lutjanidae), from the coasts of Nayarit and Sinaloa, Mexico. *Zootaxa*, Vol.20, No.1729, (March 2008), pp. 61-68, ISSN 1175-5334

Grabda, J. (1991). *Marine fish parasitology*. Wiley-VCH, ISBN 3527268987, Warszawa, Poland

Grano-Maldonado, M.; Roque, A.; Aguirre, H. & Fajer-Ávila, E.J. (2011). Egg morphology, larval development and description of the oncomiracidium of *Heterobothrium ecuadori* (Monogenea: Diclidophoridae) parasitising the bullseye pufferfish, *Sphoeroides annulatus*. *Helminthologia*, Vol.48, No.1, (March 2011), pp. 51-55, ISSN 1336-9083

Guzmán-Beltrán, L. & Zárate-Rodríguez, W.C. (2008). *Efectos patológicos de las infestaciones branquiales por el copépodo (Pseudochondracanthus sp.), parásito del pez botete diana Sphoeroides annulatus (Jenyns, 1842)*, Bachelor Degree Thesis, Universidad de La Salle, Facultad de Medicina Veterinaria, Bogotá, Colombia

Harikrishnan, R.; Balasundaram, C.; & Heo, M. (2011). Impact of plant products on innate and adaptive immune system of cultured finfish and shellfish. *Aquaculture*, Vol.317, No.1-4, (July 2011), pp. 1-15, ISSN 0044-8486

Hirazawa, N.; Mitsuboshi, T.; Hirata, T. & Shirasu, K. (2004). Susceptibility of spotted halibut *Verasper variegatus* (Pleuronectidae) to infection by the monogenean *Neobenedenia girellae* (Capsalidae) and oral therapy trials using praziquantel. *Aquaculture*, Vol.238, No.1-4, (September 2004), pp. 83-95, ISSN 0044-8486

Hirazawa, N.; Ohtaka, T. & Hata, K. (2000). Challenge trials on the anthelmintic effect of drugs and natural agents against the monogenean *Heterobothrium okamotoi* in the tiger puffer *Takifugu rubripes*. *Aquaculture*, Vol.188, No.1-2, (August 2000), pp. 1-13, ISSN 0044-8486

Hirazawa, N.; Oshima, S.; Mitsuboshi, T. & Hata, K. (2001). The anthelmintic effect of medium-chain fatty acids against the monogenean *Heterobothrium okamotoi* in the tiger puffer *Takifugu rubripes*: evaluation of doses of caprylic acid at different water temperatures. *Aquaculture*, Vol.195, No.3-4, (April 2001), pp. 211-223, ISSN 0044-8486

Ho, J.S. (1970). Revision of the genera of the Chondracanthidae, a copepod family parasitic on marine fishes. *Beaufortia*, Vol.229, No.17, pp. 105-218, ISSN 0067-4745

Ho, J.S.; Nagasawa, K.; Kim, I.H. & Ogawa, K. (2001). Occurrence of *Caligus lalandei* Barnard, 1948 (Copepoda, Siphonostomatoida) on amberjacks (*Seriola* spp.) in the western North Pacific. *Zoological Science*, Vol.18, No.3, (April 2001), pp. 423-431, ISSN 0289-0003

Hoffman, G.L. & Meyer, F.P. (1974). *Parasites of freshwater fishes: a review of their control and treatment*, TFH Publications, New Jersey, USA

Horton, T. & Okamura, B. (2001). Cymothoid isopod parasites in aquaculture: a review and case study of a Turkish sea bass (*Dicentrarchus labrax*) and sea bream (*Sparus aurata*) farm. *Diseases of Aquatic Organisms*. Vol.46, No.3, (October 2001), pp. 181-188, ISSN 0177-5103

Kabata, Z. (1985). *Parasites and diseases of fish cultured in the tropics*, Taylor & Francis, ISBN 0-85066-2850, London, UK

Kalman, J.E. (2006). Ectoparasites of Demersal Marine Fishes in Santa Monica Bay, California, U.S.A., with 31 New Host Records and Three Range Extensions. *Comparative Parasitology*, Vol.73, No.2, (July 2006), pp. 201-213, ISSN 1525-2647

Kim, K. & Cho, J. (2000). Treatment of *Microcotyle sebastis* (Monogenea: Polyopisthocotylea) infestation with praziquantel in an experimental cage simulating commercial rockfish *Sebastes schlegeli* culture conditions. *Diseases of Aquatic Organisms*, Vol.40, No.3, (April 2000), pp. 229-231, ISSN 1616-1580

Kim, K.H. & Choi, E.S. (1998). Treatment of Microcotyle sebastis (Monogenea) on the gills of cultured rockfish (*Sebastes schelegeli*) with oral administration of mebendazole and bithionol. *Aquaculture*, Vol.167, No.1-2, (August 1998), pp. 115-121, ISSN 0044-8486

Kim, K.H.; Lee, E.H.; Kwon, S.R. & Cho, J.B. (2001). Treatment of *Microcotyle sebastis* infestation in cultured rockfish *Sebastes schlegeli* by oral administration of praziquantel in combination with cimetidine. *Diseases of Aquatic Organisms*, Vol.44, No.2, (March 2001), pp. 133-136, ISSN 1616-1580

Kohn, A; Cohen, S.C. & Salgado-Maldonado, G. (2006). Checklist of Monogenea parasites of freshwater and marine fishes, amphibians and reptiles from Mexico, Central America and Caribbean. *Zootaxa*, Vol.14, No.1289, (August 2006), pp. 1-114, ISSN 1175-5334

Kritsky, D. & Boeger, W. (2002). Neotropical monogenoidea 41: New and previously described species of Dactylogyridae (Platyhelminthes) from the gill on marine and freshwater perciform fishes (Teleostei) with proposal of new genus and a hypothesis on phylogeny. *Zoosystema*, Vol.24, No.1, (Mars 2002), pp. 7-40 ISSN 1280-9551

Kritsky, D.; Tingbao, Y. & Yuan, S. (2009). Dactylogyrids (Monogenoidea, Polyonchoinea) parasitizing the gills of snappers (Perciformes, Lutjanidae): Proposal of *Haliotrematoides* n. gen. and descriptions of new and previously described species from marine fishes of the Red Sea, the eastern and Indo-west Pacific Ocean, Gulf of Mexico and Caribbean Sea. *Zootaxa*, Vol.5, No.1970, (January 2009), pp. 1-51, ISSN 1175-5334

Lamothe-Argumedo, R. (1967). Monogeneos de peces V. Redescripción de Tagia ecuadori (Meserve, 1938) Sproston, 1946. *Anales Instituto Biología, Universidad Nacional Autónoma de México, Serie Zoología*, Vol.38, pp. 35-46, ISSN 0368-8720

Leong, T. & Wong, S. (1987) Parasites of wild and diseased juvenile golden snapper, *Lutjanus johni* (Bloch), in floating cages in Penang, Malaysia. *Asian Fisheries Science*, Vol.1, No.1, (December 1987), pp. 83-90, ISSN 0116-6514

Leong, T.S. & Colorni, A. (2002). Infectious Diseases of Warmwater Fish in Marine and Brackish Water, In: *Diseases and Disorders of finfish in Cage Culture*, PTK. Woo, DW. Bruno, & LHS. Lim, (Eds.), pp. 193-230, CABI, ISBN 0851994431, UK

Lester, J.G. & Hayward, C.J. (2006). Phylum Artropoda, In: *Fish Diseases and Disorders Volume 1: Protozoan and Metazoan Infections*, PTK, Woo, (Ed.), pp 466-528, CABI, ISBN 0851990150, UK

Lester, R.J.G. (2005). Isopoda (isopods), In: *Marine Parasitology*, K. Rohde, (Ed.), pp. 138-144, CABI, ISBN 0-643-09025-8, UK

Li, H. (2005). Description of two species (including a new one) in *Euryhaliotrema* Kritsky and Boeger. 2002. *Journal of Dalian Fisheries University*, Vol.20, No.3, pp. 186-191, ISSN 1000-9957

Li, H.; AX Wu, X.Y.; Ding, X.J.; Lin, R.Q.; Xie, M.Q.; Lun, Z.R. & Zhu, X.Q. (2005). PCR-SSCP as a molecular tool for the identification of Benedeniinae (Monogenea: Capsalidae) from marine fish. *Molecular and Cellular Probes*, Vol.19, No.1, pp. 35-39, ISSN: 0890-8508

Li, H. (2006). One new species the genus *Euryhaliotrema* (Monogenea, Ancyrocephalidae)]. *Acta Zootaxonomica Sinica*, Vol.31, pp. 317-320, ISSN 1000-0739

Li, H.; Yan, X.; Yu, L.; Lan, Q. & Huang, X. (2005). Description of two new species of Ancyrocephalidae (Monogenoidea). *Journal of Guangxi Normal University*, Vol.23, pp. 86-89, ISSN 1001-6600

Lom, J. (2005). Protistan parasites and Myxozoa, Myxozoa (Myxozoans), In: *Marine Parasitology*, K. Rohde, (Ed.), pp. 41-46, CABI, ISBN 0-643-09025-8, UK

Madsen, H.C.K.; Buchmann, K. & Mellergaard, S. (2000). Treatment of trichodiniasis in eel (*Anguilla anguilla*) reared in recirculation systems in Denmark: alternatives to formaldehyde. *Aquaculture*, Vol.186, No.3-4, (June 2000) pp. 221-231, ISSN 0044-8486

McKim, J.M.; Anderson, R.L.; Benoit, D.A.; Spehar, R.L. & Stokes, G.N. (1976) Effects of pollution on freshwater fish. *Journal of the Water Pollution Control Federation*, Vol.48, pp.1544-1620 ISSN 14346163

Mendoza-Garfias, B. & Pérez-Ponce de León, G. (1998). Microcotílidos (Monogenea: Microcotylidae) parásitos de peces marinos de la Bahía de Chamela, Jalisco, México. *Anales Instituto Biología, Universidad Nacional Autónoma de México, Serie Zoología*, Vol.69, No.2, pp. 139-153, ISSN 0368-8720

Montero, F.; Crespo, S.; Padrós, F.; De la Gándara, F.; García, A. & Raga, J. (2004). Effects of the gill parasite *Zeuxapta seriolae* (Monogenea: Heteraxinidae) on the amberjack *Seriola dumerili* Risso (Teleostei: Carangidae). *Aquaculture*, Vol.232, No.1-4, (April), pp. 153-163, ISSN 0044-8486

Morales Serna, N. (2010). *Ocurrencia de copépodos parásitos en el botete diana Sphoeroides annulatus (Jenyns, 1842) en el sistema lagunar Santa María La Reforma, Sinaloa, México.* Master Degree Thesis, Ciencias del Mar y Limnología, Unidad Académica Mazatlán, UNAM, Mazatlán, México

Munday, B.L.; Sawada, Y.; Cribb, T. & Hayward, C.J. (2003). Diseases of tunas, *Thunnus* spp. *Journal of Fish Diseases*, Vol.26, No.4, (April 2003), pp. 187-206, ISSN 1365-2761

Ocampo, C.L. & Auró, O.A. (2000). *Terapia de la Enfermedades de los Peces,* (Second edition), SUA, FMVZ-UNAM, México, DF

Ogawa, K, & Yokoyama, H. (1998). Parasitic diseases of cultured marine fish in Japan. *Fish Pathology*, Vol.33, No.4, (October 1998), pp. 303-309, ISSN 1881-7335

Ogawa, K. & Inouye, K. (1997). *Heterobothrium* infection of cultured tiger puffer, *Takifugu rubripes* - infection experiments. *Fish Pathology*, Vol.32, No.1, (October 2009), pp. 21-27, ISSN 1881-7335

Ogawa, K. (1998). Egg hatching of the monogenean *Heterobothrium okamotoi,* a gill parasite of cultured tigger puffer (*Takifugu rubripes*), with a description of its oncomiracidium. *Fish Pathology*, Vol.33, No.1, (October 2009), pp. 25-30, ISSN 1881-7335

Ogawa, K. (2002). Impacts of Diclidophorid monogenean infections of fisheries in Japan. *International Journal of Parasitoloy*, Vol.32, No3, (March 2002), pp. 373-380, ISSN 0020-7519

Ogawa, K. (2005). Economic and environmental importance, effects in finfish culture, In: *Marine Parasitology*, K. Rohde, (Ed.), pp. 378-391, CABI, ISBN 0-643-09025-8, Australia

Ogawa, K.; Bondad-Reantaso, M.G.; Fukudome, M. & Wakabayashi, H. (1995). *Neobenedenia girellae* (Hargis, 1953) Yamaguti, 1963 (Monogenea: Capsalidae) from cultured marine fishes of Japan. *Journal of Parasitology*, Vol.81, No.2, (April 1995), pp. 223-227, ISSN 1937-2345

Ohashi, H.; Umeda, N.; Hirazawa, N.; Ozaki, Y.; Miura, C. & Miura, T. (2007). Antiparasitic effect of calcium and magnesium ion-free buffer treatments against a common monogenean *Neobenedenia girellae*. *Parasitology*, Vol.134, No.2, (October 2006), pp. 229-236, ISSN 0031-1820

Okabe, K., 2000. Chemotherapeutic drug (Hada-clean) of oral administrating type to control fish parasites. *Doyaku Kenkyu*, Vol.60, pp. 1-12, ISSN 0385-3519

Ontiveros-García, LA. (2008). *Efecto de las infestaciones causadas por el protozoario Amyloodinium ocellatum (Brown, 1931) sobre los parámetros sanguíneos y filamentos branquiales del pargo lunarejo, Lutjanus guttatus (Steindachner, 1869),* Master Degree Thesis, CIAD, Mazatlán, Sinaloa, México

Ortuño, J.; Esteban, M.A. & Meseguer, J. (1999). Effect of high dietary intake of vitamin C on non-specific immune response of gilthead seabream (*Sparus aurata* L.). *Fish and Shellfish Immunology*, Vol.9, No.5, (July 1999), pp. 429-443, ISSN 1050-4648

Pan J. & Zhang, J. (2006). Monogenea of marine fishes from Hainan Island, China. VIII. Two new species of *Euryhaliotrema* Kritsky & Boeger, 2002 (Dactylogyridae) from *Lutjanus argentimaculatus* (Teleostei: Lutjanidae). *Systematic parasitology*, Vol.64, No.2. (June 2006), pp. 125-130 ISSN 1573-5192

Papapanagiotou, E.P. & Trilles, J.P. (2001). Cymothoid parasite *Ceratothoa parallela* inflicts great losses on cultured gilthead sea bream, *Sparus aurata* in Greece. *Diseases of Aquatic Organisms*, Vol.45, No.3, (August 2001), pp. 237-239, ISSN 0177-5103

Paperna, I. & Overstreet, R.M. (1981) Parasites and diseases of mullets (Mugilidae), In: *Aquaculture of Grey Mullets*, OH. Oren, (Ed.), pp. 411–493, Cambridge University Press, ISBN 0521279399, Cambridge, UK

Pérez-Ponce De León, G.; García-Prieto, L.; Mendoza-Garfias, B.; León-Regagnon, V.; Pulido-Flores, G.; Aranda-Cruz, C.V. & García-Vargas, F. (1999). *Listados faunísticos de México. IX. Biodiversidad de helmintos parásitos de peces marinos y estuarinos de la Bahía de Chamela, Jalisco,* Universidad Autónoma de México, México D.F

Pérez-Urbiola, J.C.; Inohuye-Rivera, R.B. & Muhlia-Melo, A.F. (2008a) *Catálogo de parásitos de pargos Lutjanus spp. en Bahía de La Paz e Isla Cerralvo, Baja California Sur, México,* Centro de Investigaciones Biológicas del Noroeste, SC, Baja California Sur México

Pérez-Urbiola, J.C.; Inohuye-Rivera, R.B. & Muhlia-Melo, A.F. (2008b). *Protocolo de manejo de pargos Lutjanus spp. en relación a las enfermedades parasitarias de impacto en condiciones de cautiverio,* Centro de Investigaciones Biológicas del Noroeste, S.C, México

Pironet, F.N. & Jones, J.B. (2000). Treatments for ectoparasites and disease in captive western Australian dhufish. *Aquaculture International*, Vol.8, No.4, (August 2000), pp. 349-361, ISSN 1573-143X

Quispe-Maica, P.R. (2005) *Estudio de las comunidades de metazoarios parásitos del pargo lunarejo Lutjanus guttatus (Steindachner, 1869) silvestres y en jaulas flotantes en el estado de Nayarit,* Master Degree Thesis, Escuela de Ingeniería Pesquera, Universidad Autónoma de Nayarit. Centro Universitario de la Costa campus Puerto Vallarta, Universidad de Guadalajara

Reyes-Becerril, M.; Tovar-Ramírez, D.; Ascencio-Valle, F.; Civera-Cerecedo, R; Gracia-López, V. & Barbosa-Solomieu, V. (2008). Effect of dietary live yeast *Debaryomyces hansenii* on the immune and antioxidant system in juvenile leopard grouper *Mycteroperca rosacea* exposed to stress. *Aquaculture*, Vol.280, No.1-4, (August 2008), pp. 39-44, ISSN 0044-8486

Rivas-Salas, A.I. (2003) *Ciclo de vida de Lepeophtheirus simplex (Copepoda: Caligidae) parásito de botete diana Sphoeroides annulatus (Jenyns, 1842)*, Master Degree Thesis, CIAD, AC, Mazatlán, Sinaloa, México

Rojas-Alba, M. (2004). *Orígenes de la Medicina Tradicional*, Tlahui, México, DF

Ruiz, A. & Madrid, J. (1992). Studies on the biology of the parasitic isopod *Cymothoa exigua* Schioedte and Meinert, 1884 and its relationship with the snapper *Lutjanus peru* (Pisces: Lutjanidae) Nichols and Murphy, 1922 from commercial catch in Michoacán. *Ciencias Marinas*, Vol.18, No.1, pp. 19-34, ISSN 0185-3880

Sapkota, A.; Sapkota, A.R.; Kucharskib, M.; Burkec, J.; McKenzieb, S.; Walkerb, P. & Lawrenceb, R. (2008). Aquaculture practices and potential human health risks: Current knowledge and future priorities. *Environment International*, Vol.34, No.8, (November 2008), pp. 1215-1226, ISSN 0160-4120

Schmahl, G. & Mehlhorn, H. (1988). Treatment of fish parasites 4. Effects of sym. triazinone (toltrazuril) on Monogenea. *Parasitology Research*, Vol.75, No.2 (February 1988), pp. 132-143, ISSN 1432-1955

Sharp, N.J.; Diggles, B.K.; Poortenaar, C.W. & Willis, T.J. (2004). Efficacy of Aqui-S, formalin and praziquantel against the monogeneans, *Benedenia seriolae* and *Zeuxapta seriolae*, infecting yellowtail kingfish *Seriola lalandi lalandi* in New Zealand. *Aquaculture*, Vol.236, No. 1-4, (June 2004), pp 67-83, ISSN 0044-8486

Sievers, G.; Palacios, P.; Inostroza, R. & Diilz, H. (1995). Evaluation of the toxicity of 8 insecticides in *Salmo salar* and the in vitro effects against the isopode parasite, *Ceratothoa gaudichaudii*. *Aquaculture*, Vol.134, No.1-2, (July 1995), pp. 9-16, ISSN 0044-8486

Sitjà-Bobadilla, A. & Alvarez-Pellitero, P. (2009). Experimental transmission of *Sparicotyle chrysophrii* (Monogenea: Polyopisthocotylea) to gilthead seabream (*Sparus aurata*) and histopathology of the infection. *Folia Parasitologica*, Vol.56, No.2, pp. 143-151, ISSN 1803-6465

Sitjà-Bobadilla, A.; Conde de Felipe, M.; & Alvarez-Pellitero, P. (2006). *In vivo* and *in vitro* treatments against *Sparicotyle chrysophrii* (Monogenea: Microcotylidae) parasitizing the gills of gilthead sea bream (*Sparus aurata* L.). *Aquaculture*, Vol.261, No.3, (December 2006), pp. 856–864, ISSN 0044-8486

Soler-Jiménez, L.C. & Fajer-Ávila, E.J. (2012). The microecology of dactylogyrids (Monogenea: Dactylogyridae) on the gills of wild spotted rose snapper *Lutjanus guttatus* (Lutjanidae) from Mazatlan Bay, Mexico. Folia Parasitologica 59 (in press).

Stephens, F.; Cleary, J.; Jenkins, G.; Jones, B.; Raidal, S. & Thomas, J. (2003) Treatments to control *Haliotrema abaddon* in the West Australian dhufish, *Glaucosoma hebraicum*. *Aquaculture*, Vol.215, No.1-4, (January 2003), pp. 1-10, ISSN 0044-8486

Stone, J.; Sutherland, I.H.; Sommerville, C.; Richards, R.H.; & Varma, K.J. (2000). Field trials to evaluate the efficacy of emamectin benzoate in the control of sea lice, *Lepeophtheirus salmonis* (Krøyer) and *Caligus elongatus* Nordmann, infestations in Atlantic salmon *Salmo salar* L. *Aquaculture*, Vol.186, No. 3-4, (June 2000), pp. 205-219, ISSN 0044-8486

Suzuki, K.; Misaka, N. & Sakai, D.K. (2006). Efficacy of green tea extract on removal of the ectoparasitic flagellate *Ichthyobodo necator* from chum salmon, *Oncorhynchus keta*, and masu salmon, *O. masou*. *Aquaculture*, Vol.259, No.1-4, (September 2006), pp. 17–27, ISSN 0044-8486

Tang, D. Andrews, M. & Cobcroft, J.M. (2007). The first chondracanthids (Copepoda: Cyclopoida) reported from cultured finfish, with a revised key to the species of *Chondracanthus*. *Journal of Parasitology*, Vol.93, No.4, (August 2007), pp. 788-795, ISSN 1937-2345

Thoney, D.A. & Hargis, W.J. (1991). Monogenea (Platyhelminthes) as hazards for fish in confinement. *Annual Review of Fish Diseases*, Vol.1, (September 2003), pp. 133-153, ISSN 0044-8486

Tóro, R.M.; Gessner, A.A.F.; Furtado, N.A.J.C.; Ceccarelli, P.S.; Albuquerque, S.D. & Bastos, J.K. (2003). Activity of the *Pinus elliottii* resin compounds against *Lernaea cyprinacea* in vitro. *Veterinary Parasitology*, Vol.118, No.1-2, (December 2003), pp. 143–149, ISSN 0304-4017

Torres Ayala, M.D.R. (2008). *Helmintos parásitos del huachinango Lutjanus peru (Pisces: Lutjanidae) procedentes de La Cruz de Huanacaxtle, Nayarit,* Bachelor Degree Thesis, Centro Universitario de Ciencias Biológicas y Agropecuarias, Universidad de Guadalajara, Jalisco, México

Tun, T.; Yokoyama, H.; Ogawa, K; & Wakabayashi, H. (2000) Myxosporeanas and their hyperparasitic microsporeans in the intestine of emaciated tiger puffer. *Fish Pathology*, Vol.35, No.3, (October 2009), pp. 145-156, ISSN 1881-7335

Velásquez-Medina, S.P. (2004). *Evaluación experimental de desinfectantes y terapéuticos en el control de monogeneos de la subfamilia Ancyrocephalinae, parásitos branquiales del pargo lunarejo, Lutjanus gutattus Steindachner, 1869,* Degree Thesis, Universidad Jorge Tadeo Lozano, Facultad de Biología Marina, Colombia

Wang, G.X.; Jiang, D.; Zhou, Z.; Zhao, Y. & Shen, Y. (2009). In vivo assessment of anthelmintic efficacy of ginkgolic acids (C13:0, C15:1) on removal of *Pseudodactylogyrus* in European eel. *Aquaculture*, Vol.297, No.1-4, (December 2009), pp. 38-43, ISSN 0044-8486

Wellens, H. (1982). Comparison of the sensitivity of Brachydanio rerio and Leuciscus idus by testing the fish toxicity of chemicals and wastewaters. *Journal of Water and Wastewater Research*, Vol.15, No.2, pp. 49-52, ISSN 0044-3727

Whittington, I.D. (2004). The Capsalidae (Monogenea: Monopisthocotylea): a review of diversity, classification and phylogeny with a note about species complexes. *Folia Parasitologica*. Vol.51, No.2-3, pp. 109-122, ISSN 1803-6465

Whittington, I.D. (2005). Monogenea Monopisthocotylea (ectoparasic flukes), In: *Marine Parasitology* K. Rohde, (Ed.), CSIRO, ISBN 0643090258, Melbourne, Australia

Whittington, I.D. & Horton, M.A. (1996) A revision of *Neobenedenia* Yamaguti, 1963 (Monogenea: Capsalidae) including a redescription of *N. melleni* (MacCallum 1927) Yamaguti, 1963. *Journal of Natural History*, Vol.30, No.8, (February 2007), pp. 1113-1156, ISSN 0022-2933

Whittington, I.D.; Corneillie, S.; Talbot, C.; Morgan, J.A.T. & Adlard, R.D. (2001a). Infections of *Seriola quinqueradiata* Temminck and Schlegel and *S. dumerili* (Risso) in Japan by *Benedenia seriolae* (Monogenea) confirmed by morphology and 28S ribosomal DNA analysis. *Journal of Fish Diseases*, Vol.24, No.7, (September 2001), pp. 421-425, ISSN 1365-2761

Whittington, I.D.; Ernst, I.; Corneillei, S. & Talbot, C. (2001b). Sushi, fish and parasites. *Australasian Science,* Vol.22, (April 2001), pp. 22-36, ISSN 1442-679X

Wootten, R., Smith, J.W. & Needham, E.A. (1982). Aspects of the biology of the parasitic copepods *Lepeophtheirus salmonis* and *Caligus elongates* on farmed salmonids, and their treatment, *Proceedings of the Royal Society of Edinburgh Section B,* Vol.81, pp. 185-197, ISSN: 0080-455X

Wu, X.; Zhu, X.; Xie, M. & Li, A. (2006). The radiation of *Haliotrema* (Monogenea: Dactylogyridae: Ancyrocephalinae): molecular evidence and explanation inferred from LSU rDNA sequences. *Parasitology,* Vol.132, No.5, (January 2006), pp. 659-668, ISSN 0031-1820

Yamaguti, S. (1961). *Systema Helminthum. Vol. III. Nematodes of Vertebrates,* Interscience Publishers, INC. New York, USA

Yamaguti, S. (1963). *Parasitic Copepoda and Branchiura of fishes,* Interscience, USA

Yasuzaki, M.; Ogawa, K. & Yoshinaga, T. (2004). Early development of the monogenean *Heterobothrium okamotoi* on the gills of tiger puffer *Takifugu rubripes. Fish Pathology,* Vol.39, No.3, (October 2009), pp. 153- 158, ISSN 1881-7335

Zertuche-González, J.A.; Sosa-Nishizaki, O.; Vaca-Rodríguez, J.G.; del Moral-Simanek, R.; Yarish, C.; & Costa-Pierce, B.A. (April, 2008). Marine science assessment of capture-based tuna (*Thunnus orientalis*) aquaculture in the Ensenada region of Northern Baja California, Mexico, In: *University of Connecticut., DigitalCommons@UConn,* 29.07.2011, Available from http://digitalcommons.uconn.edu/ecostam_pubs/1

Zhukov, E.V (1976). New species of the genus *Haliotrema* Johnston and Tiegs, 1922, from the Gulf of Mexico fishes of the family Lutjanidae. Fauna, systematics and phylogeny of Monogenoidea, *Proceedings of Institute of Biology and Pedology, Far-East Science Centre, Academy of Sciences of the U.S.S.R., New Series 35: 33-47*

Part 2

Bacterial Diseases

4

Molecular Detection and Characterization of Furunculosis and Other *Aeromonas* Fish Infections

Roxana Beaz Hidalgo and María José Figueras
Unit of Biology and Microbiology, Faculty of Medicine and Health Sciences, IISPV, University Rovira i Virgili, Sant Llorenç, Reus Spain

1. Introduction

Species of the genus *Aeromonas,* which inhabit aquatic environments, can produce septicaemia and ulcerative and hemorrhagic fish diseases, including furunculosis, which result in mass death and important economic losses in the aquaculture sector (Austin & Austin, 2007; Beaz-Hidalgo et al., 2010; Bernoth et al., 1997; Gudmundsdóttir & Björnsdóttir, 2007; Noga, 2010; Wiklund & Dalsgaard, 1998). The species *Aeromonas salmonicida* (with 5 subspecies: *salmonicida, masoucida, smithia, achromogenes* and *pectinolytica*) and *Aeromonas hydrophila* have classically been considered the most important *Aeromonas* fish pathogens, and the subspecies *salmonicida* is thought to be the causal agent of furunculosis, a disease that was reported more than a century ago to affect trout, and later to affect other salmonids and fish species (Bernoth et al., 1997; Goodwin & Merry, 2009; Han et al., 2011; Noga, 2010; Wiklund & Dalsgaard, 1998). Nowadays, furunculosis has a worldwide distribution, having been reported in Scotland, France, Norway, Iceland, Spain, United States, Canada, Japan, Chile and Australia. The fact that many strains isolated from diseased fish do not fit the described characteristics defined for *A. salmonicida* subsp. *salmonicida* or for furunculosis has lead to the terms 'atypical strains' and 'atypical furunculosis' to be introduced when they are attributed to other *A. salmonicida* subspecies, other *Aeromonas* species or when furunculosis occurs in fish other than salmonids (Wiklund & Dalsgaard, 1998). This causes confusion because the term 'atypical' is applied in different ways by the authors.

Publications such as the specific furunculosis monography published in 1997 by Bernoth et al., chapters dealing with this infection in specific aquaculture books (Austin & Austin, 2007; Hiney & Olivier, 1999; Noga, 2010) and reviews on atypical and typical furunculosis (Wiklund & Dalsgaard, 1998) and their treatment (Gudmundsdóttir & Björnsdóttir, 2007) among others, all reflect the importance to the aquaculture sector of *Aeromonas* infections. Several PCR methods have been designed and evaluated comparatively for the fast detection and identification of typical and atypical *A. salmonicida* from infected fish tissue and the introduction of reliable molecular *Aeromonas* identification methods have enabled new species (*Aeromonas tecta* and *Aeromonas piscicola)* to be discovered and/or other known species associated to fish disease to be recognized i.e. *Aeromonas bestiarum, Aeromonas sobria, Aeromonas encheleia, Aeromonas veronii, Aeromonas eucrenophila* and *Aeromonas media* (Beaz-

Hidalgo et al., 2010; Kozińska, 2007; Li & Cai, 2011; Nawaz et al., 2006; Soriano-Vargas et al., 2010). These species would have remained masked as, mainly, *A. salmonicida* or *A. hydrophila* if biochemical identification had only been applied and would have misrepresented the real prevalence and/or diversity of the species implicated in fish infections (Beaz-Hidalgo et al., 2010; Figueras et al., 2011b).

This chapter on *Aeromonas* fish infections presents the most recent information on the taxonomy and identification of these microorganisms derived from the application of molecular techniques together with a detailed review of the currently available PCR methods that have been designed for the detection or characterization of typical or atypical *A. salmonicida* or other *Aeromonas* species in, mainly, fish tissue and water during outbreaks or regular preventive monitoring. Preventive strategies for the control of infectious fish diseases, such as those produced by *Aeromonas* are a constant challenge, and need to be regularly reviewed in order to recognize dynamic changes associated to these diseases, which may be caused both by the emergence of new pathogenic species (i.e. the recently discovered *A. piscicola* and *A. tecta*) or by environmental factors. Regarding the latter, climate change is considered to play a role in the appearance and impact of furunculosis (Tam et al., 2011) and is therefore another aspect addressed in this chapter.

2. The genus *Aeromonas*

The genus *Aeromonas* belongs to the class *Gammaproteobacteria*, order *Aeromonadales* and family *Aeromonadaceae* (Martin-Carnahan & Joseph, 2005). *Aeromonas* species are widely distributed in aquatic environments and are isolated from water, healthy or diseased fish, food products, animal and human faeces and other clinical and environmental samples (Figueras, 2005; Janda & Abbot, 2010). The first description of an *Aeromonas* species dates back to 1891, when Stainer described the bacteria *Bacillus hydrophillus fuscus* (now *A. hydrophila*) isolated from diseased frogs (Martin-Carnahan & Joseph, 2005). A few years later in 1894 Emmerich and Weibel described the species *Bacillus de Forellenseuche* (later *Bacillus salmonicida*, now *A. salmonicida*) isolated from diseased trout (Martin-Carnahan & Joseph, 2005). The formal description of the genus was made by Stainer in 1943, and in the 1970s *Aeromonas* species were classified into 2 groups on the basis of their growth temperature, motility and pigment production (Martin-Carnahan & Joseph, 2005). One group comprised mesophilic strains able to grow at 37°C, motile and non-pigmented, mainly associated with human clinical infections and represented by *A. hydrophila*. The other group comprised psycrophilic strains (optimum growth at 22-28°C), non-motile and pigmented, which were mainly fish pathogens represented by *A. salmonicida*. In the 1980s, DNA-DNA hybridization assays allowed the differentiation of various genospecies or hybridization groups (Martin-Carnahan & Joseph, 2005).

In the last edition of Bergey`s Manual (Martin-Carnahan & Joseph, 2005) the genus, that previously belonged to the family *Vibrionaceae* was placed in its own independent family *Aeromonadaceae*, it comprised 14 species i.e. *A. hydrophila* (with 2 subspecies: *hydrophila* and *ranae*), *A. bestiarum*, *A. salmonicida* (with 5 subspecies: *salmonicida*, *masoucida*, *smithia*, *achromogenes* and *pectinolytica*), *A. caviae*, *A. media*, *A. eucrenophila*, *A. sobria*, *A. veronii*, *A. encheleia*, *Aeromonas jandaei*, *Aeromonas schubertii*, *Aeromonas trota*, *Aeromonas allosaccharophila*, and *Aeromonas popoffii*. Also some species were synonymised with previously recognized species such as *A. ichthiosmia* and *A. culicicola* with *A. veronii* and *A. enteropelogenes* with *A.*

trota. Since then, the genus has expanded rapidly with the addition of 11 new species (*Aeromonas simiae, Aeromonas molluscorum, Aeromonas bivalvium, A. tecta, Aeromonas aquariorum, A. piscicola, Aeromonas fluvialis, Aeromonas taiwanensis, Aeromonas sanarellii, Aeromonas diversa and Aeromonas rivuli*) and today the genus comprises 25 validated species (Figueras et al., 2011b).

3. Furunculosis and other *Aeromonas* fish infections

Furunculosis is one of the oldest known important diseases in aquaculture first documented by Emmerich and Weibel in 1894 when they observed furuncles or ulcer lesions on the trout's skin. These characteristic ulcerated lesions are those that give rise to the name of the disease attributed to infections produced by *Bacillus de Forellenseuche*, now known as *A. salmonicida* (Austin & Austin, 2007; Bernoth et al., 1997; Martin-Carnahan & Joseph, 2005). In reality, furunculosis was the first fish infection for which the 'Koch postulates' were demonstrated more than a century ago (Austin & Austin, 2007; Bernoth et al., 1997). It was thought originally that the infection affected only salmonids, but it soon became evident that it was distributed worldwide and affected many other marine and freshwater fish species, such as the Atlantic cod (*Gadus morhua*), halibut (*Hyppoglossus hyppoglossus*), turbot (*Scopthamus maximus*), lamprey (*Petromyzon marinus*), carp (*Cyprinus carpio*), goldfish (*Carassius auratus*) and eel (*Anguilla anguilla*) among others (Austin & Austin, 2007; Bernoth et al., 1997; Godoy et al., 2010; Goldschmidt-Clermont et al., 2009; Noga, 2010; Wiklund & Dalsgaard, 1998).

Lethargy, lack of appetite, or skin hyperpigmentation can be the first signs of the infection and it may show other clinical manifestations such as the presence of the typical furuncules or ulcers, exophthalmia (swelling of the eyes), septicaemia, petequias (small haemorrhagic lesions due to broken blood capillaries), anemia, ascitis and haemorrhages in the muscle, gills, fins, nares, vent and internal organs (Austin & Austin, 2007; Bernoth et al., 1997; Hiney & Olivier, 1999; Wiklund & Dalsgaard, 1998). On the other hand, fish affected by furunculosis do not always show all these clinical symptoms and may even not show the typical furuncules or skin ulcers (Noga, 2010).

Many *A. salmonicida* carrier fish that do not show external lesions or clinical signs of the disease but that are indeed able to shed the microorganism (10^5-10^6 CFU per fish/h) and to develop the disease under conditions of stress (i.e. increase in water temperature, poor water quality, etc), producing an epidemic outbreak have been reported (Bernoth et al., 1997; Gustafson et al., 1992; Hiney & Olivier, 1999; Noga, 2010; Wiklund & Dalsgaard, 1998). It has been estimated that up to 80% of cultivated trout may carry *A. salmonicida*, thus increasing the likelihood of transmission to susceptible fish (Gustafson et al., 1992). These covertly infected fish (or carriers) play an important role in the epidemiology of the disease, therefore early detection of the pathogen is very important. Several PCR detection methods have been developed for the detection of *A. salmonicida* (see Sections 4.2.2-4.2.5, Tables 1-5), whose application is essential for an effective disease control (Altinok et al., 2008; Byers et al., 2002a, b; Gustafson et al., 1992; Onuk et al., 2010).

Diseases caused by atypical strains of *A. salmonicida* (those that do not belong to the subsp. *salmonicida*) are referred to as "atypical furunculosis", although this terminology is also applied to infections produced in non-salmonids. Furthermore, more specific names have been used to refer to infections affecting specific fish, i.e. goldfish ulcer disease, carp erythrodermatitis, and flounder ulcer disease (Austin & Austin, 2007; Noga, 2010; Wiklund & Dalsgaard, 1998). Despite some authors have tried to describe specific clinical signs for each of these named pathologies (Austin & Austin, 2007; Noga, 2010), in practice they are indistinguishable from those associated to furunculosis or other ulcerative diseases or septicaemia caused by other motile *Aeromonas* species. The latter has also been named 'motile *Aeromonas* septicaemia' and is considered a clear example of a stress-induced disease mainly affecting freshwater fish. Although motile species also inhabit brackish water, their prevalence decreases with increasing salinity (Austin & Austin, 2007; Noga, 2010). These infections are mainly attributed to *A. hydrophila*, classically considered the most important species after *A. salmonicida*. However, the importance of the species *A. hydrophila* is overestimated because biochemical identification systems erroneously identify up to 70-80% of the strains of *Aeromonas* as belonging to this species, when in fact they are many different species when identified by molecular methods (Beaz-Hidalgo et al., 2010; Figueras, 2005; 2011b; Soler et al., 2003a). Prevalent motile species associated to diseased fish following molecular identification include *A. veronii*, associated with catfish, *A. sobria* with tilapia and trout, *A. hydrophila* with trout, *A. bestiarum* with carp and the recently described *A. piscicola* recovered from diseased salmonids and turbots (Beaz-Hidalgo et al., 2010; Kozińska, 2007; Li & Cai, 2011; Martino et al., 2011; Nawaz et al., 2006; Soriano-Vargas et al., 2010). Other motile species, such as *A. encheleia, A. allosaccharophila, A. jandaei, A. media, A. eucrenophila, A. aquariorum* and *A. tecta* have also been recovered from healthy or diseased fish, though less often (Beaz-Hidalgo et al., 2010; Kozińska, 2007).

Typical or atypical furunculosis and motile *Aeromonas* septicaemias all produce similar clinical signs that are also common to those observed in other fish systemic diseases caused by other bacterial and viral pathogens (Austin & Austin, 2007; Bernoth et al., 1997; Noga, 2010; Wiklund & Dalsgaard, 1998). Therefore, a definitive diagnosis requires the isolation, culture and identification of the bacteria recovered from the lesions or internal organs of the diseased fish and/or from the dead fish. Molecular PCR methods are available (Tables 1-5) that specifically detect *A. salmonicida* without the need for culturing. Care is needed in the case of covertly infected fish because false negatives can be obtained. Some authors recommend using a pre-enrichment step before the PCR detection or to use both culturing and PCR detection in parallel (Byers et al., 2002a; Gustafson et al., 1992).

3.1 Isolation of *Aeromonas* sp. from diseased fish

Recognizing a bacterial fish pathogen requires their analysis in several organs (mainly kidney, but also spleen, skin, and ovaric fluid in breeding females) of a significant number of living fish (Byers et al., 2002b; Noga, 2010). Between 4 and 10 diseased fish are recommended to be sampled for detecting the pathogen, and between 10 and 60 fish in a population where there is a low mortality rate or the fish are apparently healthy (Noga, 2010). Detecting the pathogen in mucus, blood or faeces is also recommended because these samples do not require the fish to be sacrificed (Beaz-Hidalgo et al., 2008; Byers et al., 2002b; Gustafson et al., 1992; Kulkarni et al., 2009). When searching for asymptomatic carriers, the kidney and intestine are the recommended organs to sample. Skin and gills can also be cultured, but *A. salmonicida* may be uncovered by the presence of other dominant bacteria (Noga, 2010). False negatives in carrier fish are common since the bacteria are present in low concentrations (Byers et al., 2002b; Gustafson et al., 1992; Noga, 2010). More asymptomatic carriers can be detected by including either a pre-enrichment step before the PCR analysis, as commented before (Byers et al., 2002b; Gustafson et al., 1992) or by inducing stress using the approach named 'stress induced furunculosis' (SIF). The latter requires an intraperiotoneal injection of glucocorticoids (which generates immunosuppression) and then exposure of the fish to a heat shock (raising the temperature of the water to 18°C for 14 days). This favours bacterial proliferation and the appearance of clinical signs, and will also increase the recovery rate of the bacteria (Austin & Austin, 2007; Bernoth et al., 1997; Byers et al., 2002b; Hiney & Olivier, 1999; Noga, 2010).

Non-selective culture media used for the isolation of *Aeromonas* strains include trypticase soy agar (TSA), brain heart infusion agar (BHI) or Columbia blood agar (Austin & Austin, 2007; Bernoth et al., 1997; Hiney & Olivier, 1999). More selective media used for the recovery of *A. salmonicida* include the Furunculosis agar (tryptone, yeast extract, L-tyrosine and NaCl), TSA or BHI supplemented with L-tyrosine (0.1%) or TSA supplemented with Coomassie brilliant blue (0.01%). The latter method was designed for detecting strains that have the A-layer (or S-layer), one of the oldest known outer membrane proteins associated with *Aeromonas* infection in fish, which is implicated in the resistance against the host complement system and enables the bacteria to adhere to the host proteins, facilitating colonization (Gustafson et al., 1992). The Coomassie brilliant blue attaches to the A-layer and the colonies are then seen as deep blue in comparison with white or light blue negative A-layer colonies (Austin & Austin, 2007; Bernoth et al., 1997). However, this method is not specific for *A. salmonicida,* since other *Aeromonas* or other bacterial (*Pseudomonas, Pasteurella, Corynebacterium*) species pathogenic to fish also have the A-layer and are able to grow in this culture media. Once the bacteria are isolated they can be identified by serological, phenotypic or molecular methods, the latter two being discussed in the next section.

4. Identification of *Aeromonas* species

This section describes the established phenotypic characteristics that differentiate the genus *Aeromonas* from other related genera, the limitations of the conventional and miniaturised biochemical identification sytems for differentiating all the species, as well as the new panorama derived from the application of molecular techniques.

4.1 Phenotypic identification with conventional methods and miniaturised commercial systems

Aeromonas species are phenotypically characterized as Gram-negative bacilli, with cytochrome oxidase generally positive, an ability to grow at 0% of NaCl but not at 6%. They do not produce acid from inositol, are able to ferment glucose and most are resistant to the vibriostatic agent O/129 2,4-diamino-6,7-diisopropil-pteridine-phosphate. The genus *Aeromonas* can be differentiated from other closely-related genera like *Plesiomonas* by its fermentation of inositol and its resistance to the vibriostatic agent, and from *Vibrio* by its ability to grow at 0% but its inability to grow at 6% of NaCl (Martin-Carnahan & Joseph, 2005). Optimum growth temperature for *Aeromonas* species is 22-37°C, except for some strains of *A. salmonicida,* which is 22-25°C (Martin-Carnahan & Joseph, 2005).

As indicated previously, the species *A. salmonicida* includes 5 subspecies (*A. salmonicida* subsp. *salmonicida, A. salmonicida* subsp. *achromogenes, A. salmonicida* subsp. *smithia, A.*

salmonicida subsp. *masoucida* and *A. salmonicida* subsp. *pectinolytica*), which are practically impossible to differentiate using phenotypic or molecular methods. All of them, except the subspecies *pectinolytica* have been implicated in fish pathology (Austin & Austin, 2007; Goldschmidt-Clermont et al., 2009; Han et al., 2011; Noga, 2010; Wiklund & Dalsgaard, 1998). Being unable to assign strains implicated in fish disease to any of the subspecies *masoucida, smithia, achromogenes* or *pectinolytica* using biochemical characteristics, they have been termed "atypical A. salmonicida" to differentiate them from the "typical A. salmonicida", a term restricted to the subspecies *salmonicida* (Beaz-Hidalgo et al., 2008; Goodwin & Merry, 2009; Wiklund & Daalsgard 1998). "Typical A. salmonicida" strains are psychrophilic, non-motile and produce brown pigment. In contrast, the "atypical A. salmonicida" are a more heterogeneous group with different phenotypic features and are usually isolated from non-salmonid fish (Noga, 2010). Among the "atypical A. salmonicida" strains, there is a group of strains that show a mesophilic behaviour (growing at 37°C), are motile and do not produce pigment and that have been also identified from human clinical samples (Figueras, 2005; Martínez-Murcia et al., 2005). Some of the mentioned characteristics (motility and mesophilic behaviour) are also phenotypic traits of *A. salmonicida* subsp. *pectinolytica* (Pavan et al., 2000). To distinguish these mesophilic *A. salmonicida* strains from other mesophilic species like *A. bestiarum* or *A. piscicola* is practically impossible when using biochemical characteristics or the sequences of the 16S rRNA gene (Beaz-Hidalgo et al., 2010; Figueras et al., 2011b; Martínez-Murcia et al., 2005). However, these species can be differentiated with sequences of the housekeeping gene *rpoD* (Beaz-Hidalgo et al., 2010).

Numerous biochemical schemes have been proposed for the characterization of *Aeromonas* species, but they mainly discriminate three big phenotypic groups i.e. the "A. hydrophila" complex (including *A. hydrophila, A. bestiarum* and *A. salmonicida* and also latter *A. popoffii*), the "Aeromonas caviae" complex (including *A. caviae, A. media* and *A. eucrenophila*) and "Aeromonas sobria" complex (including *A. sobria, A. veronii, A. jandaei* and *A. trota*) (Abbott et al., 1992; 2003; Borrell et al., 1998; Kozińska et al., 2002; Martin-Carnahan & Joseph, 2005; Martínez-Murcia et al., 2005). In the descriptions of new *Aeromonas* species, the mandatory criteria of identifying differential phenotypic tests from other existing *Aeromonas* species has been fulfilled (Figueras et al., 2011b and references therein). However, biochemical tests may show intra-species variability either due to the variable nature of the bacterial phenotypic characters or due to a lack of reproducibility of the tests when they are carried out under different laboratory conditions, such as temperatures, etc. (Figueras et al., 2011b). Biochemical identifications in fact lack precision and tend to generate a lot of inconsistent results in comparison to those obtained by molecular methods (Beaz-Hidalgo et al., 2010; Figueras et al., 2011b; Kozińska et al., 2002, Kozińska, 2007; Nawaz et al., 2006).

Miniaturized commercial identification systems, such as API (20E, 20NE), Vitek, Biolog GN Microplates, BBL Crystal or MicroScan Walk/Away are also routinely used in ichthyopathology laboratories, but are not always able to identify *Aeromonas* species precisely (Figueras, 2005; Joseph & Carnahan, 1994; Kozińska et al., 2002; Park et al., 2003; Soler et al., 2003a). The mentioned systems normally tend to identify all *Aeromonas* strains as *A. hydrophila* (Figueras, 2005; Soler et al., 2003b). Some other authors have also reported poor results from the API systems and discrepancies with traditional tube testing when identifying *Aeromonas* strains isolated from fish (Godoy et al., 2010; Han et al., 2011; Joseph & Carnahan, 1994; and references therein). Some methods may even misidentify *Aeromonas* strains as *Vibrio* species (Park et al., 2003; Soler et al., 2003b). To avoid this confusion,

Chacón and colleagues designed a specific probe based on a fragment of the glycerophospholipid-cholesterol acyltransferase (GCAT) gene that is able to hybridize with all *Aeromonas* species but not with strains from other genera (Chacón et al., 2002).

4.1.1 Discrepancies between biochemical and molecular identification methods

Because phenotypic characterization can give imprecise results, one cannot guarantee that a strain has been correctly identified at the species level, especially for complex genera like *Aeromonas*, if it has not been verified later using reliable molecular methods (Figueras et al., 2011a). Using biochemical and genetic methods in parallel for species identification has allowed big discrepancies to be highlighted and the characteristic phenotypic traits that might be responsible for the confusion to be discovered (Beaz-Hidalgo et al., 2010; Kozińska, 2007; Nawaz et al., 2006).

In a recent study, Nawaz et al. (2006) demonstrated that with the Vitek-GNI system 81 strains isolated from the intestines of catfish (*Ictalurus punctatus*) collected from different geographical regions of United States were identified as *A. hydrophila* (n=23), *A. trota* (n=7), *A. veronii* (n=42), *A. caviae* (n=6) and *A. jandaei* (n=3). However, when they were evaluated with the restriction fragment length polymorphism of the 16S rRNA gene (16S rDNA-RFLP) method proposed for *Aeromonas* identification (Borrell et al., 1997; Figueras et al., 2000), all the 81 strains were identified as *A. veronii* (Nawaz et al., 2006). In this case, the errors in phenotypic identification were masking the importance that the species *A. veronii* may have in catfish pathology.

In another recent study, Beaz-Hidalgo et al. (2010) used phenotypic and genetic methods (16S rDNA-RFLP and sequencing of the *rpoD* gene) to retest 119 *Aeromonas* strains recovered from diseased fish and shellfish that had been biochemically identified in a routine ichthyopathology laboratory. They found that, of the strains considered to belong to the genus *Aeromonas* using biochemical methods, 24.4% (29/119) did not belong when using the genus-specific PCR identification method based on the GCAT gene, which is specific for the genus *Aeromonas* as previously explained (Chacón et al., 2002). Sequencing the 16S rRNA gene identified these strains as belonging to the genera *Pseudomonas* and *Vibrio*. Considering only the 86 *Aeromonas* strains isolated from diseased fish, the species recognised by biochemical identification methods were: *A. hydrophila* (n=63), *A. salmonicida* (n=12), *A. sobria* (n=2), *A. veronii* (n=2), *A. media* (n=1), *A. trota* (n=1), *A. bestiarum* (n=1) and 4 strains could not be assigned to any known species (*Aeromonas* sp.). However, 66.3% (57/86) of them had been incorrectly identified phenotypically when compared with the genetic results that showed that the order of prevalence was as follows: *A. sobria* 25.6% (22/86), *A. hydrophila* 17.4% (25/86), *A. salmonicida* 17.4% (15/86), *A. bestiarum* 16.3% (14/86), *A. piscicola* 11.6% (10/86), *A. media* 7% (6/86), *A. eucrenophila* 2.3% (2/86), *A. encheleia* 1.2% (1/86) and *A. tecta* 1.2% (1/86) (Beaz-Hidalgo et al., 2010). The strains of the species *A. veronii* (n=2) and *A. trota* (n=1) therefore belonged to totally different species. For instance, the two *A. veronii* belonged to *A. sobria* and *A. piscicola*, while *A. trota* was *A. sobria* and the 4 strains of *Aeromonas* sp. belonged to *A. sobria* (2 strains), *A. media* and *A. bestiarum*. Furthermore, genetic identification revealed that 50 of the 64 phenotypically identified *A. hydrophila* strains did in fact belong to 8 other *Aeromonas* species, including 17 strains belonging to *A. sobria* and 6 strains to new species *A. piscicola* (Beaz-Hidalgo et al., 2009; 2010). This study clearly highlighted once more that there was a false importance attributed to the species *A.*

hydrophila as the most prevalent (Beaz-Hidalgo et al., 2010; Figueras, 2005; Figueras et al., 2011b; Soler et al., 2003b). Regarding *A. salmonicida,* 85.7% (12/14) of the strains that showed to be psycrophilic, non motile and pigment producers had been correctly identified. The two strains misidentified phenotypically as *A. hydrophila* showed mesophilic behaviour (i.e. were motile and did not produce pigment). Looking at the most prevalent *Aeromonas* species by fish host, it was found that *A. sobria* (15/22 strains) and *A. hydrophila* (8/15 strains) were more associated with trout, *A. salmonicida* with Atlantic salmon and turbot (6/15 strains each) and *A. piscicola* (5/10 strains) and *A. bestiarum* (8/14 strains) with Atlantic salmon (Beaz-Hidalgo et al., 2010). The only strain of the species *A. tecta* was isolated from bogue *(Chondrostoma comun),* this being the first report in this kind of fish and the second report from fish since its recent description from trout in 2008 (Beaz-Hidalgo et al., 2010 and references therein).

These results gave evidence of a greater diversity of *Aeromonas* species involved in fish pathology than was first thought and that the species *A. salmonicida* or *A. hydrophila* were not the most prevalent.

Despite the known limitations of phenotypic identification methods, they are still being used in ichthyopathology laboratories, which contributes to an underestimation of the true diversity of *Aeromonas* species associated with diseased fish (Beaz-Hidalgo et al., 2010).

4.2 Molecular identification

This section describes the above-mentioned molecular methods used for identifying *Aeromonas* in fish pathology (i.e. 16S rDNA-RFLP, the sequences of the 16S rRNA gene or other housekeeping genes) as well as the different PCR detection and identification methods that target typical and atypical *A. salmonicida.*

4.2.1 The 16S rRNA and the housekeeping genes

The 16S rRNA gene is essential to bacteria and has been used as a specific molecular marker for their identification (Alperi et al., 2008 and references therein; Martínez-Murcia et al., 1992). An RFLP method based on this gene was developed by our group to differentiate all the *Aeromonas* species described up to 2000 (Borrel et al., 1997; Figueras et al., 2000) and has proven to be useful to different authors who have identified the strains recovered from diseased fish (Beaz-Hidalgo et al., 2010; Kozińska et al., 2002; Kozińska, 2007; Nam & Joh, 2007; Nawaz et al., 2006; Soriano-Vargas et al., 2010). However, this method is not able to differentiate closely related species that have an equal or almost equal 16S rRNA gene sequence because it produces the same RFLP pattern for all of them. This is shown in the case of the new species *A. piscicola,* which has the same RFLP pattern as *A. salmonicida* and *A. bestiarum* whose 16S rRNA sequences share 99.8-100% similarity (Beaz-Hidalgo et al., 2010; Figueras et al., 2011b). The same is true with *A. caviae* and *A. aquariorum,* which show a 99.8% similarity and the same RFLP pattern (Figueras et al., 2009, 2011b). The latter, i.e. *A. aquariorum,* was isolated originally from ornamental fish, but has not so far been isolated again from other fish. The 16S rDNA-RFLP method is also not useful for the 8% of *Aeromonas* strains that show mutations on the 16S rRNA gene in the targeted region of the endonucleases used, because these produce a different pattern from the one expected for the species (Alperi et al., 2008).

In the genus *Aeromonas* the sequences of the 16S rRNA gene are not very useful tools for identification because only a few distant species can be well discriminated, but it is useful for confirming whether or not the strains under analysis belong to the genus (Figueras et al., 2011b; Han et al., 2011).

As a complementary tool to the 16S rDNA-RFLP for identifying all *Aeromonas* species, we have introduced the use of housekeeping genes that codify essential proteins for the survival of the bacteria (Figueras et al., 2011b; Soler et al., 2004; Yáñez et al., 2003). In reality, the analysis of 5 housekeeping genes is one of the methods used in the description of new bacterial taxa that has been recommended by the *ad hoc* committee for the re-evaluation of the species definition in bacteriology (Stackebrandt et al., 2002). The first housekeeping genes described for the identification of *Aeromonas* were the *gyrB* gene, which codifies for the subunit B of the DNA gyrase (Yáñez et al., 2003) and the *rpoD* gene, which encodes the σ^{70} factor of the RNA polymerase (Soler et al., 2004). Nowadays the cost of sequencing has fallen considerably being much cheaper to use external services that provide the sequences from a given extracted DNA than to perform the sequencing at your own laboratory. Therefore we recommend sequencing the *rpoD* or *gyrB* gene to establish the identity of the isolated strain because it is a faster and more reliable identification approach than the 16S rDNA-RFLP (Figueras et al., 2011b). However, misinterpretations may still occur when comparing the sequences obtained with those available for all the species of the genus that are held at the GenBank, especially if sequences are too short or of poor quality. Furthermore the database also contains wrongly labelled strains, so comparison should always be made with the sequences of type strains (Figueras et al., 2011b). We also recommend to deposit the new sequences at the GenBank for further broadening of the existing database. Figure 1 illustrates a phylogenetic tree constructed with the *rpoD* sequences of the type strains of the 25 species that include at present the genus and several *Aeromonas* strains isolated from diverse diseased fish studied in our laboratory.

Other housekeeping genes that have been used in *Aeromonas* are the *rpoB,* which codifies for the β subunit of the RNA polymerase (Küpfer et al., 2006), the *dnaJ,* which codifies a thermal shock protein (Nhung et al., 2007), the *recA,* which codifies a protein involved in DNA repair (Sepe et al., 2008) and the *cpn60,* which codifies the chaperone Cpn60 type I that is involved in protein assembly (Miñana-Galbis et al., 2009). A Multilocus Phylogenetic Analysis (MLPA) of the genus *Aeromonas* has recently been established using the concatenated information derived from the sequences of 7 housekeeping genes: *gyrB, rpoD, recA, dnaJ, gyrA, dnaX,* and *atpD* (Martínez-Murcia et al., 2011). The resulting phylogenetic tree (4705 bp) agreed with the taxonomy of the genus as recognized to date, showing that the MLPA is a robust way of identifying unequivocally all the species. The MLPA is also called Multilocus Sequence Analysis (MLSA) by some authors, which derives from the original name Multilocus Sequence Typing (MLST) (Martínez-Murcia et al., 2011). Nowadays, MLST schemes are recognised to be the best way of establishing the epidemiological relationships among isolates using the sequences of internal fragments of multiple housekeeping genes (approximately 450-500 bp of 7 genes). For each gene the sequences of different strains are compared and each unique sequence is assigned a specific number. The strains that show the same numbers assigned to all genes belong to the same sequence type (ST) which is also identified by a specific number.

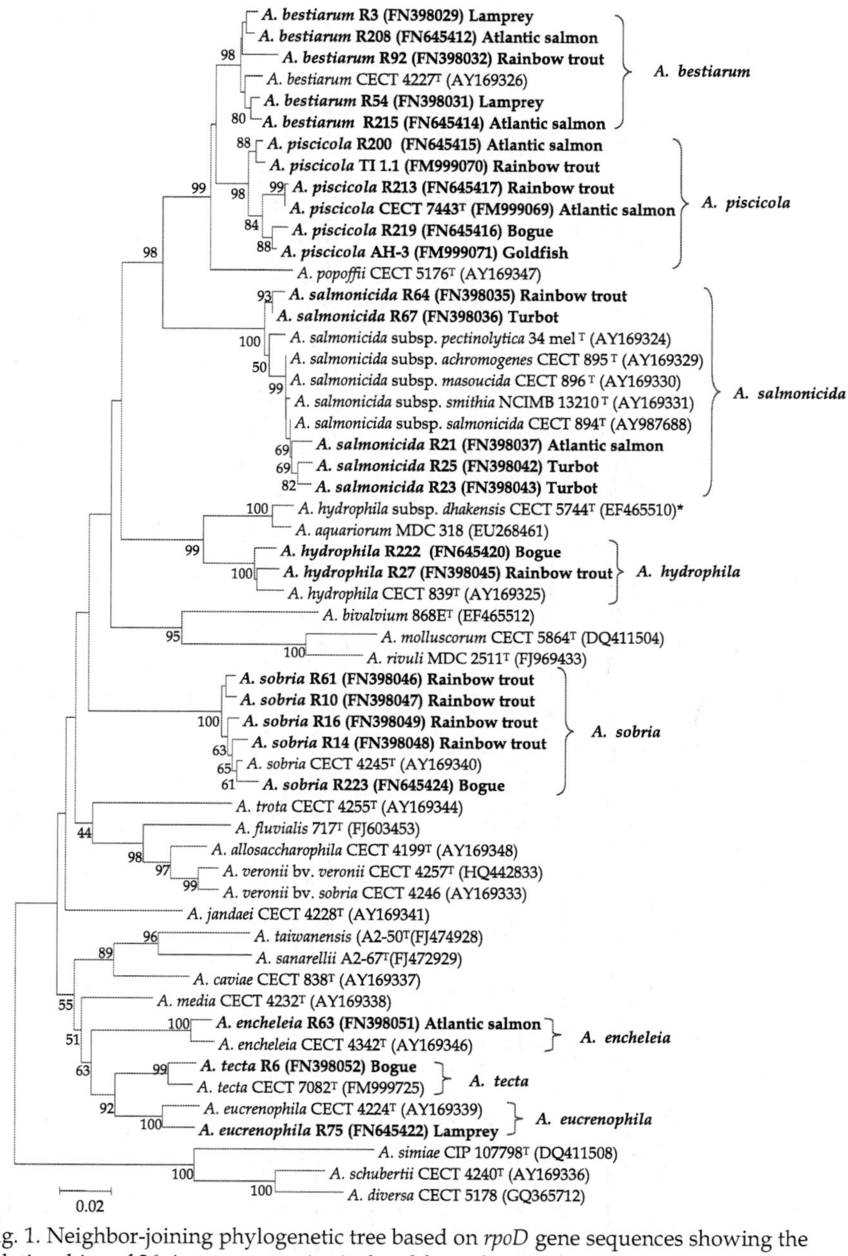

Fig. 1. Neighbor-joining phylogenetic tree based on *rpoD* gene sequences showing the relationships of 26 *Aeromonas* strains isolated from diverse diseased fish (in bold) (Beaz-

Hidalgo et al., 2010) with the 25 species included presently in the genus. The scientific names of the fish species: Atlantic salmon (*Salmo salar*), rainbow trout (*Oncorhychus mykis*), lamprey (*Petromyzum marinus*), bogue (*Boops boops*), goldfish (*Carasius auratus*), turbot (*Scopthalmus maximus*). **A. hydrophila* subsp. *dhakensis* is considered a synonym of the species *A. aquariorum* (Figueras et al., 2011b).

The first *Aeromonas* open access MLST scheme (http://pubmlst org/aeromonas) was recently constructed (Martino et al., 2011) and another that included a set of 7 housekeeping genes (Lamy, 2011) was presented at the 10th International Symposium on *Aeromonas* and *Plesiomonas*. The open access MLST uses 6 genes (*gyrB, groL, gltA, metG, ppsA* and *recA*, 3,084 nt) and was applied to a total of 96 reference and field strains that included a high proportion of strains, 79.2% (76/96) obtained from known diseased freshwater and marine fish species mostly collected in the north-eastern area of Italy (Martino et al., 2011). The phylogenetic tree constructed with the concatenated sequences grouped the 76 fish strains, in order of prevalence, in *A. veronii* (n=28), *A. sobria* (n=25), *A. salmonicida* (n=6), *A. bestiarum* (n=6), *A. allosaccharophila* (n=4), *A. media* (n=4), *A. hydrophila* (n=2) and *A. encheleia* (n=1). The strains of the prevailing species *A. veronii* and *A. sobria* came from 11 and 6 fish species, respectively. However, the majority of the strains of *A. veronii* were from catfish (7 strains belonging to 7 different STs) followed by carp (6 strains belonging to 6 different STs) while *A. sobria* prevailed in trout (13 strains belonging to 12 different STs).

Recent studies have used the above-mentioned molecular methods (the 16S rDNA-RFLP and/or the housekeeping genes *rpoD* or *gyrB*) and have identified *A. salmonicida* subsp. *salmonicida* in farmed Arctic charr (*Salvelinus alpinus*) in Austria (Goldschmit-Clermont et al., 2009), *A. bestiarum* in farmed carp in Mexico (Soriano-Vargas et al., 2010), *A. sobria* from an outbreak that occurred in tilapias in China (Li & Cai, 2011), *A. veronii* from catfish in United States (Nawaz et al., 2006) and the studies carried out by Kozińska et al. (2002), Kozińska (2007) and Beaz-Hidalgo et al. (2010) that characterized numerous *Aeromonas* species isolated from diseased fish in Poland and in Spain respectively. Jun et al. (2010), using a multiplex PCR (m-PCR) and the 16S rRNA gene, identified *A. hydrophila* as the agent responsible for an outbreak that killed 50% of a farmed population of Korean cyprinid loach (*Misgurnus anguillicaudatus*). Nam & Joh (2007) using the 16S rDNA-RFLP found that 84% (252/300) of the isolates obtained from diseased trout belonged to the species *A. sobria*, and also found other *Aeromonas* species in a lower prevalence i.e. *A. encheleia* (9.3%), *A. salmonicida* (3.7%) and *A. bestiarum* (3%). Using several housekeeping genes (*gyrB, rpoD, dnaJ, recA*) also enabled Han et al. (2011) to identify *A. salmonicida* from ulcer lessions and haemorrhages in the black rockfish (*Sebastes schlegeli*) in Korea. These authors also sequenced the *vapA* gene (encoding the A-protein, a subunit of the A-layer or S-layer) and found that the sequences from the isolated *A. salmonicida* strains clustered with *A. salmonicida* subsp. *masoucida* in a neighbour-joining phylogenetic tree that included the sequences of other atypical *A. salmonicida* strains isolated from fish and of the subspecies *salmonicida, masoucida* and *smithia* (Han et al., 2011). However, these and other authors (Lund & Mikkelsen, 2004) stated that many atypical strains do not cluster with any of the subspsepecies of *A. salmonicida* when using the *vapA* sequences, and consider this gene not to be useful in the delineation of *A. salmonicida* to the subspecies level.

Pridgeon et al. (2011) sequenced the 16S-23S rDNA intergenic spacer region and 3 housekeeping genes (*cpn60, gyrB* and *rpoD*), to identify 6 isolates as *A. hydrophila* from catfish (of which 3 were associated to an outbreak) and used suppression subtractive hybridization to investigate genome differences between a highly virulent and a less virulent strain, and they found 64 different sequences. By performing specific PCR reactions targeting these sequences, they determined that 3 of them that encoded a hypothetical protein XAUC_13870, a putative methyltranferase and a structural toxin protein RtxA were specific for the virulent strain. The specifically designed PCRs for these 3 sequences were further tested in another 7 field isolates (6 of them obtained from fish) for which the virulence had previously been determined by *in vivo* experiments in catfish. The two sequences encoding the hypothetical protein XAUC_13870 and the putative methyltranferase were present in the 4 highly virulent *A. hydrophila* strains but absent in the 5 less virulent strains, suggesting that these two sequences might be new virulence factors and therefore useful molecular markers for identifying highly virulent isolates of *A. hydrophila*.

4.2.2 PCR methods for the detection of *A. salmonicida* from fish tissue and water

Molecular PCR methods are fast, sensible and specific and have therefore been developed and applied for detecting typical and atypical *A. salmonicida* strains producing fish diseases (Tables 1-5). The methods that rely on the 16S rRNA gene have not been taken into consideration, because as commented above this gene does not enable the correct separation of *A. salmonicida* from its closely related species *A. bestiarum* and *A. piscicola* because they have an almost identical 16S rRNA gene sequence.

Table 1 describes in chronological order the most relevant PCR methods. In 1992, Gustafson et al. and Hiney et al. designed primers targeting 2 specific regions of *A. salmonicida,* which were later used by several authors for detecting the pathogen in water and fish samples (Tables 1, 2). The method described by Gustafson et al. (1992) targeted the *vapA* gene (which encodes the A-protein of the A-layer, as commented previously) of *A. salmonicida*. The authors validated their PCR method with 54 *A. salmonicida* strains (28 typical and 26 atypical) and found only one negative strain that was also not able to express the protein forming the A-layer. The authors suggested that the *vapA* gene in this strain possessed a mutation that affected both the amplification and the expression of the protein. However, they also found that 13 PCR positive strains did not express the protein (negative by Western blot) indicating that in those strains the mutations in the sequence of the *vapA* gene affected only the expression but not the PCR amplification. Further PCR analysis with primers covering the whole gene revealed that strains not expressing the A-layer had amplicons of lower molecular weight compared to strains expressing the protein, indicating that the mutations were deletions. The sensitivity of the PCR method was also validated using artificially infected samples that were prepared by inoculating a strain of *A. salmonicida* in the fish tank water and into homogenates of fish tissue and faeces of rainbow trout. The sensitivity from the direct detection in fish tissue was of 10 CFU/mg while in pure culture suspensions and water it was 1 CFU/ml (Table 1). The method was further tested using fish tissue and faecal samples of 25 naturally infected dead fish and of 25 suspected carrier fish, evaluating in parallel the results of the PCR carried out both before and after an enrichment incubation step in nutrient broth. Results for all 25 infected dead fish were positive with both methodologies (with and without a pre-enrichment). However, 5 of the 25 carrier fish were only positive after the pre-enrichment. Furthermore, the 10 water samples taken from

Reference	Target region or gene (bp)	N° of *As* strains evaluated	Type of samples	Results and/or conclusions
Gustafson et al. (1992)	*vapA* (421 bp) coding for the A protein (A or S layer).	28 typical (*Ass*), and 26 atypical *As*.	Dilutions of pure culture. Experimentally inoculated kidney, spleen and faeces homogenates and samples of sterile water. 25 dead infected fish, 25 suspected carrier fish and their tank water.	All typical *Ass* and atypical *As* strains were positive except 1 typical *Ass* strain. Detection limits were: 1 CFU in pure culture and in 100 ml of inoculated water and 10 CFU/mg in all tissue. All faecal and tissue samples of the 25 infected and 20 carriers fish were positive, 5 carrier fish were only positive after an enrichment step, therefore this step is necessary to eliminate false negatives.
Hiney et al. (1992)	DNA fragment (423-bp) recognized by Southern blot with unknown function[1].	25 *As*.	Dilutions of pure cultures.	All *As* were positive by dot blot hybridization with no cross-reaction with any of the 14 *Aeromonas* strains belonging to other 4 different species. Detection limit for the PCR assay from pure cultures was 2.4 cells.
Miyata et al. (1996)	DNA RAPD fragment (512 bp) of unknown function[2].	10 *Ass*, 2 *Asa*, 1 *Asm* and 4 atypical *As*.	Dilutions of pure cultures. Kidney of 15 experimentally infected fish.	The method detects only typical strains (the 10 *Ass* were positive) because the 2 *Asa*, the 1 *Asm* and the 4 atypical strains tested were negative[3]. The detection limit in pure cultures was 10 fg of bacterial DNA. A positive reaction was obtained from kidney samples from all the experimentally infected fish.
Høie et al. (1997)	DNA plasmid fragment of *Ass* and *Asa* of unknown function (710 bp).	2 *Ass*, 1 *Asa* and 1 *Asm*.	Dilutions of pure cultures of 4 strains of *Ass*, 1 of *Asm* and 1 of *Asa*. Experimentally inoculated kidney and gill homogenates. Kidney and gill from 109 covertly infected Atlantic salmon. Kidney from 88 wild brood Atlantic salmon.	All strains were positive except the only *Asm* tested. The PCR products were confirmed by hybridization. Detection limit in inoculated kidney and gill homogenates was 10^4 CFU/100 ml. Kidney and gill samples from covertly infected Atlantic salmon were negative by PCR and by culture. *Ass* was not detected in kidney of brood salmon by PCR but 6/88 were positive by culture. However, a PCR based on the 16S rRNA gene of *As* was positive for 29 of the 88 kidney samples analyzed[4].

Table 1. Developed PCR methods for the detection of *A. salmonicida* from diverse infected fish tissue and water.

Reference	Target region or gene (bp)	N° of *As* strains evaluated	Type of samples	Results and/or conclusions
Oakey et al. (1998)	5 RAPD fragments of unknown function were used as probes in Southern blot experiments.	10 *Ass*, 6 *Asa*, 4 *Asm*, 1 *Ass*	Dilutions of pure cultures.	All 5 Southern blot probes hybridize with *Ass* and with some of the subspecies, therefore the method does not discriminate typical from atypical *As* strains.
Nilsson et al. (2006)	DNA fragment (749 pb) of the insertion sequence (ISasa4) of the *tapA* gene encoding a protein of the type IV pili.	29 atypical *As*, 24 typical (*Ass*)	Dilutions of pure cultures.	This PCR differentiates most of the atypical *As* (27/29 were positive) from typical *As* (24/24 were negative. The detection limit was approx. 250 fg of template.
Beaz-Hidalgo et al. (2008)	*fstA* (422 bp) encoding a ferric siderophore receptor.	66 *Ass*, 2*Asa*, 1 *Asm*	Dilutions of pure and mixed bacterial cultures. Experimentally inoculated (kidney and skin) tissue homogenates, mucus and blood. Mucus and blood of 31 wild salmon.	All *As* strains tested were positive, so it does not differentiate typical from atypical strains. Detection limits from pure cultures was 20-200 cells/ml and from mixed cultures 60-600 cells/ml. The latter was approx. the same detection limit for 100 mg of infected tissue with pure and mixed cultures. Detection limits from mucus was 10^2 cells/ml and in blood 10^5 cells/ml. 6/31 samples of wild salmon were positive.
	gyrB (760 bp) encoding the B subunit of the DNA gyrase.		Dilutions of pure cultures.	The *gyrB* primers amplified all 69 *As* strains but showed less sensitivity when compared with the *fstA* primers, so no other experiments for *gyrB* were performed.

As, A. salmonicida; Ass, A. salmonicida subsp. salmonicida; Asa, A. salmonicida subps. achromogenes; Asm, A. salmonicida subsp. masoucida; CFU, colony forming units.
[1]A Blast analysis we carried out revealed that it shares 99% of similarity with the gene *mobA* (accession number: AJ508382) which is a mobilization protein of 1263 bp localized in the plasmid PAsa1 of *A. salmonicida* subsp. *salmonicida* (strain JF2267).[2]The sequence of this fragment is not available for comparison. [3]This method has been used in several studies (Table 2) to differentiate the atypical strains based on the negative PCR reaction obtained with this method. [4]We found that these 16S rRNA primers do not differentiate *A. salmonicida* from *A. bestiarum* and *A. piscicola*.

Table 1. Developed PCR methods for the detection of *A. salmonicida* from diverse infected fish tissue and water. (continued) the tanks of the suspected carrier fish were only investigated using the enrichment step and all were positive. Gustafson et al. (1992) concluded that the enrichment step increases sensitivity and avoids possible false negative results (Table 1).

Hiney and co-workers (1992) discovered an apparently specific DNA fragment (423 bp) from a genomic DNA library of strain 7222V of *A. salmonicida* by differential hybridization with a strain of *A. hydrophila.* They tested the specificity of this fragment by dot blot hybridization using 25 *A. salmonicida* strains, other *Aeromonas* species (8 *A. hydrophila,* 3 *A. caviae,* 2 *A. sobria,* 1 *A. media*) and 11 strains belonging to other genera. The results demonstrated that the probe only reacted with *A. salmonicida.* This DNA fragment was isolated, sequenced and a specific PCR was designed obtaining a sensitivity of 2.4 cells of *A. salmonicida* in pure cultures (Table 1). Despite the authors sequenced the amplified fragment, they were not able, at that time, to determine the identity of this sequence. In the search for establishing its identity, now that the complete genome of *A. salmonicida* is known and that more sequences are available for comparison at the GenBank, we found that it shares 99% similarity with the sequence of the gene *mobA* (access number: AJ508382), which is a mobilization protein of 1263 bp localized in the plasmid PAsa1 of *A. salmonicida* (Fehr et al., 2006). The PCR method designed by Hiney et al. (1992) has been used and evaluated in posterior studies which are listed in Table 2.

In 1996, Miyata et al. sequenced a DNA fragment (512 bp) of unknown function obtained from a RAPD amplification of *A. salmonicida* and developed a PCR for the early detection of furunculosis, evaluating the method with several typical and atypical strains of *A. salmonicida* and experimentally infected fish. They found amplification only for the 10 *A. salmonicida* subsp. *salmonicida* strains (Table 1). The method could therefore be considered specific for detecting only typical furunculosis or typical *A. salmonicida* strains (i.e. those belonging only to the subsp. *salmonicida*). The assay was also positive in kidneys of 15 experimentally infected amago salmon (*Oncorhynchus rhodurus* subsp. *macrostomus*). So far, the sequence of this fragment has not been deposited in the GenBank and since this target has also been used in several later studies (Table 2), it is important to obtain the sequence in order to know its function.

Other studies performed by Høie et al. (1997) and Oakey et al. (1998) also developed PCRs for targeting DNA fragments of *A. salmonicida* of unknown functions, but these methods were not able to differentiate the subspecies. To our knowledge, these protocols have not been used since in any later studies and the main results of the two studies are summarised in Table 1.

Two new PCR methods were described in the first decade of the 21st century (Beaz-Hidalgo et al., 2008; Nilsson et al., 2006). The one of Nilsson and co-workers (2006) is based on the presence of an insertion element (ISE) in the *tapA* gene (encoding a pili subunit protein of the type IV pili) of *A. salmonicida,* which they named ISAsa4. They demonstrated that ISAsa4 was present in multiple copies (>30) but only in atypical strains of *A. salmonicida,* because none of the 24 typical strains tested were PCR positive, whereas 27/29 atypical strains were (Table 1). The identity of the 53 typical and atypical strains was confirmed by their positive amplification using the PCR method described by Gustafson et al. (1992) (which detects both typical and atypical strains) and by the non-amplification obtained for the 29 atypical strains using the method described by Miyata et al. (1996) (which only detects typical strains i.e. *A. salmonicida* subsp. *salmonicida*). On the basis of their results, Nilsson and co-workers

(2006) proposed this method as useful for differentiating most atypical strains. Furthermore, by Southern blot, using the ISasa4 as a probe, and after sequencing the *tapA* gene, these authors also observed that the atypical strains showed a high heterogeneity in this region/gene in comparison with the typical strains. Similar results were obtained previously when the amino acid sequences of the *vapA* gene from typical and atypical strains were compared, because identical sequences were found in typical strains, which contrasted with the significant variability observed in atypical strains (Lund & Mikkelsen, 2004). We believe that this high variability could indicate that the group of atypical strains may in fact not only embrace other subspecies of *A. salmonicida* but also other misidentified *Aeromonas* species. This PCR method directed at atypical *A. salmonicida* strains was used recently by Godoy et al. (2010) for characterising strains isolated from freshwater Atlantic salmon (*Salmo salar*) with successful results (Table 2), which we will discuss later.

The method developed by Beaz-Hidalgo et al. (2008) targets a fragment of 422 bp of the *fstA* gene of *A. salmonicida* that encodes a siderophore receptor and was validated by testing 69 strains of this species using Gustafson's method in parallel. All of the 69 strains produced the expected amplicon of the *fstA* gene, although 4 strains did not amplify with the method designed by Gustafson et al. (1992). This is probably due to the presence of mutations in the targeted DNA sequence of the *vapA* gene as Gustafson et al. (1992) discussed. The detection limit of the *fstA* PCR in dilutions of pure and mixed cultures (*A. salmonicida, Vibrio anguillarum* and *A. hydrophila*) was within the range of 20 to 600 cells/ml. In artificially infected kidney and skin samples the range was relatively similar, 60-600 cells per 100mg of tissue (Table 1). The proposed PCR for *A. salmonicida* was considered a non-destructive diagnostic tool when used in blood or mucus. Detection limits of experimentally infected mucus and blood samples were 2.5×10^2 and 1.5×10^5 CFU/ml respectively. Bacteria present in seeded blood was only detected at a high concentration (1.5×10^5 CFU/ml), which is probably due to interferences with heparin or other unknown blood components that compete with the bacterial DNA in the PCR amplification (Beaz-Hidalgo et al., 2008). This problem was reported by other authors previously (Høie et al., 1997 and references therein) and could maybe be avoided including a pre-enrichment step (Byers et al., 2002b; Gustafson et al., 1992) or evaluating the quality of the extracted DNA for the presence of inhibitors (Mooney et al., 1995). However, none of these approaches were tested by Beaz-Hidalgo et al. (2008). In the search for possible asymptomatic carrier fish, the *fstA* protocol was further assayed in the mucus and blood of 31 wild salmon that showed no signs of furunculosis, obtaining 4 *A. salmonicida* PCR-positive samples of mucus and 6 of blood recovered from 6 (19%) of the 31 salmon assayed (Table 1). On the other hand, culture methods were only able to isolate *A. salmonicida* from one blood sample, indicating that the PCR method had a higher sensitivity. It was concluded that the method was fast, specific and sensitive to *A. salmonicida* in both infected and asymptomatic carrier fish (Beaz-Hidalgo et al., 2008). The possible application to mucus and blood, which can be obtained without the need to perform necropsies or to sacrifice the fish, was a differential and advantageous characteristic over other molecular methods that were designed to detect the pathogen only from internal organs. Beaz-Hidalgo et al. (2008) within the same study also designed an additional method that relied on the amplification of the *gyrB* housekeeping gene of *A. salmonicida* and that produced the expected amplicon for the 69 strains tested. However, no further experiments were carried out with the *gyrB* PCR because its detection limit was higher than that obtained for the *fstA* gene (Table 1). Both the *fstA* and *gyrB* primers designed by Beaz-

Hidalgo et al. (2008) have been employed in later studies (Table 3) and have also been used in m-PCR methods developed for the simultaneous detection of *A. salmonicida* and other bacterial species that cause fish diseases (Table 4, section 4.2.5). In our laboratory, we have recently evaluated 10 strains of each of the species *A. piscicola, A. bestiarum* and *A. salmonicida* (including the type strains of all the *A. salmonicida* subspecies) using the PCR method described by Beaz-Hidalgo et al. (2008), which targets a fragment of 422 bp of the *A. salmonicida fstA* gene. This assay was carried out to know whether the primers only reacted with *A. salmonicida* or could also react with the closely-related species *A. piscicola* and *A. bestiarum.* To our knowledge those primers have never been tested in the recently described species *A. piscicola* (Beaz-Hidalgo et al., 2009) and only in the type strain of *A. bestiarum* (Beaz-Hidalgo et al., 2008). The results confirmed the specificity of the method for *A. salmonicida* because only the 10 strains of this species showed a unique band of expected size of the *fstA* gene. In contraposition none of the *A. piscicola* and *A. bestiarum* strains showed this specific band (422 bp) despite some bands of others sizes were amplified in some strains (unpublished results).

Another potential new specific target of *A. salmonicida* is a hypothetical protein named AssHPA, which was discovered by chance by Kingombe et al. (2010) while they were developing a m-PCR for the detection of three enterotoxins (*act, alt* and *ast*). Since the AssHPA amplicon (148 bp) was present in the nine *A. salmonicida* strains isolated from dead fish, the authors suggested that its usefulness as a specific target for the identification of this species could be evaluated. However, it was also present in two other strains, one identified as *A. bestiarum* and one identified as belonging to the *A. caviae/A. media* phenotypic complex. Considering this, we believe that the specificity of this PCR reaction might be compromised by this cross reactivity with other *Aeromonas* species and so requires further investigation with a greater number of strains of other species, including *A. piscicola.* The latter should also be evaluated with all the molecular methods available that can differentiate *A. salmonicida* strains (typical and atypical). This would guarantee that *A. piscicola,* which is commonly infecting fish, would not interfere by producing false positive reactions.

4.2.3 Application or comparative evaluation of the described PCR methods

Several studies have evaluated individually, or compared, the methods described in Table 1 and the results and conclusions are summarized in Table 2. In this sense, Morgan et al. (1993) used the PCR method designed by Hiney et al. (1992) to confirm the presence of viable not culturable (VNC) cells of a strain of *A. salmonicida* after 21 days incubation in sterile and untreated lake water microcosms (Table 2). Flow cytometry was used to determine the number of viable cells comparing it with the amount recovered after cultivation on agar plates. In the sterile microcosm culturable *A. salmonicida* cells were evaluated on trypticase soy agar (TSA) plates. However, to differentiate this species from other indigenous bacteria present in untreated lake water, the strain tested was marked using a kanamycin resistant plasmid and the *xylE* gene (that encodes a D-xylose transporter). Growth of colonies on TSA agar plates supplemented with kanamycin revealed the presence of the tested *A. salmonicida* strain in the untreated water. The bacteria were also detected using the PCR designed by Hiney at al. (1992), which targets the *vapA* gene and in the case of the untreated water also the specific PCR targeting the *xylE* gene only present in the marked *A. salmonicida* cells. After 21 days, counts of viable cell measured by flow cytometry were 4 logarithms higher than culturable cells in the sterile microcosm, and a positive PCR detection was obtained at all the times tested (Table 2). In the untreated microcosm, the number of culturable *A. salmonicida* cells had declined below the detection limit after 14 days, although both viable cells and positive PCR amplicons (confirmed by hybridization) were observed until the end of the experiments.

The same DNA fragment discovered by Hiney et al. (1992) was also used by O´Brien et al. (1994) to investigate the presence of *A. salmonicida* in samples of faeces, effluent and tank water of a hatchery of Atlantic salmon smolts. Their results showed that although cultures of water and particulate matter of effluents of tanks with apparently healthy fish were negative for *A. salmonicida,* the pathogen was detected by PCR. The estimated sensitivity was approximately 1000 fg of *A. salmonicida* DNA (or 200 genome equivalents) per g of sample by PCR detection, but this sensitivity increased 10 times by dot blot hybridization (Table 2).

Byers and colleagues (2002a) compared three PCR methods designed by Gustafson et al. (1992), Hiney et al. (1992) and Miyata et al. (1996) in order to verify the identity of 308 *A. salmonicida* strains, which included type and reference collection strains as well as strains recovered from 38 teleost fish. They determined their sensitivity in inoculated fish tissue homogenates and evaluated the method on mucus and other samples (gill, muscle lesion, intestine, spleen and kidney) from experimentally infected salmonids (Table 2). They found that the methods of Hiney et al. (1992) and Gustafson et al. (1992) correctly identified 92.5% and 93.5% of the 308 strains (typical and atypical) and their simultaneous use gave a positive result for 99.4% of the strains (Table 2). In agreement with what it was described by Miyata et al. (1996) their PCR was able to identify 100% of the *A. salmonicida* subsp. *salmonicida* strains but none of the atypical strains. The latter were positive only with the methods of Hiney et al. (1992) and Gustafson et al. (1992). Thus they concluded that the simultaneous application of these three methods appeared to be useful for distinguishing typical and atypical isolates of *A. salmonicida.* Byers et al., (2002a) found that some PCR negative isolates for the method of Hiney et al. (1992) targeting the *vapA gene* of the A-layer, were still able to produce this layer and suggested that this was probably due to a mutation within the primer site that did not affect the expression of the gene as it was already described in the original study of Hiney and co-workers. They found that the methods of Hiney et al. (1992) and Gustafson et al. (1992) were more sensitive than that of Miyata et al. (1996) (Table 2). The obtained sensitivity in pure cultures ranged from 0.2 to 2 pg of DNA in 50 µl, while in the different inoculated fish tissues it ranged from 10^3 to 10^5 CFU per g. Furthermore, they detected the presence of the pathogen in tissue of all the experimentally infected fish (Byers et al., 2002a). In a second complementary study, Byers et al. (2002b) investigated the presence of *A. salmonicida* in covertly infected rainbow trout, Atlantic salmon and Arctic charr again using the same 3 PCR methods in parallel with conventional cultures. However, they only evaluated the two most sensitive methods (Gustafson et al., 1992; Hiney et al., 1992) for the direct PCR detection of *A. salmonicida* in experimentally infected fish. They concluded that culturing was a more reliable method than the PCR assays, and estimated that the lowest detection limit of the 2 PCR methods assayed (Gustafson et al., 1992; Hiney et al., 1992) was 4 x 10^5 CFU/g of tissue. In addition, for evaluating the presence of *A. salmonicida* in covertly infected fish, they also recommended the use of a pre-enrichment culture step of the tissue before the PCR assay (Byers et al. 2002b). Another interesting result was that mucus samples yielded PCR products more often

Reference	Objective	PCR method/name of primers	Type of samples	Results and/or conclusion
Morgan et al. (1993)	To study the viable but not culturable state (VNC) of *As.*	Hiney et al. (1992)/AP	Experimentally inoculated sterile and untreated lake freshwater.	The number of bacteria recovered from inoculated sterile lake water after 21 days by conventional culture was only 4 CFU/100 ml. However 5.6 x 10^4 cells/ml viable cells were confirmed by flow cytometry. In untreated lake water colonies were not obtained after 14 days. However, the bacteria were detected over the 21-day experiment by PCR. *A. salmonicida* possess a VNC state in inoculated freshwater.
O´Brien et al. (1994)	To detect *A. salmonicida* in a hatchery of Atlantic salmon.	Hiney et al. (1992)/AP	Faeces, effluent and tank water samples from a hatchery of Atlantic salmon smolts.	The method enables detection of *A. salmonicida* from all samples. Parallel culture techniques were negative in the effluent and tank water samples. The estimated detection limit of the PCR products by gel electrophoresis was approx. 200 *A. salmonicida* genome equivalents/g of sample but dot blot hybridization was 10 times more sensitive.
Byers et al. (2002a)	To compare the results obtained for the detection and identification of 308 *A. salmonicida* strains using three existing PCR methods in parallel.	Gustafson et al. (1992)/PAAS Hiney et al. (1992)/AP Miyata et al. (1996)/MIY	Dilutions of pure cultures. Experimentally inoculated tissue homogenates (mucus, gill, kidney and intestine). Mucus, gill, muscle lesion, intestine, spleen and kidney from experimentally infected fish.	Of the 308 *A. salmonicida* typical and atypical strains tested the AP amplified 285 and the PAAS 288. Their parallel application improved the individual results and detected 306 (99.4%) strains. The MIY method only amplified the typical *Ass* strains. The AP and PAAS methods provided better sensitivity in pure cultures (0.2 pg-2 pg/50 µl) and in tissue (10^3-10^5 CFU/g) than the MIY method (200 pg-20000 pg/50 µl in culture and 10^6-10^7 CFU/g in tissue). Tissues from all the experimentally infected fish were positive.

Table 2. Studies that apply the PCR methods described in Table 1.

Reference	Objective	PCR method/name of primers	Type of samples	Results and/or conclusion
Byers et al. (2002b)	To compare the results obtained for the detection and identification of *A. salmonicida* in experimentally and covertly infected rainbow trout, Atlantic salmon and Artic charr using three existing PCR methods in parallel.	Gustafson et al. (1992)/PAAS Hiney et al. (1992)/AP Miyata et al. (1996)/MIY	Tissue (mucus, gill, spleen, kidney, intestine) of experimentally infected salmonids as well as covertly infected (or carriers) salmonids.	All experimentally infected fish were positive with AP and PAAS methods. The 3 PCR methods identified correctly the 32 isolates recovered from covertly infected fish by culture. Detection limit in covertly infected salmonids with the AP and PAAS methods was 10^5 CFU/g. An enrichment step is required to increase the sensitivity.
Skugor et al. (2009)	To detect and determine levels of *A. salmonicida* in experimentally infected vaccinated and unvaccinated Atlantic salmon.	Hiney et al. (1992)/AP	Liver and spleen from experimentally infected vaccinated and unvaccinated Atlantic salmon.	*A. salmonicida* was detected in the liver and spleen of all analyzed fish but the load was substantially lower in vaccinated compared to unvaccinated fish.
Godoy et al. (2010)	To identify by PCR 5 biochemically characterized *A. salmonicida* (typical or atypical) strains isolated from a furunculosis outbreak in freshwater farmed Atlantic salmon.	Gustafson et al. (1992)/PAAS Miyata et al. (1996)/MIY Nilsson et al. (2006)/ISasa4	External lesions, kidney, liver, spleen, heart of moribund Atlantic salmon.	Since the 5 strains only amplified with the PAAS primers (that does not discriminate between typical and atypical) and with those described by Nilsson et al. (2006) (specific for atypical strains) but did not with the MIY primers (specific for typical strains), the strains were considered as atypical *A. salmonicida*.

Table 2. Studies that apply the PCR methods described in Table 1. (continued)

than tissues like the intestine (Byers et al. 2002b). The authors suggested that this might indicate either that the pathogen is more abundant in certain body parts or that some types of infected tissue may inhibit the PCR more than others. In our view, it should be considered that the mucus that covers all the surface of the fish is the most accessible part of the body to *A. salmonicida,* which could be another reason for these results.

Another study in which several molecular methods were used in parallel was the one performed by Godoy et al. (2010) that aimed at re-identifying 5 biochemically characterised *A. salmonicida* strains isolated from a furunculosis outbreak that occurred in Chile in a freshwater farm of Atlantic salmon. The study tried to determine if the infection was produced by typical or atypical *A. salmonicida* strains, since atypical furunculosis had been detected previously in seawater in Chile. They used the primers designed by Gustafson et al. (1992), which recognize both typical and atypical *A. salmonicida* strains, those of Nilsson et al. (2006), which targeted the repeated insertion element ISasa4 specific only of the atypical strains (non *salmonicida* subspecies), and those designed by Miyata et al. (1996), which only recognize typical strains (*A. salmonicida* subsps. *salmonicida*). The five isolates were positive for the PCR that targeted the insertion element ISAsa4 (Nilsson et al., 2006) and the PCR that targeted the DNA region of Gustafson et al. (1992), but were negative for that of Miyata et al. (1996). These results indicated that the freshwater salmon isolates were atypical strains of *A. salmonicida*. Sequencing nearly the complete 16S rRNA gene was also carried out and revealed that the five isolates obtained from diseased Atlantic salmon reared in freshwater shared 100% similarity with *A. salmonicida* subsp. *achromogenes* and *A. salmonicida* subsp. *masoucida,* whereas with *A. salmonicida* subsps. *salmonicida* the similarity was 99.85%. In our view Godoy et al. (2010) forgot to consider that the species *A. bestiarum* shows an identical 16S rRNA gene sequence to *A. salmonicida* subsp. *achromogenes* and *A. salmonicida* subsp. *masoucida* (Martinez-Murcia el al., 2005). Furthermore, some strains of the recently proposed new species *A. piscicola* also share a 100% 16S rRNA gene similarity with strains of *A. bestiarum* and *A. salmonicida,* including the subspecies *achromogenes* and *masoucida.* When the authors tested another 12 atypical *A. salmonicida* strains isolated from cultured marine fish, they found that the ISasa4 PCR was negative. Godoy et al. (2010) suggested that this might be due to a lower number of copies of this insertion element being present in the atypical *A. salmonicida* strains they isolated from the marine environment, in comparison with those isolated from freshwater, or due to variations in the targeted sequence. However, none of these suggestions have been investigated experimentally or proven by sequencing the specific region.

4.2.4 Studies that develop other PCR technologies using previously described targets

Other authors have used the PCR primers or a part of the specific target region described in previous studies for developing other PCR technologies, such as a nested PCR (Mooney et al., 1995), quantitative PCR methods (Balcázar et al., 2007; Goodwin & Merry, 2009) or a loop mediated isothermal amplification (LAMP) (Kulkarni et al., 2009), which are all summarized in Table 3 and described in this section. In the study carried out by Mooney et al. (1995), they applied the method described by Hiney et al. (1992) for the detection of *A. salmonicida* in the blood of 61 wild Atlantic salmon, but all samples were negative. In order to verify this, the authors developed an improved extraction procedure and a test to evaluate the quality of DNA by performing a PCR suitability test. The latter consisted of a PCR targeting a region of the fish host genome in order to determine if the sample was free of PCR inhibitors that might interfere with the PCR detection of *A. salmonicida*. Furthermore they developed a nested PCR procedure (involving two consecutive PCRs) to improve sensitivity. So, after the conventional PCR (using the primers of Hiney et al., 1992) was carried out, specific primers that further amplified a sub-region (278 bp) of the original amplicon (423 bp) were designed and used in a second PCR. In this way, Mooney et al. (1995) were able to obtain amplification bands at the expected size in 87% of the 61 fish blood samples (Table 3).

Balcázar et al. (2007) developed a quantitative real-time PCR (Q-PCR) for the detection of *A. salmonicida* in inoculated fish tissue and in naturally infected fish. The Q-PCR targeted 131 bp situated within the 423 bp of the *mobA* gene that Hiney et al. (1992) amplified with the advantage of being able to quantify the PCR product (Tables 1, 3). The specificity of the method was confirmed by the positive results obtained for the 16 *A. salmonicida* isolates tested (belonging to 3 subspecies: *salmonicida, achromogenes* and *masoucida*) and the negative results for 10 strains belonging to other *Aeromonas* species and for 16 strains of different bacterial genera. Sensitivity of the Q-PCR was similar in pure culture and in inoculated tissue and was within the range of 0.5 pg to 50 ng of DNA that was established to be equivalent to 16 CFU (Table 3). In order to further validate the designed Q-PCR, it was applied to fish from natural outbreaks of furunculosis, and the levels of *A. salmonicida* per g of tissue (kidney, liver and spleen) obtained ranged from of 5.12×10^2 to 1.05×10^4 CFU.

In a more recent study, Goodwin & Merry (2009) used the three pairs of primers previously described by Gustafson et al. (1992), Hiney et al. (1992) and Miyata et al. (1996) and adapted them for a Q-PCR to analyse 62 ulcer swab samples obtained from carp to investigate the possible association of typical and atypical *A. salmonicida* in the ulcerative disease of the carp. Using the conventional PCR, they demonstrated that atypical strains were present in 84% (52/62) of the carp ulcers analysed. Like in other studies (Byers et al., 2002a) the distinction between typical and atypical strains was based on the positive amplification obtained using the primers of Hiney et al. (1992) and Gustafson et al. (1992) and the negative amplification using those of Miyata et al. (1996). Sixteen percent (10/62) of the samples were found to be PCR negative, which coincided with the ulcer swab samples taken from carp that live in water with high temperatures. These authors suggested that these increasing temperatures are likely to negatively affect the survival of the pathogen or to help the carp immune system to eliminate it (Goodwin & Merry 2009). Using Hiney's primers, Goodwin & Merry (2009) calculated a range of 9.83×10^2- 7.74×10^7 genome copies of *A. salmonicida* per µg of host DNA extracted from the ulcer swab samples of the carp (Table 3). In parallel, the authors also grew cultures of the swabs but failed to isolate the bacteria and they argued that this is most likely due the overgrowth of other bacteria, particularly fast-growing motile aeromonads (which might also be responsible for the disease) as occurs with gills and skin where dominant bacteria mask the presence of lower prevalence bacteria. The authors concluded that the Q-PCR protocol might be a useful new tool to study the *A. salmonicida* epidemiology of the carp ulcer (Goodwin & Merry 2009).

Another totally different method is the loop-mediated isothermal amplification (LAMP) assay developed by Kulkarni et al. (2009) for the detection of furunculosis in Atlantic cod (*Gadus morhua*). For the development of the method, the authors designed 5 sets of primers targeting the *gyrB* gene region described by Beaz-Hidalgo et al. (2008) as useful for detecting *A. salmonicida*. The LAMP technique was first described by Notomi et al. (2000) to detect the

Reference	Objective	Type of samples	Results and/or conclusion
Mooney et al. (1995)	To increase the sensitivity of the PCR of Hiney et al. (1992) improving the DNA extraction (developing a procedure for testing its quality) and performing a nested-PCR.	Blood samples from 61 wild Atlantic salmon.	Conventional PCR did not detect *A. salmonicida* in the blood of wild salmon but the improved method was positive for 53 of the 61 evaluated fish. These results were confirmed by hybridization. Sensitivity was < 100 *A. salmonicida* genome equivalents/fish sample.
Balcázar et al. (2007)	To develop a quantitative real time-PCR (Q-PCR) for the detection of *A. salmonicida* from fish tissue using a smaller fragment (131 bp) of the region targeted by Hiney et al. (1992).	Tissue of naturally infected fish recovered from outbreaks.Dilutions of pure cultures. Experimentally inoculated fish tissue homogenates (liver, kidney, spleen and intestine).	*A. salmonicida* was detected and quantified in all naturally infected fish (in concentrations ranging from 5.12 x10^2 CFU/g to 1.05 x 10^4 CFU/g). Detection limits in pure cultures and in inoculated tissue (liver, kidney, intestine, spleen) was approximately 16 CFU per Q-PCR reaction.
Goodwin & Merry (2009)	To establish the incidence of typical and atypical *A. salmonicida* in ulcerative lesions of the carp using primers designed by Gustafson et al. (1992), Hiney et al (1992) and Miyata et al. (1996) and to used them in a Q-PCR.	Swabs from ulcerative lesions of diseased carp	52 of the 62 strains recovered from ulcers were considered atypical and therefore those are the *As* strains associated to the carp ulcerative disease. The range of genome of copies of *A. salmonicida* in ulcer swab samples was 9.83 x 10^2 to 7.74 x 10^7 /µg of DNA.
Kulkarni et al. (2009)	To develop a LAMP protocol for the rapid, sensitive and specific detection of furunculosis in Atlantic cod using 5 *A. salmonicida* specific primers for the *gyrB* gene targeted by Beaz-Hidalgo et al. (2008).	Spleen obtained from experimentally infected and uninfected fish. Dilutions of pure cultures. Experimentally inoculated mucus from a healthy fish.	The LAMP assay was specific for *A. salmonicida* and was positive for all the experimentally infected fish samples. LAMP was more sensitive than conventional PCR detection in pure cultures (1 pg of DNA/ml vs. 100 pg of DNA/ml) and in infected mucus (10 pg of DNA/ml vs. 1000 pg of DNA/ml).

LAMP, loop-mediated isothermal amplification.

Table 3. Studies that improve the PCR methods described in Table 1 or use the target regions and/or primers to design other methods.

hepatitis B virus and has the advantages of being faster (45 min) than conventional PCR and of not needing a thermocycler or any sophisticated equipment. Furthermore, on-site detection of the pathogen can be carried out by direct visualization of the fluorescent staining (SYBR Safe green) of LAMP products under a handheld UV light (Kulkarni et al., 2009). In their work, Kulkarni et al. (2009) compared the LAMP assay with the conventional PCR using the *gyrB* and *fstA* primers previously designed by Beaz-Hidalgo et al. (2008). They tested serial DNA dilutions of a pure culture of *A. salmonicida* and achieved a sensitivity of 1 pg of bacterial DNA/ml with LAMP, which is a 100-fold improvement over conventional PCR detection. Both LAMP and conventional PCR were assayed inoculating mucus samples, and the detection limits of 10 pg of bacterial DNA/ml and 1000 pg of bacterial DNA/ml, respectively, were achieved (Table 3). The use of the LAMP technique has rarely been reported in the literature, the study by Kulkarni et al., (2009) being the first one to use it for the detection of an *Aeromonas* fish pathogen. Further validation studies are needed for it to be applied in diagnostic laboratories and in on-site programmes (Kulkarni et al., 2009).

4.2.5 Multiplex PCR (m-PCR) methods to detect simultaneously *A. salmonicida* and other fish pathogenic bacteria

The advantage of using an m-PCR is its ability to detect several targeted sequences simultaneously, enabling the presence of several pathogens to be evaluated at once, being less time consuming and more cost effective than any individual PCR. González et al. (2004) developed an m-PCR that could be adapted to a microarray that targeted *A. salmonicida* and 4 other important marine fish pathogenic bacteria i.e. *V. anguillarum*, *Photobacterium damselae* subsp. *damselae*, *Vibrio parahaemolyticus* and *Vibrio vulnificus* (Table 4). For detecting *A. salmonicida,* 2 regions of the *vapA* gene encoding the A-layer were selected, a region of 177 bp different from that of Gustafson et al. (1992) and a smaller fragment (101 bp) of the same region (423 bp) described by Hiney et al. (1992). They tested a total of 75 strains, representing 28 species of several genera, including 3 strains of *A. salmonicida* subsp. *salmonicida* and 3 strains belonging to other *Aeromonas* species (Table 4). The specificity of the assay was 100%, but the authors indicated that false negatives might arise as result of naturally occurring mutations in the region targeted by the primers. The sensitivity obtained with pure bacterial cultures (4-5 CFU) was similar to that obtained with other previously discussed PCR methods (Altinok et al., 2008; Gustafson et al., 1992; Hiney et al., 1992; Onuk et al., 2010).

Altinok et al. (2008) described another m-PCR for the detection of *A. hydrophila, A. salmonicida* subsp. *salmonicida* and 3 other major bacterial fish pathogens (*Flavobacterium columnare, Renibacterium salmoninarum* and *Yersinia ruckeri*) (Table 4). For detecting *A. salmonicida,* a modification of the Hiney et al. (1992) primers were used, while for detecting *A. hydrophila* the primers were those previously described by Nielsen et al. (2001) targeting the 16S rRNA gene. However, when we searched for any possible cross reactivity of *A. hydrophila* with sequences of other *Aeromonas* species, we found that complementary sequences to the primer regions were also present in the species *A. molluscorum* (only known from shellfish) and *A. encheleia* (originally isolated from European eels in Valencia, Spain and found in association with fish diseases in other studies). This seems to indicate that those primers cannot be considered specific for detecting *A. hydrophila,* and this needs to be tested experimentally using several strains of the mentioned interfering species. The method

Reference	Simultaneous detection of:	Target region or gene (bp)	N° of *Aeromonas* strains and type of samples evaluated	Results and/or conclusion
González et al. (2004)	*A. salmonicida* *Photobacterium damselae* subsp. *damselae*, *Vibrio vulnificus* *Vibrio anguillarum* *Vibrio parahaemolyticus*	A different region of 177 bp of the *vapA* gene targeted by Gustafson et al. (1992) and a smaller fragment (101 bp) of the region (423 bp) targeted by Hiney et al. (1992)[1].	3 *A. salmonicida* subsp. *salmonicida*, 1 *A. caviae*, 1 *A. hydrophila* and 1 *A. sobria*. Dilutions of pure cultures.	Detection limit for *A. salmonicida* in pure cultures was <20 fg of genomic DNA, which is equivalent to 4-5 CFU. The m-PCR products were used to develop a microarray and clear hybridization signals for both regions were obtained.
Altinok et al. (2008)	*A. salmonicida* subsp. *salmonicida* *A. hydrophila* *Flavobacterium columnare* *Renibacterium salmoninarum* *Yersinia ruckeri*	A slightly smaller fragment (416 bp) of the region (423 bp) targeted by Hiney et al. (1992)[1] for *A. salmonicida*. The region of the 16S rRNA gene proposed by Nielsen et al. (2001) for *A. hydrophila*.	5 *A. salmonicida* subsp. *salmonicida*, 4 *A. hydrophila* and 2 *A. sobria*. Dilutions of pure cultures. Tissue (liver, spleen, kidney, gill, skin and fins) of diseased farmed rainbow trout.	Detection limit in pure cultures was 1 CFU. Of the 558 diseased rainbow trout obtained from 31 farms, 112 were positive for the detection of any of the 3 pathogens including 35 positive for *A. hydrophila* and 22 for *A. salmonicida*.
Onuk et al. (2010)	*A. salmonicida* subsp. *salmonicida* *Flavobacterium psychrophilum* *Yersinia ruckeri*	Identical region (*fstA*, 422 bp) as Beaz-Hidalgo et al. (2008).	9 *A. salmonicida* subsp. *salmonicida*. Dilutions of pure cultures. Experimentally inoculated homogenates of liver tissue obtained from a healthy salmon.	Detection limits were 30 CFU in pure cultures and 250 CFU in infected liver. All *A. salmonicida* subsp. *salmonicida* were positive.
Kulkarni et al. (2010)	*A. salmonicida* *Francisella piscicida* *V. anguillarum* (three major pathogens in farmed Atlantic cod).	Identical region (*gyrB*, 760 bp) as Beaz-Hidalgo et al. (2008).	1 *A. salmonicida* Dilutions of pure cultures.	Detection limit of 10µg-50 ng DNA/ml in pure cultures. The only reference strain of *A. salmonicida* tested was positive.

[1]Hiney et al. (1992) indicated that this sequence had an unknown function. However, a Blast analysis we carried out revealed that it shares 99% of similarity with the gene *mobA* (accession number: AJ508382), which is a mobilization protein of 1263 bp localized in the plasmid PAsa1 of *A. salmonicida* subsp. *salmonicida* (strain JF2267).

Table 4. Studies that develop multiplex-PCR (m-PCR) to detect simultaneously *Aeromonas* spp. (mainly *A. salmonicida*) and other fish pathogens.

appeared to be specific for the 5 strains of *A. salmonicida* evaluated, with no cross-reaction with the other bacteria (Altinok et al., 2008). The detection limits in dilutions of pure culture were 2 CFU for *A. hydrophila* and 1 CFU for *A. salmonicida,* respectively, and the assay detected *A. hydrophila* in 35 fish and *A. salmonicida* in 22 fish of the 558 diseased rainbow trout analysed (Table 4).

In a very recent study, Onuk et al. (2010) has developed an m-PCR for the simultaneous detection of *A. salmonicida* and two other bacteria (*Flavobacterium psychrophilum* and *Y. ruckeri*) able to produce contagious infections in salmonids (Table 4). The detection limit was 30 CFU of *A. salmonicida* from culture suspensions and 250 CFU from inoculated homogenated liver tissue (Table 4). For the design of the m-PCR, they assayed in parallel the *vapA* primers of Hiney et al. (1992) and those of the *fstA* of Beaz-Hidalgo et al. (2008), showing that the latter had better specificity as none of the 9 strains analyzed showed any false negative reactions or non-specific amplifications. False negatives were, however, obtained for 2 of the 9 *A. salmonicida* isolates (2.2%) using the primers of Hiney et al. (1992). The failure of some *A. salmonicida* isolates to amplify with the latter primers agrees with results obtained by Byers et al. (2002a), who reported no amplification in 23 of the 308 isolates (7.5%) examined with these primers.

For monitoring the presence of *A. salmonicida, Francisella piscicida* and *V. anguillarum,* considered the three most important pathogens in cultured Atlantic cod, Kulkarni et al. (2010) developed an m-PCR. For detecting *A. salmonicida* they used the primers of the *gyrB* gene designed by Beaz-Hidalgo et al. (2008). The m-PCR was specific for the detection of the single strain of *A. salmonicida* tested with no cross-reaction with the other pathogens tested simultaneously and had a good detection limit when assayed in bacterial suspensions (Table 4).

5. Impact of climate change in *Aeromonas* infections

Several studies have reported that climate change can affect the aquaculture sector and its production, as it may increase the vulnerability of cultured fish to diseases due to an increase in water temperature and/or to a decrease in water quality (Alborali, 2006; Karvonen et al., 2010; Marcogliese, 2008; Marcos-López et al., 2010; Mohanty et al., 2010; Tam et al., 2011). It is well known that small changes in water temperature alter both the fish metabolism and physiology that may have consequences for their growth, fecundity or feeding behaviour (Alborali, 2006; Marcogliese, 2008; Mohanty et al., 2010). An increase in water temperature causes fish to suffer thermal stress, making them more susceptible to infections (by opportunistic pathogens such as *Aeromonas* spp.) and degrading their environmental habitat, lowering oxygen concentrations and altering the levels of nutrients (Alborali, 2006; Karvonen et al., 2010; Marcogliese, 2008; Marcos-López et al., 2010; Tam et al., 2011). However, it also has to be considered that at a higher water temperature many bacteria replicate at a higher rate and they might therefore be more abundant, favouring the spread of infectious diseases (Marcos-López et al., 2010). Motile opportunistic *Aeromonas* species have an optimal growth temperature under laboratory conditions of 25-30°C, and the ideal water temperature for *A. salmonicida* to survive is within the range 12.8°C to 21.1°C (Tam et al., 2011 and references therein). Classically, outbreaks of *Aeromonas* septicaemias and furunculosis are linked to a rise in temperature, usually occurring during spring and summer (Tam et al., 2011). Increased temperature may also lengthen the transmission

Target region or gene (bp)	Primer set- Sequence of oligonucleotides	Conditions	Reference
vapA (421)	AP-1- 5´-GGC TGA TCT CTT CAT CCT CAC CC-3´ AP-2- 5´-CAG AGT GAA ATC TAC CAG CGG TGC-3´	94°C 1s 57°C 25s 73°C 25s (x 35)	Gustafson et al. (1992)
Unknown region[1] (423)	PAAS-1- 5´-CGT TGG ATA TGG CTC TTC CT-3´ PAAS-2- 5´-CTC AAA ACG GCT GCC TAC CA-3´	94°C 1min 55°C 1min 72°C 1min (x 30)	Hiney et al. (1992)
Unknown region (512)	MIY-1- 5´-AGC CTC CAC GCG CTC ACA GC-3´ MIY-2- 5´-AAG AGG CCC CAT AGT GTG GG-3´	94°C 30s 60°C 30s 72°C 1min (x 30) 72°C 5min	Miyata et al. (1996)
IS sequence ISasa4 (749)	ISasa4F- 5´-CCT GCA CCG CCT CAT TTC TC-3´ ISasa4R- 5´-GAA AAC CCA GTG ATC TGA GC-3´	94°C 2.5min 94°C 30s 67°C 30s 72°C 30s (x 30) 72°C 9.5min	Nilsson et al. (2006)
gyrB (422)	Asg1- 5´-TGG CAT GGA ACA TTC CTC CT-3´ Asg2- 5´-GTC GCC TGC TTT TTC CAG CA-3´	95°C 3min 95°C 30s 57°C 30s 72°C 1min (x 40) 72°C 5min	Beaz-Hidalgo et al. (2008)[2]
fstA (760)	Fer3- 5´-CGG TTT TGG CGC AGT GAC G-3´ Fer4- 5´- AGG CGC TCG GGT TGG CTA TCT-3´	92°C 3min 92°C 1min 60°C 1min 72°C 1min (x 30) 72°C 5min	Beaz-Hidalgo et al. (2008)

[1] Blast analysis we carried out revealed that it shares 99% of similarity with the gene *mobA* (accession number: AJ508382), which is a mobilization protein of 1263 bp localized in the plasmid PAsa1 of *A. salmonicida* subsp. *salmonicida* (strain JF2267). [2]Amplification conditions were not indicated in that study so those indicated were from Kulkarni et al. (2010). All these methods detected typical and atypical strains except Miyata et al. (1996) that detected only typical and Nilsson et al. (2006) that detected only atypical.

Table 5. Primers and conditions of the most commonly used PCR methods for the detection of typical and atypical *A. salmonicida* strains.

season, leading to a higher prevalence of the disease and to more widespread epidemics (Karvonen et al., 2010). There might also be ecological alterations between species or strains, resulting in the emergence of new pathogenic strains affecting a wider range of hosts (Karvonen et al., 2010; Marcos-López et al., 2010). Furthermore, bacteria might show a greater virulence, for instance *A. hydrophila* has shown a greater virulence in largemouth bass (*Micropterus salmoides*) at warmer temperatures because of either reduced resistance of the host or to an increased expression of virulence factors (Marcogliese, 2008 and references therein).

In a recent study, Tam et al. (2011) carried out a regional impact assessment of climate change in relation to furunculosis investigating fish populations in two lakes in Canada.

Lake water temperatures from 1963 to 2001 were used for the model to project future lake temperatures (2011-2100) considering different scenarios and calculating the vertical and surface water temperatures over different time scales. They recognized a significant rise in air temperature since 1963 and in the mid 1990s they detected the occurrence of furunculosis when the mean air temperature was 12.8°C and summer mean water temperature was 15°C. Between 2011 and 2100 they predict that the summer mean water temperatures will rise from 15-15.5°C to 16.5-17.4°C, conditions being within the range of the survival temperatures of *A. salmonicida* (12.8-21.1°C). Furthermore, they noticed that the estimated range of temperature between 16.5 and 17.4°C is relatively near to the 18°C at which virulence of *A. salmonicida* is better expressed (Daher et al., 2011). Tam et al., (2011) considered that the effects of climate change might also be transferred to other anthropogenic impacts, such as contamination that impacts on water quality. This is another factor that might favour the development of furunculosis.

In conclusion, we could say that there is a general agreement that as a consequence of global warming, water temperature will increase and as a result endemic diseases like furunculosis will become more prevalent and more difficult to control in immunodepressed fish populations (Marcos-López et al., 2010). In the long term, aquaculture must respond to climate change by minimizing discharges into water ecosystems, try to mitigate the negative impacts of climate change in water quality and avoid fish crowding to minimize disease transmission (Marcos-López et al., 2010).

6. Conclusions and perspectives

Classically the species of *Aeromonas* implicated in fish disease that have been considered important in ichthyopathology were *A. salmonicida* and *A. hydrophila*. However, recently this panorama of species has expanded with the discovery of new species like *A. piscicola* and *A. tecta*, which have been isolated mainly from diseased salmonids and turbot. These new species might have an important role in fish pathology that needs to be explored in the future. Other species such as *A. veronii* or *A. sobria* seem to have a specific role in the pathology of catfish and trout, respectively. All these species should have remained masked under *A. hydrophila* when only biochemical identification methods were applied. Therefore these methods should be avoided as they provide results that misrepresent the real prevalence and/or diversity of the species.

The species *A. salmonicida* includes a broad diversity of strains, some able and others unable to produce pigment or be motile under laboratory conditions. This heterogeneous behaviour has led to the introduction of the terms 'typical' and 'atypical' strains (subspecies different from *salmonicida*), which in our view is very confusing. However, strains of both groups are known to cause furunculosis and ulcerative diseases in a variety of fish hosts, in which they produce similar clinical characteristics. Furthermore, there has been evidence over the years that the identification of the strains of *A. salmonicida* as belonging to the different subspecies (other than *salmonicida*) is both phenotypically and genetically complex. Therefore, it is probably time to realize that the separation of this species into different subspecies does not fulfil the aim of helping to clarify the identity of the isolates on the basis of stable phenotypic and genetic characters, but just makes the situation more difficult and confusing. We believe that after genetic confirmation, it is probably better to avoid the use of the terms 'typical' and 'atypical' and to refer to the strains simply as *A. salmonicida*.

In ichthyopathology, correct identification is essential for determining the true etiology of the disease during outbreaks at aquaculture facilities and this is the basis for establishing adequate treatment and prevention programmes. Fast and reliable detection of *Aeromonas* is a key element to minimise the impact of the infection. Moreover, the continuous monitoring of *Aeromonas* in fish farms (both the water and the fish) is needed because these bacteria are autochthonous of the aquatic environment and can act as opportunistic pathogens.

Phenotypic methods are unreliable for identifying *Aeromonas* species. Therefore it is necessary to direct efforts towards the use of suitable and reliable molecular techniques.

There is a broad panorama of PCR-based methods developed to detect *A. salmonicida* in fish tissue (mucus, blood, and other tissues) that have shown to provide a good specificity and the tendency in the future will be to continue using m-PCR methods that will enable to screen several pathogens simultaneously. Sequencing the housekeeping genes (i.e. *rpoD* and *gyrB*) of the strains recovered from fish have proven to be useful for identifying the *Aeromonas* species and its routine use will revert on the clarification of the diversity of species involved in fish disease.

Also more knowledge will be gathered in the future from the complete genomes of both bacteria and the infected host fish. In the genus *Aeromonas* there is only one complete genome of the 3 available belonging to a strain recovered from diseased fish (trout) of the species *A. salmonicida* (strain A449). It can be expected that soon other ones will be available. These genomes can provide important information about the expressed genes in the host, in response to vaccination or infection that can be useful for selecting resistant fish populations in the future.

It is clear that one of the predicted effects of climate change will be an increase in the water temperature, a tendency that has already been observed and that can change the fish's immunocompetence and susceptibility to disease and affect fisheries and aquaculture. There is evidence to suggest that emergence, distribution and transmission of many pathogenic bacteria like *Aeromonas* will increase under the effect of global warming but other anthropogenic impacts such as those derived from water contamination might also be important drivers in exacerbating the problem. Few studies have tried to predict the impact of climate change on fish *Aeromonas* infections and therefore this needs to be further explored in order to search for timely corrective measures that can be implemented to counter the effects of global warming.

7. Acknowledgments

The authors thank Dr. Jesús López Romalde, Dr. Juan Luis Barja and Dr. Alicia Estevez Toranzo from the University of Santiago de Compostela for providing information and the *Aeromonas* fish strains used in some of our articles reported in this chapter.

8. References

Abbott, S.L.; Cheung, W.K. & Janda, J.M. (2003) The genus *Aeromonas*: biochemical characteristics, atypical reactions, and phenotypic identification schemes. *Journal of Clinical Microbiology* 41, pp. 2348-2357.

Abbott, S.L.; Cheung, W.K.; Kroske-Bystrom, S.; Malekzadeh, T. & Janda, J.M. (1992). Identification of *Aeromonas* strains to the genospecies level in the clinical laboratory. *Journal of Clinical Microbiology* 30, pp. 1262-1266.

Alborali, L. (2006) Climatic Variations Related to Fish Diseases and Production *Veterinary Research Communications*, 30, pp. 93–97.

Alperi, A., Figueras, M.J.; Inza, I. & Martínez-Murcia, A.J. (2008). Analysis of 16S rRNA gene mutations in a subset of *Aeromonas* strains and their impact in species delineation. *International Microbiology* 11, pp. 185-194.

Altinok, I.; Capkin, E. & Kayis, S. (2008). Development of multiplex PCR assay for simultaneous detection of five bacterial fish pathogens. *Veterinary Microbiology* 131, pp. 332-338.

Austin, B.; Austin, D.A. (2007). *Bacterial fish pathogens, disease of farmed and wild fish* (4th edition) Springer-Praxis, ISBN 978-1-4020-6068-7, Chichester, UK.

Balcázar, J.L.; Vendrell, D.; de Blas, I.; Ruiz-Zarzuela, I.; Gironés, O. & Múzquiz, J.L. (2007). Quantitative detection of *Aeromonas salmonicida* in fish tissue by real-time PCR using self-quenched, fluorogenic primers. *Journal of Medical Microbiology* 56, pp. 323-328.

Beaz-Hidalgo, R.; Alperi, A.; Buján, N.; Romalde, J.L. & Figueras, M.J. (2010). Comparison of phenotypical and genetic identification of *Aeromonas* strains isolated from diseased fish. *Systematic and Applied Microbiology* 33, pp. 149-153.

Beaz-Hidalgo, R.; Alperi, A.; Figueras, M.J. & Romalde, J.L. (2009). *Aeromonas piscicola* sp. nov., isolated from diseased fish. *Systematic and Applied Microbiology* 32, pp. 471-479.

Beaz-Hidalgo, R.; Magi, G.E.; Balboa, S.; Barja, J.L. & Romalde, J.L. (2008). Development of a PCR protocol for the detection of *Aeromonas salmonicida* in fish by amplification of the *fstA* (ferric siderophore receptor) gene. *Veterinary Microbiology* 128, pp. 386-394.

Bernoth, E.M.; Ellis, A.; Midtlyng, P.; Olivier, G. & Smith, P. (1997). *Furunculosis Multidisciplinary Fish Disease Research* (1st edition), Academic Press, ISBN 0-12-093040-4.

Borrell, N.; Acinas, S.G.; Figueras, M.J. & Martínez-Murcia, A.J. (1997). Identification of *Aeromonas* clinical isolates by restriction fragment length polymorphism of PCR-amplified 16S rRNA genes. *Journal of Clinical Microbiology* 35, pp. 1671-1674.

Borrell, N., Figueras, M.J. & Guarro, J. (1998). Phenotypic identification of *Aeromonas* genomospecies from clinical and environmental sources. *Canadian Journal of Microbiology* 44, pp. 7-12.

Byers, H.K.; Gudkovs, N. & Crane, M.S. (2002a). PCR-based assays for the fish pathogen *Aeromonas salmonicida* I. Evaluation of three PCR primer sets for detection and identification. *Diseases of Aquatic Organisms* 49, pp. 129-138.

Byers, H.K.; Cipriano, R.C.; Gudkovs, N. & Crane, M.S. (2002b). PCR-based assays for the fish pathogen *Aeromonas salmonicida* II. Further evaluation and validation of three PCR primer sets with infected fish. *Diseases of Aquatic Organisms* 49, pp. 139-144.

Chacón, M.R.; Castro-Escarpulli, G.; Soler, L.; Guarro, J. & Figueras, M.J. (2002). A DNA probe specific for *Aeromonas* colonies. *Diagnostic Microbiology and Infectious Disease* 44, pp. 221–225.

Daher, R.K.; Filion, G.; Tan, S.G.; Dallaire-Dufresne, S.; Paquet, V.E, & Charette, S.J. (2011). Alteration of virulence factors and rearrangement of pAsa5 plasmid caused by the growth of *Aeromonas salmonicida* in stressful conditions. *Veterinary Microbiology* 152, pp. 353-360.

Fehr, D.; Casanova, C.; Liverman, A.; Blazkova, H.; Orth, K.; Dobbelaere, D.; Frey, J. & Burr, S.E. (2006). AopP, a type III effector protein of *Aeromonas salmonicida*, inhibits the NF-kappaB signalling pathway. *Microbiology* 152, pp. 2809-28018.

Figueras, M. J. (2005). Clinical relevance of *Aeromonas*. *Reviews in Medical Microbiology* 16, pp. 145-153.

Figueras, M.J.; Alperi, A.; Saavedra, M.J.; Ko, W.C.; Gonzalo, N.; Navarro, M. & Martínez-Murcia A (2009). Clinical relevance of the recently described *Aeromonas aquariorum*. *Journal of Clinical Microbiology* 47, pp. 3742-3746.

Figueras, M.J.; Alperi, A.; Beaz-Hidalgo, R.; Stackebrandt, E.; Brambilla, E.; Monera, A. & Martínez-Murcia, A.J. (2011a). *Aeromonas rivuli* sp. nov., isolated from the upstream region of a karst water rivulet in Germany. *International Journal of Systematic and Evolutionary Microbiology* 61, pp. 242-248.

Figueras, M.J.; Beaz-Hidalgo, R.; Collado, L. & Martínez-Murcia, A.J. (2011b) Point of view on the recomendations for new bacterial species description and their impact on the genus *Aeromonas* and *Arcobacter*. *The Bulletin of Bergey´s International Society for Microbial Systematics* 2, part 1, pp. 1-16.

Figueras, M.J.; Soler, L.; Chacón, M.R.; Guarro, J. & Martínez-Murcia, A.J. (2000). Use of restriction fragment length polymorphism of the PCR-amplified 16S rRNA gene for the identification of *Aeromonas* spp. *Journal of Clinical Microbiology* 38, pp. 2023-2025.

Godoy, M.; Gherardelli, V.; Heisinger, A.; Fernández, J.; Olmos, P.; Ovalle, L.; Ilardi, P. & Avendaño-Herrera, R. (2010). First description of atypical furunculosis in freshwater farmed Atlantic salmon, *Salmo salar* L., in Chile. *Journal of Fish Diseases* 33, pp. 441-449.

Goldschmidt-Clermont, E.; Hochwartner, O.; Demarta, A.; Caminada, A.P. & Frey, J. (2009). Outbreaks of an ulcerative and haemorrhagic disease in Arctic char *Salvelinus alpinus* caused by *Aeromonas salmonicida* subsp. *smithia*. *Diseases of Aquatic Organisms* 86, pp. 81-86.

González, S.F.; Krug, M.J.; Nielsen, M.E.; Santos, Y. & Call, D.R. (2004). Simultaneous detection of marine fish pathogens by using multiplex PCR and a DNA microarray. *Journal of Clinical Microbiology* 42, pp. 1414-1419.

Goodwin, A.E. & Merry, G.E. (2009). Are all koi ulcer cases associated with infection by atypical *Aeromonas salmonicida*? Polymerase chain reaction assays of koi carp skin swabs submitted by hobbyists. *Journal of Aquatic Animal Health* 21, pp. 98-103.

Gudmundsdóttir, B.K. & Björnsdóttir, B. (2007). Vaccination against atypical furunculosis and winter ulcer disease of fish. *Vaccine* 25, pp. 5512-5523.

Gustafson, C.E.; Thomas, C.J, & Trust, J.T. (1992). Detection of *Aeromonas salmonicida* from fish by using polymerase chain reaction amplification of the virulence surface array protein gene. *Applied and Environmental Microbiology* 58, pp. 3816-3825.

Han, H.J.; Kim, D.Y.; Kim, W.S.; Kim, C.S.; Jung, S.J.; Oh, M.J. & Kim, D.H. (2011). Atypical *Aeromonas salmonicida* infection in the black rockfish, *Sebastes schlegeli* Hilgendorf, in Korea. *Journal of Fish Diseases* 34, pp. 47-55.

Hiney, M.; Dawson, M.T.; Heery, D.M.; Smith, P.R.; Gannon, F. & Powell, R. (1992). DNA probe for *Aeromonas salmonicida. Applied and Enviromental Microbiology* 58, pp. 1039-1042.

Hiney, M. & Olivier, G. (1999) Furunculosis (*Aeromonas salmonicida*). in, *Fish Diseases and Disorders, Vol. 3. Viral, Bacterial and Fungal Infections.* PTK Woo and DW Bruno (Eds), pp. 341-425. CABI Publishing, ISBN 9781845935542, New York, USA.

Høie, S.; Heum, M. & Thoresen, O.F. (1997). Evaluation of a polymerase chain reaction-based assay for the detection of *Aeromonas salmonicida* subsp. *salmonicida* in Atlantic salmon *Salmo salar. Diseases of Aquatic Organisms* 30, pp. 27-35.

Janda, J.M. & Abbott, S.L. (2010). The genus *Aeromonas*, taxonomy, pathogenicity, and infection. *Clinical Microbiology Reviews* 23, pp. 35-73.

Joseph, S.W. & Carnahan, A. (1994). The isolation, identification and systematics of the motile *Aeromonas* species. Ann Rev Fish Dise 4, pp. 315-343.

Jun, J.W.; Hyung, K.J.; Gomez, D.K.; Choresca, C.H.; Han, J.E.; Shin, S.P. & Chang, P.S. (2010). Occurrence of tetracycline-resistant *Aeromonas hydrophila* infection in Korean cyprinid loach (*Misgurnus anguillicaudatus*) *African Journal of Microbiology Research* 4, pp. 849-855.

Karvonen, A.; Rintamäki, P.; Jokela, J. & Valtonen, E.T. (2010). Increasing water temperature and disease risks in aquatic systems, climate change increases the risk of some, but not all, diseases. *International Journal for Parasitology* 40, pp. 1483-1488.

Kingombe, C.I.; D'Aoust, J.Y.; Huys, G.; Hofmann, L.; Rao, M. & Kwan, J. (2010). Multiplex PCR method for detection of three *Aeromonas* enterotoxin genes. *Applied and Environmental Microbiology* 76, pp. 425-433.

Kozińska, A. (2007) Dominant pathogenic species of mesophilic aeromonads isolated from diseased and healthy fish cultured in Poland. *Journal of fish diseases* 30, pp. 293-301.

Kozińska, A.; Figueras, M.J.; Chacón, M.R. & Soler, L. (2002). Phenotypic characteristics and pathogenicity of *Aeromonas* genomospecies isolated from common carp (*Cyprinus carpio* L.) *Journal of Applied Microbiology* 93, pp. 1034-1041.

Kulkarni, A.; Caipang, C.M.; Brinchmann, M.F. & Kiron, V. (2010). Simultaneous detection of pathogens causing francisellosis, furunculosis and vibriosis in Atlantic cod by multiplex polymerase chain reaction. *Aquaculture Research* 41, pp. 1533-1538.

Kulkarni, A.; Caipang, C.M.; Brinchmann, M.F.; Korsnes, K. & Kiron, V. (2009). Loop-mediated isothermal amplification, an assay for the detection of atypical furunculosis caused by *Aeromonas salmonicida* in Atlantic cod, *Gadus morhua. Journal of Rapid Methods and Automation in Microbiology* 17, pp. 476-489.

Küpfer, M.; Kuhnert, P.; Korczak, B.M.; Peduzzi, R. & Demarta, A. (2006). Genetic relationships of *Aeromonas* strains inferred from 16S rRNA, *gyrB* and *rpoB* gene sequence. *International Journal of Systematic and Evolutionary Microbiology* 56, pp. 2743-2751.

Lamy, B. (2011). Development and evaluation of the use of a Multilocus Sequence Scheme for the whole genus *Aeromonas*. 10th International Symposium on *Aeromonas* and *Plesiomonas*. Galvestone, Texas, USA, May 2011.

Li, Y. & Cai, S.H. (2011). Identification and pathogenicity of *Aeromonas sobria* on tail-rot disease in juvenile tilapia *Oreochromis niloticus*. *Current Microbiology* 62, pp. 623-627.

Lund, V. & Mikkelsen, H. (2004). Genetic diversity among A-proteins of atypical strains of *Aeromonas salmonicida*. *Diseases of Aquatic Organisms* 61, pp. 257-262.

Marcos-López, M.; Gale, P.; Oidtmann, B.C. & Peeler, E.J. (2010). Assessing the impact of climate change on disease emergence in freshwater fish in the United Kingdom. *Transboundary and Emerging Diseases* 57, pp. 293-304.

Marcogliese, D.J. (2008). The impact of climate change on the parasites and infectious diseases of aquatic animals. *Revue Scientifique et Technique-Office International des Epizooties* 27, pp. 467-484.

Martin-Carnahan, A. & Joseph, S.W. (2005). Family I. *Aeromonadaceae* Colwell, MacDonell and DeLey 1986. In *Bergey's Manual of Systematic Bacteriology*, 2nd Ed., vol. 2. Brennan DJ, Krieg NR, Staley JT, Garrity GN (Eds.), pp.556–578, Springer-Verlag, ISBN 0-387-95040-0, New York, USA.

Martínez-Murcia, A.J.; Benlloch, S. & Collins, M.D. (1992). Phylogenetic interrelationships of members of the genera *Aeromonas* and *Plesiomonas* determined by 16S ribosomal DNA sequencing, lack of congruence with results of DNA-DNA hybridizations. *International Journal of Systematic Bacteriology* 42, pp. 412-421.

Martínez-Murcia, A.; Morena, A.; Saavedra, M.J.; Ocina, R.; López-Álvarez, M.; Lara, E. & Figueras, M.J. (2011). Multilocus Phylogenetic Analysis of the Genus *Aeromonas*. *Systematic and Applied Microbiology* 34, pp. 189-199.

Martínez-Murcia, A.J.; Soler, L.; Saavedra, M.J.; Chacón, M.R.; Guarro, J.; Stackebrandt, E. & Figueras, M.J. (2005). Phenotypic, genotypic, and phylogenetic discrepancies to differentiate *Aeromonas salmonicida* from *Aeromonas bestiarum*. *International Microbiology* 8, pp. 259-269.

Martino, M.E.; Fasolato, L.; Montemurro, F.; Rosteghin, M.; Manfrin, A.; Patarnello, T.; Novelli, E. & Cardazzo, B. (2011). Determination of microbial diversity of *Aeromonas* strains on the basis of multilocus sequence typing, phenotype, and presence of putative virulence genes. *Applied and Environmental Microbiology* 77, pp. 4986-5000.

Miñana-Galbis, D.; Urbizu-Serrano, A.; Farfán, M.; Fusté, M.C. & Lorén, J.G. (2009). Phylogenetic analysis and identification of *Aeromonas* species based on sequencing of the *cpn60* universal target. *International Journal of Systematic and Evolutionary Microbiology* 59, pp. 1976-1983.

Miyata, M.; Inglis, V. & Aoki, T. (1996) Rapid identification of *Aeromonas salmonicida* subspecies *salmonicida* by the polymerase chain reaction. *Aquaculture* 141, pp. 13-24.

Mohanty, B.; Mohanty, S.; Sahoo, J. & Sharma, A. (2010). Climate Change, Impacts on Fisheries and Aquaculture. In *Climate Change and Variability*, Suzanne Simard (Ed.), ISBN, 978-953-307-144-2, InTech, Retrieved from, http,//www.intechopen.com/articles/show/title/climate-change-impacts-on-fisheries-and-aquaculture.

Mooney, J.; Powell, E.; Clabby, C. & Powell, R. (1995). Detection *of Aeromonas salmonicida* in wild Atlantic salmon using a specific DNA probe test. *Diseases of Aquatic Organisms* 21, pp. 131-135.

Morgan, J.A.W.; Rhodes, G. & Pickup, R.W. (1993). Survival of non-culturable *Aeromonas salmonicida* in lake water. *Applied and Environmental Microbiology* 59, pp. 874-880.

Nam, I.Y. & Joh, K. (2007). Rapid detection of virulence factors of *Aeromonas* isolated from a trout farm by hexaplex-PCR. *Journal of Microbiology* 45, pp. 297-304.

Nawaz, M.; Sung, K.; Khan, S.A.; Khan, A.A. & Steele, R. (2006). Biochemical and molecular characterization of tetracycline-resistant *Aeromonas veronii* isolates from catfish. *Applied and Environmental Microbiology* 27, pp. 6461–6466.

Nhung, P.H.; Hata, H.; Ohkusu, K.; Noda, M.; Shah, M.M.; Goto, K. & Ezaki, T. (2007). Use of the novel phylogenetic marker *dnaJ* and DNA-DNA hybridization to clarify interrelationships within the genus *Aeromonas*. *International Journal of Systematic and Evolutionary Microbiology* 57, pp. 1232-1237.

Nielsen, M.E.; Høi, L.; Schmidt, A.S.; Qian, D.; Shimada, T.; Shen, J.Y. & Larsen, J.L. (2001) Is *Aeromonas hydrophila* the dominant motile *Aeromonas* species that causes disease outbreaks in aquaculture production in the Zhejiang Province of China? *Diseases of Aquatic Organism* 46, pp. 23-29.

Nilsson, W.B.; Gudkovs, N. & Strom, M.S. (2006). Atypical strains of *Aeromonas salmonicida* contain multiple copies of insertion element ISAsa4 useful as a genetic marker and a target for PCR assay. *Diseases of Aquatic Organisms* 70, pp. 209-217

Noga, E.J. (2010). *Fish Diseases* (2nd edition) Willey-Blackwell, ISBN 978-0-8138-0697-6, Singapore.

Notomi, T.; Okayama, H.; Masubuchi, H.; Yonekawa, T.; Watanabe, K.; Amino, N. & Hase, T. (2000). Loop-mediated isothermal amplification of DNA. *Nucleic Acids Research* 28, pp. E63.

Oakey, H.; Gibson, L.F. & George, A.M. (1998). Co-migration of RAPD-PCR amplicons from *Aeromonas hydrophila*. *FEMS Microbiology Letters* 164, pp. 35–38.

O'Brien, D.; Mooney, J.; Ryan, D.; Powell, E.; Hiney, M.; Smith, P.R. & Powell, R. (1994) Detection of *Aeromonas salmonicida,* causal agent of furunculosis in salmonid fish, from the tank effluent of hatchery-reared Atlantic salmon smolts. *Applied and Environmental Microbiology* 60, pp. 3874-3877.

Onuk, E.E.; Ciftci, A.; Findik, A. & Durmaz, Y. (2010). Development and evaluation of a multiplex PCR assay for simultaneous detection of *Flavobacterium psychrophilum, Yersinia ruckeri* and *Aeromonas salmonicida* subsp. *salmonicida* in culture fisheries. *Journal of Veterenary Science* 11,pp. 235-241.

Park, T.S.; Oh, S.H.; Lee, E.Y.; Lee, T.K.; Park, K.H.; Figueras, M.J. & Chang, C.L. (2003). Misidentification of *Aeromonas veronii* biovar sobria as *Vibrio alginolyticus* by the Vitek system. *Letters of Applied Microbiology* 37, pp. 349-353.

Pavan, M.E.; Abbott, S.L.; Zorzópulos, J. & Janda, J.M. (2000). *Aeromonas salmonicida* subsp. *pectinolytica* subsp. *nov.*, a new pectinase-positive subspecies isolated from a heavily polluted river. *International Journal of Systematic and Evolutionary Microbiology* 50, pp. 1119-1124.

Pridgeon, J.; Klesius, P.H.; Mu, X.; Carter, D.; Fleming, K.; Xu, D.; Srivastava, K. & Reddy, G. (2011). Identification of unique DNA sequences present in highly virulent 2009 Alabama isolates of *Aeromonas hydrophila*. *Veterinary Microbiology* 152, pp. 117-125.

Sepe, A.; Barbieri, P.; Peduzzi, R. & Demarta, A. (2008) Evaluation of *recA* sequencing for the classification of *Aeromonas* strains at the genotype level *Letters of Applied Microbiology* 46, pp. 439-444.

Shaw, J.G. (2011). Molecular chracterization of the flagellar systems of *Aeromonas* species. 10th International Symposium on *Aeromonas* and *Plesiomonas*. Galvestone, Texas, USA, May 2011.

Skugor, S.; Jørgensen, S.M.; Gjerde, B. & Krasnov, A. (2009). Hepatic gene expression profiling reveals protective responses in Atlantic salmon vaccinated against furunculosis. *BMC Genomics* 10,503.

Soler L, Figueras MJ, Chacón MR, Guarro J, Martínez-Murcia AJ (2003a). Comparison of three molecular methods for typing *Aeromonas popoffii* isolates. *Antonie van Leeuwenhoek* 83, pp. 341–349.

Soler, L.; Marco, F.F.; Vila, J.; Chacón, M.R.; Guarro, J. & Figueras, M.J. (2003b) Evaluation of two miniaturized systems microscan W/A and BBL crystal E/NF for identification of clinical isolates of *Aeromonas* spp. *Journal of Clinical Microbiology* 41, pp. 5732-5734.

Soler, L.; Yañez, M.A.; Chacón, M.R.; Aguilera-Arreola, M.G.; Catalán, V.; Figueras, M.J. & Martínez-Murcia, A.J. (2004). Phylogenetic analysis of the genus *Aeromonas* based on two housekeeping genes. *International Journal of Systematic and Evolutionary Microbiology* 54, pp. 1511-1519.

Soriano–Vargas, E.; Castro–Escarpulli, G.; Aguilera–Arreola, M.G.; Vega–Castillo, F. & Salgado–Miranda, C. (2010). Aislamiento e identificación de *Aeromonas bestiarum* a partir de carpa común de cultivo (*Cyprinus carpió* L.) procedentes de Santa María Chapa de Mota, Estado de México, México. *Veterinaria México* 41, pp. 111-115.

Stackebrandt, E.; Frederiksen, W.; Garrity, G.M.; Grimont, P.A.D.; Kämpfer, P.; Maiden, M.C.J.; Nesme, X.; Rosselló-Móra, R.; Swings, J.; Trüper, H.G.; Vauterin, L.; Ward, A.C. & Whitman, W.B. (2002). Report of the ad hoc committee for the reevaluation of the species definition in bacteriology. *Internationa Journal of Systematic and Evolutionary Microbiology 52*, pp. 1043–1047.

Tam, B.; Gough, W.A. & Tsuji, L. (2011). The impact of warming on the appearance of furunculosis in fish of the James Bay region, Quebec, Canada. *Regional Environmental Change* 11, pp. 123–132.

Wiklund, T. & Dalsgaard, I. (1998). Occurrence and significance of atypical *Aeromonas salmonicida* in non-salmonid and salmonid fish hosts. *Diseases of Aquatic Organisms* 32, pp. 49-69.

Yañez, M.A.; Catalán, V.; Apraiz, D.; Figueras, M.J. & Martínez-Murcia, A.J. (2003). Phylogenetic analysis of members of the genus *Aeromonas* based on *gyrB* gene sequences. *International Journal of Systematic and Evolutionary Microbiology* 53, pp. 875-883.

5

An Overview of Virulence-Associated Factors of Gram-Negative Fish Pathogenic Bacteria

Jessica Méndez, Pilar Reimundo, David Pérez-Pascual, Roberto Navais, Esther Gómez, Desirée Cascales and José A. Guijarro
Department of Microbiology, IUBA, University of Oviedo, Oviedo, Asturias Spain

1. Introduction

Bacterial diseases are among the most important causes of losses among fish stocks in the aquaculture industry, affecting the economic development of the sector in many countries. The ability of bacteria to cause disease depends to a large extent on the expression of virulence factors, which help them to invade the host, produce pathological effects and evade host defences. The study of these factors is essential for the development of new immunoprophylactic and chemotherapeutic reagents to fight the bacterial infections, since the development of antibiotic resistance by bacteria has led to these diseases becoming one of the major problems in the sector. In the last decade, the application of *in vivo* and *in vitro* molecular techniques to fish pathogenic bacteria, together with the availability of adequate models for studying the disease, have allowed the discovery and characterization of novel virulence determinants, as well as a deeper insight into well-known pathogenic mechanisms. In reference to bacterial diseases, Gram-negative bacteria have long been recognized as one of the main problems in the aquaculture industry. They can cause systemic infections in which they invade the fish and damage internal organs or can cause external infections affecting the gills or causing fin rot and body ulcers. This review describes current understanding of the virulence factors shown to be involved in the virulence of Gram-negative bacteria causing disease in fish.

2. Bacterial adherence and colonization

2.1 Adhesins

Bacterial adherence to the host may involve either specific interactions between a receptor and a ligand or hydrophobic interactions. The receptors are usually specific carbohydrate or peptide residues on the eucaryotic cell surface and the ligands called adhesins are bacterial-surface proteins or polysaccharides.

Many pathogenic Gram-negative bacteria display long adhesive fibers, called type IV *pili*, in order to mediate cellular attachment to host tissue receptors. *Pili* are also involved in several other bacterial processes, including bacterial auto-aggregation, target tissue specificity and natural competence for DNA uptake.

The fish pathogenic *bacterium Aeromonas salmonicida* subsp. *salmonicida* contains in its genome a complete set of genes for two type IV *pilus* systems, Tap and Flp (Boyd *et al.*, 2008). *In vivo* experiments in rainbow trout (*Oncorhynchus mykiss*) showed that the Tap *pilus* contributes moderately to virulence. A *tapA* mutant constructed by allelic exchange was found to be slightly less pathogenic than wild type when delivered by intraperitoneal injection (Masada *et al.*, 2002). Boyd *et al.* (2008) showed that the Tap *pilus* also made a moderate contribution to virulence in Atlantic salmon (*Salmo salar*), while the Flp *pilus* made little or no contribution. In addition to type IV *pili*, *A. salmonicida* has also a type I or Fim *pilus* system (similar to Pap *fimbriae* of *Escherichia coli*) encoded by several genes clustered into an operon. When this operon was deleted in *A. salmonicida* A449, the virulence of this strain was not affected in direct live challenges of Atlantic salmon. However, an *ex vivo* adherence and invasion assay using freshly excised salmon gastrointestinal tract showed that, compared to the parental strain, the ability of a *fim* mutant strain to adhere to the salmon gastrointestinal tract was reduced but, once adhered, its capacity to invade was unaffected (Dacanay *et al.*, 2010).

In *A. salmonicida*, the S-layer also acts as an adhesin promoting high levels of adherence to non-phagocytic fish cell lines (Garduño *et al.*, 2000). The capsular polysaccharide is involved in the ability of *A. salmonicida* and *Aeromonas hydrophila*, the leading cause of fatal hemorrhagic septicemia in rainbow trout, to both adhere to and invade fish cell lines, being an important factor for intracellular invasion (Merino *et al.*, 1996, 1997a). The capsule also plays an important role in the pathogenicity of *Photobacterium damselae* subsp. *piscicida* (formerly *Pasteurella piscicida),* the etiological agent of pasteurellosis in fish. The induction of its expression increased the degree of virulence for fish of the non-pathogenic strains and conferred resistance to serum killing (Magariños *et al.*, 1996).

2.2 Motility

Bacteria have developed at least six different types of motility (Henrichsen, 1972). The evolutionary maintenance of motility, despite its high energetic cost, indicates its importance for the survival of bacteria in changing milieus. Additionally, motility is considered essential during host tissue colonization in pathogenic bacteria.

For instance, the fish pathogen *Listonella anguillarum* (formerly *Vibrio anguillarum*), the causal agent of vibriosis, required the presence of the *flagellum* to produce the disease when rainbow trout were infected by immersion in bacteria-containing water. However, this appendix and the bacteria's resulting motility were not required for pathogenicity following intraperitoneal injection (O'Toole *et al.*, 1996). Ormonde *et al.* (2000) suggested that active motility is essential for this *bacterium* to enter the fish host, maybe in a process driven by a fish skin and intestinal mucus chemotactic response. In this work, the authors created a strain with an intact but paralyzed *flagellum* by disruption of the *motY* gene, a flagellar motor gene. The virulence of the *motY* mutant was 750-fold lower than that of the wild type strain when the *bacterium* was supplied by immersion, using rainbow trout as the infection model; however, no loss in virulence was seen when the *motY* mutant was injected intraperitoneally. Taken together, these results suggest that motility and not the flagellin proteins are required for *L. anguillarum* to invade rainbow trout successfully (Ormonde *et al.*, 2000). In *A. hydrophila*, the motility is important for the adhesion and invasion by the bacteria of fish cell lines, as was demonstrated by Merino *et al.* (1997b). Gavín *et al.* (2002)

concluded that the lateral flagella are mainly responsible for the adhesion process in this species and also for its ability to form biofilm. However, in the case of *Yersinia ruckeri*, the causal agent of yersiniosis, a non-motile biogroup identified as *Y. ruckeri* serovar I biotype 2 has been described, which is able to cause the disease in rainbow trout (Austin *et al.*, 2003; Fouz *et al.*, 2006). In fact, Evenhuis *et al.* (2009) have demonstrated that the lack of the *flagellum* and flagellar secretion machinery do not affect *Y. ruckeri* virulence.

In general, there are few cases establishing a direct relation between fish pathogenic bacteria motility and disease development. The reason may be that mutations in motility-related genes usually involve pleiotropic phenotypes. Therefore, it is difficult to ascribe the absence of virulence exclusively to the lack of motility. However, all authors recognise the importance of motility for host tissue colonization during the first stages of the infection.

3. Growth and invasion

3.1 Iron acquisition

Iron is an essential element for most bacteria due to its participation as a cofactor in numerous biochemical cellular processes. Thus, bacteria have developed high-affinity iron-transport systems that are considered to be important factors in pathogenicity. Many bacteria are able to synthesize low molecular mass molecules, or siderophores, which bind ferric iron (Ratledge & Dover, 2000; Ratledge, 2007). These iron compounds are then recognized by specific outer membrane receptors and introduced into the cell by an energy-transduction complex named TonB system (Andrews *et al.*, 2003; Ratledge, 2007; Ratledge & Dover, 2000; Wandersman & Delepelaire, 2004). A second iron-acquisition mechanism is based on a direct interaction between the host iron-containing proteins transferrin, lactoferrin and ferritin and specific receptors on the bacterial cell surface (Butler, 2003; Ratledge, 2007). Some pathogenic bacteria are also able to acquire iron from free haem or haem proteins, such as hemoglobin or hemopexin within host tissues (Genco & Dixon, 2001; Tong & Guo, 2009).

Evidence for the role of different iron scavenging mechanisms in promoting Gram-negative bacterial infections within fish hosts have been reported in literature on several occasions. For example, the isolation and further analysis of a mutant showing impaired growth under iron-limited conditions using a Tn*4351*-mutagenesis system led to the identification of a TonB system which includes ExbB, ExbD1, ExbD2 and TonB proteins in *Flavobacterium psychrophilum*, the causative agent of cold water disease in salmonids (Álvarez *et al.*, 2008). Moreover, *in vivo* assays conducted in rainbow trout fry determined that the ExbD2 protein plays an important role in virulence of *F. psychrophilum*, since the *exbD2* mutant strain was approximately 450-fold attenuated compared to the wild strain and conferred a high level of protection after vaccination (Álvarez *et al.*, 2008).

Two different TonB systems, TonB1 and TonB2, were identified and characterized in the pathogen *Vibrio algynoliticus*, which causes vibriosis in marine fish (Wang *et al.*, 2008). Construction and further analysis of mutants in these systems led to the conclusion that they are involved in iron uptake from ferrichrome and vibrioferrin and that they are also essential for virulence, since the corresponding mutant strains showed an 11 to 25-fold increase in LD_{50} value in zebra fish (*Danio rerio*) in comparison to that of the wild type strain (Wang *et al.*, 2008). Besides, the TonB1 gene cluster was shown to specifically contribute to haemin and haemoglobin utilization in this fish pathogen (Wang *et al.*, 2008). In the same way, two different iron-regulated TonB mechanisms were identified in *L. anguillarum* (Stork *et al.*, 2004). The analysis of these systems revealed that TonB2 protein but not TonB1 is involved in the transport of the siderophores anguibactin and enterobactin and that TonB2 protein is essential for virulence of *L. anguillarum,* since the LD_{50} value for the *tonB2* mutant strain was more than 100-fold higher than that of the parental strain (Stork *et al.*, 2004).

Following suppressive subtraction hybridization experiments, the existence of a siderophore biosynthesis gene cluster in *P. damselae* subsp. *piscicida* was reported (Osorio *et al.*, 2006). Insertional mutation of an *irp1* gene included within this cluster, which codes for a putative non-ribosomal peptide synthetase, led to impaired growth under iron-limited conditions, loss of siderophore production and a 100-fold decrease in degree of virulence in turbot fingerlings (*Scophthalmus maximus*) (Osorio *et al.*, 2006).

The application of an *in vivo* expression technology system to the study of the fish pathogen *Y. ruckeri* permitted the identification of, among others, four clones involved in the biosynthesis and transport of a cathecol siderophore named ruckerbactin (Fernández *et al.*, 2004). An isogenic mutant in *rucC* (involved in ruckerbactin biosynthesis) was conducted and it was demonstrated that this strain was impaired for growth under iron-depleted conditions with respect to the wild type strain. Moreover, *in vivo* assays carried out in rainbow trout indicated that the LD_{50} value for the *rucC* mutant was 100-fold higher than that of the wild type strain (Fernández *et al.*, 2004).

The iron-uptake processes and their importance as determinants in pathogenicity have been widely studied in *L. anguillarum.* In 1980, it was demonstrated in *in vivo* challenge experiments on juvenile Coho salmon that heat-mediated curation of a 65 Kb plasmid of this bacterium, named pJM1, was correlated with an attenuation of virulence of about 3 logarithms (Crosa *et al.*, 1980). The pJM1-type plasmids encode the genes responsible for the biosynthesis and transport of the siderophore anguibactin, an iron-sequestering system that represents a major virulence factor in *L. anguillarum* (Crosa, 1989; Lemos *et al.*, 1988). One of these plasmid-encoded genes, *angR*, is involved in the regulation of both the expression of the iron transport genes *fatDCBA* and the production of the siderophore anguibactin (Chen *et al.*, 1996; Salinas *et al.*, 1989). Construction and subsequent analysis of site-directed *angR* mutants and deletion derivatives indicated that an intact AngR protein is required for full virulence and anguibactin production in *L. anguillarum* but not for regulation of iron-transport gene expression (Wertheimer *et al.*, 1999). Virulence tests carried out on juvenile trout showed that the *angR* mutation results in a dramatic attenuation of virulence of about five logarithms in *L. anguillarum* (Wertheimer *et al.*, 1999). Additional iron uptake systems have been found in this *bacterium.* For example, isolation of mutants defective in haem utilization led to the identification of the gene *huvA,* which encodes an iron-regulated outer membrane protein involved in a specific haem uptake mechanism (Mazoy *et al.*, 2003). The analysis of the *huvA* mutant strain revealed inability to grow in the presence of haem as the sole iron source as well as a decrease in the degree of virulence for turbot fingerlings in experimental infections in which fish were previously overloaded with haemin (Mazoy *et al.*, 2003).

It has been clearly demonstrated that the ability of pathogenic bacteria to scavenge iron from the fish host is of vital importance to the outcome of the disease. Thus, iron acquisition mechanisms should be deeply studied in bacterial fish pathogens and may be considered as optimal targets for the development of new antimicrobial agents in aquaculture.

3.2 Extracellular products (ECPs)

Separation of the extracellular products (ECPs) secreted by bacteria from the cell fraction has been a common strategy for studying the virulence factors of fish-pathogenic bacteria. The analysis of the pathological effects of their different components on fish can give us substantial information about host-pathogen interactions. Among these components, molecules with different activities such as haemolytic, cytolytic, proteolytic and lipolytic, etc. have been identified.

ECPs and extracellular proteases have been well studied in the genus *Aeromonas* and there are numerous reports showing their implication in the virulence of this pathogen. The species *A. hydrophila* produces extracellular substances that are capable of causing pathological effects when injected into rainbow trout and tilapia (*Tilapia nilotica*) (Allan & Stevenson, 1981; Khalil & Mansour, 1997; Santos *et al.*, 1988). Allan and Stevenson (1981) showed that the *in vivo* effect observed in rainbow trout, as well as the proteolytic and haemolytic activities, were lost when ECPs were heated. These authors also suggested haemolytic activity as a significant lethality factor.

The ECPs of *A. salmonicida* seem to play a relevant role in the pathogenesis of fish furunculosis. Ellis *et al.* (1981) showed that all the lesions associated with this disease were reproduced when ECPs were injected intraperitoneally or intramuscularly into rainbow trout. A few years later, it was indicated that the presence of protease and haemolysin activities in the ECPs of *A. salmonicida* was correlated with the development of lesions but not with the lethal toxicity of the ECPs in rainbow trout. Thus, an unidentified component of ECP was responsible for killing fish (Ellis *et al.*, 1988). Concerning the two major extracellular enzymes of *A. salmonicida*, glycerophospholipid: cholesterol acyltransferase (GCAT) and a serine protease (AspA), it has been surprisingly revealed that no major decrease in virulence in Atlantic salmon occurred when their encoding genes were mutated (Vipond *et al.*, 1998).

The role in virulence of the activities present in ECPs within the family *Vibrionaceae* has also been closely studied. In *L. anguillarum*, different exoenzymes (i.e. haemolysins, cytotoxins, and dermatotoxins) can contribute to the development of infections (Kodama *et al.*, 1984). Nevertheless, metalloproteases and undetermined low molecular weight substances are the main toxins responsible for the lethality of their ECPs (Toranzo & Barja, 1993). This was illustrated with the work of Santos *et al.* (1991) which showed that, although all *L. anguillarum* isolates tested were virulent for trout, salmon and turbot, rainbow trout was the most susceptible fish species to experimentally induced vibriosis. In contrast, the ECPs (with proteolytic, haemolytic, cytotoxic activities and permeability factors) exhibited similar lethal doses for turbot, salmon and trout. Therefore, differences in susceptibility to vibriosis were not completely due to a differential sensitivity of fish to the extracellular products of this *bacterium*. In *L. anguillarum*, a membrane-bound lytic murein transglycosylase D (*mltD*) mutant was generated and its extracellular protease activity decreased markedly together with a total loss of haemolytic activity compared with the wild type strain (Xu *et al.*, 2011). The MltD protein was characterized and it showed haemolytic, phospholipase, gelatinase and diastase activities. Surprisingly, virulence of the *mltD* mutant was enhanced compared with that of the wild type when it was inoculated intraperitoneally into zebra fish. This could be partially explained by the hypothesis that peptidoglycan (PG) fragments, released during growth, can contribute to the pathogenesis of multiple bacterial infections (Xu *et al.*, 2011).

With regards to other *Vibrionaceae* species, it has been demonstrated that the ECPs from all the strains of *P. damselae* sp. *damselae* are strongly lethal for fish (Fouz *et al.*, 1993). These ECP samples possessed low proteolytic activity but remarkable phospholipase and haemolytic activities for turbot red blood cells and were cytotoxic for fish. Finally, a correlation could be established between the levels of enzymatic and cytotoxic activities of ECPs and the degree of virulence for fish (Fouz *et al.*, 1993). Additionally, it was shown that the ECPs from 16 strains of *P. damselae* subsp. *damselae* that were strongly lethal for redbanded seabream (*Pagrus auriga)* exhibited lipase, phospholipase and esterase-lipase activities among others (Labella *et al.*, 2010). They also displayed a strong cytotoxic effect on four fish cell lines, although this effect disappeared when ECPs were heated at 100°C. The virulence of the strains tested could not be related to the haemolytic activity or to the production of the toxin damselysin. Therefore, another unknown type of toxin could play an important role in the virulence mechanisms of this *bacterium* (Labella *et al.*, 2010).

There are also many works with other Gram-negative bacteria whose ECPs have been tested on their hosts. As described by Romalde & Toranzo (1993), ECPs (including proteolytic haemolytic, cytotoxic, and lipolytic activities) could play a role in the pathogenicity of *Y. ruckeri* because when injected into fish they lead to the appearance of symptoms related to yersiniosis. The same occurs with *F. psychrophilum*. A crude extracellular preparation (CEP) from a strain of this *bacterium* was capable of causing serious muscle necrosis in rainbow trout after intramuscular injection. The CEP degraded gelatin, but the addition of protease inhibitors to the CEP simultaneously terminated its ability to degrade this protein *in vitro* and to produce muscle necrosis in rainbow trout. Both effects were restored following the addition of zinc chloride to the protease inhibitor-treated CEP, suggesting that this strain of *F. psychrophilum* secretes a protein complex with zinc metalloprotease-like activity (Ostland *et al.*, 2000).

In relation to the degradation of non-proteinaceous components from the extracellular matrix of fish tissue, the activity of a chondroitin AC lyase present in *Flavobacterium columnare* could be the cause of the necrotic lesions characteristic of the columnaris disease (Suomalainen *et al.*, 2006).

3.2.1 Extracellular proteases

Production of extracellular proteolytic enzymes is a property shared by non-pathogenic and pathogenic microorganisms. These enzymes are indispensable factors in their life cycles and may be lethal to the host when produced by pathogenic bacteria (Miyoshi & Shinoda, 2000). The role of proteases in pathogenesis is not clear, but it seems that they are involved in colonization and invasion during host-pathogen interaction, apart from providing nutrients for the microorganism.

This idea is also strengthened by the findings of a wide variety of studies, such as that on *Moritella viscosa*, which causes winter ulcer disease in salmonids (Bjornsdottir *et al.*, 2009). The metallopeptidase MvP1 of this pathogen caused extensive tissue necrosis and haemorrhages at the site of injection but was non-lethal to salmon at concentrations up to 0.22 μg/g fish. The authors suggested that MvP1 could aid in the invasion and dissemination of the *bacterium* in the host by causing tissue destruction (Bjornsdottir *et al.*, 2009). The same function is attributed to the protease Yrp1 of *Y. ruckeri*, included within the serralysin metalloendopeptidase subfamily. The mutation of the *yrp1* gene caused the loss of the proteolytic activity as well as attenuation in virulence when the mutant was injected intraperitoneally into rainbow trout (Fernández *et al.*, 2002).

Additional reports relating to the activity of extracellular metalloproteases in virulence can be found. The participation of an extracellular zinc metalloprotease in the first steps of the infectious process by promoting invasion is suggested in *L. anguillarum*. Norqvist *et al.* (1990) isolated a mutant with a low level of protease activity. This mutant behaved also as an invasiveness-defective strain. When compared with wild type strain, this mutant had a 1,000-fold higher LD_{50} value after immersion infection of rainbow trout. In the work of Yang *et al.* (2007), it is also proposed that an extracellular zinc metalloprotease, EmpA, is a putative virulence factor of the fish pathogen *L. anguillarum*.

Proteolytic enzymes may play an important role in invasiveness and establishment of infection by overcoming initial host defences and by providing nutrients for cell proliferation, as has been suggested for *Aeromonas* spp. (Leung & Stevenson, 1988). The first report in which the mutation of an extracellular protease of *A. hydrophila* was related to a decrease in virulence was the work of Cascón *et al.* (2000). The protease (AhpB) hydrolyzed casein and elastin and showed a high sequence similarity to other metalloproteases. The mutation of *ahpB* resulted in 100-fold attenuation in virulence for rainbow trout. Indeed, in local infections proteases can cause necrotic or haemorrhagic tissue damage through digestion of structural components of the ground substance and form oedematous lesions through generation of an inflammatory response (Miyoshi & Shinoda, 2000). As Abolghait *et al.* (2010) indicated, intramuscular infection of goldfish (*Carassius auratus*) with wild type *A. hydrophila* led to the development of a characteristic large ulcer at the injection site while the PepO (a thermoregulated outer membrane M13 family zinc endopeptidase) deficient mutant strain lost its ulcerogenic property *in vivo*. However, this mutant strain caused a higher mortality in goldfish than the wild type *A. hydrophila*. This paradox could be partially clarified by the evidence that *pepO*-mutagenesis changed the extracellular proteome, suggesting that PepO may regulate the secretion and/or the expression of some of *A. hydrophila* virulence factors present in ECPs. What is more, metalloproteases from *A. hydrophila* could have a more harmful effect and even be lethal to the host. Extensive haemorrhages in the abdominal cavity were caused after injection with the protease obtained from the culture filtrate of this pathogen (Kanai & Wakabayashi, 1984).

Combining protease isolation and gene interruption in *A. salmonicida*, as early as 1985, Sakai showed that a protease-deficient mutant (NTG-1) lost its virulence and proteolytic activity in Sockeye salmon (*Oncorhynchus nerka*) and rainbow trout. In addition, the activity of an extracellular metallo-caseinase, AsaP1, was linked with lethal toxicity and a strong pathogenic effect (Gunnlaugsdóttir & Gudmundsdóttir, 1997) in *A. salmonicida* subsp. *achromogenes* for Atlantic salmon fingerlings. Besides, the lethal dose of an AsaP1-defective strain was 10-fold higher in Artic charr (*Salvelinus alpinus*) and 5-fold higher in Atlantic salmon than that of the wild type strain (Arnadottir *et al.*, 2009).

However, the involvement of extracellular metalloproteases in the virulence of fish pathogenic bacteria is not a general rule, given that the metalloprotease Vvp of *V. vulnificus,* is not an essential lesion factor. Thus, when the ECPs from a Vvp-defective mutant were injected into fish, similar lesions to those caused by the wild type strain appeared in eels (*Anguilla anguilla*) (Valiente *et al.*, 2008). A similar case is that of the Fpp2 proteolytic enzyme of *F. psychrophilum*, which seems not to be involved in the infection process, having a putative nutritional role (Pérez-Pascual *et al.*, 2011). In contrast, according to Zhang *et al.* (2009b), the AprX of *Pseudomonas fluorescens* (an extracellular alkaline metalloprotease of the serralysin family) could be involved in the infection process, since an *aprX* mutant strain exhibited significantly attenuated ability to disseminate and survive within Japanese flounder (*Paralichthys olivaceus*) blood and tissues. These results, together with the observation that purified recombinant AprX was highly toxic in cultured flounder gill cells, demonstrated that AprX is a virulence factor that contributes to bacterial infection.

3.2.2 Haemolysins and phospholipases

Bacterial haemolysins are cytolytic exotoxins, generally considered as important virulence factors. These toxins cause damage to erythrocytes and other cell types, such as leukocytes or neutrophils, by two different models of action which involve a pore-forming protein or a phospholipase enzyme (Rowe & Welch, 1994). However, their specific mechanism of action as well as their specific contribution to pathogenicity varies in each organism.

There are many reports about bacterial fish pathogens, especially *Vibrio* species, which possess haemolysin proteins. Thus, Rodkhum *et al.* (2005) identified four genes, *vah2, vah3, vah4* and *vah5,* which encode four haemolysins in *L. anguillarum.* These genes were cloned and the corresponding proteins subsequently purified. It was demonstrated that all of them displayed haemolytic activity against erythrocytes of rainbow trout. In addition, construction of a mutant strain for each haemolysin gene and subsequent LD_{50} assays in rainbow trout determined that the four mutant strains were less virulent than the wild type strain (Rodkhum *et al.*, 2005). Rock & Nelson (2006) characterized a haemolysin gene cluster that encodes the *L. anguillarum* haemolysin Vah1, a putative phospholisase (Plp) and a putative lactonizing lipase (LlpA). Mutation in the *plp* gene resulted in a significant increase in haemolytic activity but not in virulence. On the contrary, mutations in the *vah1* and *llpA* genes did not affect haemolytic activity. Moreover, the *vah1* mutant strain showed virulence attenuation in juvenile Atlantic salmon. The data obtained in this study suggest that *plp* constitutes a negative regulator of the haemolysin genes *vah1* and *llpA* and that *vah1* plays a role in the pathogenicity of this *bacterium*. A repeat-in-toxin (RTX) gene cluster (*rtxACHBDE*) related to haemolysis in *L. anguillarum* was identified by Li *et al.* (2008) a few years later. Haemolytic-deficient mutant strains were obtained by interrupting *vah1* and an *rtx* gene, leading to the conclusion that the *rtx* gene cluster represents a new haemolytic mechanism in *L. anguillarum.* It was proved that Vah1 and RtxA proteins displayed cytotoxic effects on Atlantic salmon kidney cells, whereas the *vah1 rtxA* double mutant strain lost this cytotoxic activity. Besides, the *rtxA* mutant strain showed reduced virulence in juvenile Atlantic salmon, suggesting that this gene is a relevant virulence factor for *L. anguillarum.*

The thermostable direct haemolysin gene (*tdh*) from *V. alginolyticus* was identified and sequenced by Cai *et al.* (2007). The encoded protein was expressed and purified in *E. coli* and it was shown that the Tdh had haemolytic activity and it was toxic for crimson snapper (*Lutjanus erythopterus*) (Cai *et al.*, 2007). In the same way, Jia *et al.* (2010) expressed and purified the thermolabile haemolysin (TLH) from *V. alginolyticus*. The experiments carried out with the protein revealed that it had both phospholipase and haemolytic activities against flounder erythrocytes. What is more, toxicity of TLH to zebra fish was demonstrated when injected intraperitoneally, evidencing the importance of this protein in the pathologic process provoked by this *bacterium*.

The *V. harveyi* haemolysin gene (*vhhA*) was overexpressed, purified and characterized by Zhong *et al.* (2006). Evident phospholipase and haemolytic activities against turbot erythrocytes were determined for this protein. In addition, the protein was proved cytotoxic for flounder gill cells as well as lethal for flounder (Zhong *et al.*, 2006). Later, Sun *et al.* (2007) carried out the construction of a *vhhA* site-directed mutant strain. It was observed that this mutant strain exhibited loss of haemolytic and phospholipase activities, together with a complete lack of virulence for turbot (*Scophthalmus maximus*). A haemolysin gene cluster, which encodes an YhlB haemolysin activation protein and a YhlA haemolysin, was identified in *Y. ruckeri* by Fernández *et al.* (2007). Insertional mutant strains of the two proteins were constructed and it was concluded that both mutant strains exhibit reduced haemolytic activity relative to that of the parental strain. What is more, LD_{50} assays carried out in rainbow trout determined that *yhlA* and *yhlB* mutant strains showed a 10-fold and 100-fold decrease in virulence respectively, which clearly implicated this system in the virulence of the *bacterium* (Fernández *et al.*, 2007).

The mutation of the *esrB* gene of *Edwardsiella tarda*, part of the two-component system EsrA-EsrB, resulted in severe attenuation in virulence but this was accompanied by significantly enhanced haemolytic activity and cell-invasion capability (Wang *et al.*, 2010b). The authors observed that a haemolysis-associated protein, EthA, was up-regulated in this mutant strain. The construction and subsequent analysis of an *ethA* mutant strain determined that it exhibited decreased capacities of internalization into epithelial papilloma of carp cells. Nevertheless, no significant differences were recorded between the LD_{50} values obtained in zebra fish and Japanese flounder for the mutant and the wild type strains, thus questioning its contribution to lethality in the fish host (Wang *et al.*, 2010b).

Several proteins have been shown to be involved in the haemolytic process in *A. hydrophila*. Thus, Li *et al.* (2011) evaluated the relationship between the presence of the aerolysin (*aerA*), cytotoxic enterotoxin (*alt*) and serine protease (*ahp*) genes, and virulence of *A. hydrophila* isolates in zebra fish. The authors conclude that the *aerA+alt+ahp+* isolates were more virulent to zebra fish than other single or two-virulence-factor combination strains. Apart from haemolysins, phospholipases constitute other important membrane-active agents in many pathogenic bacteria (Bai *et al.*, 2010). In *A. hydrophila* AH-3, the *plc* gene codes for a 65-kDa protein involved in phospholipase C activity. An insertion mutant in this gene showed a 10-fold increase in its LD_{50} in rainbow trout and mice, suggesting that Plc protein is a virulence factor (Merino *et al.*, 1999). Furthermore, Plc is also a cytotoxic factor against epithelioma papulosum of carp (*Cyprinus carpium*) monolayers and slightly haemolytic for rainbow trout erythrocytes (Merino *et al.*, 1999).

As a conclusion, it has been clearly demonstrated that ECPs from Gram-negative fish-pathogenic bacteria have a variety of damaging factors as haemolysins, cytotoxins, proteases, phospholipases, etc. which result in the appearance of symptoms of the disease and even more frequently in a toxic effect causing the death of fish.

3.3 Protein secretion systems

Protein secretion is involved in different processes in the bacterial life cycle, including organelle biogenesis, nutrient acquisition and virulence-factor expression. In Gram-negative bacteria, where secretion involves translocation across inner and outer membranes, up to six different secretion pathways for protein export to the extracellular environment have been identified. Some secreted proteins are exported across the inner and outer membranes in a single step via the type I, type III, type IV or type VI pathways (Tseng *et al.*, 2009). Other proteins are first exported into the periplasmic space via the universal Sec or two-arginine (Tat) pathways and then translocated across the outer membrane via the type II, type V or less commonly, the type I or type IV machinery (Tseng *et al.*, 2009). In Gram-negative fish pathogenic bacteria we can find several examples in which all of these secretion systems, with the exception of the Sec and the general type II pathways, have been related to virulence.

The bacterial twin-arginine translocation system (Tat) is involved in the translocation of proteins in a folded state using a proton gradient as an energy source (Müller, 2005). In *V. alginolyticus,* the Tat pathway plays pleiotropic roles in growth, motility and secretion of some virulent factors such as the extracellular alkaline serine protease (Asp), an important exotoxin in this *bacterium* as was previously indicated. Fish infection and cytotoxicity assays showed that the Tat system is also required for the virulence of this bacterium in zebra fish and against an epithelioma papulosum cyprinid cell line (He *et al.*, 2011).

The type I or ATP-binding cassette (ABC) transporter pathway exports substrates such as toxins, proteases and lipases (Binet *et al.*, 1997) directly across the inner and outer membranes without periplasmic intermediates. The Yrp1 protease of *Y. ruckeri* (described in section 3.2.1) is secreted by an ABC protein secretion system composed of three genes termed *yrpD, yrpE* and *yrpF,* and a protease inhibitor *inh.* Site-directed insertion mutations into the *yrpE* gene led to the loss of protease activity and attenuation in virulence when bacteria were injected intraperitoneally into rainbow trout (Fernández *et al.*, 2002).

The type III pathway is composed of a complex protein structure spanning both the inner and the outer membranes and it is used exclusively by pathogenic bacteria to deliver virulence factors into host cells, directly interfering with and altering host processes. In fish, two species of the genus *Aeromonas (A. salmonicida* and *A. hydrophila)* have been reported to have a functional type III secretion system involved in virulence (Burr *et al.*, 2002; Yu *et al.*, 2004). In *A. salmonicida,* a mutation in the *ascV* gene that encodes an inner membrane component of the type III secretion apparatus results in the lack of toxicity against RTG-2 rainbow trout gonad cells (Burr *et al.*, 2002). In *A. hydrophila* AH-1, insertional inactivation of two of the type III secretion system genes (*aopB* and *aopD*) led to decreased cytotoxicity in carp epithelial cells, increased phagocytosis and reduced virulence in blue gourami (*Trichogaster trichopterus* Pallas) (Yu *et al.*, 2004). *E. tarda* also has a type III secretion system that is essential for pathogenesis formed by three genes *eseD, eseB* and *eseC.* Infection experiments in fish showed that the *eseD* mutant exhibited slower proliferation and a 10-fold decrease in virulence in fish (Wang *et al.*, 2010a). Finally, in *V. alginolyticus,* a type III secretion system (T3SS) is required to cause rapid death of infected carp fish cells (Zhao *et al.*, 2010).

The type IV secretion system is related to the transport of macromolecules, such as proteins and DNA. It allows the secretion of nucleoprotein complexes, multi-subunit toxins or monomeric proteins. Recently, this pathway has been divided into two subgroups: type IVa, similar to the VirB secretion system of *Agrobacterium tumefaciens* and type IVb, assembled from Tra homologues of the Incl ColIb-P9 plasmid of *Shigella flexneri* (Sexton & Vogel, 2002). Proteins using this pathway can be secreted into the extracellular milieu or directly into a host cell. In *Y. ruckeri*, the *traHIJKCLMN* operon, which encodes a putative type IVb secretion system, is involved in the virulence of the *bacterium* (Méndez *et al.*, 2009). LD_{50} determinations with rainbow trout fry indicated that a mutation in the *traI* gene resulted in virulence attenuation. It is suggested that this system contributes to bacterial pathogenicity through the secretion of some effector molecules into the host cells (Méndez *et al.*, 2009).

The type V secretion system presents the largest family of protein-translocating outer membrane porins in Gram-negative bacteria and the simplest secretion apparatus (Yen *et al.*, 2002). Proteins using this system are translocated across the outer membrane via a transmembrane pore formed by a self-encoded β-barrel structure. The Pfa1 autotransporter of the fish pathogen *P. fluorescens* is associated with virulence since a mutation in this gene significantly attenuates the virulence of the *bacterium* and impairs its ability in biofilm production, interaction with host cells, modulation of host immune response and dissemination in host blood (Hu *et al.*, 2009).

The type VI secretion machinery is a recently characterized secretion system that appears to constitute a phage-tail-spike-like injectisome that has the potential to introduce effector proteins directly into the cytoplasm of host cells (Tseng *et al.*, 2009). In *E. tarda,* a type VI secretion system was identified as EVP (*E. tarda* virulence protein) gene cluster and consisted of 16 components (*evpP–evpO*). Compared with the parental strain, in-frame deletion of *evpP* in *E. tarda* EIB202 led to a significantly increased LD_{50} in zebra fish and Japanese flounder, decreased haemolytic activities, failure to adhere to mucus and reduced serum resistance (Wang *et al.*, 2009b). Moreover, the *evpP* deletion mutant exhibited incapacity to internalize in epithelioma papulosum of carp cell model *in vitro*, demonstrating that EvpP in type VI secretion machinery plays a critical role in the invasion mechanism of *E. tarda* and merits investigation as a potential target for attenuated live vaccine construction (Wang *et al.*, 2009b).

Finally, different uptake systems such as a cysteine transporter and a zinc transporter were shown to be involved in the ability of *Y. ruckeri* to infect fish (Dahiya & Stevenson, 2010a; Méndez *et al.*, 2011).

4. Endotoxins

An endotoxin is a toxin which is a structural molecule of Gram-negative bacteria and which is recognized by the immune system of the host. The prototypical examples of endotoxins are complex molecules called lipopolysaccharides (LPS), which are a constituent of the cell wall outer membrane of Gram-negative bacteria. LPS structure consists of three components: an outer polysaccharide region that is highly variable among different bacteria, known as "O antigen", a highly conserved polysaccharide chain called "core" and an inner fatty-acid-rich region known as "lipid A".

Lower vertebrates such as frog and fish are stated to be resistant to endotoxic shock, whereas higher animals are very sensitive to it (Berczi *et al.*, 1966a, 1966b; Wedemeyer *et al.*, 1969). However, in recent years endotoxins/LPS have frequently been shown to be responsible for the pathogenesis of several bacterial fish diseases. The Gram-negative fish pathogens in which endotoxins have been identified as disease-related virulence mechanisms belong mainly to three genera: *Aeromonas, Edwardsiella* and *Vibrio*. Gram-negative bacteria have the unique characteristic of smooth and rough variation, which is mainly based on the presence or absence of an O-specific chain (Lukácová *et al.*, 2008). The attachment of O-antigen to core lipid-A results in smooth phenotype, while core lipid-A lacking O-antigen is referred to as rough phenotype (Swain *et al.*, 2010). Genes involved in the biosynthesis of the O-antigen have been associated with virulence in different studies. For example, *A. hydrophila* AH-3 mutants in which the gene that codes for UDP *N*-acetylgalactosamine 4-epimerase (*gne*) is affected, showed the O- phenotype (LPS without O-antigen molecules) and were less virulent for fish compared to the wild type strain (Canals *et al.*, 2006). The results obtained in a later work (Canals *et al.*, 2007) confirmed that LPS is essential in *A. hydrophila* pathogenicity. In another study carried out with the same strain, a mutation in *galU*, a gene that codes for UDP-glucose pyrophosphorylase, caused two types of LPS structures (Vilches *et al.*, 2007). The *galU* mutation reduced the survival of this strain in serum to less than 1%, decreased the ability to adhere and reduced the virulence of *A. hydrophila* AH-3 in a septicaemia model in fish (Vilches *et al.*, 2007). In a recent work, two rough attenuated variants of *A. hydrophila*, derived from two smooth virulent strains by continuous cultivation in brain–heart infusion agar over a period of 8 years, were discovered to be attenuated, since they produced neither disease nor mortality in the injected fishes (Swain *et al.*, 2010).

In *Edwardsiella ictaluri*, an LPS O side-chain (O antigen) mutant strain was isolated using transposon mutagenesis (Lawrence *et al.*, 2001). The mutant failed to express O side-chains and was highly attenuated in channel catfish (*Ictalurus punctatus*) fingerlings compared with the parent wild type strain.

In the virulent strain *E. tarda* EIB202, a mutation in the gene *waaL*, which codes for a putative O-antigen ligase, resulted in the absence of O-antigen side chains in the LPS production (Xu *et al.*, 2010). This *waaL* mutant was attenuated in virulence, showed an impaired ability in internalization of epithelioma papulosum cyprinid cells and a poor ability to proliferate *in vivo*. The *waaL* mutant also exhibited a decreased resistance to serum and polymyxinB and an increased sensitivity to H_2O_2, indicating that the LPS was involved in endurance to oxidative stress in hosts during infection. In another study in *E. tarda* (Wang *et al.*, 2010c), a comparison of pathogenicity of purified LPS and lipid A from virulent and avirulent strains demonstrated that LPS was one of the virulence factors of the *E. tarda* isolates.

In *V. vulnificus*, there are classically two biotypes with the virulence for eels being specific to strains belonging to biotype 2, a homogeneous LPS-based O serogroup. It was demonstrated that the O side chain of this LPS determines the selective virulence of biotype 2 for eels (Amaro *et al.*, 1997). In this study, biotype 1 strains were destroyed by the bactericidal action of non-immune eel serum, biotype 2 strains were resistant and rough mutants of biotype 2 lacking the O polysaccharide side-chain were sensitive and avirulent for eels. Another study in *V. vulnificus* biotype 2 showed that the *gne* gene is essential for O-antigen biosynthesis and virulence in eels (Valiente *et al.*, 2008). Mutation in *gne* increased the sensitivity to microcidal peptides, to eel serum and to phagocytosis/opsonophagocytosis. Moreover, significant attenuation of virulence for eels was observed. The change in the attenuated-virulence phenotype produced by the *gne* mutation was correlated with the loss of the O-antigen LPS.

In *L. anguillarum*, the presence of the O1 antigen side chain was shown to be crucial for the resistance to the bactericidal action of serum from rainbow trout (Welch & Crosa, 2005). In this work, a mutant in *rmlC*, a gene involved in the biosynthesis of dTDP-rhamnose (a common constituent of bacterial LPS O side chains) was isolated. The *rmlC* mutant was shown to be defective in the production of the O antigen. In addition to this, a mutant obtained by allelic exchange in *rmlD*, another ORF in the dTDP-rhamnose biosynthetic cluster, showed the same O1-deficient phenotype and was highly attenuated compared to the wild type strain.

5. Regulation of virulence gene expression

Pathogenic bacteria are submitted to continuous environmental changes, which may vary significantly during infection process. Therefore, the pathogenicity of the bacteria depends on their ability to survive in stressful environmental conditions. To confront these surrounding variables, bacteria present a complex regulation of gene expression.

5.1 Two component regulatory system (TCS)

Bacteria efficiently survive under changeable conditions by utilizing different signal transduction systems. One of the most extended systems is known as the two component regulatory system (TCS) (Gao *et al.*, 2007; Robinson *et al.*, 2000). The typical TCS consists of a sensor kinase that responds to specific signals, phosphorylating the second component of the system (Mitrophanov & Groisman, 2008). The EsrA-EsrB TCS is well characterized as a virulence regulatory system in *E. tarda*. Tan *et al.* (2005) showed that this TCS regulates a type III secretion system related to the pathogenicity of the *bacterium*. When *esrA* and *esrB* genes were disrupted, proteins that compound the type III secretion system (EseB, EseC and EseD) were missing or considerably reduced (Tan *et al.*, 2005). The EsrA-EsrB TCS system also controls other regulator called EsrC. The *esrC* mutant showed an increase of 50 % in its virulence using blue gourami as the infection model. In this sense, the authors concluded that the EsrC regulator plays an important role in the virulence of *E. tarda*, forming a regulation cascade complex with the TCS EsrA-EsrB, which regulates the expression of the secreted proteins encoded by the type III secretion system and the *evp* cluster (Zheng *et al.*, 2005). The EsrA-EsrB TCS is also involved in the regulation of the type VI secretion system of *E. tarda* exercising a positive effect on the transcription of *evpP* (Wang *et al.*, 2009b). Recently, it has been demonstrated that this EsrA-EsrB TCS governs the expression of EthA haemolysin of *E. tarda*, which belongs to Eth haemolysin system in this *bacterium*, which comprises EthA and EthB subunits (Wang *et al.*, 2010b).

Another novel TCS is the BarA-UvrY described for *Y. ruckeri*. In this system, a mutant strain in the response regulator *uvrY* gene showed less ability to infect epithelioma papulosum cyprini cells, more sensitivity to H_2O_2 and was unable to maintain a high bacterial load in rainbow trout kidney (Dahiya & Stevenson, 2010b).

5.2 GntR family regulators

The GntR regulators are a metabolite-responsive family that represents one of the most abundant groups of Helix-turn-helix (HTH) transcription factors. These proteins contain a characteristic DNA-binding HTH domain at their N-terminus (Hoskisson & Rigali, 2009). Besides the mentioned regulation system of *E. tarda* virulence factors, Wang *et al.* (2009a) have characterized an EthB subunit regulation belonging to the Eth haemolysin system. The *ethB* gene encodes the activation/secretion machinery required for the maturation and translocation of EthA haemolysin (Hirono *et al.*, 1997). In the study of Wang *et al.* (2009a), they identified the *ethB* regulator EthR, a transcription regulator of the GntR family, which controls *ethB* expression by direct interaction with the *ethB* promoter region. Disruption of the regulated expression of *ethR* significantly decreases bacterial virulence using Japanase flounder as infection model by intraperitoneal injection (Wang *et al.*, 2009a).

5.3 Sigma factors

Sigma factors are another type of regulation system. These consist of a class of proteins constituting essential dissociable subunits of prokaryotic RNA polymerase and they are involved in promoter recognition and transcription initiation. These regulation systems are diverse and they have been shown to regulate expression of virulence genes as well as virulence-associated genes (Kazmierczak *et al.*, 2005). Related to the fish pathogenic bacteria, the alternative *rpoN* sigma factor, classified into the σ^{54} subfamily, was proposed as a virulence factor regulator in *L. anguillarum*. The disruption of the *rpoN* gene generated an aflagellated mutant, and in consequence, a non-motile strain. The infectivity of *rpoN* mutant was similar to that of the wild type strain following intraperitoneal injection of fish; however, it was reduced significantly when fish were immersed in bacteria-containing water (O'Toole *et al.*, 1997). The authors concluded that RpoN regulated the expression of the polar *flagellum*, an important virulence factor, which is necessary during host colonization in the first stages of the water-borne infection (O'Toole *et al.*, 1997). In *V. alginolitycus*, the alternative *rpoS* sigma factor has been defined as a virulence factor expression regulator, comprising the extracellular protease activity and cytotoxicity of extracellular products (Tian *et al.*, 2008). In the same way, the authors confirmed the implication of RpoS sigma factor in the regulatory network of the LuxS *quorum* sensing system; the disruption of *rpoS* gene showed a decrease of extracellular autoinducer-2 level, involved in the LuxS system (Tian *et al.*, 2008).

5.4 *Quorum* sensing

Bacteria are social organisms that display complex cooperative behaviour, such as conjugation, biofilm formation, antibiotic synthesis, sporulation, secretion of virulence factors and bioluminescence. Many of these kinds of behaviour are regulated by a process known as *quorum* sensing. *Quorum* sensing is a cell-to-cell communication system that enables bacteria to synchronize gene expression with population density. The prototype of *quorum* sensing in Gram-negative bacteria is the LuxI/LuxR circuit of *V. fischeri* (Engebrecht *et al.*, 1983; Fuqua & Winans, 1994).

In *A. hydrophila*, the involvement of *quorum* sensing in pathogenicity could be demonstrated (Bi *et al.*, 2007). In this species, the *ahyR* gene encodes the LuxR-type response regulator. An *ahyR* mutant was highly attenuated relative to the wild type strain. The analysis of exoenzyme activity revealed that the *ahyR* mutant could not produce exoproteases, amylases, haemolysins and Dnases, while the wild type strain of *A. hydrophila* had a high level of exoenzyme activity. The S layer of *A. hydrophila* could not be detected in the mutant either.

A. salmonicida also possesses the LuxIR-type *quorum* sensing system, termed AsaIR. It has been shown that the autoinducer synthase AsaI plays a role in the virulence of *A. salmonicida* subsp. *achromogenes* (Schwenteit *et al.*, 2011). A knockout mutant of AsaI did not produce a detectable *quorum* sensing signal and its virulence in fish was significantly decreased. In addition to this, the expression of two virulence factors (the toxic protease AsaP1 and a cytotoxic factor), was reduced in the mutant. AsaP1 production was also inhibited by synthetic *quorum* sensing inhibitors.

In *E. tarda,* it has been suggested that some virulence factors are regulated by the *quorum* sensing system. An *E. tarda* strain isolated from deceased flounder was found to produce N-acyl-homoserine-lactones (AHLs) as *quorum* sensing signal molecules that control the expression of a 55-kDa virulent-strain-specific protein (Morohoshi *et al.*, 2004). In the same *bacterium,* it has also been shown that the overexpression of EthR, a transcriptional regulator of the GntR family, drastically altered the expression patterns of *ethB* and *luxS* in the host environment during infection, causing vitiation in the tissue dissemination and survival ability of the *bacterium,* and significant attenuation of the overall bacterial virulence (Wang *et al.*, 2009a). EthR positively modulates *luxS* expression and autoinductor AI-2 production by binding to the *luxS* promoter region (Wang *et al.*, 2009a). In another study (Zhang *et al.*, 2009a), it was found that 5411 and 5906, two small peptides which share sequence identities with the C-terminal of *E. tarda* LuxS, inhibited AI-2 activity and could vitiate the infectivity of the pathogenic *E. tarda* strain TX1. The inhibitory effect of 5411 and 5906 on AI-2 activity was proven to be exerted on LuxS, with which these peptides specifically interact. The expression of 5411 and 5906 in TX1 produced effects that are similar to those caused by interruption of LuxS expression, such as the alteration of biofilm production and a decrease in the expression of certain virulence-associated genes. It could also be demonstrated that fish expressing 5411 directly from tissues exhibited enhanced resistance against TX1 infection (Zhang *et al.*, 2009a).

In *L. anguillarum,* a species known to produce AHLs as *quorum* sensing signal molecules, the use of furanone C-30, a *quorum* sensing inhibitor, caused a significant reduction of mortality in rainbow trout during challenge with the *bacterium* (Rasch *et al.*, 2004). Although neither growth, survival, proteome, motility nor respiration of the *bacterium* were affected by the concentrations of furanone C-30 used in the challenge experiments, it could not be discounted that the disease suppression effect of furarone C-30 is caused by action directly on the fish defence systems, since it is known that AHLs exert a direct immune modulatory effect on the host. Furthermore, the absence of any clear up or down regulation of *L. anguillarum* proteins would suggest that the furanone is affecting the fish host rather than the *quorum* sensing system.

In *V. alginolyticus,* it has been shown that the LuxS *quorum* sensing system plays an important role in regulating the expression of virulence factors (Ye *et al.*, 2008). The *luxS* mutants of both a standard strain and a fish-clinical isolate had reduced lethality in red seabream (*Pagrus major*). The two *luxS* mutants exhibited a lower growth rate and defective flagellar biosynthesis. They also showed a significant decrease in protease production and an increase in both extracellular polysaccharide production and biofilm development. Moreover, in *V. alginolyticus* the LuxO-LuxR regulatory system is involved in the regulation of the exotoxin alkaline serine protease Asp. A LuxR-deficient mutant showed a reduction in *asp* transcript and the disruption of the *luxO* gene caused an increase in this virulence factor. Furthermore, the interruption of *luxO* resulted in a higher level of *luxR* mRNA, indicating that LuxO negatively regulates the expression of *luxR,* which in turn activates the expression of *asp* (Rui *et al.*, 2009). Recently, Liu *et al.* (2011) have found that Hfq, an sRNA chaperone, plays an important role in post-translational regulation of the alkaline serine protease Asp, besides other environmental stress responses. The interruption of the *hfq* gene caused attenuation of virulence in zebra fish and grouper infection models (Liu *et al.*, 2011).

6. Perspectives and future

The increasing level of production in the aquaculture industry, in terms of both quantity and the number of fish species cultured, has resulted in the appearance of new bacterial diseases. In this sense, the development of vaccines and diagnostic techniques is probably the major challenge in the field. The study of the biology of fish pathogenic bacteria and in particular, of their virulence factors, is essential in order to achieve these goals. Gene mutation is currently the major and, probably, the best means of determining the involvement of a gene in the pathogenic mechanisms of bacteria. Most of the studies described in this review are based on this fact. However, the relationship between virulence and a particular phenotype as a consequence of gene mutation should be carefully established, since in most cases this alteration produces pleiotropic effects. The recent development of specific and sensitive techniques such as microarrays, genome sequencing, *in vivo* expression technology, etc. will provide massive information about virulence-related genes as well as their expression and regulation. This will lay the foundations for tackling and solving the infection problems in the aquaculture industry.

7. References

Abolghait, S. K., Akeda, Y., Kodama, T., Cantarelli, V. V., Iida, T. & Honda, T. (2010). *Aeromonas hydrophila* PepO outer membrane endopeptidase activates human big endothelin-3 *in vitro* and induces skin ulcer in goldfish (*Carassius auratus*). *Vet Microbiol* 145, 113-121.

Allan, B. J. & Stevenson, R. M. (1981). Extracellular virulence factors of *Aeromonas hydrophila* in fish infections. *Can J Microbiol* 27, 1114-1122.

Álvarez, B., Álvarez, J., Menéndez, A. & Guijarro, J. A. (2008). A mutant in one of two *exbD* loci of a TonB system in *Flavobacterium psychrophilum* shows attenuated virulence and confers protection against cold water disease. *Microbiology* 154, 1144-51.

Amaro, C., Fouz, B., Biosca, E. G., Marco-Noales, E. & Collado, R. (1997). The lipopolysaccharide O side chain of *Vibrio vulnificus* serogroup E is a virulence determinant for eels. *Infect Immun* 65, 2475-2479.

Andrews, S. C., Robinson, A. K. & Rodríguez-Quinones, F. (2003). Bacterial iron homeostasis. *FEMS Microbiol Rev* 27, 215-37.

Arnadottir, H., Hvanndal, I., Andresdottir, V., Burr, S. E., Frey, J. & Gudmundsdottir, B. K. (2009). The AsaP1 peptidase of *Aeromonas salmonicida* subsp. *achromogenes* is a highly conserved deuterolysin metalloprotease (family M35) and a major virulence factor. *J Bacteriol* 191, 403-410.

Austin, D. A., Robertson, P. A. W. & Austin, B. (2003). Recovery of a new biogroup of *Yersinia ruckeri* from diseased rainbow trout (*Onchorynchus mykiss*, Walbaum). *Syst Appl Microbiol* 26, 127-131.

Bai, F., Sun, B., Woo, N. Y. S. & Zhang, X.-H. (2010). *Vibrio harveyi* hemolysin induces ultrastructural changes and apoptosis in flounder (*Paralichthys olivaceus*) cells. *Biochem Biophys Res Commun* 395, 70-75.

Berczi, I., Baintner, K., Jr & Antal, T. (1966a). Comparative assay of endotoxins by oral and parenteral administration. *Zentralblatt Veterinarmedizin Reihe B* 13, 570-575.

Berczi, I., Bertok, I. & Bereznai, T. (1966b). Comparative studies on the toxicity of bacterial endotoxins. *Annual review of Microbiology* 1070-1091.

Bi, Z. X., Liu, Y. J. & Lu, C. P. (2007). Contribution of AhyR to virulence of *Aeromonas hydrophila* J-1. *Res Vet Sci* 83, 150-156.

Binet, R., Létoffé, S., Ghigo, J. M., Delepelaire, P. & Wandersman, C. (1997). Protein secretion by Gram-negative bacterial ABC exporters--a review. *Gene* 192, 7-11.

Bjornsdottir, B., Fridjonsson, O. H., Magnusdottir, S., Andresdottir, V., Hreggvidsson, G. O. & Gudmundsdottir, B. K. (2009). Characterisation of an extracellular vibriolysin of the fish pathogen *Moritella viscosa*. *Vet Microbiol* 136, 326-334.

Boyd, J. M., Dacanay, A., Knickle, L. C., Touhami, A., Brown, L. L., Jericho, M. H., Johnson, S. C. & Reith, M. (2008). Contribution of type IV pili to the virulence of *Aeromonas salmonicida* subsp. *salmonicida* in Atlantic salmon (Salmo salar L.). *Infect Immun* 76, 1445-1455.

Burr, S. E., Stuber, K., Wahli, T. & Frey, J. (2002). Evidence for a type III secretion system in *Aeromonas salmonicida* subsp. *salmonicida*. *J Bacteriol* 184, 5966-5970.

Butler, A. (2003). Iron acquisition: straight up and on the rocks? *Nat Struct Biol* 10, 240-1.

Cai, S. H., Wu, Z. H., Jian, J. C. & Lu, Y. S. (2007). Cloning and expression of gene encoding the thermostable direct hemolysin from *Vibrio alginolyticus* strain HY9901, the causative agent of vibriosis of crimson snapper (*Lutjanus erythopterus*). *J Appl Microbiol* 103, 289-296.

Canals, R., Jiménez, N., Vilches, S., Regué, M., Merino, S. & Tomás, J. M. (2006). The UDP N-acetylgalactosamine 4-epimerase gene is essential for mesophilic *Aeromonas hydrophila* serotype O34 virulence. *Infect Immun* 74, 537-548.

Canals, R., Jiménez, N., Vilches, S., Regué, M., Merino, S. & Tomás, J. M. (2007). Role of Gne and GalE in the virulence of *Aeromonas hydrophila* serotype O34. *J Bacteriol* 189, 540-550.

Cascón, A., Yugueros, J., Temprano, A., Sánchez, M., Hernanz, C., Luengo, J. M. & Naharro, G. (2000). A major secreted elastase is essential for pathogenicity of *Aeromonas hydrophila*. *Infect Immun* 68, 3233-3241.

Chen, Q., Wertheimer, A. M., Tolmasky, M. E. & Crosa, J. H. (1996). The AngR protein and the siderophore anguibactin positively regulate the expression of iron-transport genes in *Vibrio anguillarum*. *Mol Microbiol* 22, 127-34.

Crosa, J. H. (1989). Genetics and molecular biology of siderophore-mediated iron transport in bacteria. *Microbiol Rev* 53, 517-30.

Crosa, J. H., Hodges, L. L. & Schiewe, M. H. (1980). Curing of a plasmid is correlated with an attenuation of virulence in the marine fish pathogen *Vibrio anguillarum*. *Infect Immun* 27, 897-902.

Dacanay, A., Boyd, J. M., Fast, M. D., Knickle, L. C. & Reith, M. E. (2010). *Aeromonas salmonicida* Type I *pilus* system contributes to host colonization but not invasion. *Dis Aquat Org* 88, 199-206.

Dahiya, I. & Stevenson, R. M. (2010a). The ZnuABC operon is important for *Yersinia ruckeri* infections of rainbow trout, *Onchorynchus mykiss* (Walbaum). *J Fish Dis* 33, 331-340.

Dahiya, I. & Stevenson, R. M. (2010b). The UvrY response regulator of the BarA-UvrY two-component system contributes to *Yersinia ruckeri* infection of rainbow trout (*Onchorynchus mykiss*). *Arch Microbiol* 192, 541-7.

Ellis, A. E., Burrows, A. S. & Stapleton, K. J. (1988). Lack of relationship between virulence of *Aeromonas salmonicida* and the putative virulence factors: A-layer, extracellular proteases and extracellular haemolysins. *Journal of Fish Diseases* 11, 309-323.

Ellis, A. E., Hastings, T. S. & Munro, A. L. S. (1981). The role of extracellular products in the pathology of furunculosis. *Journal of Fish Diseases* 4, 41-51.

Engebrecht, J., Nealson, K. & Silverman, M. (1983). Bacterial bioluminescence: isolation and genetic analysis of functions from *Vibrio fischeri*. *Cell* 32, 773-781.

Evenhuis, J. P., Lapatra, S. E., Verner-Jeffreys, D. W., Dalsgaard, I. & Welch, T. J. (2009). Identification of flagellar motility genes in *Yersinia ruckeri* by transposon mutagenesis. *Appl Environ Microbiol* 75, 6630-6633.

Fernández, L., Márquez, I. & Guijarro, J. A. (2004). Identification of specific *in vivo*-induced (*ivi*) genes in *Yersinia ruckeri* and analysis of ruckerbactin, a catecholate siderophore iron acquisition system. *Appl Environ Microbiol* 70, 5199-207.

Fernández, L., Prieto, M. & Guijarro, J. A. (2007). The iron- and temperature-regulated haemolysin YhlA is a virulence factor of *Yersinia ruckeri*. *Microbiology* 153, 483-9.

Fernández, L., Secades, P., López, J. R., Márquez, I. & Guijarro, J. A. (2002). Isolation and analysis of a protease gene with an ABC transport system in the fish pathogen *Yersinia ruckeri*: insertional mutagenesis and involvement in virulence. *Microbiology* 148, 2233-43.

Fouz, B., Barja, J. L., Amaro, C., Rivas, C. & Toranzo, A. E. (1993). Toxicity of the extracellular products of *Vibrio damsela* isolated from diseased fish. *Current Microbiology* 27, 341-347.

Fouz, B., Zarza, C. & Amaro, C. (2006). First description of non-motile *Yersinia ruckeri* serovar I strains causing disease in rainbow trout, *Onchorynchus mykiss* (Walbaum), cultured in Spain. *J Fish Dis* 29, 339-346.

Fuqua, W. C. & Winans, S. C. (1994). A LuxR-LuxI type regulatory system activates *Agrobacterium* Ti plasmid conjugal transfer in the presence of a plant tumor metabolite. *J Bacteriol* 176, 2796-2806.

Gao, R., Mack, T. R. & Stock, A. M. (2007). Bacterial response regulators: versatile regulatory strategies from common domains. *Trends Biochem Sci* 32, 225-234.

Garduño, R. A., Moore, A. R., Olivier, G., Lizama, A. L., Garduño, E. & Kay, W. W. (2000). Host cell invasion and intracellular residence by *Aeromonas salmonicida*: role of the S-layer. *Can J Microbiol* 46, 660-668.

Gavín, R., Rabaan, A. A., Merino, S., Tomás, J. M., Gryllos, I. & Shaw, J. G. (2002). Lateral flagella of *Aeromonas* species are essential for epithelial cell adherence and biofilm formation. *Mol Microbiol* 43, 383-397.

Genco, C. A. & Dixon, D. W. (2001). Emerging strategies in microbial haem capture. *Mol Microbiol* 39, 1-11.

Gunnlaugsdóttir, B. & Gudmundsdóttir, B. K. (1997). Pathogenicity of atypical *Aeromonas salmonicida* in Atlantic salmon compared with protease production. *J Appl Microbiol* 83, 542-551.

He, H., Wang, Q., Sheng, L., Liu, Q. & Zhang, Y. (2011). Functional characterization of *Vibrio alginolyticus* twin-arginine translocation system: its roles in biofilm formation, extracellular protease activity, and virulence towards fish. *Curr Microbiol* 62, 1193-1199.

Henrichsen, J. (1972). Bacterial surface translocation: a survey and a classification. *Bacteriol Rev* 36, 478-503.

Hirono, I., Tange, N. & Aoki, T. (1997). Iron-regulated haemolysin gene from *Edwardsiella tarda*. *Mol Microbiol* 24, 851-856.

Hoskisson, P. A. & Rigali, S. (2009). Chapter 1: Variation in form and function the helix-turn-helix regulators of the GntR superfamily. *Adv Appl Microbiol* 69, 1-22.

Hu, Y.-hua, Liu, C.-sheng, Hou, J.-hui & Sun, L. (2009). Identification, characterization, and molecular application of a virulence-associated autotransporter from a pathogenic *Pseudomonas fluorescens* strain. *Appl Environ Microbiol* 75, 4333-4340.

Jia, A., Woo, N. Y. S. & Zhang, X.-H. (2010). Expression, purification, and characterization of thermolabile hemolysin (TLH) from *Vibrio alginolyticus*. *Dis Aquat Org* 90, 121-127.

Kanai, K., and H. Wakabayashi. 1984. Purification and some properties of protease from *Aeromonas hydrophila*. Bull. Jpn. Soc. Sci. Fish. 50:1367-1374.

Kazmierczak, M. J., Wiedmann, M. & Boor, K. J. (2005). Alternative sigma factors and their roles in bacterial virulence. *Microbiol Mol Biol Rev* 69, 527-543.

Khalil, A. H. & Mansour, E. H. (1997). Toxicity of crude extracellular products of *Aeromonas hydrophila* in tilapia, *Tilapia nilotica*. *Lett Appl Microbiol* 25, 269-273.

Kodama, H., Moustafa, M., Ishiguro, S., Mikami, T. & Izawa, H. (1984). Extracellular virulence factors of fish *Vibrio*: relationships between toxic material, hemolysin, and proteolytic enzyme. *Am J Vet Res* 45, 2203-2207.

Labella, A., Sánchez-Montes, N., Berbel, C., Aparicio, M., Castro, D., Manchado, M. & Borrego, J. J. (2010). Toxicity of *Photobacterium damselae* subsp. *damselae* strains isolated from new cultured marine fish. *Dis Aquat Org* 92, 31-40.

Lawrence, M. L., Banes, M. M. & Williams, M. L. (2001). Phenotype and Virulence of a Transposon-Derived Lipopolysaccharide O Side-Chain Mutant Strain of *Edwardsiella ictaluri*. *Journal of Aquatic Animal Health* 13, 291-299.

Lemos, M. L., Salinas, P., Toranzo, A. E., Barja, J. L. & Crosa, J. H. (1988). Chromosome-mediated iron uptake system in pathogenic strains of *Vibrio anguillarum*. *J Bacteriol* 170, 1920-5.

Leung, K. Y. & Stevenson, R. M. (1988). Tn5-induced protease-deficient strains of *Aeromonas hydrophila* with reduced virulence for fish. *Infect Immun* 56, 2639-2644.

Li, J., Ni, X. D., Liu, Y. J. & Lu, C. P. (2011). Detection of three virulence genes *alt*, *ahp* and *aerA* in *Aeromonas hydrophila* and their relationship with actual virulence to zebrafish. *J Appl Microbiol* 110, 823-830.

Li, W. K., Chen, Y. S., Wann, S. R., Liu, Y. C. & Tsai, H. C. (2008). *Lactococcus garvieae* endocarditis with initial presentation of acute cerebral infarction in a healthy immunocompetent man. *Intern Med* 47, 1143-6.

Liu, H., Wang, Q., Liu, Q., Cao, X., Shi, C. & Zhang, Y. (2011). Roles of Hfq in the stress adaptation and virulence in fish pathogen *Vibrio alginolyticus* and its potential application as a target for live attenuated vaccine. *Appl Microbiol Biotechnol* 91, 353-364.

Lukácová, M., Barák, I. & Kazár, J. (2008). Role of structural variations of polysaccharide antigens in the pathogenicity of Gram-negative bacteria. *Clin Microbiol Infect* 14, 200-206.

Magariños, B., Bonet, R., Romalde, J. L., Martínez, M. J., Congregado, F. & Toranzo, A. E. (1996). Influence of the capsular layer on the virulence of *Pasteurella piscicida* for fish. *Microb Pathog* 21, 289-297.

Masada, C. L., LaPatra, S. E., Morton, A. W. & Strom, M. S. (2002). An *Aeromonas salmonicida* type IV pilin is required for virulence in rainbow trout. *Dis Aquat Org* 51, 13-25.

Mazoy, R., Osorio, C. R., Toranzo, A. E. & Lemos, M. L. (2003). Isolation of mutants of *Vibrio anguillarum* defective in haeme utilisation and cloning of *huvA*, a gene coding for an outer membrane protein involved in the use of haeme as iron source. *Arch Microbiol* 179, 329-38.

Méndez, J., Fernández, L., Menéndez, A., Reimundo, P., Pérez-Pascual, D., Navais, R. & Guijarro, J. A. (2009). A chromosomally located *traHIJKCLMN* operon encoding a putative type IV secretion system is involved in the virulence of *Yersinia ruckeri*. *Appl Environ Microbiol* 75, 937-945.

Méndez, J., Reimundo, P., Pérez-Pascual, D., Navais, R., Gómez, E. & Guijarro, J. A. (2011). A novel *cdsAB* operon is involved in the uptake of L-cysteine and participates in the pathogenesis of *Yersinia ruckeri*. *J Bacteriol* 193, 944-951.

Merino, S., Aguilar, A., Nogueras, M. M., Regue, M., Swift, S. & Tomás, J. M. (1999). Cloning, sequencing, and role in virulence of two phospholipases (A1 and C) from mesophilic *Aeromonas* sp. serogroup O:34. *Infect Immun* 67, 4008-4013.

Merino, S., Aguilar, A., Rubires, X., Abitiu, N., Regué, M. & Tomás, J. M. (1997a). The role of the capsular polysaccharide of *Aeromonas hydrophila* serogroup O:34 in the adherence to and invasion of fish cell lines. *Res Microbiol* 148, 625-631.

Merino, S., Aguilar, A., Rubires, X., Simon-Pujol, D., Congregado, F. & Tomás, J. M. (1996). The role of the capsular polysaccharide of *Aeromonas salmonicida* in the adherence and invasion of fish cell lines. *FEMS Microbiol Lett* 142, 185-189.

Merino, S., Rubires, X., Aguilar, A. & Tomás, J. M. (1997b). The role of flagella and motility in the adherence and invasion to fish cell lines by *Aeromonas hydrophila* serogroup O:34 strains. *FEMS Microbiol Lett* 151, 213-217.

Mitrophanov, A. Y. & Groisman, E. A. (2008). Signal integration in bacterial two-component regulatory systems. *Genes Dev* 22, 2601-2611.

Miyoshi, S. & Shinoda, S. (2000). Microbial metalloproteases and pathogenesis. *Microbes Infect* 2, 91-8.

Morohoshi, T., Inaba, T., Kato, N., Kanai, K. & Ikeda, T. (2004). Identification of *quorum-sensing* signal molecules and the LuxRI homologs in fish pathogen *Edwardsiella tarda*. *J Biosci Bioeng* 98, 274-281.

Müller, M. (2005). Twin-arginine-specific protein export in *Escherichia coli*. *Res Microbiol* 156, 131-136.

Norqvist, A., Norrman, B. & Wolf-Watz, H. (1990). Identification and characterization of a zinc metalloprotease associated with invasion by the fish pathogen *Vibrio anguillarum*. *Infect Immun* 58, 3731-3736.

Ormonde, P., Hörstedt, P., O'Toole, R. & Milton, D. L. (2000). Role of motility in adherence to and invasion of a fish cell line by *Vibrio anguillarum*. *J Bacteriol* 182, 2326-2328.

Osorio, C. R., Juiz-Rio, S. & Lemos, M. L. (2006). A siderophore biosynthesis gene cluster from the fish pathogen *Photobacterium damselae* subsp. *piscicida* is structurally and functionally related to the *Yersinia* high-pathogenicity island. *Microbiology* 152, 3327-41.

Ostland, V. E., Byrne, P. J., Hoover, G. & Ferguson, H. W. (2000). Necrotic myositis of rainbow trout, *Onchorynchus mykiss* (Walbaum): proteolytic characteristics of a crude extracellular preparation from *Flavobacterium psychrophilum*. *Journal of Fish Diseases* 23, 329-336.

O'Toole, R., Milton, D. L., Hörstedt, P. & Wolf-Watz, H. (1997). RpoN of the fish pathogen *Vibrio (Listonella) anguillarum* is essential for flagellum production and virulence by the water-borne but not intraperitoneal route of inoculation. *Microbiology (Reading, Engl)* 143 (Pt 12), 3849-3859.

O'Toole, R., Milton, D. L. & Wolf-Watz, H. (1996). Chemotactic motility is required for invasion of the host by the fish pathogen *Vibrio anguillarum*. *Mol Microbiol* 19, 625-637.

Pérez-Pascual, D., Gómez, E., Álvarez, B., Méndez, J., Reimundo, P., Navais, R., Duchaud, E. & Guijarro, J. A. (2011). Comparative analysis and mutation effects of *fpp2-fpp1* tandem genes encoding proteolytic extracellular enzymes of *Flavobacterium psychrophilum*. *Microbiology* 157, 1196-1204.

Rasch, M., Buch, C., Austin, B., Slierendrecht, W. J., Ekmann, K. S., Larsen, J. L., Johansen, C., Riedel, K., Eberl, L., & other authors. (2004). An inhibitor of bacterial *quorum* sensing reduces mortalities caused by Vibriosis in rainbow trout (*Onchorynchus mykiss*, Walbaum). *Syst Appl Microbiol* 27, 350-359.

Ratledge, C. (2007). Iron metabolism and infection. *Food Nutr Bull* 28, S515-23.

Ratledge, C. & Dover, L. G. (2000). Iron metabolism in pathogenic bacteria. *Annu Rev Microbiol* 54, 881-941.

Robinson, V. L., Buckler, D. R. & Stock, A. M. (2000). A tale of two components: a novel kinase and a regulatory switch. *Nat Struct Biol* 7, 626-633.

Rock, J. L. & Nelson, D. R. (2006). Identification and characterization of a hemolysin gene cluster in *Vibrio anguillarum*. *Infect Immun* 74, 2777-2786.

Rodkhum, C., Hirono, I., Crosa, J. H. & Aoki, T. (2005). Four novel hemolysin genes of *Vibrio anguillarum* and their virulence to rainbow trout. *Microb Pathog* 39, 109-119.

Romalde, J. L. & Toranzo, A. E. (1993). Pathological activities of *Yersinia ruckeri*, the enteric redmouth (ERM) bacterium. *FEMS Microbiol Lett* 112, 291-299.

Rowe, G. E. & Welch, R. A. (1994). Assays of hemolytic toxins. *Meth Enzymol* 235, 657-667.

Rui, H., Liu, Q., Wang, Q., Ma, Y., Liu, H., Shi, C. & Zhang, Y. (2009). Role of alkaline serine protease, asp, in *Vibrio alginolyticus* virulence and regulation of its expression by *luxO-luxR* regulatory system. *J Microbiol Biotechnol* 19, 431-438.

Sakai, D. K. (1985). Loss of virulence in a protease-deficient mutant of *Aeromonas salmonicida*. *Infect Immun* 48, 146-152.

Salinas, P. C., Tolmasky, M. E. & Crosa, J. H. (1989). Regulation of the iron uptake system in *Vibrio anguillarum*: evidence for a cooperative effect between two transcriptional activators. *Proc Natl Acad Sci U S A* 86, 3529-33.

Santos, Y., Laixier, R., Bandin, I., Lamas, J. & Toranzo, A. E. (1991). Susceptibility of turbot (*Scophthalmus maximus*), coho salmon (*Oncorhynchus kisutch)*, and rainbow trout (*O.*

mykiss) to strains of *Vibrio anguillarum* and their exotoxins. *Journal of Applied Ichthyology* 7, 160-167.

Santos, Y., Toranzo, A. E., Barja, J. L., Nieto, T. P. & Villa, T. G. (1988). Virulence properties and enterotoxin production of *Aeromonas* strains isolated from fish. *Infect Immun* 56, 3285-3293.

Schwenteit, J., Gram, L., Nielsen, K. F., Fridjonsson, O. H., Bornscheuer, U. T., Givskov, M. & Gudmundsdottir, B. K. (2011). *Quorum* sensing in *Aeromonas salmonicida* subsp. *achromogenes* and the effect of the autoinducer synthase AsaI on bacterial virulence. *Vet Microbiol* 147, 389-397.

Sexton, J. A. & Vogel, J. P. (2002). Type IVB secretion by intracellular pathogens. *Traffic* 3, 178-185.

Stork, M., Di Lorenzo, M., Mourino, S., Osorio, C. R., Lemos, M. L. & Crosa, J. H. (2004). Two *tonB* systems function in iron transport in *Vibrio anguillarum*, but only one is essential for virulence. *Infect Immun* 72, 7326-9.

Sun, B., Zhang, X.-H., Tang, X., Wang, S., Zhong, Y., Chen, J. & Austin, B. (2007). A single residue change in *Vibrio harveyi* hemolysin results in the loss of phospholipase and hemolytic activities and pathogenicity for turbot (*Scophthalmus maximus*). *J Bacteriol* 189, 2575-2579.

Suomalainen, L.-R., Tiirola, M. & Valtonen, E. T. (2006). Chondroitin AC lyase activity is related to virulence of fish pathogenic *Flavobacterium columnare*. *J Fish Dis* 29, 757-763.

Swain, P., Behera, T., Mohapatra, D., Nanda, P. K., Nayak, S. K., Meher, P. K. & Das, B. K. (2010). Derivation of rough attenuated variants from smooth virulent *Aeromonas hydrophila* and their immunogenicity in fish. *Vaccine* 28, 4626-4631.

Tan, Y. P., Zheng, J., Tung, S. L., Rosenshine, I. & Leung, K. Y. (2005). Role of type III secretion in *Edwardsiella tarda* virulence. *Microbiology (Reading, Engl)* 151, 2301-2313.

Tian, Y., Wang, Q., Liu, Q., Ma, Y., Cao, X. & Zhang, Y. (2008). Role of RpoS in stress survival, synthesis of extracellular autoinducer 2, and virulence in *Vibrio alginolyticus*. *Arch Microbiol* 190, 585-594.

Tong, Y. & Guo, M. (2009). Bacterial heme-transport proteins and their heme-coordination modes. *Arch Biochem Biophys* 481, 1-15.

Toranzo, A. E. & Barja, J. L. (1993). Virulence factors of bacteria pathogenic for coldwater fish. *Annual Review of Fish Diseases* 3, 5-36.

Tseng, T.-T., Tyler, B. M. & Setubal, J. C. (2009). Protein secretion systems in bacterial-host associations, and their description in the Gene Ontology. *BMC Microbiol* 9 Suppl 1, S2.

Valiente, E., Padrós, F., Lamas, J., Llorens, A. & Amaro, C. (2008). Microbial and histopathological study of the vibriosis caused by *Vibrio vulnificus* serovar E in eels: the metalloprotease Vvp is not an essential lesional factor. *Microb Pathog* 45, 386-393.

Vilches, S., Canals, R., Wilhelms, M., Saló, M. T., Knirel, Y. A., Vinogradov, E., Merino, S. & Tomás, J. M. (2007). Mesophilic *Aeromonas* UDP-glucose pyrophosphorylase (GalU) mutants show two types of lipopolysaccharide structures and reduced virulence. *Microbiology (Reading, Engl)* 153, 2393-2404.

Vipond, R., Bricknell, I. R., Durant, E., Bowden, T. J., Ellis, A. E., Smith, M. & MacIntyre, S. (1998). Defined deletion mutants demonstrate that the major secreted toxins are not essential for the virulence of *Aeromonas salmonicida*. *Infect Immun* 66, 1990-1998.

Wandersman, C. & Delepelaire, P. (2004). Bacterial iron sources: from siderophores to hemophores. *Annu Rev Microbiol* 58, 611-47.

Wang, B., Mo, Z. L., Xiao, P., Li, J., Zou, Y. X., Hao, B. & Li, G. Y. (2010a). EseD, a putative T3SS translocon component of *Edwardsiella tarda*, contributes to virulence in fish and is a candidate for vaccine development. *Mar Biotechnol* 12, 678-685.

Wang, F., Zhang, M., Hu, Y. Z, Zhang, W. W & Sun, L. (2009a). Regulation of the *Edwardsiella tarda* hemolysin gene and *luxS* by EthR. *J Microbiol Biotechnol* 19, 765-773.

Wang, Q., Liu, Q., Cao, X., Yang, M. & Zhang, Y. (2008). Characterization of two TonB systems in marine fish pathogen *Vibrio alginolyticus*: their roles in iron utilization and virulence. *Arch Microbiol* 190, 595-603.

Wang, X., Wang, Q., Xiao, J., Liu, Q., Wu, H., Xu, L. & Zhang, Y. (2009b). *Edwardsiella tarda* T6SS component *evpP* is regulated by *esrB* and iron, and plays essential roles in the invasion of fish. *Fish Shellfish Immunol* 27, 469-477.

Wang, X., Wang, Q., Xiao, J., Liu, Q., Wu, H. & Zhang, Y. (2010b). Hemolysin EthA in *Edwardsiella tarda* is essential for fish invasion *in vivo* and *in vitro* and regulated by two-component system EsrA-EsrB and nucleoid protein HhaEt. *Fish Shellfish Immunol* 29, 1082-1091.

Wang, Y., Zhang, X. H. & Austin, B. (2010c). Comparative analysis of the phenotypic characteristics of high- and low-virulent strains of *Edwardsiella tarda*. *J Fish Dis* 33, 985-994.

Wedemeyer, G., Ross, A. J. & Smith, L. (1969). Some Metabolic Effects of Bacterial Endotoxins in Salmonid Fishes. *Journal of the Fisheries Research Board of Canada* 26, 115-122.

Welch, T. J. & Crosa, J. H. (2005). Novel role of the lipopolysaccharide C1 side chain in ferric siderophore transport and virulence of *Vibrio anguillarum*. *Infect Immun* 73, 5864-5872.

Wertheimer, A. M., Verweij, W., Chen, Q., Crosa, L. M., Nagasawa, M., Tolmasky, M. E., Actis, L. A. & Crosa, J. H. (1999). Characterization of the *angR* gene of *Vibrio anguillarum*: essential role in virulence. *Infect Immun* 67, 6496-6509.

Xu, L., Wang, Q., Xiao, J., Liu, Q., Wang, X., Chen, T. & Zhang, Y. (2010). Characterization of *Edwardsiella tarda waaL*: roles in lipopolysaccharide biosynthesis, stress adaptation, and virulence toward fish. *Arch Microbiol* 192, 1039-1047.

Xu, Z., Wang, Y., Han, Y., Chen, J. & Zhang, X.-H. (2011). Mutation of a novel virulence-related gene *mltD* in *Vibrio anguillarum* enhances lethality in zebra fish. *Res Microbiol* 162, 144-150.

Yang, H., Chen, J., Yang, G., Zhang, X.-H. & Li, Y. (2007). Mutational analysis of the zinc metalloprotease EmpA of *Vibrio anguillarum*. *FEMS Microbiol Lett* 267, 56-63.

Ye, J., Ma, Y., Liu, Q., Zhao, D. L., Wang, Q. Y. & Zhang, Y. X. (2008). Regulation of *Vibrio alginolyticus* virulence by the LuxS -sensing system. *J Fish Dis* 31, 161-169.

Yen, M. R., Peabody, C. R., Partovi, S. M., Zhai, Y., Tseng, Y. H. & Saier, M. H. (2002). Protein-translocating outer membrane porins of Gram-negative bacteria. *Biochim Biophys Acta* 1562, 6-31.

Yu, H. B., Rao, P. S. S., Lee, H. C., Vilches, S., Merino, S., Tomas, J. M. & Leung, K. Y. (2004). A type III secretion system is required for *Aeromonas hydrophila* AH-1 pathogenesis. *Infect Immun* 72, 1248-1256.

Zhang, M., Jiao, X. D., Hu, Y. H. & Sun, L. (2009a). Attenuation of *Edwardsiella tarda* virulence by small peptides that interfere with LuxS/autoinducer type 2 sensing. *Appl Environ Microbiol* 75, 3882-3890.

Zhang, W. W., Hu, Y. H., Wang, H. L. & Sun, L. (2009b). Identification and characterization of a virulence-associated protease from a pathogenic *Pseudomonas fluorescens* strain. *Vet Microbiol* 139, 183-188.

Zhao, Z., Chen, C., Hu, C.-Q., Ren, C.-H., Zhao, J.-J., Zhang, L.-P., Jiang, X., Luo, P. & Wang, Q.-B. (2010). The type III secretion system of *Vibrio alginolyticus* induces rapid apoptosis, cell rounding and osmotic lysis of fish cells. *Microbiology (Reading, Engl)* 156, 2864-2872.

Zheng, J., Tung, S. L. & Leung, K. Y. (2005). Regulation of a type III and a putative secretion system in *Edwardsiella tarda* by EsrC is under the control of a two-component system, EsrA-EsrB. *Infect Immun* 73, 4127-4137.

Zhong, Y., Zhang, X.-H., Chen, J., Chi, Z., Sun, B., Li, Y. & Austin, B. (2006). Overexpression, purification, characterization, and pathogenicity of *Vibrio harveyi* hemolysin VHH. *Infect Immun* 74, 6001-6005.

Part 3

Antibiotics and Probiotics

6

Antibiotics in Aquaculture – Use, Abuse and Alternatives

Jaime Romero[1], Carmen Gloria Feijoó[2] and Paola Navarrete[1]
[1]*Universidad de Chile, Biotechnology Laboratory, Institute for Nutrition and Food Technology (INTA)*
[2]*Universidad Andrés Bello, Laboratory of Developmental Biology, Santiago*
Chile

1. Introduction

According to the UN Food and Agriculture Organization, aquaculture is growing more rapidly than all other animal food-production sectors (www.fao.org). Its contribution to global supplies of several species of fish, crustaceans and mollusks increased from 3.9% of total production by weight in 1970 to 33% in 2005. It has been estimated that fisheries and aquaculture supplied the world with about 110 million metric tons of food fish per year (FAO, State of World Fisheries and Aquaculture 2010.), providing a per capita supply of 16.7 kg (live weight equivalent). Of this supply, 47% is derived from aquaculture production. However, this production is hampered by unpredictable mortalities that may be due to negative interactions between fish and pathogenic bacteria. To solve this problem, farmers frequently use antibiotic compounds to treat bacterial diseases (Cabello 2006).

Aquaculture is becoming a more concentrated industry, with fewer, but much larger, farms. Infectious diseases are always a hazard and may cause significant stock losses and problems with animal welfare. Intensive aquaculture (shrimp and fish farming) has led to growing problems with bacterial diseases, the treatment of which now requires the intensive use of antimicrobials. Although various authors have emphasized the putative negative effects of using antimicrobial agents in fish farms (Alderman and Hastings, 1998; Cabello, 2006), few studies on antimicrobial resistance in the aquaculture industry have been performed *in situ*. (Fernández -Alarcón 2010, Miranda & Zemelman 2002).

Because a wide variety of chemicals are currently used in aquaculture production, control measures have been introduced over the years. These include disinfectants (e.g., hydrogen peroxide and malachite green), antibiotics (e.g., sulfonamides and tetracyclines) and anthelmintic agents (e.g., pyrethroid insecticides and avermectins) (Rawn et al. 2009). However, disease control is an active research field, and alternatives to antibiotic treatments have been explored. The public health hazards related to antimicrobial use in aquaculture include the development and spread of antimicrobial-resistant bacteria and resistance genes and the presence of antimicrobial residues in aquaculture products and the environment.

The aim of this chapter is to present information about current knowledge regarding antibiotic use in aquaculture systems. This will include basic information, for example, mechanisms of action and resistance, the role of antibiotics in disease control and the putative negative impact of the use of antimicrobial agents in fish farms, and also some alternative strategies that could reduce the use of these chemicals.

2. Use of antimicrobials in aquaculture

2.1 Controlling diseases using antibiotics

Antimicrobial agents can be defined as substances that have the capacity to kill or inhibit the growth of microorganisms. After their formal discovery by Fleming in 1928, antibiotics have become essential drugs for human and animal health and welfare. Antibiotics can be derived from natural sources or have synthetic origins. Antibiotics should be safe (non-toxic) to the host, allowing their use as chemotherapeutic agents for the treatment of bacterial infectious diseases. In addition to their use in human medicine, antimicrobials are also used in food animals and aquaculture, and their use can be categorized as therapeutic, prophylactic or metaphylactic. Therapeutic use corresponds to the treatment of established infections. Metaphylaxis is a term used for group-medication procedures that aim to treat sick animals while also medicating others in the group to prevent disease. Prophylaxis means the preventative use of antimicrobials in either individuals or groups to prevent the development of infections. In aquaculture, antibiotics at therapeutic levels are frequently administered for short periods of time via the oral route to groups of fish that share tanks or cages . All drugs legally used in aquaculture must be approved by the government agency responsible for veterinary medicine, for example, the Food and Drug Administration (FDA) in the USA). For instance, in the USA the following antimicrobials are authorized for use in aquaculture: oxytetracycline, florfenicol, and Sulfadimethoxine/ormetoprim. These regulatory agencies may set rules for antibiotic use, including permissible routes of delivery, dose forms, withdrawal times, tolerances, and use by species, including dose rates and limitations. The most common route for the delivery of antibiotics to fish occurs through mixing the antibiotic with specially formulated feed. However, fish do not effectively metabolize antibiotics and will pass them largely unused back into the environment in feces. It has been estimated that 75 percent of the antibiotics fed to fish are excreted into the water (Burridge et al., 2010).

In most of the countries with an important aquaculture industry, government agencies exert some controlling actions. For example, in Norway the use of antimicrobials requires a veterinarian's prescription, and hence, their use is therapeutic. They are sold in pharmacies or in feed plants authorized by the Norwegian Medicines Agency. In Norway, it is mandatory to report the amount of antibiotics used and retain records of prescriptions.

Intensive fish farming has promoted the growth of several bacterial diseases, which has led to an increase in the use of antimicrobials (Defoirdt et al., 2011, 2007). Current levels of antimicrobial use in aquaculture worldwide are not easy to determine because different countries have different distribution and registration systems. Nevertheless, Burridge et al. (2010) reported that the amount of antibiotics and other compounds used in aquaculture differed significantly between countries. Defoirdt et al., (2011) previously estimated that approximately 500–600 metric tons of antibiotics were used in shrimp farm production in Thailand in 1994; he also emphasized the large variation between different countries, with antibiotic use ranging from 1 g per metric ton of production in Norway to 700 g per metric ton in Vietnam.

2.2 Antibiotics – Mechanisms of action

Antimicrobial drugs may have different types of chemical structures, and they act on different parts of bacterial machinery. In general, antibiotics work by one of two mechanisms (Figure 1):

i. A bactericidal effect, i.e., the antibiotic generally kills the bacteria by interfering with either the formation of the bacterium's cell wall or its cell contents. Examples include penicillin, fluoroquinolones, and metronidazole.
ii. A bacteriostatic effect, i.e., the antibiotic stops bacteria from multiplying by interfering with bacterial protein production, DNA replication, or other aspects of bacterial cellular metabolism. Examples include tetracyclines, sulfonamides, chloramphenicol, and macrolides.

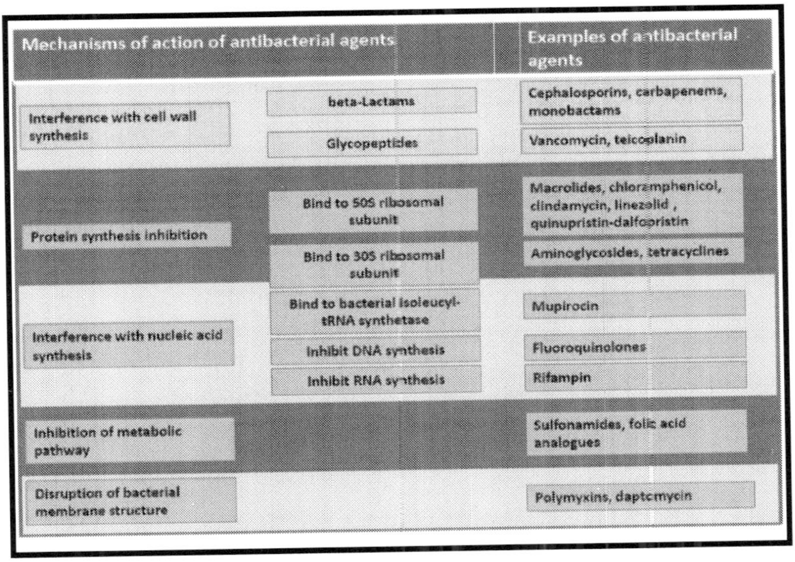

Fig. 1. Diagram showing the different mechanisms of action of antibiotics.

Some of the antibiotics that inhibit bacterial cell wall synthesis include Beta-lactams (penicillins, cephalosporins) and glycopeptides. Beta-Lactam drugs block the synthesis of the bacterial cell wall by interfering with the enzymes required for the synthesis of the peptidoglycan layer. In contrast, vancomycin and teicoplanin work by binding to the terminal D-alanine residues of growing peptidoglycan chains, thereby preventing the cross-linking steps required for stable cell wall synthesis.

Antibacterial drugs that work by inhibiting protein synthesis include macrolides, aminoglycosides, tetracyclines and chloramphenicol. These antibacterial drugs take advantage of the structural differences between bacterial and eukaryotic ribosomes to selectively inhibit bacterial growth. Macrolides, aminoglycosides, and tetracyclines bind to the 30S subunit of the ribosome, whereas chloramphenicol binds to the 50S subunit.

Fluoroquinolones exert their antibacterial effects by disrupting DNA synthesis and causing lethal double-strand DNA breaks during DNA replication. For example, the bactericidal action of ciprofloxacin results from the inhibition of topoisomerase II (DNA gyrase) and topoisomerase IV (both Type II topoisomerases), which are required for bacterial DNA replication, transcription, repair, and recombination. Sulfonamides and trimethoprim (TMP) block the pathway for folic acid synthesis, which ultimately inhibits DNA synthesis. The common antibacterial drug combination of TMP, a folic acid analogue, plus sulfamethoxazole (SMX), a sulfonamide, inhibits 2 steps in the enzymatic pathway for bacterial folate synthesis. For example, sulfadimethoxine and ormetoprim are two different antibiotics compounded into one drug. Sulfadimethoxine is a long-acting sulfonamide and ormetoprim is a diaminopyrimidine structurally related to trimethoprim. These antibiotic drugs act in synergy because they block two sequential steps in bacterial folic acid synthesis, thus inhibiting bacterial thymidine synthesis. Sulfadimethoxine blocks the conversion of para-aminobenzoic acid to dihydrofolic acid by inhibiting the enzyme dihydrofolate synthetase. Ormetoprim blocks the conversion of dihydrofolic acid to tetrahydrofolic acid by inhibiting dihydrofolate reductase. The net effect is that of a potentiated sulfa whose action is not merely bacteriostatic but bactericidal.

Disruption of bacterial membrane structure may be a fifth, although less well characterized, mechanism of action. It is postulated that polymyxins accumulate in the bacterial cell membrane and exert their inhibitory effects by increasing bacterial membrane permeability. The cyclic lipopeptide daptomycin apparently inserts its lipid tail into the bacterial cell membrane, causing membrane depolarization and, eventually, the death of the bacterium (Carpenter & Chambers, 2004).

2.3 Resistance mechanisms and transference

The use of antimicrobial drugs in aquaculture has particular differences from their use in terrestrial animals. In aquaculture, antimicrobials are regularly added to the feed, which is then placed in the water where the fish are kept. In some cases, antimicrobials may be added directly to the water. These procedures result in a selective pressure in the exposed environments (usually water). The use of antimicrobials in aquaculture may involve a broad environmental application that affects a wide variety of bacteria.

Several bacterial species may survive unfavorable conditions or environmental changes after selecting mutations that improve their fitness in the new conditions. Furthermore, bacteria take advantage of mobile genetic elements, such as plasmids and transposable elements. With these elements, bacteria can access a large pool of itinerant genes that move from one bacterial cell to another and can spread through bacterial populations. Some of these genes may provide the ability to resist antibiotic effects. Antibiotic resistance takes two forms:

i. Inherent or intrinsic resistance, i.e. the species is not normally susceptible to a particular drug. This may be due to the inability of the antibacterial agent to enter the bacteria cell and reach its target site, or a lack of affinity between the antibacterial and its target (site of action), or the absence of the target in the cell. It has been suggested that some species of bacteria are innately resistant to whole classes of antimicrobial agents. In such cases, all strains of that bacterial species are resistant to all members of the antibacterial classes.

ii. Acquired resistance. This type of resistance represents the major cause for concern because of the transmissible nature of the resistance mechanisms. In this case, the bacterial species is normally susceptible to a particular drug, but some strains express drug resistance. Initially susceptible populations of bacteria become resistant to an antibacterial agent and proliferate and spread under the selective pressure induced by the use of that agent. Genes responsible for antibiotic resistance can be transferred between bacteria by three processes that involve lateral DNA transfer:

1. Transformation, i.e., bacteria acquire genes from the uptake of (foreign) DNA from the external environment;
2. Transduction, i.e., bacteria obtain genes through infection with viral DNA. This alternative has the potential to play an important role in resistance transference because of the high concentrations of viruses (bacteriophages) in aquatic habitats, seawater and the marine sediment.
3. Conjugation, i.e., bacteria gain genes by cell-to-cell mating. In this process, a plasmid is passed from one organism to another through a pilus. This may occur between members of same species or between bacteria from different genera or families. The spread of genes coding for antibiotic resistance is facilitated by mobile genetic elements called transposons, which can move from plasmids to the bacterial chromosome and in the reverse direction. A large family of discrete mobile genetic units called cassettes has been described; these elements act similarly to transposons. Cassettes may contain only one antibiotic resistance gene and a family of receptor elements called integrons that provide both the site into which gene cassettes are integrated and the enzyme responsible for gene movement (integrase). This enzyme can move these resistance cassettes in and out of the integron, thereby substantially increasing the horizontal mobility of antibiotic resistance genes and allowing bacteria to quickly adapt to environmental changes.

Several mechanisms of antimicrobial resistance are readily spread to a variety of bacterial genera. The microorganism may acquire genes encoding enzymes, such as beta-lactamases, which destroy beta–lactams (penicillins). Other antibiotic-inactivating enzymatic reactions include phosphorylation, adenylation, and acetylation. Recently, Kumarasamy et al., (2010) described the beta-lactamase NDM-1 as an example of how significant a single enzyme can be. This metallo-beta-lactamase is the cause of a dramatic and frightening rise in antibiotic resistance among enteric bacteria isolated from patients in India, Pakistan and the U.K. Bacteria may also acquire efflux pumps that excrete the antibacterial agent from the cell before it can reach its target site and exert its effect; these molecular pumps may energetically transfer antibiotics out of the cell. Bacteria may acquire several genes for a metabolic pathway that ultimately produces an altered bacterial cell wall that no longer contains the binding site for the antimicrobial agent, or bacteria may acquire mutations that limit the access of antimicrobial agents to the intracellular target site via the downregulation of porin genes. Ribosomes (RNA or proteins) may become altered due to mutations and chemical-physical changes that prevent antibiotic attachment. Therefore, normally susceptible bacterial populations may become resistant to antimicrobial agents through mutation and selection or by acquiring genetic information that encodes resistance from other bacteria.

Lateral DNA transfer mechanisms allow bacteria to acquire resistance to multiple classes of antibiotics. Bacteria with multidrug resistance (defined as resistance to > 3 antibacterial drug classes) have become a cause for serious concern, particularly in healthcare institutions where they tend to occur most commonly. Similarly, an important consequence of the large amounts of antibiotics used for farm animals and fish in aquaculture is the selection of pathogenic bacteria resistant to multiple drugs. Multidrug resistance in bacteria may be generated by one of two mechanisms. First, these bacteria may accumulate multiple genes, each coding for resistance to a single drug, within a single cell. This accumulation typically occurs on resistance plasmids. Second, multidrug resistance may also occur through the increased expression of genes that code for multidrug efflux pumps that excrete a wide range of drugs.

3. Antibiotic effects on host microbiota

The intestinal tracts of healthy fish harbor a microbiota that has been investigated by several authors due to its assumed importance in digestion, nutrition and disease control (Navarrete et al. 2008). Studies in germ-free zebrafish have revealed that gut microbiota could be involved in important processes such as epithelial proliferation, the promotion of nutrient metabolism and innate immune responses (Bates et al. 2006). An important aspect of these results was the specificity of the host response, which depends on the bacterial species that colonize the digestive tract (Rawls et al. 2004). Possible modifications in gastrointestinal microbiota due to antibiotic treatment could alter this presumably beneficial host-microbiota relationship. Therefore, understanding how antibacterial compounds modify the gastrointestinal microbiota of farmed fish could help to improve the management of hatcheries to reduce antibiotic use and enhance the safety of farmed fish. However, few studies have focused on determining the effects of antibiotic treatment on the microbial ecology of the fish gut. In general, published studies have mainly focused on describing the frequency of antibiotic resistance during and after the use of antibiotics (Kerry et al. 1997), the susceptibility of fish pathogens isolated from fish and fish farms to antibiotics (Giraud et al., 2006; Kerry et al., 1997; Akinbowale et al., 2007) and molecular determinants of antibiotic resistance (Miranda et al., 2003; Miranda & Zemelman, 2002). The impact of a specific antibiotic treatment on bacterial diversity will be reviewed in the next paragraphs. A special effort was made to describe the dominant bacterial components, especially the newly arising microbiota.

Navarrete et al (2008) evaluated the effects of oxytetracycline (OTC) treatment on bacterial populations present in the intestines of healthy juvenile salmon. Oxytetracycline was administered via medicated feed to Atlantic salmon held in experimental tanks and their intestinal microbiota were analyzed after culture. Isolates were analyzed by restriction fragment length polymorphism (RFLP) and sequencing of 16S rDNA amplicons. Microbiota from the intestines of untreated fish were more diverse and their main components were *Pseudomonas, Acinetobacter, Bacillus, Flavobacterium, Psycrobacter* and *Brevundimonas/ Caulobacter/Mycoplana.* In contrast, the microbiota of the OTC treated group were characterized by less diversity and were only composed of *Aeromonas,* clustering with *A. sobria* and *A. salmonicida.* The frequency of resistant bacteria, defined as those capable of colony formation on TSA medium containing 30 μg ml^{-1} OTC, indicated that no resistant bacteria were detected (< 10^2 CFU per gram) in the three tanks before OTC treatment. In treated fish, resistant bacteria accounted for 60%, 33% and 25% of isolates from the samples collected on day 11, 21, and 28, respectively.

All resistant bacteria isolated from the treated group showed an identical RFLP pattern to that obtained for *Aeromonas* spp. 16S rDNA sequence analysis confirmed that these resistant phylotypes belonged to *Aeromonas* spp. The presence of class A family *Tet* tetracycline resistance genes (*tet*A, *tet*B, *tet*C, *tet*D, *tet*E and *tet*H) was assessed by PCR and *Hae*III digestion of amplicons (Jacobs & Chenia, 2006; Schnabel & Jones, 1999). The *tet*E determinant was detected most frequently among the isolates (78%), while 22% of the isolates possessed tetD/H determinants.

Figure 2 shows a shift in the composition of the intestinal microbiota of the OTC-treated salmon, with several phylotypes disappearing and an *Aeromonas* population appearing (Figure 2). Bacteria belonging to this genus have been widely isolated from the gut microbiota of fish (Huber et al., 2004; Romero& Navarrete, 2006) and are considered to be a normal bacterial component. However, some species of *Aeromonas*, including *A. salmonicida, A. hydrophila, A. caviae* and *A. sobria*, are also regarded as common pathogens of fish because they may cause furunculosis and hemorrhagic septicemia. More recently, Ringø et al., (2004) proposed that the digestive tract could represent a port of entry for invading bacteria, especially *Aeromonas*. Compared with the OTC-treated salmon, a more diverse bacterial composition was observed in the untreated salmon (Fig. 2B). Some authors have suggested that, to maintain a successful culture environment in an aquatic hatchery, it is necessary to maintain a diverse microbial community that includes innocuous and beneficial bacteria (Schulze et al 2006). Therefore, the reduction in the diversity of the intestinal microbiota observed after OTC treatment could facilitate the proliferation or invasion of opportunistic microorganisms, as indicated by the rise of some phylotypes that became prevalent several weeks after treatment. Antibiotic treatment can eradicate susceptible microorganisms and promote opportunists that may occupy ecological niches previously unavailable to them. The occurrence of OTC-resistant bacteria, including *Aeromonas* species, in salmon farming has been demonstrated previously (Jacobs & Chenia 2007). Mobile resistance determinants have also been detected in this genus (Miranda et al. 2003). The presence of bacteria harboring resistance determinants could be related to the widespread use of antibiotics in aquaculture (Cabello 2006). Some authors have even suggested that common components of the microbiota could disperse resistance genes via horizontal gene transfer because of the high density and proximity of resident bacteria in the gastrointestinal tract microenvironment (Navarrete et al. 2008).

In their study with rainbow trout, Austin and Al-Zahrani (1988) used erythromycin, oxolinic acid (OA), oxytetracycline (OTC), penicillin G and sulfafurazole to study the effects of antimicrobial compounds on aerobic heterotrophic gut microbiota. These authors observed that oxolinic acid, oxytetracycline and sulfafurazole, which are used to combat infections by gram-negative bacterial pathogens, caused an increase in bacterial numbers throughout the digestive tract, with maximal numbers in the lower intestine. Conversely, erythromycin and penicillin G, which are used to treat some diseases caused by Gram-positive bacteria, caused a rapid reduction in bacterial numbers within the gastrointestinal tract. Overall, this evidence suggests that antibiotic treatment may change the composition of the intestinal microbiota of farmed fish, causing a reduction in bacterial diversity. This evidence supports the current concern that antibiotic treatment can eradicate microorganisms of the normal microbiota, facilitating the proliferation of opportunistic bacteria by depleting competition.

4. Antibiotic effects on environmental bacteria

The effects of fish farming on bacterial density, biomass, and community structure and the possible connections between these factors and antibiotic resistance have been investigated

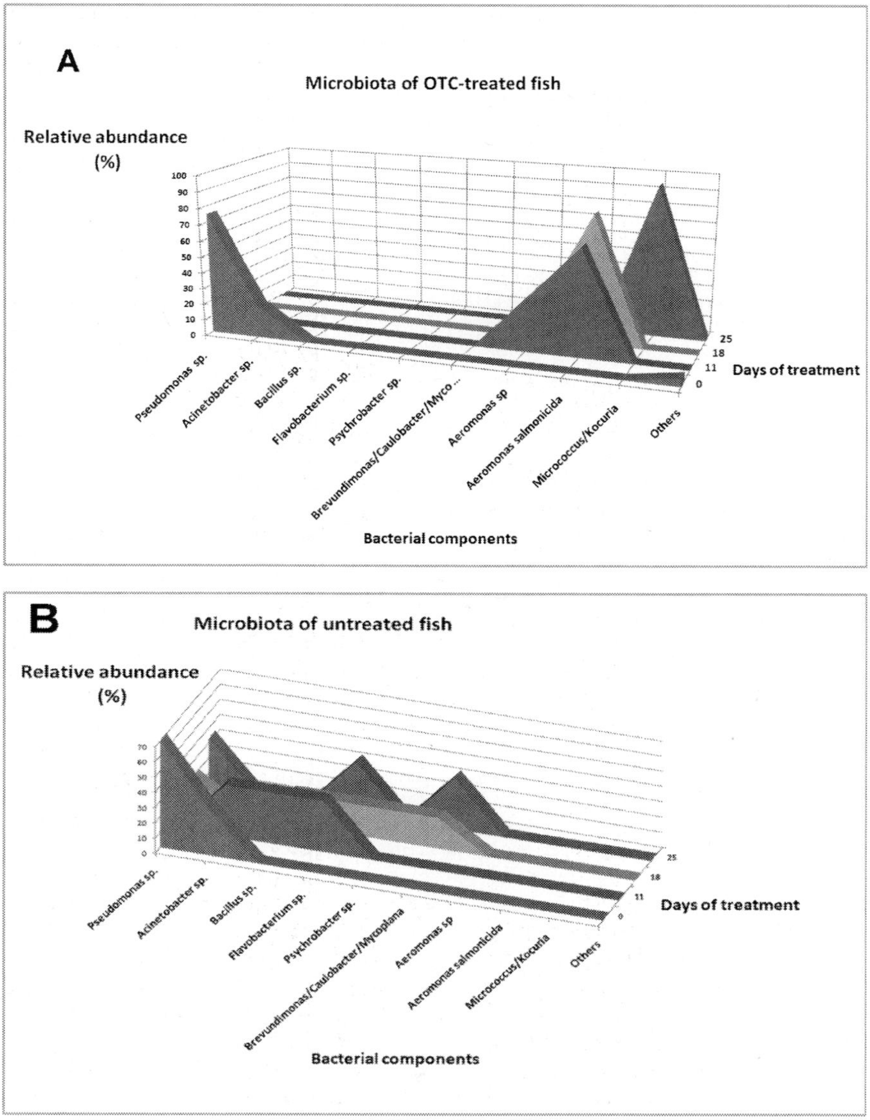

Fig. 2. Relative abundance of bacterial components in the gut microbiota. Bacteria were isolated from the intestines of untreated (control) and OTC-medicated salmon at different sampling times (day 0, i.e., one day before treatment, and days 11, 18 and 25 after treatment). A: OTC-medicated salmon, B: untreated salmon (control tanks). Adapted from Navarrete (2008).

in several studies; however, the overall results are still controversial. When an antibiotic treatment begins (usually via medicated feed) the gut microbiota and environmental bacteria can come in contact with the antibiotics present in fish farm and hatchery wastes. In fact, treatment of salmonids with various antibiotics (including OTC) has been shown to result in significant increases in the proportion of the gut microbiota showing resistance to the administered antibiotics (Austin and Al-Zahrani 1988). The same is true for environmental bacteria coming into contact with wastes containing antibiotics such as OTC. The use of oxytetracycline in fish farming has been demonstrated to coincide with an increased frequency of oxytetracycline-resistant microorganisms (DePaola et al 1995). Miranda & Zemelman (2002) also investigated the prevalence of oxytetracycline resistance in freshwater salmon farms, finding that the highest proportions of resistant bacteria were found in effluent samples (8-69%), and these were significantly higher than those from the influent samples (0-16%), which exhibited the lowest proportions of resistant bacteria. Clearly, these data suggest that the numbers of oxytetracycline-resistant bacteria are usually higher in fish farms undergoing antimicrobial therapy because susceptible microorganisms are inhibited, thus allowing colonization by resistant bacteria.

Oxytetracycline, one of the most commonly used antibiotics in fish farms and hatcheries, is very poorly absorbed through the intestinal tract of fish. It has to be administered at a high dosage rate of 100–150 mg per kg fish per day for 10–15 days. This treatment consequently causes the slow excretion of large amounts of this antibiotic, thus increasing the selective pressure which might lead to the selection of oxytetracycline-resistant bacteria in the gut (Austin and Al-Zahrani 1988; Navarrete 2008). Furthermore, it has been estimated that 70-80% of OTC is intact in the feces (Samuelsen, 2006; Ellingsen et al., 2002); although it appears to be degraded in seawater, it may persist in sediments. Following several OTC treatments, OTC levels in farm sediments reached 11 µg/g and the half-lives for OTC persistence were estimated to range from 9 to 415 days (Smith and Samuelsen 1996). In marine salmon farms, the proportion of OTC-resistant environmental bacteria can reach 25%, compared with less than 5% in sites not affected by marine salmon farms.

In another report, Akinbowale et al (2007) used polymerase chain reaction (PCR) amplification to detect the genetic determinants responsible for tetracycline resistance (*tet*R genes) in oxytetracycline-resistant bacteria from aquaculture sources in Australia. Samples included different fish (skin, intestine), rearing tanks, and water from prawn farming. This study suggested that bacteria from aquaculture sources in Australia may contribute to the reservoir of resistance genes because one or more tet genes were detected in 75% of OTC resistant isolates. tetM (50%) was the most common determinant, followed by tetE (45%), tetA (35%) and tetD (15%). Furthermore, some OTC resistant isolates were able to transfer their R-plasmid to *Escherichia coli* recipients of chicken, pig and human origins *in vitro*. Therefore, most of the studies indicate that increased levels of antibiotic resistance can be expected to occur for as long as antibiotics are used in aquaculture. However, if the use of a given antibiotic in aquaculture is discontinued or if the frequency with which it is used is reduced, it appears likely that the advantage of possessing resistance to the antibiotic would disappear. Nevertheless, more recent observations could not confirm these initial speculations.

The available studies have focused on characterizing antibiotic-resistant bacterial isolates from aquaculture farms. However, it should be considered that almost 99% of environmental microbes are uncultivable (Amann et al., 1995); therefore, this approach has important limitations when attempting to determine the prevalence of resistance genes in the environment.

Using a culture-independent approach, Seyfried et al. (2010) assessed whether OTC use in aquaculture facilities increased the detection frequency of *tet*R genes relative to facilities with no recent OTC treatment. Using a conventional (qualitative) PCR strategy, these authors screened water and sediment from four noncommercial fish farms. *tet*R was detected at significantly higher frequencies in water from farms with recent OTC use compared with water from farms without recent OTC use. Although OTC use was associated with increased prevalence and diversity of *tet*R genes in water samples, it was not found to be correlated with the prevalence of *tet*R genes in sediment samples. Sediment samples from facilities with no recent OTC use had significantly higher frequencies of *tet*R gene detection than did samples from facilities with recent OTC use. These findings suggest that OTC treatment in aquaculture facilities and the farms themselves may be sources of tetR gene introduction to the environment.

A more comprehensive understanding of drug-resistance in aquaculture requires an unambiguous and quantitative analysis of resistance genes using a culture-independent approach. Tamminen et al. (2011) investigated the prevalence of *tet* resistance genes in sediments from aquaculture farms and their surroundings and analyzed the stability of these resistance genes over a period of several years. The prevalence of *tet*- genes was monitored by quantitative polymerase chain reaction (qPCR), and the total amounts of tetracycline and oxytetracycline in the samples were also measured. None of the farms were using tetracycline at the time of the sampling, and one of the farms had stopped all antibiotic use six years prior to the first sampling. Two of the farms were sampled over four successive summers, and two were sampled once. The authors reported greater copy numbers of *tet*A, *tet*C, *tet*H, and *tet*M at the farms compared to pristine sites. However, no resistance genes were found in samples collected 200 m from any of the farms. Furthermore, the analysis of tetracyclines indicated that none of the samples contained therapeutic concentrations at any of the sampling times, suggesting that the prevalence of tetracycline-resistance genes may be caused by the persistence of these genes in the absence of selection pressure. An increase in antibiotic-resistance genes in the absence of the antibiotic itself has also been attributed to co-selection with other antibiotics.

Miranda & Rojas (2007) described the prevalence of florfenicol-resistant bacteria in a farm under florfenicol therapy two weeks before the date, designated LF1, and a farm with no recent history of antibacterial therapy, designated LF2. Samples from surface water, pellets, *Salmo salar* fingerlings and control and under-cage sediments were collected from each salmon farm. A low (< 9%) percentage of water and fingerling samples from both farms showed florfenicol resistance. However, 27% of sediment samples from LF1 showed florfenicol resistance (under-cage), and this was significantly higher than the prevalence in LF2 samples, which was < 1%(under-cage). In a complementary study, Fernández-Alarcón et al. (2010) detected the florfenicol resistant gene in florfenicol-resistant isolates from these farms and other locations by using specific PCR amplification. The isolates carrying the *flo*R gene showed a high incidence of multi-drug resistance, with all strains resistant to at least 5 of the following antibacterial drugs: ampicillin, cefotaxime, streptomycin, kanamycin, gentamicin, chloramphenicol, florfenicol, oxytetracycline, nalidixic acid, oxolinic acid, flumequine, furazolidone and trimethoprim-sulfamethoxazole. This observation indicated that a single antibiotic has the potential to co-select for a diversity of resistance genes.

Molecular approaches and massive sequencing methods could be important tools to elucidate the diversity of antibiotic-resistance genes present in the environment. The resistome concept has been used to describe the diversity of antibiotic resistance that exists naturally in a particular environment (Fernández-Alarcón 2010). However, the resistome of aquaculture environments has been poorly described. A comprehensive understanding of the influence of anthropogenic activities on the environment, for example, the long-term effects of antibiotic use on aquaculture facilities and their surroundings, will required more studies using molecular approaches. These approaches should allow the diversity of antibiotic resistance genes in an environment to be analyzed, even when no antibiotics are used and also permit the effects of antibiotics on bacterial populations to be evaluated. Recent studies have emphasized the observation that a single antibiotic has the potential to co-select for a diversity of resistances. To assess this potential risk, future studies should focus on the ability of different antibiotics used in aquatic environments to co-select for multiple resistances.

5. Selection of fish pathogens resistant to antibiotics

The major concerns with treating fish with antibiotics are the potential impact of antimicrobials on the aquatic environment, both marine and fresh water, and the wider theoretical risks associated with the development of antimicrobial resistance by fish pathogens. The spread of antimicrobial resistance due to exposure to antimicrobial agents is well documented in both human and veterinary medicine. It is also well documented that fish pathogens and other aquatic bacteria can develop resistance as a result of antimicrobial exposure. Examples include *Aeromonas salmonicida, Aeromonas hydrophila, Edwardsiella tarda, Yersinia ruckeri, Photobacterium damselae* and *Vibrio anguillarum.*

Aeromonas salmonicida, which causes disease in fish from temperate and colder areas, easily develops resistance when exposed to antimicrobials. Sulfonamide was used in the 1950 to control this pathogen; however, resistance to this drug emerged with a prevalence > 75%. More recently, multiresistant *Aeromonas salmonicida* isolates have been described in various parts of the world, and transferable resistance plasmids are commonly detected in these strains. Similar findings have been reported in other countries for other bacterial fish pathogens following the use of antibiotics to control diseases in cultured salmonids. When antibiotic resistance occurs, the effectiveness of the antibiotics for treating fish diseases is compromised.

Currently, antibiotics are only partially effective due to the emergence of resistant bacteria; therapeutic treatments may have limited success at controlling infectious bacterial diseases. For instance, Karunasagar et al. (1994) reported mass mortality in *Penaeus monodon* larvae caused by *Vibrio harveyi* strains with multiple resistances to cotrimoxazole, chloramphenicol, erythromycin and streptomycin. Genetic determinants of antibiotic resistance that have been described in aquaculture environments are regularly located on mobile genetic elements. Indeed, resistance genes have been found located on transferable plasmids and integrons in pathogenic bacteria such as *Aeromonas* spp., *Edwardsiella* spp. and *Vibrio* spp. (Defoirdt et al 2011).

In summary, the high proportions of antibiotic-resistant bacteria that persist in sediments and farm environments may provide a threat to fish farms because they can act as sources of antibiotic-resistance genes for fish pathogens in the vicinity of the farms. Because resistant bacteria may transfer their resistance elements to bacterial pathogens, the implementation of efficient strategies to contain and manage resistance-gene emergence and spread is critical. Inefficiencies in the antibiotic treatment of fish illnesses may lead to significant economic losses in the future.

6. Antibiotics and their effects on fish stress responses

The use of antimicrobial drugs in aquaculture has well-known positive effects on the control of bacterial infections; however, several side effects that affect both the fish and the environment are associated with excessive use. If one takes into account that 70 to 80% of the antibiotics administered to fish as medicated pelleted feed are released into the aquatic environment via urinary and fecal excretion and/or as unused medicated food (Martinsen & Horsberg, 1995; Smith & Samuelsen, 1996; Samuelsen, 2006), it is not hard to imagine the extent to which antibiotics can affect the aquatic habitat. The effects of antibiotics on the environment are mainly due to the overuse of these drugs by the aquaculture industry and the presence of drug residues in fish products (Saglam & Yonar, 2009). Unfortunately, there are only a few studies that analyze the side effects of antibiotic use on fish themselves. There is evidence that some antibiotics can induce nephrotoxicity (Hentschel et al., 2005), but the most well documented side effect is immunomodulation (Rijkers et al., 1981; Grondel et al., 1985; Wishkovsky et al., 1987; Tafalla et al., 2002).

In the case of nephrotoxicity, a study conducted by the Bonventre group (Hentschel et al., 2005) determined that, as in rats and humans, gentamicin, an aminoglycoside antibiotic, induces acute renal failure in fish. Their results showed that gentamicin induced pericardial edema in a time- and dose-dependent manner, which resulted in the fish being unable to maintain fluid homeostasis. In addition, a histological analysis of treated larvae pronephros demonstrated the existence of lysosomal phospholipidosis, flattening of the brush border, accumulation of debris in the tubular lumen, and tubular and glomerular distention. Moreover, they observed peritubular accumulation of leukocytes with occasional infiltration into the glomerulus in the medicated larvae, a typical feature of human acute renal failure.

Immunomodulation is a consequence of a change in the quantity and/or function of the cells or molecules involved in the immune responses. In addition to natural immunoregulatory mechanisms, a number of drugs and environmental chemicals are known to induce alterations in the immune system. It is important to remember that the immunomodulatory effects of the antibiotics used in aquaculture are variable and depend on the drug, the protocol used in the analysis, and the fish species. At present, a few reports have been published that describe the effects of oxytetracycline, florfenicol and, to a lesser extent, oxolinic acid on the fish immune system. There is evidence suggesting that oxytetracycline can suppress immune functions in carp, rainbow trout, turbot and Atlantic cod. The first studies were carried out *in vitro* and showed that oxytetracycline suppressed mitogenic and allogenic leukocyte responses in fish and that low concentrations of this antibiotic delayed the mitogenic response, but did not reduce it (Grondel et al., 1985). *In vitro* experiments in rainbow trout and in turbot demonstrated that oxytetracycline also suppressed macrophage phagocytic capacity in these two species (Wishkovsky et al., 1987;

Tafalla et al., 1999). Moreover, in turbot, both respiratory burst activity and phagocytosis were significantly suppressed by incubation with oxytetracycline *in vitro*. However, phagocytosis seems to be more sensitive to oxytetracycline because it was inhibited by all the doses used (Tafalla et al., 1999). Unexpectedly, these two macrophage functions were not suppressed when the effects of oxytetracycline were examined in vivo by the same authors, probably as a result of the low antibiotic levels present in the head kidney due to poor absorption rates. In many cases, contradictory results have been reported depending on the fish species and especially the antibiotic dose and the route of administration, making it very difficult to compare results from different studies. One study showing a discrepancy between different administration routes was conducted by Rijkers and collaborators. They analyzed the in vivo effects of oxytetracycline, administered either orally in the feed or by intraperitoneal injection, on immune function in the carp. To determinate the effects on cellular immunity, allogeneic scale transplantation was carried out. In the case of oral administration, they determined that there was no effect on the median survival time (MST) of the scales. However, the injected fish showed significantly prolonged MST, 11–20 days compared to control fish with 8.5 days. Thus, cellular immunity was not affected by oral administration of oxytetracycline, but injections did have a dramatic immunosuppressive effect. To study the effect of oxytetracycline on humoral immunity, animals were injected with red blood cells from rabbits and sheep, and the number of rosette-forming cells in the spleen was determined. Regardless of the route of administration, the serum immunoglobulin levels were drastically decreased by oxytetracycline (Rijkers et al., 1981). Lundên and collaborators studied the effects of several antibiotics, including oxolinic acid, oxytetracycline, florfenicol and trimethoprim in combination with sulfadiazine, on different aspects of immune function in rainbow trout (*Oncorhynchus mykiss*) (Lunden et al., 1998; Lunden et al., 1999; Lunden & Bylund, 2002). The results for oxolinic acid and oxytetracycline, both in vitro and in vivo, indicated that these antibiotics suppressed the mitogenic response of head kidney lymphoid cells. Moreover, the suppression of the response was stronger in T cells than in B cells.

Reports concerning florfenicol are even more controversial, but there is a consensus that its effects appear to be less pronounced than those demonstrated for oxytetracycline and oxolinic acid. Lundên and collaborators analyzed immune responses after vaccination with simultaneous oral antibiotic treatment in rainbow trout (Lunden et al., 1999). They found that florfenicol did not have any significant effect on antibody production or on circulating leukocyte levels. They only detected a decrease in the number of phagocytic cells 5–6 weeks after vaccination and a slightly reduction in fish survival after challenge with *A. salmonicida*. The same research group used a different approach to show that oxolinic acid and florfenicol suppressed the respiratory burst of phagocytic cells in rainbow trout both in vitro and in vivo (Lunden & Bylund., 2002). Again, contradictory results were obtained due to the use of different protocols to analyze the same biological question. Caipang and collaborators (Caipang et al., 2009) demonstrated that florfenicol and oxolinic acid were able to modulate different compounds of the immune system in another fish species, namely Atlantic cod. They found that, at therapeutic concentrations, the activity of alkaline phosphatase, but not that of myeloperoxidase, was altered. Transcript levels of the pro-inflammatory cytokines IL-1β and IL-8 were upregulated, as was the bactericidal permeability-increasing protein (BPI); the expression of g-type lysozyme was downregulated. In the case of the oxidative stress-related genes catalase and phospholipid hydroperoxide glutathione peroxidase (GSH-

Px), they observed differential effects of florfenicol and oxolinic acid on the expression levels (Caipang et al., 2009).

The antimicrobial effects on the immune system described above were detected using assays in which the drug was administered orally or injected into the fish; these studies did not examine the effects caused by exposure to antibiotics that remain in the water column and/or in the sediment. It is common to dose farmed fish with antibiotics in their food to protect against disease and, as fish pens are typically located in rivers or lakes, the toxic feces, uneaten food pellets, dead fish, and antibiotic residues are distributed over the entire ecosystem. This is an important point because, as active compounds, antibiotics must be considered potential environmental micropollutants and thus a source of artificial environmental stress for fish. Furthermore, in addition to the presence of the antibiotic itself, the existence of its degradation products is also a cause for concern. Unfortunately, no studies of these problems have been published, and only limited information is available regarding the presence, or absence, of antibiotics (mainly oxytetracycline, florfenicol and oxolinic acid) in the sediment around fish farm nets in a few countries (Carson et al., 2002; Lalumera et al., 2004; Pouliquen et al., 2007). The use of drugs in aquaculture has different legal constraints in each country, and supervision of compliance with regulations also varies between different countries. Likewise, companies have different ways of working, depending on the country where they are. Some information can be obtained from the industry websites. An example of this is Marine Harvest, which is one of the most important salmon farming businesses, which is headquartered in Norway and has subsidiaries in several countries. In the recent Sustainability reports submitted by Marine Harvest (http://marineharvest.com/Global/Sustainability/UoP%20final%20LO.pdf), they stated the amount of antibiotics used in all countries where the company conducts salmonid farming. In spite of important disparities in the amount of antibiotic used in the different countries, a global tendency to reduce antibiotic to disease control can be observed.

It is imperative to investigate how long different antibiotics can persist in the water and whether this time period is sufficient to alter the wellbeing of the farmed fish. This invaluable information will help determine the conditions that promote fish health and survival. Pouliquen and collaborators (Pouliquen et al., 2009) have quantified the amount of oxolinic acid, flumequine, oxytetracycline, and florfenicol present in four fish farms in France. They only detected florfenicol in one sediment sample, possibly due to the limited use of this antibiotic, which is mainly used in winter treatment and restricted to fry or young trout. In the case of oxytetracycline, they found relatively few sediment samples were contaminated with this antibiotic, presumably due to its weak stability. OTC undergoes photolysis and hydrolysis in the water column and in the first few centimeters of each layer of sediment. However, oxytetracycline can still be detected in sediments under aerobic conditions after 30 days. Compared with florfenicol and oxytetracycline, oxolinic acid and flumequine were more frequently detected due to their greater consumption and persistence. It is important to note that this work was done in a country with very low levels of fish farming; the situation is very different in countries like Ireland, Canada, Chile and Norway, where aquaculture is a highly developed industry. The difference is apparent, even in the case of Norway, which uses fewer antibiotics than the other three countries. Samuelsen and collaborators determined the amount of oxytetracycline in the sediment of three selected cages at a fish farm in Norway over a period of 18 months after 10 days of medication with the antibiotic. Most of the oxytetracycline disappeared during the first weeks, but it persisted in the sediment at lower concentrations for quite some time after the initial medication. The half-life of oxytetracycline in the sediment was measured as 125, 144 and 87 days under each of the cages. At the end of the treatment protocol, all three sediments contained 100% oxytetracycline-resistant bacteria (Samuelsen et al., 1992). The next step will be to analyze whether there are any physiological changes in fish that live in pens where antibiotic residues were found in the surrounding sediment.

Due to the absence of any published report on the effects of chronic exposure to any antibiotic on stress levels and/or immune function, we decided to investigate the effect of two of the most commonly used antibiotics, namely oxytetracycline and florfenicol, on zebrafish (*Danio rerio*). This teleost fish has several advantages as a model organism, including an especially well-studied biology, rapid development, ease of handling, and more importantly, the ability to perform in vivo analyses. Due to its versatility, the zebrafish has become one of the preferred animal models for performing ecotoxicology and toxicology research (Froehlicher et al., 2009). We took advantage of the existence of a stress-responsive transgenic line, Hsp70::GFP, that expresses green fluorescent protein when the larvae are exposed to stressors like extreme temperature and heavy metals. The primary response of fish to stress involves the activation of the hypothalamic–pituitary–interrenal axis. This activation leads to increased levels of adrenocorticotropic hormone and the glucocorticoid hormone, cortisol, (Barton, 2002). Moreover, plasma cortisol is used as a routine indicator of the magnitude and duration of the stress response. Previous research in rainbow trout has demonstrated that Hsp70 and the glucocortocoid receptor (GR) maintain a balanced sequential pro-inflammatory and anti-inflammatory cytokine expression profile that is required for effective immune responses. Heat stress and cortisol treatment have also been shown to stimulate the association of Hsp70 with the glucocorticoid receptor (Stolte et al., 2009), indicating that Hsp70 is a good molecular marker to address for both cellular and molecular stress. To determinate whether the selected antibiotics could modulate stress levels in vivo, we developed an assay in which we incubated zebrafish larvae for 48 hours in the highest sublethal concentration of each antibiotic. In the case of oxytetracycline, we used 750 ppm because this concentration is not lethal nor does it produce obvious phenotypic changes during 4 days of larvae incubation. Next, we analyzed whether chronic exposure to 750 ppm of oxytetracycline could activate the Hsp70 promoter. We incubated HSP70::GFP transgenic larvae for 48 hrs in 750 ppm oxytetracycline per triplicate and analyzed the resulting fluorescence. At 6 hours post treatment (hpt), we found that there was no obvious difference between the control and experimental larvae; the only GFP detected was the basal expression in the lens. At 12 hpt, fluorescence in the trunk and tail was limited; only a few muscle fibers expressed GFP. After 24 hrs of treatment the GFP expression became more ubiquitous, but it was not until 48 hpt when strong fluorescence was detected (Figure 3) (Feijoo, unpublished data). These results indicate that the activation of the Hsp70 promoter is gradual, and the full response is obtained after 48 hours of treatment. Prolonged exposure to oxytetracycline strongly upregulated the stress levels of the zebrafish and, thus, may alter immune responses and the ability to cope with pathogen infections. Further experiments are needed to corroborate the effect of oxytetracycline-triggered stress on the immune response, such as determining the level of pro-inflammatory cytokines, including IL-1, IL-8, and IL-6, and chronic inflammation markers such as iNos and the respiratory burst. Nonetheless, previous studies supported our hypothesis because they indicated that the immune response is inhibited or depressed as a consequence of stress (Ortuño et al., 2001). Moreover,

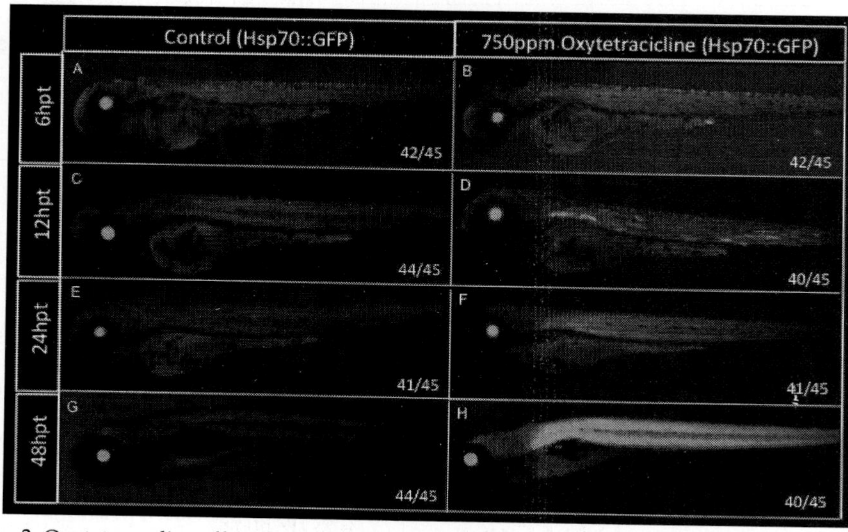

Fig. 3. Oxytetracycline effects on stress levels. Lateral view of Hsp70::GFP transgenic larva incubated in 750 ppm oxytetracycline or control medium for 6 hrs (A, B), 12 hrs (C, D), 24 hrs (E, F) and 48 hrs (G, H). hpt, hours post treatment.

chronic elevation of plasma cortisol levels in fish was shown to result in a dose-dependent increase in mortality due to common bacterial and fungal diseases (Pickering & Pottinger, 1989).

7. Antibiotics and public health

The use of antibiotics in aquaculture depends on the local regulations, which vary widely between different countries. The emerging view that antibiotics should be used with more care has prompted more strict regulations on the use of antibiotics in aquaculture and on the presence of antibiotic residues in aquaculture products. In some countries, regulations on the use of antibiotics are strict, and only a few antibiotics are licensed for use in aquaculture. However, a large proportion of global aquaculture production takes place in countries that have permissive regulations. Furthermore, many governments have set obligatory Maximum Residue Levels (MRLs) for aquaculture products. The public health risk associated with antimicrobial residues depends on the quantity of the antimicrobial encountered or consumed, i.e. the exposure. In a FAO/OIE/WHO consultation on scientific issues related to non-human usage of antimicrobials held in Geneva in December, 2003, it was concluded that antimicrobial residues in foods represent a significantly less important human health risk than the risk related to antimicrobial-resistant bacteria in food.

The presence of antibiotic-resistant bacteria in foods of animal origin is a potential health threat because resistance can be transferred among bacteria, and antibiotic-resistant pathogens may not respond to antibiotic treatments. In a microbiological study of market products, Duran & Marshall (2005) examined several brands of ready-to-eat shrimp that were obtained from grocery stores. A total of 1,564 isolates corresponding to 162 bacterial species were recovered while screening for resistance to the following 10 antibiotics: ampicillin, ceftriaxone, chloramphenicol, clindamycin, erythromycin, nalidixic acid, streptomycin, tetracycline, trimethoprim, and vancomycin. These authors reported that 42% of the isolates and 81% of the species showed resistance to antibiotics. Several human pathogens were observed among the resistant isolates, including *Escherichia coli, Salmonella, Shigella* and *Vibrio* spp.

Antimicrobial-resistant bacteria in aquaculture present a risk to public health. The appearance of acquired resistance in fish pathogens and other aquatic bacteria means that such resistant bacteria can act as a reservoir of resistance genes from which genes can be further disseminated and may ultimately end up in human pathogens. Plasmid-borne resistance genes have been transferred by conjugation from the fish pathogen *A. salmonicida* to *Escherichia coli,* a bacterium of human origin, some strains of which are pathogenic for humans. In other examples, plasmid-borne drug resistance genes have also been transferred from the fish pathogen *Vibrio anguillarum* to the causative bacterium of cholera in humans, *Vibrio cholera* (Nakajima et al. 1983).

It is important to consider that most antibiotics used for treating infections are produced by environmental microorganisms, meaning that the genes for antibiotic resistance must also have emerged in non-clinical/artificial habitats (Martínez 2008). A better understanding of the ecological role of antibiotics and antibiotic resistance in natural environments may eventually help to predict and counteract the emergence and evolution of resistance.

8. Alternative treatments: Probiotics, essential oils and phage therapy

The rise in bacterial antibiotic resistance and antibiotic residues has become global concerns, and there is a need to develop alternative therapies for bacterial pathogens in animal production, especially in aquaculture. Vaccination is an ideal method for preventing infectious diseases, but it is not a treatment for existing infections, and commercially available vaccines are still very limited in the aquaculture field. Several alternatives to the use of antibiotics have been used successfully in aquaculture. The use of innocuous microorganisms to avoid bacterial infection in aquatic organisms has been tested in aquaculture. Here, we include a brief review of the use of probiotics in aquaculture and discuss some of the suggested mechanisms by which they might control aquatic pathogens. Another source of alternative treatments is essential oils, which are natural components from plants that are generally recognized as safe substances (GRAS). Due to their antimicrobial properties, these oils may constitute alternative prophylactic and therapeutic agents in aquaculture. Furthermore, phage therapy has gained much attention for its advantages in preventing and controlling pathogen infections; since 1999, phages have been used successfully in aquaculture facilities.

8.1 The use of probiotics as an alternative to antibiotics in aquaculture

Elie Metchnikoff was the first to report the beneficial effects of fermented milk products containing microorganisms; however Kollath was the first to suggest the term "probiotics" to designate organic or inorganic substances that are essential to the healthy development of life (Desriac et al., 2010). The definition of probiotics has changed many times, and The

International Scientific Association for Probiotics and Prebiotics recently adopted the definition proposed by the World Health Organization, "probiotics are live microorganisms that, when administrated in adequate amounts, confer a health benefit on the host" (Reid et al., 2003). A definition of probiotics applied to aquaculture has to be adapted to include reference to some specific characteristics of aquatic organisms. For example, the microbiota of aquatic organisms interacts constantly with their environment, which has a much greater influence on the health of the fish (Cahill, 1990; Romero & Navarrete, 2006). Opportunistic pathogens can proliferate in seawater outside their host and reach a high load in the environment (Moriarty, 1998). This implies that the health status of aquatic organisms is strongly influenced by its environment. This also means that probiotics can be active in the environment as well as in the host. Probiotics are often defined as applications of entire microorganisms or components of microorganisms that are beneficial to the health of the host (Irianto & Austin, 2002). However, based on the important influence of the environment on aquatic organisms health, the following more complete definition for probiotics has been proposed: "a live microbial adjunct which has a beneficial effect on the host by modifying the host-associated or ambient microbial community, by ensuring improved use of the feed or enhancing its nutritional value, by enhancing the host response towards disease, or by improving the quality of its ambient environment" (Verschuere et al., 2000). As defined by these authors, probiotics may include microorganisms that prevent the multiplication of pathogens in the gut, on structural surfaces, and in the growing environment, improve the water quality of the culture, contribute to food digestion, or stimulate host immune responses.

Several microorganisms has been evaluated as probiotics in aquaculture, and these have been extensively reviewed elsewhere (Verschuere et al., 2000; Irianto & Austin, 2002; Balcázar et al., 2006a; Kesarcodiwatson et al., 2008). The most studied are lactic acid bacteria (LAB) (*Lactobacillus, Carnobacterium, Enterococcus, Lactococcus, Micrococcus, Streptococcus,* and *Weissella*) (Balcazar et al., 2008; Balcázar et al., 2007; Hagi & Hoshino, 2009; Pérez-Sánchez et al., 2011; Vazquez et al., 2005; Villamil et al., 2002), *Bacillus* (Ai et al., 2011; Antony et al., 2011; Balcázar & Rojas-Luna, 2007; Bandyopadhyay & Das Mohapatra, 2009; Ochoa-Solano & Olmos-Soto, 2006; Liu et al., 2009; Nakayama et al., 2009; Newaj-Fyzul et al., 2007; Olmos et al., 2011; Salinas et al., 2005; Sun et al., 2010; Vaseeharan & Ramasamy, 2003), *Vibrio* (Fjellheim et al., 2007; Thompson et al., 2010), *Pseudomonas* (Abd El-Rhman et al., 2009; Chythanya, 2002; Das et al., 2006; Preetha et al., 2007; Ström-Bestor, Wiklund, 2011), and *Aeromonas* (Irianto et al., 2003; Lategan, 2004; Lategan et al., 2006; Lategan et al., 2004). Yeasts (*Saccharomyces, Debaryomyces*) (Abdeltawwab et al., 2008; Reyes-Becerril et al., 2008; Tovar-Ramírez et al., 2010), bacterial spore formers (Hong et al., 2005) and recently *Actinobacteria* have also generated interest due to their high metabolic potential (Das et al., 2010; You et al., 2007). These probiotics have been used in different aquatic organisms, such as teleost fish (Merrifield et al., 2010; Dimitroglou et al., 2011), prawns (Van Hai et al., 2009), shrimp (Farzanfar, 2006; Ninawe & Selvin, 2009), and bivalve molluscs (Kesarcodi-Watson et al., 2008; Prado et al., 2010) and have been shown to be successful, not only for their ability to prevent disease, but also for improving digestion and growth. Many of these applications have been targeted at the early stages of development of the aquatic organisms, such as the larval stages, because these stages are more susceptible to infections (Dierckens et al., 2009; Vine et al., 2006; Avella et al., 2011; Bricknell, Dalmo, 2005; Fjellheim et al., 2007; Fjellheim et al., 2010; Nhan et al., 2010; Planas et al., 2006; Tinh et al., 2008; Zhou et al., 2009).

The mechanisms of action of probiotics in aquaculture have been extensively reviewed (Balcázar et al., 2006; Gómez et al., 2007; Irianto, Austin, 2002; Kesarcodiwatson et al., 2008; Nayak, 2010; Prado et al., 2010; Tinh et al., 2008; Verschuere et al., 2000). Some of the proposed mechanisms that provide protection against pathogens involve the production of inhibitory compounds, competition for essential nutrients and adhesion sites, the enhancement of disease resistance and the modulation of host immune responses (Nayak, 2010; Magnadottir, 2010; Verschuere et al., 2000; Desriac et al., 2010; Balcázar et al., 2006). However, some of these probiotic activities, such as the production of inhibitory compounds in the aquatic environment, remain controversial because there is no scientific evidence for these mechanisms of action *in vivo*.

Interestingly, the disruption of quorum sensing in bacterial pathogens has recently been suggested as a novel strategy for use in aquaculture (Defoirdt et al., 2004; Defoirdt et al., 2008; Defoirdt et al., 2011; Merrifield et al., 2010; Tinh et al., 2008; Vine et al., 2006; Natrah et al., 2011a). Quorum sensing is a process that involves bacterial cell-to-cell communications with the participation of low molecular weight signaling molecules that elicit population-density-dependent responses. The signal molecules AHL (N-acyl homoserine lactone) and/or AI-2 (autoinducer 2) have been found to be involved in the regulation of virulence factors in many pathogenic bacteria, including fish pathogens (Federle, Bassler, 2003; Morohoshi et al., 2004; Bruhn et al., 2005; Defoirdt et al., 2005; Natrah et al., 2011b; Rasch et al., 2007; Ruwandeepika et al., 2011). Bacteria that are able to degrade quorum sensing molecules might be useful as biocontrol agents in aquaculture. To date, several bacteria of aquatic origin have been studied for their quorum quenching properties (Chu et al., 2011; Defoirdt et al., 2004; Defoirdt et al., 2008; Merrifield et al., 2010; Nakayama et al., 2009; Nhan et al., 2010; Tinh et al., 2007; Tinh et al., 2008). Marine *Bacillus* and *Halobacillus salinus* have been shown to quench the quorum sensing system of *Vibrio harveyi* (Musthafa et al., 2011; Teasdale et al., 2009). A *Bacillus* (QSI)-1-like bacteria isolated from the intestine of the fish *Carassius auratus gibelio* can degrade AHLs *in vitro* and reduced the amount of AHLs and the extracellular protease activity of *Aeromonas hydrophila* in coculture with the fish pathogen (Chu et al., 2011). Carp fed a diet supplemented with (QSI)-1 showed good survival after an experimental infection with *Aeromonas hydrophila* compared with those fed a control diet (Chu et al., 2011). Recombinant AHL-lactonase from *Bacillus* sp., when co-injected with the fish pathogen *Aeromonas hydrophila* in common carp, decreased the mortality rate and delayed the time of death of the fish (Chen et al., 2010). The *Actinobacteria Streptomyces albus* has been also suggested as a promising candidate for aquaculture because it was able to attenuate biofilm formation and inhibit the quorum-sensing system of *Vibrio harveyi*, *Vibrio vulnificus*, and *Vibrio anguillarum* (You et al., 2007).

An important aspect to be considered is that probiotics have to be innocuous or not pathogenic to the host, other aquatic organisms or human consumers. For instance, probiotics have to be free of plasmid-encoded antibiotic resistance genes. In addition, considerable further research in terms of food and environmental safety is needed.

Most of the literature about probiotics consists of reports on the study and application of a single bacterial strain. However, the conditions in which the aquatic organisms are cultured are continuously changing and have strong effects on host health. It is has been suggested that a probiotic which contained several bacteria could be more efficient at controlling bacterial pathogens (Verschuere et al., 2000; Chapman et al., 2011) This is based on the assumption that it would be difficult for a single bacteria to be able to remain dominant in a continuously changing environment. However, different strains could interact under a variety of conditions and be able to maintain their dominance in a dynamic way (Verschuere et al., 2000). An optimal combination of strains with different mechanisms of action may result in successful anti-pathogenic effects (Fjellheim et al., 2010). For example, rainbow trout fed a multistrain formulation containing *Enterobacter cloacae* and *Bacillus mojavensis* and then challenged with *Yersinia ruckeri* had a higher survival rate compared with control fish (Capkin & Altinok, 2009).

There is a consensus in aquaculture that some microorganisms are beneficial to aquaculture organisms in terms of reducing the incidence of disease. However, despite the promising potential benefits demonstrated in the current literature, the probiotic mechanisms which mediate host benefits are poorly understood. Future studies should be focused on evaluating the mechanisms by which probiotics interact with the host and pathogens. To study the probiotic-host interaction, the use of axenic and gnotobiotic aquatic organisms have been proposed, which eliminate interference from the environmental and host microbiota, seems the most appropriate approach. Axenic organisms are raised or treated to eliminate microbes derived from parents, gametes or environment; hence they are considered germ free animals. Gnotobiotic organisms are animals harboring a known microbe or microbiota, formerly, they could be derived from a germ free animal, colonized by a known microbe. To date, very limited information has been reported about the development of axenic aquatic organisms. To our knowledge, only zebrafish (Bates et al., 2007; Rawls et al., 2004) and more recently sea bass (Dierckens et al., 2009) larvae that are free of microorganisms have been reported.

8.2 Essential oils: Antibacterial properties and potential applications in aquaculture systems

Essential oils (EOs) are volatile liquid fractions that contain the substances responsible for the aromas of plants; they are obtained from different organs, such as flowers, buds, seeds, leaves, twigs, bark, herbs, wood, fruits and roots (Bakkali et al., 2008). There are several methods for extracting essential oils, including expression, fermentation, enfleurage, extraction, and the use of liquid carbon dioxide or microwaves, but the steam distillation method is most commonly used for the commercial production of EOs (Burt, 2004). The chemical composition of EOs depends on the extraction method and can also vary according to conditions such as climate, soil composition, origin, season, plant organ, age and vegetative cycle stage (Abu-Darwish et al., 2011; Angioni et al., 2006; Ben Marzoug et al., 2011; Chung et al., 2011; Ennajar et al., 2011; Karakaya, 2011; Masotti et al., 2003). The composition of EOs is very complex and can include more than sixty components; however, a few major components constitute up to 85% of the EOs and generally determine their biological properties (Bakkali et al., 2008), whereas other components are present only as traces (Bader et al. 2010; Rali et al., 2007). The components include two groups of distinct biosynthetic origin that are synthesized in secondary metabolism (Nagegowda, 2010; Pichersky et al., 2006). The main group is composed of terpenes and terpenoids and the other of aromatic and aliphatic constituents, all characterized by low molecular weights (Bakkali, et al., 2008).

The primary roles of essential oils in plants are believed to be as pollinator attractors and defenses against pathogens and pests due to their antibacterial, antiviral, antifungal, and insecticidal effects. The essential oils of many plants contain phenolic compounds, and these comprise the majority of plant antimicrobial components (Consentino et al, 1999; Ultee et al, 2002). Among the most studied are thymol from thyme and oregano; cinnamaldehyde from cinnamon; eugenol from clove, carvacrol from oregano and anethole from anise, whose antibacterial properties have been examined in several studies (Hammer et al., 1999; Bagamboula, 2004; Burt, 2004; Calsamiglia et al., 2007; Delaquis et al., 2002; Kim et al., 1995; Lambert et al., 2001; García-García et al., 2011). Several investigations have been conducted to determine the efficiency of essential oils and their individual components in extending the shelf-life of different foods (Busatta et al., 2008; Mejlholm, Dalgaard, 2002; Kiskó, Roller, 2005; Lambert et al., 2001; Lin et al., 2004; Ponce et al., 2003; Ultee et al., 2002; Fratianni et al., 2010; Doulgeraki et al., 2011; Mastromatteo et al., 2010; Mahmoud et al., 2004; Kostaki et al., 2009; Shekarforoush et al., 2007). These reports have focused on the *in vitro* evaluation of the effectiveness of EOs against food-borne pathogens (Pasqua et al., 2005; Dorman, Deans, 2000; Nascimento et al., 2000; Irkin et al., 2010; de Oliveira et al., 2011; Si et al., 2006) and spoilage microorganisms (Barbosa et al., 2009; Bevilacqua et al., 2010; Pasqua et al., 2005; Dorman et al., 2000; Mejlholm & Dalgaard 2002; Mastromatteo et al., 2010). The molecular basis of the antibacterial action of EOs is poorly understood. It has been suggested that they can disrupt the permeability of the bacterial cell membrane (Bouhdid et al., 2009; Pascua et al., 2007; Turina et al., 2006; Devi et al., 2010), leading to the disruption of the proton motive force, electron flow and active transport (Lambert et al. 2001; Ultee et al. 2002; Cox et al. 2000; Helander et al. 1998; Fisher et al., 2009). Other proposed mechanisms are related to the coagulation of cell contents (Lambert et al. 2001, Becerril et al., 2007), and recently, evidence has been provided for the inhibition of quorum sensing (Brackman et al., 2008; Kahn et al., 2009), the induction of heat shock proteins and the prevention of flagella development (Burt et al., 2007). Interestingly, cinnamaldehyde was recently found to interfere with autoinducer-2 (AI-2), which is involved in quorum sensing in *Vibrio* spp., by decreasing the DNA-binding ability of LuxR, resulting in several marked phenotypic changes, including reduced virulence. Because inhibitors of AI-2-based quorum sensing are rare, and given the role of AI-2 in several processes, these compounds may provide useful leads for antipathogenic drugs (Brackman et al., 2008).

Limited information is available concerning the *in vivo* antibacterial effects of EOs; however, interest is rising in the use of these natural compounds as an alternative to antibiotic growth promoters to improve gut health and control the pathogens carried in the guts of livestock, swine and poultry (Bampidis et al., 2006; Benchaar et al. 2008; Busquet et al., 2006; Calsamiglia et al., 2007; Jang et al., 2007; Fraser et al., 2007; McIntosh et al., 2003; Muhl & Liebert, 2007; Cross et al., 2007; Maenner et al., 2011; Windisch et al., 2008; Yang et al., 2007). The *in vivo* effects of dietary treatment with EOs to control bacterial pathogens or to modify the microbiota composition are very controversial, and the effects largely depend on factors such as the type or combination of EOs used, the EO concentration, the host in which EOs are tested, and the composition and susceptibilities of the bacterial groups present in the gut.

The dietary addition of a commercial blend of essential oils, including thymol at 25 and 50 mg/kg diet, showed a decrease in *E. coli* CFU in the ileo-cecal digesta in growing broiler chickens, whereas the *Lactobacillus* population was not affected (Jang et al. 2007). In contrast, a dietary treatment with 1000 mg/kg feed of thyme, oregano, marjoram, rosemary or yarrow herb (with an assumed 100 g oil/kg herb) did not affect the total viable counts of lactic acid bacteria, coliforms, anaerobes or *Clostridium perfringens* in chickens from 7 to 28 days of age (Cross et al. 2007; Muhl & Liebert, 2007). In cows, viable bacteria, cellulolytic bacteria and protozoa were not influenced by supplementation with a mixture of EOs, including thymol (Benchaar et al. 2008). The inclusion of thymol [1% (w/w)] in a pig diet caused clear changes in the small intestine microbial community, notably decreasing *Actinobacillus* spp. to undetectable levels (Janczyk et al., 2008)

EOs have been studied in aquaculture as preserving agents in seafood (Kostaki et al., 2009; Lin et al., 2004; Mahmoud et al., 2004; Mejlholm, Dalgaard, 2002; Pyrgotou et al., 2010), and only recently have they been used *in vivo* as antibacterial agents to control bacterial infections (Yeh et al., 2009; Randrianarivelo et al., 2010). For example, the shelf-life of carp fillets has been extended by dipping fillets into a solution containing both carvacrol and thymol, leading to reduced growth and numbers of bacteria (Kim et al., 1995; Mejlholm, Dalgaard, 2002; Mahmoud et al., 2006). Trout fillet treatment with oregano EO also extended the shelf life by 7 to 8 days for fresh trout fillets (Pyrgotou et al., 2010).

Examples of the *in vivo* use of EOs in aquaculture systems are scarce but promising. The few studies that exist have reported the effects of EOs on shrimp and some fish. A recent report by Yeh et al. (2009) reported that EOs from *Cinnamomum kanehirae* (stout camphor tree) showed antibacterial effects against different pathogens of aquatic animals, and shrimp (*Litopenaeus vannamei*) treated with hot-water extracts from twigs of *Cinnamomum kanehirae* showed a significant decrease in their sensitivity to *Vibrio alginolyticus* (Yeh et al., 2009). When the EO of *Cinnamosma fragrans,* an endemic plant to Madagascar, was added directly to the water tank, there was an increase in the survival of the shrimp larvae (*Penaeus monodon*) concomitant with a decrease in bacterial concentration (Randrianarivelo et al., 2010). Similarly, two EOs of *C. fragrans* (B8: linalool-type and B143: 1,8-cineole-type) reduced the total heterotrophic aerobic bacteria and the *Vibrio* concentrations in the rearing water of *P. monodon* shrimp larvae (Sarter et al., 2011)

The effects of supplementing diets with acetone extract (1% w/w) from four medicinal plants (Bermuda grass, *Cynodon dactylon;* beal, *Aegle marmelos;* wintercherry, *Withania somnifera;* and ginger, *Zingiber officinale*) were evaluated in tilapia (*Oreochromis mossambicus).* The results showed that tilapia fed the plant extracts showed improvements in their growth, and their non-specific immune responses were stimulated, as indicated by increases in leucocrit value, phagocytic index and lysozyme activity. The acetone extract from *W. somnifera* showed the strongest inhibition of *Vibrio* spp. and *Photobacterium damselae* growth. A challenge test with *V. vulnificus* showed 100% mortality in *O. mossambicus* fed the control diet by day 15, whereas the fish fed the experimental diets registered only 63–80% mortality at the end of challenge experiment (30 days) (Immanuel et al., 2009). Interestingly, channel catfish (*Ictalurus punctatus*) fed with the natural oregano EO extracted from *Origanum heracleoticum* had the lowest mortality following an *Aeromonas hydrophila* infection compared with fish fed a combination of carvacrol and thymol, which are the principal active components of oregano EO (Zheng et al., 2009). These results highlight the fact that the use of the entire EO is more effective than treating with a combination of its principal components.

It is important to evaluate the effect of EOs on the normal gut microbiota. Recently, Navarrete et al. (2010) evaluated the effects of a diet supplemented with *Thymus vulgaris* essential oil (TVEO) on the composition of the rainbow trout intestinal microbiota using molecular profiling methods based on 16S rRNA gene analysis (restriction fragment length polymorphism (RFLP)) and PCR-temporal temperature gradient electrophoresis (PCR-TTGE). No significant changes (P > 0.05) were detected in the RFLP and TTGE profiles of TVEO-treated trout compared with controls, indicating that the dominant microbiota was not affected by the EO. In addition, *in vitro* determination of the antibacterial activity of TVEO was performed using several bacteria isolated from the gut of healthy trout and some fish pathogens. The inhibitory concentrations for all bacteria tested were higher than the TVEO levels used in trout, which may explain the *in vivo* results. The MICs were similar to those previously reported (Cosentino et al. 1999; Burt, 2004). Notably, some pathogenic bacteria, such as *Lactococcus piscium*, were clearly more susceptible to TVEO than those isolates belonging to the indigenous microbiota. It should be noted that the level of TVEO required to inhibit bacterial pathogens (480 mg/L) is higher than the level used in the *in vivo* study (20 mg TVEO/kg feed). Therefore, more *in vivo* studies are needed to evaluate the effects of higher TVEO concentrations on gut bacteria. It will also be necessary to evaluate whether these higher concentrations can alter feed flavor or induce toxic responses in the fish (Stroh et al. 1998). To enhance shelf life and avoid the degradation of EO in supplemented feed, the encapsulation of EOs could be a plausible means of delivering active EOs into the fish gut, as this would reduce interactions with the food matrix and possibly reduce any toxic effects. This approach has recently been evaluated, and the application of this method to aquaculture is highly promising (Wang et al., 2009; Piva et al., 2007; Pérez-Conesa et al., 2011; van Vuuren et al., 2010; Donsi et al., 2011).

8.3 Phage therapy as a potential therapy in aquaculture

Bacteriophages are defined as viruses that can infect, multiply in and kill susceptible bacteria. They are both ubiquitous and abundant in the environment, especially in seawater, in which the total numbers of viruses frequently exceeds the bacterial concentration by a factor of 10 (Børsheim, 1993). Since their discovery in 1915, phages have been studied for their therapeutic properties and ability to control infectious bacteria; however these studies were later abandoned due to the introduction of cheap, broad-spectrum antibiotics. Recently, after the increase in bacterial antibiotic resistance, phage therapy has reappeared as an effective alternative to the use of antibiotics. To date, several early studies have shown the therapeutic and prophylactic effects of phage therapy in animals and humans (Mathur et al., 2003). Phages have several advantages over other therapeutic agents: 1) they have specific narrow host ranges, meaning that they do not harm the normal intestinal microbiota; 2) they can self-replicate in the bacterial target, which eliminates the need for multiple administrations and 3) there have been no reports of side effects (Mathur et al., 2003; Nakai, Park, 2002).

The use of phage therapy in aquaculture began with the work of Nakai et al. (1999) and has been recently reviewed (Almeida et al., 2009). The first studies evaluated the ability of phages to prevent the infection of yellowtail (*Seliora quinqueradiata*) and ayu (*Plecoglossus altivelis*) with *Lactococcus garvieae* and *Pseudomonas plecoglossicida*, respectively. Phages can easily enter the fish body through the skin and gills; phages can be detected in the kidney after dipping fish in phage solution. Bath administration of phages will be effective for those fish in which infection is initiated by bacterial colonization on the skin and gills (Nakai &

Park, 2002). Administration of phages through feed will be advantageous for infections in which the oral route is the major route for pathogen transmission, as is the case for *L. garvieae* infection of yellowtail and *P. plecoglossicida* infection of ayu (Nakai & Park, 2002).

An anti-*Lactococcus garvieae* phage was intraperitoneally or orally administered to yellowtail and protected the fish from experimental *L. garvieae* infection, suggesting a potential use for the phage in controlling this disease (Nakai et al., 1999). The oral administration of phage-impregnated feed to ayu (*Plecoglossus altivelis*) resulted in protection against infection with *Pseudomonas plecoglossicida,* which quickly disappeared from the kidneys of the phage-treated fish (Park et al., 2000). In the presence of phages, bacterial counts in freshwater were also low. These results suggest the feasibility of using phages to control the disease caused by *P. plecoglossicida,* the causative agent of bacterial hemorrhagic ascites disease in cultured ayu fish (Park et al., 2000). Furthermore, in another study, a mixture of two phages induced the highest protection in ayu (*P. altivelis*) against water-borne infection with *P. plecoglossicida* compared with the administration of one phage alone (Park & Nakai, 2003). Interestingly, neither phage-resistant organisms nor phage-neutralizing antibodies were detected in diseased fish or apparently healthy fish, respectively.

Phage therapy has also been successfully used to protect against *Vibrio* infection in a shrimp and prawn hatchery. A bacteriophage of *Vibrio harveyi* was isolated from shrimp farm water from the west coast of India and showed a broad lytic activity against *V. harveyi* isolates. Shrimp larvae infected with *V. harveyi* showed a higher rate of survival in the presence of the bacteriophage compared with the control. Treatment with the bacteriophage in field trials where there was a natural outbreak of *V. harveyi* improved larval survival and reduced the *V. harveyi* counts in hatchery tanks (Vinod et al., 2006). Shivu et al. (Shivu et al., 2007) also isolated four phages from oyster tissue and a shrimp hatchery that had lytic activity against *V. harveyi.* These bacteriophages were effective in controlling the population of *V. harveyi* in hatchery systems and enhanced the survival of shrimp (*Penaeus monodon*). Bacteriophages with lytic activity against *V. harveyi* were isolated from prawn farm samples. Purified phages of the family *Siphoviridae* had a clear lytic ability and no apparent transducing properties, indicating they are appropriate for phage therapy (Crothers-Stomps et al., 2010). Although several phages isolated from hatcheries and water from aquaculture systems showed highly lytic activity against *V. harveyi* (Shivu et al., 2007), some phages that infect *V. harveyi* occur in a hatchery and co-exist with *V. harveyi* cells and are unable to control the outbreak of luminescent bacterial disease in a shrimp system Chrisolite et al. (2008).

More recently, Imbeault et al. (2006) confirmed the earlier results of Park and Nakai (Park & Nakai, 2003) showing that a bacteriophage mixture was successfully used to prevent *Aeromonas salmonicida* infection (furunculosis) in brook trout. They showed that more than one phage could infect *A. salmonicida* and that a mutant resistant to one phage was sensitive to one or more other phages. Resistant bacteria had a shorter generation time than the original strain and their replication success was very low (Imbeault et al., 2006). This highlights the necessity of using more than one phage to avoid bacterial resistance. In contrast, Atlantic salmon was not protected by the application of bacteriophages from *A. salmonicida* furunculosis, which could be due to the limited duration of survival of bacteriophages in the target animal, limiting the effectiveness of the treatment (Verner–Jeffreys et al., 2007).

In order to control *Flavobacterium psychrophilum*, the causative agent of systemic bacterial coldwater disease (CWD), a number of lytic phages of *F. psychrophilum* that infect trout, were isolated and showed a broad host range on *F. psychrophilum*, suggesting that they could be used in phage therapy (Stenholm et al., 2008). Another recent study that involved searching for phages using the enrichment method on pond water collected from Japanese ayu farms showed that the phage PFpW-3 had high infectivity for *F. psychrophilum* and demonstrated sufficient survivability in the stability tests. This study may be the basis for further evaluation of phage therapy in the treatment of CWD in Japanese ayu farms (Kim et al., 2010).

Walakira et al. isolated two lytic bacteriophages specific for *Edwardsiella ictaluri* which causes enteric septicemia of catfish (Walakira et al, 2008). Each *E. ictaluri* strain tested was susceptible to phage infection with variable efficiency, but the phages showed no evidence of lysogeny, and no plaques were detected on other bacterial species, demonstrating their potential use as biotherapeutic and diagnostic agents associated with enteric septicemia of catfish.

Besides the broad range of advantages, the following limitations of the use of phages have been identified (Mathur et al., 2003): 1) the induction of toxin genes, 2) the rapid release of bacterial endotoxins due to the lytic effect of the phage, 3) the risk that phages might mediate genetic exchange among bacteria, i.e. transduction or phage conversion, 4) the maintenance of phages by regular propagation needs expertise and an established set up, 5) the development of antibodies against phages may also lead to their decreased effectiveness, 6) the cost 7) some temperate phages contribute to bacterial virulence, 8) the rapid appearance of phage-resistant bacteria, 9) the presence of a disease outbreak with unknown bacteria in which high specificity may be a problem and 10) phage therapy cannot be used for intracellular bacteria because the phages are continuously cleared by the spleen, liver and other filtering organs (recticulo-endothelial system).

Despite the limitations mentioned, neutralizing antibodies have not been detected to date. Conversely, the low immunogenicity of phages in fish might provide an advantage for phage therapy in fish (Nakai et al, 2002). In addition, although phage-resistant mutants are uncommon, the use of a multiphage therapy could reduce the resistance frequency and avoid resistance (Levin & Bull, 2004).

Recently, phage therapy has generated a great deal of interest, and this interest has led to the formation of a strong network for the development of phage therapy in aquaculture (AQUAPHAGE, http://lib.bioinfo.pl/projects/view/26098). The implementation of this initiative involves a common project focused on the identification and exploitation of phages specific for bacterial pathogens that constitute serious threats for both freshwater and marine aquaculture. The target bacteria are *Listonella anguillarum* and *V. harveyi*, pathogens of Mediterranean aquaculture species (European sea bass and gilthead sea bream); *Flavobacterium psychrophilum*, a serious trout pathogen; and *Aeromonas salmonicida*, an obligate bacterial pathogen of Atlantic salmon.

9. Conclusion

Intensive fish farming has promoted the spread of several bacterial diseases, which in turn has led to the increased use of antimicrobials. Concerns about the consequences of antibiotic use on public health have encouraged the development of strict regulations controlling the use of antibiotics and have led to only a few antibiotics being licensed for use in aquaculture.

The high proportions of antibiotic-resistant bacteria that persist in sediments and farm environments may provide a threat to fish farms because they can serve as sources of antibiotic-resistance genes for fish pathogens in the vicinity of the farms. Because resistant bacteria may be transferred to humans and are capable of transferring their resistance elements to opportunistic human pathogens, the implementation of efficient strategies to contain and manage resistance-gene emergence and spread is critical. In addition to the potential effects on human health, inefficiencies in antibiotic treatment of fish illnesses lead to significant economic losses.

One strategy for reducing antibiotic use in aquaculture is to implement rearing practices that minimize the level of stress on the fish and that reduce the likelihood that infections requiring antibiotic treatment will occur.

Several alternatives to antibiotics have been developed; including probiotics, phage therapy and essential oils, and some of these have been successfully used to control bacterial infections in aquaculture facilities. These microorganisms, compounds, and/or their components are gaining increasing interest because of their relatively safe status, wide acceptance by consumers and their potential for multipurpose uses. Although the application of these alternatives to aquaculture is very promising, further studies are needed to gain more insight about their mechanisms of actions, to improve their stability and to evaluate their impact on the environment and the host microbiota.

10. Acknowledgments

This work has been supported by FONDECYT grants No. 1110253, 11090102 and 3100075 from CONICYT-Chile.

11. References

Abd El-Rhman, Azza M; Khattab, Yassir a E; Shalaby, Adel M E (2009): „Micrococcus luteus and Pseudomonas species as probiotics for promoting the growth performance and health of Nile tilapia, Oreochromis niloticus.". In: Fish & shellfish immunology. Elsevier Ltd 27 (2), pp. 175-80.

Abdeltawwab, M; Abdelrahman, A; Ismael, N (2008): „Evaluation of commercial live bakers' yeast, Saccharomyces cerevisiae as a growth and immunity promoter for Fry Nile tilapia, Oreochromis niloticus (L.) challenged in situ with Aeromonas hydrophila". In: Aquaculture. 280 (1-4), pp. 185-189.

Abu-Darwish, Mohammad S et al. (2011): „Seasonal variation in essential oil yield and composition from Thymus vulgaris L. during different growth stages in the south of Jordan". In: Natural Product Research. Taylor & Francis pp. 1-8.

Ai, Qinghui et al. (2011): „Effects of dietary supplementation of Bacillus subtilis and fructooligosaccharide on growth performance, survival, non-specific immune response and disease resistance of juvenile large yellow croaker, Larimichthys crocea". In: Aquaculture. Elsevier B.V. 317 (1-4), pp. 155-161.

Akinbowale, O L; Peng, H; Barton, M D (2007): „Diversity of tetracycline resistance genes in bacteria from aquaculture sources in Australia.". In: Journal of applied microbiology. 103 (5), pp. 2016-25.

Alderman, D J; Hastings, T S (1998): „Antibiotic use in aquaculture: development of antibiotic resistance - potential for consumer health risks •" In: International Journal of Food Science and Technology. (July 1997), pp. 139-155

Almeida, Adelaide et al. (2009): „Phage therapy and photodynamic therapy: low environmental impact approaches to inactivate microorganisms in fish farming plants.". In: Marine drugs. 7 (3), pp. 268-313.

Amann, R I; Ludwig, W; Schleifer, K H (1995): „Phylogenetic identification and in situ detection of individual microbial cells without cultivation.". In: Microbiological reviews. 59 (1), pp. 143-69.

Angioni, Alberto et al. (2006): „Chemical composition, seasonal variability, and antifungal activity of Lavandula stoechas L. ssp. stoechas essential oils from stem/leaves and flowers.". In: Journal of agricultural and food chemistry. 54 (12), pp. 4364-70.

Antony, Swapna P. et al. (2011): „Antimicrobial peptide gene expression in tiger shrimp, Penaeus monodon in response to gram-positive bacterial probionts and white spot virus challenge". In: Aquaculture. Elsevier B.V. 316 (1-4), pp. 6-12.

Austin, B.; Al-Zahrani, a. M. J. (1988): „The effect of antimicrobial compounds on the gastrointestinal microflora of rainbow trout, Salmo gairdneri Richardson". In: Journal of Fish Biology. 33 (1), pp. 1-14.

Avella, Matteo a. et al. (2011): „Use of Enterococcus faecium to improve common sole (Solea solea) larviculture". In: Aquaculture. Elsevier B.V. 315 (3-4), pp. 384-393.

Bader, Ammar; Cioni, Pier Luigi; Flamini, Guido (2010): „GC-MS analysis of the essential oils of ripe fruits, roots and flowering aerial parts of Elaeoselinum asclepium subsp. meoides growing in Sicily.". In: Natural product communications. 5 (7), pp. 1111-1114.

Bagamboula, C (2004): „Inhibitory effect of thyme and basil essential oils, carvacrol, thymol, estragol, linalool and p-cymene towards Shigella sonnei and S. flexneri". In: Food Microbiology. 21 (1), pp. 33-42.

Bakkali, F et al. (2008): „Biological effects of essential oils--a review.". In: Food and chemical toxicology : an international journal published for the British Industrial Biological Research Association. 46 (2), pp. 446-75.

Balcazar, J et al. (2008): „Characterization of probiotic properties of lactic acid bacteria isolated from intestinal microbiota of fish". In: Aquaculture. 278 (1-4), pp. 188-191.

Balcázar, José Luis et al. (2006): „The role of probiotics in aquaculture.". In: Veterinary microbiology. 114 (3-4), pp. 173-86.

Balcázar, José Luis et al. (2007): „In vitro competitive adhesion and production of antagonistic compounds by lactic acid bacteria against fish pathogens.". In: Veterinary microbiology. 122 (3-4), pp. 373-80.

Balcázar, José Luis; Rojas-Luna, Tyrone (2007): „Inhibitory activity of probiotic Bacillus subtilis UTM 126 against vibrio species confers protection against vibriosis in juvenile shrimp (Litopenaeus vannamei).". In: Current microbiology. 55 (5), pp. 409-12.

Bampidis, V a et al. (2006): „Effect of dried oregano leaves versus neomycin in treating newborn calves with colibacillosis.". In: Journal of veterinary medicine. A, Physiology, pathology, clinical medicine. 53 (3), pp. 154-6.

Bandyopadhyay, Partha; Mohapatra, Pradeep K Das (2009): „Effect of a probiotic bacterium Bacillus circulans PB7 in the formulated diets: on growth, nutritional quality and immunity of Catla catla (Ham.).". In: Fish physiology and biochemistry. 35 (3), pp. 467-78.

Barton, Bruce a (2002): „Stress in fishes: a diversity of responses with particular reference to changes in circulating corticosteroids.". In: Integrative and comparative biology. 42 (3), pp. 517-25.

Bates, J M (2006): „, Distinct signals from the microbiota promote different aspects of zebrafish gut differentiation". In: Dev. Biol. 297 , p. 374e386.

Bates, Jennifer M et al. (2007): „Intestinal alkaline phosphatase detoxifies lipopolysaccharide and prevents inflammation in zebrafish in response to the gut microbiota.". In: Cell host & microbe. 2 (6), pp. 371-82.

Ben Marzoug, Hajer Naceur et al. (2011): „Eucalyptus oleosa essential oils: chemical composition and antimicrobial and antioxidant activities of the oils from different plant parts (stems, leaves, flowers and fruits).". In: Molecules (Basel, Switzerland). 16 (2), pp. 1695-709.

Benchaar, C et al. (2008): „A review of plant-derived essential oils in ruminant nutrition and production". In: Animal Feed Science and Technology. 145 (1-4), pp. 209-228.

Borsheim, Knut Yngve (1993): „Native marine bacteriophages". In: FEMS Microbiology Letters. 102 (3-4), pp. 141-159.

Bricknell, Ian; Dalmo, Roy A (2005): „The use of immunostimulants in fish larval aquaculture.". In: Fish & shellfish immunology. 19 (5), pp. 457-72.

Bruhn, Jesper B et al. (2005): „Quorum sensing signal molecules (acylated homoserine lactones) in gram-negative fish pathogenic bacteria.". In: Diseases of aquatic organisms. 65 (1), pp. 43-52.

Burridge, Les et al. (2010): „Chemical use in salmon aquaculture: A review of current practices and possible environmental effects". In: Aquaculture. Elsevier B.V. 306 (1-4), pp. 7-23.

Burt, Sara (2004): „Essential oils: their antibacterial properties and potential applications in foods--a review.". In: International journal of food microbiology. 94 (3), pp. 223-53.

Busatta, C et al. (2008): „Application of Origanum majorana L. essential oil as an antimicrobial agent in sausage.". In: Food microbiology. 25 (1), pp. 207-11.

Busquet, M et al. (2006): „Plant extracts affect in vitro rumen microbial fermentation.". In: Journal of dairy science. 89 (2), pp. 761-71.

Cabello, Felipe C (2006): „Heavy use of prophylactic antibiotics in aquaculture: a growing problem for human and animal health and for the environment.". In: Environmental microbiology. 8 (7), pp. 1137-44.

Cahill, Marian (1990): „Bacterial flora of fishes: A review". In: Microbial Ecology. Springer New York 19 (1), pp. 21-41.

Caipang, Christopher Marlowe a et al. (2009): „In vivo modulation of immune response and antioxidant defense in Atlantic cod, Gadus morhua following oral administration of oxolinic acid and florfenicol.". In: Comparative biochemistry and physiology. Toxicology & pharmacology : CBP. Elsevier Inc. 150 (4), pp. 459-64.

Calsamiglia, S et al. (2007): „Invited review: Essential oils as modifiers of rumen microbial fermentation.". In: Journal of dairy science. 90 (6), pp. 2580-95.

Capkin, E; Altinok, I (2009): „Effects of dietary probiotic supplementations on prevention/treatment of yersiniosis disease.". In: Journal of applied microbiology. 106 (4), pp. 1147-53.

Carpenter, Christopher F; Chambers, Henry F (2004): „Daptomycin: another novel agent for treating infections due to drug-resistant gram-positive pathogens.". In: Clinical infectious diseases : an official publication of the Infectious Diseases Society of America. 38 (7), pp. 994-1000.

Carson, M.C. Bullock, G.; Bebak-Williams, J. (2002): „Determination of oxytetracycline residues in matrixes from a freshwater recirculating aquaculture system.". In: journal of AOAC International. 85 (2), pp. 341-348.

Chapman, C M C; Gibson, G R; Rowland, I (2011): „Health benefits of probiotics: are mixtures more effective than single strains?". In: European journal of nutrition. 50 (1), pp. 1-17.

Chen, Ruidong et al. (2010): „High yield expression of an AHL-lactonase from Bacillus sp. B546 in Pichia pastoris and its application to reduce Aeromonas hydrophila mortality in aquaculture.". In: Microbial cell factories. 9 , p. 39.

Chrisolite, B et al. (2008): „Distribution of luminescent Vibrio harveyi and their bacteriophages in a commercial shrimp hatchery in South India". In: Aquaculture. 275 (1-4), pp. 13-19.

Chu, W et al. (2011): „Isolation and characterization of new potential probiotic bacteria based on quorum-sensing system.". In: Journal of applied microbiology. 110 (1), pp. 202-8.

Chung, Ill-Min et al. (2011): „Composition of the essential oil constituents from leaves and stems of Korean Coriandrum sativum and their immunotoxicity activity on the Aedes aegypti L.". In: Immunopharmacology and Immunotoxicology. Informa Healthcare pp. 1-5.

Chythanya, R (2002): „Inhibition of shrimp pathogenic vibrios by a marine Pseudomonas I-2 strain". In: Aquaculture. 208 (1-2), pp. 1-10.

Cosentino, S et al. (1999): „In-vitro antimicrobial activity and chemical composition of Sardinian Thymus essential oils.". In: Letters in applied microbiology. 29 (2), pp. 130-5.

Cross, D E et al. (2007): „The effect of herbs and their associated essential oils on performance, dietary digestibility and gut microflora in chickens from 7 to 28 days of age.". In: British poultry science. 48 (4), pp. 496-506.

Crothers-Stomps, C et al. (2010): „Isolation of lytic bacteriophage against Vibrio harveyi.". In: Journal of applied microbiology. 108 (5), pp. 1744-50.

Das, B et al. (2006): „Antagonistic activity of cellular components of Pseudomonas species against Aeromonas hydrophila". In: Aquaculture. 253 (1-4), pp. 17-24.

Das, Surajit; Ward, Louise R.; Burke, Chris (2010): „Screening of marine Streptomyces spp. for potential use as probiotics in aquaculture". In: Aquaculture. Elsevier B.V. 305 (1-4), pp. 32-41.

Defoirdt, Tom et al. (2004): „Disruption of bacterial quorum sensing: an unexplored strategy to fight infections in aquaculture". In: Aquaculture. 240 (1-4), pp. 69-88.

Defoirdt, Tom et al. (2005): „The impact of mutations in the quorum sensing systems of Aeromonas hydrophila, Vibrio anguillarum and Vibrio harveyi on their virulence towards gnotobiotically cultured Artemia franciscana.". In: Environmental microbiology. 7 (8), pp. 1239-47.

Defoirdt, Tom et al. (2007): „Alternatives to antibiotics to control bacterial infections: luminescent vibriosis in aquaculture as an example.". In: Trends in biotechnology. 25 (10), pp. 472-9.

Defoirdt, Tom et al. (2008): „Quorum sensing and quorum quenching in Vibrio harveyi: lessons learned from in vivo work.". In: The ISME journal. 2 (1), pp. 19-26.

Defoirdt, Tom; Sorgeloos, Patrick; Bossier, Peter (2011): „Alternatives to antibiotics for the control of bacterial disease in aquaculture.". In: Current opinion in microbiology. 14 (3), pp. 251-8.

Delaquis, Pascal J et al. (2002): „Antimicrobial activity of individual and mixed fractions of dill, cilantro, coriander and eucalyptus essential oils.". In: International journal of food microbiology. 74 (1-2), pp. 101-9.

Depaola, a; Peller, J T; Rodrick, G E (1995): „Effect of oxytetracycline-medicated feed on antibiotic resistance of gram-negative bacteria in catfish ponds.". In: Applied and environmental microbiology. 61 (9), p. 3513.

Desriac, Florie et al. (2010): „Bacteriocin as weapons in the marine animal-associated bacteria warfare: inventory and potential applications as an aquaculture probiotic.". In: Marine drugs. 8 (4), pp. 1153-77.

Dierckens, K et al. (2009): „Development of a bacterial challenge test for gnotobiotic sea bass (Dicentrarchus labrax) larvae.". In: Environmental microbiology. 11 (2), pp. 526-33.

Dimitroglou, Arkadios et al. (2011): „Microbial manipulations to improve fish health and production--a Mediterranean perspective.". In: Fish & shellfish immunology. Elsevier Ltd 30 (1), pp. 1-16.

Donsì, Francesco et al. (2011): „Design of nanoemulsion-based delivery systems of natural antimicrobials: Effect of the emulsifier.". In: Journal of biotechnology.

Doulgeraki, Agapi I; Paramithiotis, Spiros; Nychas, George-John E (2011): „Characterization of the Enterobacteriaceae community that developed during storage of minced beef under aerobic or modified atmosphere packaging conditions.". In: International journal of food microbiology. Elsevier B.V. 145 (1), pp. 77-83.

Duran, G.M; Marshall, Douglas (2005): „Ready-to-eat shrimp as an international vehicle of antibiotic-resistant bacteria.". In: Journal of Food Protection. 68 (11), pp. 2395-2401.

Ellingsen, Odd F et al. (2002): „Dosage regime experiments with oxolinic acid and flumequine in Atlantic salmon (Salmo salar L) held in seawater". In: Aquaculture. 209 (1-4), pp. 19-34.

Ennajar, Monia et al. (2011): „Influence of the process, season, and origin on volatile composition and antioxidant activity of Juniperus phoenicea L. leaves essential oils.". In: Journal of food science. 76 (2), pp. C224-30.

FAO Fisheries and Aquaculture Department Food And Agriculture Organization Of The United Nations. 2010. The State of World Fisheries and Aquaculture.

Farzanfar, Ali (2006): „The use of probiotics in shrimp aquaculture.". In: FEMS immunology and medical microbiology. 48 (2), pp. 149-58.

Federle, Michael J; Bassler, Bonnie L (2003): „Interspecies communication in bacteria". In: Journal of Clinical Investigation. 112 (9), pp. 1291-1299.

Fernández-Alarcón, C et al. (2010): „Detection of the floR Gene in a Diversity of Florfenicol Resistant Gram-Negative Bacilli from Freshwater Salmon Farms in Chile.". In: Zoonoses and public health. 57 , pp. 181-188.

Fjellheim, a et al. (2007): „Vibrionaceae dominates the microflora antagonistic towards Listonella anguillarum in the intestine of cultured Atlantic cod (Gadus morhua L.) larvae". In: Aquaculture. 269 (1-4), pp. 98-106.

Fjellheim, Anders Jón et al. (2010): „Selection of candidate probionts by two different screening strategies from Atlantic cod (Gadus morhua L.) larvae.". In: Veterinary microbiology. 144 (1-2), pp. 153-9.

Fraser, G R et al. (2007): „Assessment of the effects of cinnamon leaf oil on rumen microbial fermentation using two continuous culture systems.". In: Journal of dairy science. 90 (5), pp. 2315-28.

Fratianni, Florinda et al. (2010): „Preservation of chicken breast meat treated with thyme and balm essential oils.". In: Journal of food science. 75 (8), pp. M528-35.

Froehlicher, Mirjam et al. (2009): „Zebrafish (Danio rerio) neuromast: promising biological endpoint linking developmental and toxicological studies.". In Aquatic toxicology (Amsterdam, Netherlands). 95 (4), pp. 307-19.

García-García, Rebeca; López-Malo, Aurelio; Palou, Enrique (2011): „Bactericidal action of binary and ternary mixtures of carvacrol, thymol, and eugenol against Listeria innocua.". In: Journal of food science. 76 (2), pp. M95-100.

Giraud, Etienne et al. (2006): „Survey of antibiotic resistance in an integrated marine aquaculture system under oxolinic acid treatment.". In: FEMS microbiology ecology. 55 (3), pp. 439-48.

Gómez, Geovanny; Balcázar, José Luis; Shen, M A (2007): „Probiotics as Control Agents in Aquaculture". In: Journal of Ocean University of China. 6 (1), pp. 76-79.

Grondel, J.L. Gloudemans, A.G.M; Muiswinkel, Willem B van (1985): „The influence of antibiotics on the immune system. II. Modulation of fish leukocyte responses in culture". In: Veterinary Immunology and Immunopathology. 6 (3), pp. 251-260.

Hagi, Tatsuro; Hoshino, Takayuki (2009): „Screening and Characterization of Potential Probiotic Lactic Acid Bacteria from Cultured Common Carp Intestine". In: Bioscience, Biotechnology, and Biochemistry. 73 (7), pp. 1479-1483.

Hai, Ngo Van; Buller, Nicky; Fotedar, Ravi (2009): „The use of customised probiotics in the cultivation of western king prawns (Penaeus latisulcatus Kishinouye, 1896).". In: Fish & shellfish immunology. Elsevier Ltd 27 (2), pp. 100-4.

Hammer, K a; Carson, C F; Riley, T V (1999): „Antimicrobial activity of essential oils and other plant extracts.". In: Journal of applied microbiology. 86 (6), pp. 985-90.

Hentschel, Dirk M et al. (2005): „Acute renal failure in zebrafish: a novel system to study a complex disease.". In: American journal of physiology. Renal physiology. 288 (5), pp. F923-9.

Hong, Huynh a; Duc, Le Hong; Cutting, Simon M (2005): „The use of bacterial spore formers as probiotics.". In: FEMS microbiology reviews. 29 (4), pp. 813-35.

Huber, I. et al. (2004): „Phylogenetic analysis and in situ identification of the intestinal microbial community of rainbow trout (Oncorhynchus mykiss, Walbaum)". In: Journal of Applied Microbiology. 96 (1), pp. 117-132.

Imbeault, Sandra et al. (2006): „Using Bacteriophages to Prevent Furunculosis Caused by Aeromonas salmonicida in Farmed Brook Trout". In: Journal of Aquatic Animal Health. 18 (3), pp. 203-214.

Immanuel, G et al. (2009): „Dietary medicinal plant extracts improve growth, immune activity and survival of tilapia Oreochromis mossambicus.". In: Journal of fish biology. 74 (7), pp. 1462-75.

Irianto, A; Austin, B (2002): „Review Probiotics in aquaculture". In: Journal of Fish Diseases. (1997), pp. 633-642.

Irianto, a; Robertson, P a W; Austin, B (2003): „Oral administration of formalin-inactivated cells of Aeromonas hydrophila A3-51 controls infection by atypical A. salmonicida in goldfish, Carassius auratus (L.)". In: Journal of Fish Diseases. 26 (2), pp. 117-120.

Jacobs, Liezl; Chenia, Hafizah Y (2006): „Characterization of integrons and tetracycline resistance determinants in Aeromonas spp . isolated from South African aquaculture systems". In: International Journal of Food Microbiology.

Janczyk, Pawel et al. (2008): „Effect of thymol on microbial diversity in the porcine jejunum.". In: International journal of food microbiology. 126 (1-2), pp. 258-61.

Jang, I et al. (2007): „Effect of a commercial essential oil on growth performance, digestive enzyme activity and intestinal microflora population in broiler chickens". In: Animal Feed Science and Technology. 134 (3-4), pp. 304-315.

Karakaya, Sibel et al. (2011): „Antioxidant and Antimicrobial Activities of Essential Oils Obtained from Oregano (Origanum vulgare ssp. hirtum) by Using Different Extraction Methods". In: Journal of Medicinal Food. Mary Ann Liebert, Inc., publishers 14 (6), pp. 645-652.

Karunasagar, Iddya et al. (1994): „Mass mortality of Penaeus monodon larvae due to antibiotic-resistant Vibrio harveyi infection". In: Aquaculture. 128 (3-4), pp. 203-209.

Kerry, J (1997): „Changes in oxytetracycline resistance of intestinal microflora following oral administration of this agent to Atlantic salmon (Salmo salar L.) smolts in a marine environment". In: Aquaculture. 157 (3-4), pp. 187-195.

Kesarcodiwatson, a et al. (2008): „Probiotics in aquaculture: The need, principles and mechanisms of action and screening processes". In: Aquaculture. 274 (1), pp. 1-14.

Kim, Jeongmok; Marshall, Maurice R.; Wei, Cheng-i (1995): „Antibacterial activity of some essential oil components against five foodborne pathogens". In: Journal of Agricultural and Food Chemistry. 43 (11), pp. 2839-2845.

Kim, Ji Hyung et al. (2010): „Isolation and identification of bacteriophages infecting ayu Plecoglossus altivelis altivelis specific Flavobacterium psychrophilum.". In: Veterinary microbiology. 140 (1-2), pp. 109-15.

Kiskó, Gabriella; Roller, Sibel (2005): „Carvacrol and p-cymene inactivate Escherichia coli O157:H7 in apple juice.". In: BMC microbiology. 5 , p. 36.

Kostaki, Maria et al. (2009): „Combined effect of MAP and thyme essential oil on the microbiological, chemical and sensory attributes of organically aquacultured sea bass (Dicentrarchus labrax) fillets.". In: Food microbiology. Elsevier Ltd 26 (5), pp. 475-82.

Kumarasamy, Karthikeyan K et al. (2010): „Emergence of a new antibiotic resistance mechanism in India, Pakistan, and the UK: a molecular, biological, and epidemiological study.". In: The Lancet infectious diseases. 10 (9), pp. 597-602.

Lalumera, Giorgia Mary et al. (2004): „Preliminary investigation on the environmental occurrence and effects of antibiotics used in aquaculture in Italy.". In: Chemosphere. 54 (5), pp. 661-8.

Lambert, R J et al. (2001): „A study of the minimum inhibitory concentration and mode of action of oregano essential oil, thymol and carvacrol.". In: Journal of applied microbiology. 91 (3), pp. 453-62.

Lategan, M (2004): „Biocontrol of saprolegniosis in silver perch Bidyanus bidyanus (Mitchell) by Aeromonas media strain A199". In: Aquaculture. 235 (1-4), pp. 77-88.

Lategan, M.J. et al. (2006): „An inhibitory substance produced by Aeromonas media A199, an aquatic probiotic". In: Aquaculture 254 (1-4), pp. 115-124.

Lategan, M.J. Torpy, F.R.; Gibson, L.F. (2004): „Control of saprolegniosis in the eel Anguilla australis Richardson, by Aeromonas media strain A199". In: Aquaculture. 240 (1-4), pp. 19-27.

Levin, Bruce R; Bull, James J (2004): „Population and evolutionary dynamics of phage therapy.". In: Nature reviews. Microbiology. 2 (2), pp. 166-73.

Lin, Y T; Labbe, R G; Shetty, Kalidas (2004): „Inhibition of Listeria monocytogenes in Fish and Meat Systems by Use of Oregano and Cranberry Phytochemical Synergies". In: Society. 70 (9), pp. 5672-5678.

Liu, C-H et al. (2009): „Improvement in the growth performance of white shrimp, Litopenaeus vannamei, by a protease-producing probiotic, Bacillus subtilis E20, from natto.". In: Journal of applied microbiology. 107 (3), pp. 1031-41.

Lunden, T. et al. (1998): „Influence of oxytetracycline and oxolinic acid on the immune response of rainbow trout (Oncorhynchus mykiss)". In: Fish and Shellfish Immunology. 8 (3), pp. 217-230.

Lunden, Tuula et al. (1999): „Effect of florfenicol on the immune response of rainbow trout (Oncorhynchus mykiss)". In: Veterinary immunology and immunopathology. 67 (4), pp. 317-325.

Lunden, Tuula; Bylund, Göran (2002): „Effect of sulphadiazine and trimethoprim on the immune response of rainbow trout (Oncorhynchus mykiss).". In: Veterinary immunology and immunopathology. 85 (1-2), pp. 99-108.

Maenner, K; Vahjen, W; Simon, O (2011): „Studies on the effects of essential-oil-based feed additives on performance, ileal nutrient digestibility, and selected bacterial groups in the gastrointestinal tract of piglets.". In: Journal of animal science. 89 (7), pp. 2106-12.

Magnadottir, Bergljot (2010): „Immunological control of fish diseases.". In: Marine biotechnology (New York, N.Y.). 12 (4), pp. 361-79.

Mahmoud, Barakat S.M et al. (2004): „Bacterial microflora of carp (Cyprinus carpio) and its shelf-life extension by essential oil compounds". In: Food Microbiology. 21 (6), pp. 657-666.

Martínez, José L (2008): „Antibiotics and antibiotic resistance genes in natural environments.". In: Science (New York, N.Y.). 321 (5887), pp. 365-7.

Martinsen, B; Horsberg, T E (1995): „Comparative single-dose pharmacokinetics of four quinolones, oxolinic acid, flumequine, sarafloxacin, and enrofloxacin, in Atlantic salmon (Salmo salar) held in seawater at 10 degrees C.". In: Antimicrobial agents and chemotherapy. 39 (5), pp. 1059-64.

Mastromatteo, Marianna et al. (2010): „Shelf life of ready to use peeled shrimps as affected by thymol essential oil and modified atmosphere packaging.". In: International journal of food microbiology. Elsevier B.V. 144 (2), pp. 250-6.

Mathur, M D; Vidhani, S; Mehndiratta, P L (2003): „Bacteriophage therapy: an alternative to conventional antibiotics.". In: The Journal of the Association of Physicians of India. 51 (June), pp. 593-6.

Mcintosh, F M et al. (2003): „Effects of Essential Oils on Ruminal Microorganisms and Their Protein Metabolism". In: Society. 69 (8), pp. 5011-5014.

Mejlholm, O; Dalgaard, P (2002): „Antimicrobial effect of essential oils on the seafood spoilage micro-organism Photobacterium phosphoreum in liquid media and ® sh products". In: Letters in Applied Microbiology. pp. 27-31.

Merrifield, Daniel L. et al. (2010): „The current status and future focus of probiotic and prebiotic applications for salmonids". In: Aquaculture. Elsevier B.V. 302 (1-2), pp. 1-18.

Miranda, C , Rojas R. (2007): „ Occurrence of florfenicol resistance in bacteria associated with two Chilean salmon farms with different history of antibacterial usage ". In: Aquaculture 266 39–46

Miranda, C , Zemelman R. (2002): „Bacterial resistance to oxytetracycline in Chilean salmon farming". In: Aquaculture. 212 (1-4), pp. 31-47.

Miranda, Claudio D et al. (2003): „Diversity of Tetracycline Resistance Genes in Bacteria from Chilean Salmon Farms". In: Society. 47 (3), pp. 883-888.

Moriarty, D J W (1998): „Control of luminous Vibrio species in penaeid aquaculture ponds". In: Aquaculture. 164 (1-4), pp. 351-358.

Morohoshi, T et al. (2004): „Identification of quorum-sensing signal molecules and the LuxRI homologs in fish pathogen Edwardsiella tarda.". In: Biosci Bioeng. 98 (4), pp. 274-276.

Muhl, a; Liebert, F (2007): „Growth and parameters of microflora in intestinal and faecal samples of piglets due to application of a phytogenic feed additive.". In: Journal of animal physiology and animal nutrition. 91 (9-10), pp. 411-8.

Musthafa, K Syed et al. (2011): „Antipathogenic potential of marine Bacillus sp. SS4 on N-acyl-homoserine-lactone-mediated virulence factors production in Pseudomonas aeruginosa (PAO1)". In: Journal of Biosciences. 36 (1), pp. 55-67.

Nagegowda, Dinesh a (2010): „Plant volatile terpenoid metabolism: biosynthetic genes, transcriptional regulation and subcellular compartmentation.". In: FEBS letters. Federation of European Biochemical Societies 584 (14), pp. 2965-73.

Nakai, T et al. (1999): „Protective effects of bacteriophage on experimental Lactococcus garvieae infection in yellowtail.". In: Diseases of aquatic organisms. Faculty of Applied Biological Science, Hiroshima University, Higashihiroshima, Japan. nakaitt@ipc.hiroshima-u.ac.jp 37 (1), pp. 33-41.

Nakai, Toshihiro; Park, Se Chang (2002): „Bacteriophage therapy of infectious diseases in aquaculture.". In: Research in microbiology. 153 (1), pp. 13-8.

Nakajima, T. et al. (1983): „Transmission of R plasmids in Vibrio anguillarum to Vibrio cholera". In: Microbiology and Immunology. 27 , pp. 195-198.

Nakayama, T; Lu, H; Nomura, N (2009): „Inhibitory effects of Bacillus probionts on growth and toxin production of Vibrio harveyi pathogens of shrimp.". In: Letters in applied microbiology. 49 (6), pp. 679-84.

Natrah, F M I; Defoirdt, Tom; et al. (2011): „Disruption of bacterial cell-to-cell communication by marine organisms and its relevance to aquaculture.". In: Marine biotechnology (New York, N.Y.). 13 (2), pp. 109-26.

Natrah, F M I; Ruwandeepika, H a Darshanee; et al. (2011): „Regulation of virulence factors by quorum sensing in Vibrio harveyi.". In: Veterinary microbiology. Elsevier B.V.

Navarrete, Paola et al. (2010): „Effect of Thymus vulgaris essential oil on intestinal bacterial microbiota of rainbow trout, Oncorhynchus mykiss (Walbaum) and bacterial isolates". In: Aquaculture Research. pp. 667-678.

Navarrete, Paola et al. (o. J.): „Oxytetracycline treatment reduces bacterial diversity of intestinal microbiota of Atlantic salmon (Salmo salar)". In:

Nayak, S K (2010): „Probiotics and immunity: a fish perspective.". In: Fish & shellfish immunology. Elsevier Ltd 29 (1), pp. 2-14.

Newaj-Fyzul, a et al. (2007): „Bacillus subtilis AB1 controls Aeromonas infection in rainbow trout (Oncorhynchus mykiss, Walbaum).". In: Journal of applied microbiology. 103 (5), pp. 1699-706.

Nhan, D T et al. (2010): „Quorum quenching bacteria protect Macrobrachium rosenbergii larvae from Vibrio harveyi infection.". In: Journal of applied microbiology. 109 (3), pp. 1007-16.

Ninawe, A S; Selvin, Joseph (2009): „Probiotics in shrimp aquaculture: Avenues and challenges". In: Critical Reviews in Microbiology. Informa Healthcare 35 (1), pp. 43-66.

Ochoa-Solano, Leonel; Olmos-Soto, Jorge (2006): „The functional property of Bacillus for shrimp feeds.". In: Food microbiology. 23 (6), pp. 519-25.

Olmos, Jorge et al. (2011): „Functional Feed Assessment on Litopenaeus vannamei Using 100% Fish Meal Replacement by Soybean Meal, High Levels of Complex Carbohydrates and Bacillus Probiotic Strains.". In: Marine drugs. 9 (6), pp. 1119-32.

Ortuño, J. Esteban, M.A.; Meseguer, J. (2001): „Effects of short-term crowding stress on the gilthead seabream (Sparus aurata L.) innate immune response". In: Fish and Shellfish Immunology. 11 (2), pp. 187-197.

Park, S C et al. (2000): „Isolation of bacteriophages specific to a fish pathogen, Pseudomonas plecoglossicida, as a candidate for disease control.". In: Applied and environmental microbiology. 66 (4), pp. 1416-22.

Park, Se Chang; Nakai, Toshihiro (2003): „Bacteriophage control of Pseudomonas plecoglossicida infection in ayu Plecoglossus altivelis". In Diseases Of Aquatic Organisms. 53 , pp. 33-39.

Pérez-Conesa, D et al. (2011): „Inactivation of Listeria monocytogenes and Escherichia coli O157:H7 biofilms by micelle-encapsulated eugenol and carvacrol.". In: Journal Of Food Protection. 74 (1), pp. 55-62.

Pérez-Sánchez, T et al. (2011): „Identification and characterization of lactic acid bacteria isolated from rainbow trout, Oncorhynchus mykiss (Walbaum), with inhibitory activity against Lactococcus garvieae.". In: Journal of fish diseases. 34 (7), pp. 499-507.

Pichersky, E; Noel, J; Dudareva, N (2006): „Biosynthesis of plant volatiles: Nature's Diversity and ingenuity". In: Science. 10 , pp. 808-811.

Pickering, A. D.; Pottinger, T.G. (1989): „Stress responses and disease resistance in salmonid fish: Effects of chronic elevation of plasma cortisol". In: Fish physiology and biochemistry. 7 (1-4), pp. 253-258.

Piva, a et al. (2007): „Lipid microencapsulation allows slow release of organic acids and natural identical flavors along the swine intestine.". In: Journal of animal science. 85 (2), pp. 486-93.

Planas, M et al. (2006): „Probiotic effect in vivo of Roseobacter strain 27-4 against Vibrio (Listonella) anguillarum infections in turbot (Scophthalmus maximus L.) larvae". In: Aquaculture. 255 (1-4), pp. 323-333.

Ponce, a.G. et al. (2003): „Antimicrobial activity of essential oils on the native microflora of organic Swiss chard". In: LWT - Food Science and Technology. 36 (7), pp. 679-684.

Pouliquen, H et al. (2007): „Comparative hydrolysis and photolysis of four antibacterial agents (oxytetracycline oxolinic acid, flumequine and florfenicol) in deionised water, freshwater and seawater under abiotic conditions". In: Aquaculture. 262 (1), pp. 23-28.

Prado, Susana; Romalde, Jesús L; Barja, Juan L (2010): „Review of probiotics for use in bivalve hatcheries.". In: Veterinary microbiology. 145 (3-4), pp. 187-97.

Preetha, R et al. (2007): „Optimization of carbon and nitrogen sources and growth factors for the production of an aquaculture probiotic (Pseudomonas MCCB 103) using response surface methodology.". In: Journal of applied microbiology. 102 (4), pp. 1043-51.

Pyrgotou, Nikoletta et al. (2010): „Quality assessment of salted, modified atmosphere packaged rainbow trout under treatment with oregano essential oil.". In: Journal of food science. 75 (7), pp. M406-11.

Rali, Topul; Wossa, Stewart W; Leach, David N (2007): „Comparative chemical analysis of the essential oil constituents in the bark, heartwood and fruits of Cryptocarya massoy (Oken) Kosterm. (Lauraceae) from Papua New Guinea.". In: Molecules Basel Switzerland. 12 (2), pp. 149-154.

Randrianarivelo, R et al. (2010): „Novel alternative to antibiotics in shrimp hatchery: effects of the essential oil of Cinnamosma fragrans on survival and bacterial concentration of Penaeus monodon larvae.". In: Journal of applied microbiology. 109 (2), pp. 642-50.

Rasch, M et al. (2007): „Quorum sensing signals are produced by Aeromonas salmonicida and quorum sensing inhibitors can reduce production of a potential virulence factor". In: Diseases of Aquatic Organisms. 78 (2), pp. 105-113.

Rawls, John F; Samuel, Buck S; Gordon, Jeffrey I (2004): „Gnotobiotic zebrafish reveal evolutionarily conserved responses to the gut microbiota.". In: Proceedings of the National Academy of Sciences of the United States of America. 101 (13), pp. 4596-601.

Rawn, Dorothea F. K. et al. (2009): „Analysis of fin and non-fin fish products for azamethiphos and dichlorvos residues from the Canadian retail market". In: International Journal of Food Science & Technology. 44 (8), pp. 1510-1516.

Reid, Gregor et al. (2003): „Potential Uses of Probiotics in Clinical Practice". In: Society. 16 (4), pp. 658-672.

Reyes-Becerril, Martha et al. (2008): „Oral delivery of live yeast Debaryomyces hansenii modulates the main innate immune parameters and the expression of immune-relevant genes in the gilthead seabream (Sparus aurata L.).". In: Fish & shellfish immunology. 25 (6), pp. 731-9.

Rijkers, G.T. Oosterom, R. Van; Muiswinkel, W.B. Van (1981): „The immune system of cyprinid fish. Oxytetracycline and the regulation of humoral immunity in carp (Cyprinus carpio)". In: Veterinary immunology and immunopathology. 2 (3), pp. 281-290.

Ringø, Einar et al. (2004): „Damaging effect of the fish pathogen Aeromonas salmonicida ssp. salmonicida on intestinal enterocytes of Atlantic salmon (Salmo salar L.).". In: Cell and tissue research. 318 (2), pp. 305-11.

Romero, Jaime; Navarrete, Paola (2006): „16S rDNA-based analysis of dominant bacterial populations associated with early life stages of coho salmon (Oncorhynchus kisutch).". In: Microbial ecology. 51 (4), pp. 422-30.

Ruwandeepika, H A Darshanee et al. (2011): „Quorum sensing regulation of virulence gene expression in Vibrio harveyi in vitro and in vivo during infection of gnotobiotic brine shrimp larvae". In: Environmental Microbiology Reports. Blackwell Publishing Ltd p. no-no.

Saglam, Naim; Yonar, M Enis (2009): „Effects of sulfamerazine on selected haematological and immunological parameters in rainbow trout (Onchorhynchus mykiss , Walbaum, 1792)". In: Aquaculture Research. 40 (4), pp. 395-404.

Salinas, Irene et al. (2005): „Dietary administration of Lactobacillus delbrüeckii and Bacillus subtilis, single or combined, on gilthead seabream cellular innate immune responses.". In: Fish & shellfish immunology. 19 (1), pp. 67-77.

Samuelsen, O (2006): „Pharmacokinetics of quinolones in fish: A review". In: Aquaculture. 255 (1-4), pp. 55-75.

Samuelsen, Ole B. Torsvik, V.; Ervik, A. (1992): „Long-range changes in oxytetracycline concentration and bacterial resistance towards oxytetracycline in a fish farm sediment after medication". In: Science of the Total Environment. 114 , pp. 25-36.

Sarter, Samira et al. (2011): „Antimicrobial Effects of Essential Oils of Cinnamosma fragrans on the Bacterial Communities in the Rearing Water of Penaeus monodon Larvae". In: Vector-Borne and Zoonotic Diseases. Mary Ann Liebert, Inc., publishers 11 (4), pp. 433-437.

Schnabel, E L; Jones, a L (1999): „Distribution of tetracycline resistance genes and transposons among phylloplane bacteria in Michigan apple orchards.". In: Applied and environmental microbiology. 65 (11), pp. 4898-907.

Schulze, a et al. (2006): „Bacterial diversity in a marine hatchery: Balance between pathogenic and potentially probiotic bacterial strains". In: Aquaculture. 256 (1-4), pp. 50-73.

Seyfried, Erin E et al. (2010): „Occurrence of tetracycline resistance genes in aquaculture facilities with varying use of oxytetracycline.". In: Microbial ecology. 59 (4), pp. 799-807.

Shekarforoush, S et al. (2007): „Effects of storage temperatures and essential oils of oregano and nutmeg on the growth and survival of Escherichia coli O157:H7 in barbecued chicken used in Iran". In: Food Control. 18 (11), pp. 1428-1433.

Shivu, Mockshanath M et al. (2007): „Molecular characterization of Vibrio harveyi bacteriophages isolated from aquaculture environments along the coast of India.". In: Environmental microbiology. 9 (2), pp. 322-31.

Smith, Peter; Samuelsen, Ole B (1996): „Estimates of the significance of out-washing of oxytetracycline from sediments under Atlantic salmon sea-cages". In: Aquaculture. 144 (1-3), pp. 17-26.

Stenholm, Anne Rønnest; Dalsgaard, Inger; Middelboe, Mathias (2008): „Isolation and characterization of bacteriophages infecting the fish pathogen Flavobacterium psychrophilum.". In: Applied and environmental microbiology. 74 (13), pp. 4070-8.

Stolte, Ellen H et al. (2009): „The immune response differentially regulates Hsp70 and glucocorticoid receptor expression in vitro and in vivo in common carp (Cyprinus carpio L.).". In: Fish & shellfish immunology. Elsevier Ltd 27 (1), pp. 9-16.

Stroh, J et al. (1998): „Evaluation of the acute toxicity to juvenile Pacific coho salmon and rainbow trout of some plant essential oils, a formulated product, and the carrier.". In: Bulletin of environmental contamination and toxicology. 60 (6), pp. 923-30.

Ström-Bestor, M; Wiklund, T (2011): „Inhibitory activity of Pseudomonas sp. on Flavobacterium psychrophilum, in vitro.". In: Journal of fish diseases. 34 (4), pp. 255-64.

Sun, Yun-Zhang et al. (2010): „Probiotic applications of two dominant gut Bacillus strains with antagonistic activity improved the growth performance and immune responses of grouper Epinephelus coioides.". In: Fish & shellfish immunology. Elsevier Ltd 29 (5), pp. 803-9.

Tafalla, C. et al. (2002): „In vivo and in vitro effect of oxytetracycline treatment on the immune response of turbot, Scophthalmus maximus (L.)". In: Journal of Fish Diseases. 22 (4), pp. 271-276.

Tamminen, Manu et al. (2011): „Tetracycline resistance genes persist at aquaculture farms in the absence of selection pressure.". In: Environmental science & technology. 45 (2), pp. 386-91.

Teasdale, Margaret E et al. (2009): „Secondary metabolites produced by the marine bacterium Halobacillus salinus that inhibit quorum sensing-controlled phenotypes in gram-negative bacteria.". In: Applied and environmental microbiology. 75 (3), pp. 567-72.

Thompson, J et al. (2010): „An in vitro and in vivo assessment of the potential of Vibrio spp. as probiotics for the Pacific white shrimp, Litopenaeus vannamei.". In: Journal of applied microbiology. 109 (4), pp. 1177-87.

Tinh, Nguyen Thi Ngoc et al. (2007): „N-acyl homoserine lactone-degrading microbial enrichment cultures isolated from Penaeus vannamei shrimp gut and their probiotic properties in Brachionus plicatilis cultures.". In: FEMS microbiology ecology. 62 (1), pp. 45-53.

Tinh, Nguyen Thi Ngoc et al. (2008): „A review of the functionality of probiotics in the larviculture food chain.". In: Marine biotechnology (New York, N.Y.). 10 (1), pp. 1-12.

Tovar-Ramírez, D. et al. (2010): „Dietary probiotic live yeast modulates antioxidant enzyme activities and gene expression of sea bass (Dicentrarchus labrax) larvae". In: Aquaculture. Elsevier B.V. 300 (1-4), pp. 142-147.

Ultee, A; Bennik, M H J; Moezelaar, R (2002): „The Phenolic Hydroxyl Group of Carvacrol Is Essential for Action against the Food-Borne Pathogen Bacillus cereus". In: Society. 68 (4), pp. 1561-1568.

Van Hai, Ngo; Buller, Nicky; Fotedar, Ravi (2009): „The use of customised probiotics in the cultivation of western king prawns (Penaeus latisulcatus Kishinouye, 1896).". In: Fish & shellfish immunology. Elsevier Ltd 27 (2), pp. 100-4.

Van Vuuren, Sandy F et al. (2010): „Encapsulation of essential oils within a polymeric liposomal formulation for enhancement of antimicrobial efficacy.". In: Natural product communications. 5 (9), pp. 1401-1408.

Vaseeharan, B; Ramasamy, P (2003): „Control of pathogenic Vibrio spp. by Bacillus subtilis BT23, a possible probiotic treatment for black tiger shrimp Penaeus monodon.". In: Letters in applied microbiology. 36 (2), pp. 83-7.

Vazquez, J; Gonzalez, M; Murado, M (2005): „Effects of lactic acid bacteria cultures on pathogenic microbiota from fish". In: Aquaculture. 245 (1-4), pp. 149-161.

Verner-Jeffreys, David W. et al. (2007): „Furunculosis in Atlantic salmon (Salmo salar L.) is not readily controllable by bacteriophage therapy". In: Aquaculture. 270 (1-4), pp. 475-484.

Verschuere, L et al. (2000): „Probiotic bacteria as biological control agents in aquaculture.". In: Microbiology and molecular biology reviews : MMBR. 64 (4), pp. 655-71.

Villamil, L et al. (2002): „Evaluation of Immunomodulatory Effects of Lactic Acid Bacteria in Turbot (Scophthalmus maximus)". In: Society. 9 (6), pp. 1318-1323.

Vine, Niall G; Leukes, Winston D; Kaiser, Horst (2006): „Probiotics in marine larviculture.". In: FEMS microbiology reviews. 30 (3), pp. 404-27.

Vinod, M et al. (2006): „Isolation of Vibrio harveyi bacteriophage with a potential for biocontrol of luminous vibriosis in hatchery environments". In: Aquaculture. 255 (1-4), pp. 117-124.

Walakira, J K et al. (2008): „Identification and characterization of bacteriophages specific to the catfish pathogen, Edwardsiella ictaluri.". In: Journal of applied microbiology. 105 (6), pp. 2133-42.

Wang, Q et al. (2009): „In vitro evaluation of the activity of microencapsulated carvacrol against Escherichia coli with K88 pili.". In: Journal of applied microbiology. 107 (6), pp. 1781-8.

Windisch, W et al. (2008): „Use of phytogenic products as feed additives for swine and poultry.". In: Journal of animal science. 86 (14 Suppl), pp. E140-8.

Wishkovsky, A. Roberson, B.; Hetrick, F. M. (1987): „In vitro suppression of the phagocytic response of fish macrophages by tetracyclines". In: Journal of fish biology. 31 , pp. 61-65.

Yang, W Z et al. (2007): „Effects of garlic and juniper berry essential oils on ruminal fermentation and on the site and extent of digestion in lactating cows.". In: Journal of dairy science. 90 (12), pp. 5671-81.

Yeh, Ruo-Yun et al. (2009): „Evaluation of the antibacterial activity of leaf and twig extracts of stout camphor tree, Cinnamomum kanehirae, and the effects on immunity and disease resistance of white shrimp, Litopenaeus vannamei.". In: Fish & shellfish immunology. 27 (1), pp. 26-32.

You, JianLan et al. (2007): „Inhibition of Vibrio biofilm formation by a marine actinomycete strain A66.". In: Applied microbiology and biotechnology. 76 (5), pp. 1137-44.

Zheng, Z.L. et al. (2009): „Evaluation of oregano essential oil (Origanum heracleoticum L.) on growth, antioxidant effect and resistance against Aeromonas hydrophila in channel catfish (Ictalurus punctatus)". In: Aquaculture. Elsevier B.V. 292 (3-4), pp. 214-218.

Zhou, X; Wang, Y; Li, W (2009): „Effect of probiotic on larvae shrimp (Penaeus vannamei) based on water quality, survival rate and digestive enzyme activities". In: Aquaculture. Elsevier B.V. 287 (3-4), pp. 349-353.

7

The Use of Antibiotics in Shrimp Farming

M.C. Bermúdez-Almada* and A. Espinosa-Plascencia
Research Center for Food and Development, AC, Hermosillo, Sonora México

1. Introduction

Global aquaculture has grow dramatically over the past 50 years to around 52.5 million tones in 2008 worth US$98.5 billon and accounting for around 50 per cent of the world's food fish supply. Asia dominates this production, accounting for 89 per cent by volume and 79 per cent by value, with China by far the largest producer (32.7 million tonnes in 2008). The rapid growth in the region has been driven by a variety of factors, including pre-existing aquaculture practices, population and economic growth, relaxed regulatory framework and expanding export opportunities.

Aquaculture development in Europe and North America was rapid during the 1980s-1990s but has since stagnated, probably owing to regulatory restrictions on sites and other competitive factors, although as markets for fish and seafood they have continue to grow (Bostock et al., 2010).

In contrast to other animal production sectors, aquaculture is highly dynamic and characterized by an enormous diversity of species raised both in natural and artificial systems (Walter & Winton, 2010). Aquaculture began in Asia with the cultivation of freshwater fish, and use of the cultivation techniques currently extend to all continents, with a great diversity in the species raised (Subasinghe et al., 2009) Approximately 350 different species are raised in farms, including 34 fish, 8 crustaceans, and 12 species of mollusks that have annual production levels exceeding 100,000 tons (Walter & Winton, 2010).

Shrimp cultivation areas have expanded the most. However, this industry faces major problems with bacterial diseases, and large quantities of chemical and antibiotic products are frequently used to counteract this (Le et al., 2005; Tu et al., 2008).

2. Shrimp disease, one of the main threats facing aquaculture

In aquaculture, bacterial diseases have emerged as a serious problem and represent the most important challenge facing this industry (Morales, 2004; Holmstrom et al., 2003). Bacterial microorganisms can also cause destructive infections, such as the diseases caused by bacteria of the *Vibrio* genus and the bacteria that cause necrotizing hepatopancreatitis (NHP). These are the main diseases responsible for infections in shrimp farms (Roque et al.,

* Corresponding Author

2001). The causative agent of necrotizing hepatopancreatitis (NHP) is a gram-negative, pleomorphic, obligate intracellular pathogen. The predominant is a non-flagellated, rod-shaped, Rickettsia-like form, that occasionally, exhibits a transverse constricted zone indicative of replication by binary fission (Vincent & Lotz, 2007).

Infections due to necrotizing hepatopancreatitis have been virtually eliminated in the Americas, but the International Office of Epizootics (IOE) has stated some concerns due to the potential that this disease possesses that could be extended to the entire world, because the mortalities caused in shrimp due to necrotizing hepatopancreatitis may include up to 95% of the organisms infected in the pond, causing large losses during the crop cycle (Vincent & Lotz, 2007).

The first infection by necrotizing hepatopancreatitis in a shrimp crop was reported on a farm in the state of Texas (United States of America) in 1985, and outbreaks have subsequently occurred in Peru, Ecuador, Venezuela, Brazil, Panama, Costa Rica, and Mexico, causing significant mortality. This disease is also known as granulomatous hepatopancreatitis or Texas necrotizing hepatopancreatitis, and the infection has been identified in species of *Litopenaeus vannamei, Litopenaeus setiferus, Litopenaeus stylirostris, Farfantepenaeus aztecas,* and *Farfantepenaeus californiensis* (Vincent & Lotz, 2007).

The symptoms that shrimp infected with necrotizing hepatopancreatitis exhibit include a reduction in food ingestion, lethargy, an empty intestine, a flaccid body, darkening of the gills, an expansion of the chromatophores around the swimmerets leading to a darkened appearance, and marked atrophy of the hepatopancreas (Lightner, 1996).

Bacteria of the *Vibrio* genus constitute the majority of the bacteria isolated from among those that cause disease and death, either when equilibrium is broken or the immune system is suppressed, and these can be due to some factors. Physico-chemical changes like salinity and temperature, and hypoxia have been reported to affect the immune response of shrimp and its susceptibility to pathogen bacteria. Water with low and high pH level, as well as those with low dissolved oxygen have been reported to decrease the total haemocyte count (THC) and phenol oxidase (PO) activity (Li & Chen, 2008).

Systemic vibriosis, bacterial erosion, Zoea II syndrome, and "white ball" are some of the diseases affecting shrimp produced by bacteria of the *Vibrio* genus (Gómez-Gil et al., 2001). Systemic vibriosis, also known as seagull syndrome, affects all farmable species of shrimp, as they may be susceptible to infection under stressful conditions. *Vibrio* species: *harveyi, vulnificus, parahaemolyticus,* and *alginolyticus* are all related to this disease (Morales, 2004).

Bacterial erosion of the shell is present in all penaeid shrimp, juveniles and adults alike. It manifests with the appearance of brown or black stains in areas that have been eroded through the action of chitinolytic bacteria, such as *Vibrio* sp., *Aeromonas* sp., *Spirillum* sp., and *Flavobacterium* sp. The disease is self-limiting and generally disappears when the shrimp molt. If left untreated, it becomes more serious and may become a systemic infection (Morales, 2004).

Zoea II Syndrome causes high mortality rates in the juvenile stage of shrimp. In Ecuador, the causal agent of this disease was found to be *Vibrio harveyi*, although other authors have reported the possible presence of intracellular bacteria. This infection was detected for the first time in 1993 in farms of *Litopenaeus vannamei* in Ecuador, Mexico, and the United States (Morales, 2004).

"White ball" disease causes the appearance of small balls, arising from desquamated hepatopancreatic cells or to hypertrophied and rounded hepatocytes that appear as spherical formations. It is believed that these balls are caused by toxins produced mainly by *Vibrio* spp. The species of *Vibrio* related to this syndrome are *Vibrio alginolyticus* and *Vibrio harveyi* (Gómez-Gil et al., 2001; Vandenberghe et al., 1999)

A significant limitation to the industry is loss of stock through bacterial disease. Traditional methods to combat disease with antibiotics have been questioned and alternatives have been sought. The modern aquaculture industry demands alternative prophylacts that may help to keep a microbiologically healthy environment, resulting in better production and higher profits. Within this context, the probiotics seem to be a very promising alternative for the management of disease in aquaculture (Sáenz, et al., 2009) Another form is with the reduction in stocking density.

3. Antibiotics used in the cultivation of shrimp

Antibiotics are commonly used in aquaculture during the production cycle, both in the larval and growth phases. The use of antibiotics in aquaculture is associated with environmental and human health problems, including bacterial resistance, persistence of the disease in the aquatic environment, and effects on the biogeochemical composition of the sediment. The accumulation of antibiotic residues in the edible tissues of shrimp may also alter human intestinal flora and cause food poisoning or allergy problems (Ma et al., 2006).

The antibiotics most frequently used in aquaculture to combat bacterial diseases include oxytetracycline, florfenicol, sarafloxacin, and enrofloxacin (Roque et al., 2001; Soto-Rodríguez et al., 2006). Globally, other antibiotics such as chlortetracycline, quinolones, ciprofloxacin, norfloxacin, oxolinic acid, perfloxacin, sulfamethazine, gentamicin, and tiamulin are used (Holmstrom et al., 2003).

3.1 Oxytetracycline

Oxytetracycline is widely employed to treat bacterial infections in aquaculture farms, such as vibriosis and furunculosis (Capone et al., 1996; Prescott et al., 2000; Reed et al., 2006; Wang et al., 2004). It belongs to the tetracycline group, which exerts antimicrobial action against both Gram (-) and (+) bacteria, ricksettsias, mycoplasmas, and others (Gómez-Gil et al., 2001). Tetracyclines are produced by *Streptomyces* spp., which possess determinants for resistance to this class of antibiotics.

Oxytetracycline is a bacteriostatic antibiotic that exerts its antimicrobial effect against protein synthesis, by bonding directly to the S7 protein of the 30S subunit of the bacterial ribosome, thereby impeding the bonding of aminoacyl-tRNA (aminoacyl transfer RNA) to the A-site of the ribosome. This prevents the addition of amino acids to the growing peptide chain (Chambers, 2004; Isidori et al., 2005; Jara, 2007). In order for oxytetracycline to interact with its target site, it needs to pass through the external membrane via passive diffusion through the OmpF and OmpC pores, and through the cytoplasm membrane via an energy-dependent process (Jara, 2007).

3.2 Enrofloxacin

Enrofloxacin was developed as an antimicrobial agent during the 1980s for exclusive use in veterinary medicine and has proven to be effective in the treatment of bacterial diseases that affect aquaculture organisms. Enrofloxacin is a derivative of nalidixic acid. It has a basic dihydroquinoline (4-quinolone ring) chemical core with an ethyl group at the 4th position, favoring its absorption and availability. It is primarily lipophilic and has a low molecular weight, favoring tissue penetration. The mechanism of enrofloxacin acts at the level of the cellular nucleus, inhibiting DNA synthesis. During the multiplication phase of the bacteria, the DNA folds and unfolds alternately. This process is controlled by the enzyme DNA gyrase, which is inhibited by enrofloxacin, causing a collapse of bacterial metabolism and preventing the genetic information from being copied, thus causing the bacteriocidal effect (Williams et al., 2002).

The information related to this antibiotic for the most widely grown shrimp species such as *Litopenaeus vannamei* is scarce, but pharmacokinetic studies on enrofloxacin have been carried out using other species, such as crab (*Scylla serrata*), tilapia (*Oreochromis niloticus*), black shrimp (*Penaeus monodon*), Chinese shrimp (*Penaeus chinensis*), and European seabass (*Dicentrarchus labrax*) (Intorre et al., 2000; Tu et al., 2008; Wen et al., 2007; Xu et al., 2006). It is important to note that the pharmacokinetic results for enrofloxacin obtained for these species should not be extrapolated to other aquatic species, because each organism possesses a different metabolism, and the cultivation conditions may have a significant influence over the kinetic behavior displayed by the antibiotic.

None of the fluoroquinolones included enrofloxacin is approved for use in shrimp in the United States. Their potential use in other countries, such as Mexico, as well as the potential for extra-label use in the United States provides a need for efficient methods to monitor food supplies for the presence of fluoroquinolones residues (Schneider et al., 2005).

3.3 Ciprofloxacin

Ciprofloxacin is the main metabolite of Enrofloxacin and is active against a broad spectrum of aerobic Gram (-) bacteria, including enteric pathogens such as *Pseudomonas* and *Serratia marcescens*. It is also active against Gram (+) pathogens, even when these bacteria have developed resistance to other antibiotics, such as penicillin (Wen et al., 2007). It is not active against anaerobic bacteria and may be used occasionally, in combination with other antibacterial agents, for the treatment of mycobacterial infections.

The antibacterial effects of ciprofloxacin arise from its inhibition of Topoisomerase IV and bacterial DNA gyrase, which act by cleaving the DNA of the bacterial chromosome and rejoining the ends once a superhelix is formed (Banerjee et al., 2007). When these enzymes are inhibited, bacterial cell multiplication is interrupted.

3.4 Florfenicol

This fluorinated antibiotic, derived from thiamphenicol, is a potent and broadly acting bacteriostatic agent. It is effective in the treatment of infections caused by *Pasteurella piscicida, Aeromonas salmonicida, Vibrio anguillarum*, and *Edwardsiella tarda*. Its chemical structure is very similar to that of chloramphenicol, and florfenicol is effective against bacteria that have developed the ability to deactivate other drugs, such as thiamphenicol and chloramphenicol. Pharmacokinetically, florfenicol use has been reported among some species of fish such as Atlantic salmon (*Salmo salar*), in which a bioavailability of more than 95% is present, exhibiting a good distribution among all of the organs and tissues. Its half-life in fish is less than 15 h (Yanong & Curtis, 2005). However, published information for shrimp is scarce, meaning that the kinetic behavior of this compound among these crustaceans has not yet been completely elucidated.

3.5 Sarafloxacin

This is a white or slightly yellow crystalline solid with the chemical name 6-fluoro-1-(4-fluorophenyl)-4-oxo-7-piperazin-1-ylquinoline-3-carboxylic acid. Its solubility is 0.034 mol/L at pH 1, and its molecular weight is 335 (King et al., 2000; Oliphant et al., 2002). Two fluoroquinolones have been approved by the Food and Drug Administration (FDA) in the United States for use in the production of animal-derived food products. Sarafloxacin was approved in August 1995 for the treatment of infections caused by *Escherichia coli* in poultry (turkeys and chickens) but, along with other fluoroquinolones, has not been authorized by the Food and Drug Administration for use as a therapeutic agent in aquaculture (Roybal et al., 2002; Nakata et al.,2005).

The normal dosage used in farm animals is 10 mg kg^{-1}, and the drug is administered in drinking water. A Maximum Residue Level (MRL) has not been established for cow's milk or chicken muscle (King et al., 2000; Oliphant et al., 2002).

3.5.1 1st Generation quinolones

These are used exclusively as urinary antiseptics because they do not have sufficient serum levels and are actively eliminated with urine. All are administered orally.

3.5.2 2nd Generation quinolones

These are monofluoride quinolones. Compared to 1st generation quinolones, these exhibit more potent activity, a greater bacterial spectrum, a longer half-life, and, with the exception of norfloxacin and enrofloxacin, they achieve good serum levels, making it possible to treat systemic infections. Ciprofloxacin, Ofloxacin and Perfloxacin can be administered both orally and parenterally.

3.5.3 3rd Generation quinolones

These are bi- and trifluoride fluoroquinolones. Some have a greater half-life and others have a greater antibacterial spectrum, in some cases including Gram (−) bacteria (mainly streptococci), intracellular bacteria (Chlamydia, mycoplasma, mycobacteria, etc.), and anaerobic bacteria.

3.5.4 4th Generation quinolones

This group has improved activity against Gram (+) bacteria and anaerobic bacteria (King et al., 2000; Oliphant et al., 2002).

4. Studies on the accumulation and elimination of enrofloxacin and oxytetracycline antibiotics among *Litopenaeus vannamei* shrimp

The exposure of the consumer to antibiotic residues in seafood is of great importance for health. Information related to residues of oxytetracycline, enrofloxacin and its metabolite ciprofloxacin in *Litopenaeus vannamei* tissues is very scarce, even though this species is among the most highly valued for its commercialization and oxytetracycline and enrofloxacin are frequently employed when farming this species.

However, pharmacokinetic and bioavailability studies of antimicrobial agents in farmed shrimp are important in order to determine optimal dosage regimens and formulations, to establish safe withdrawal periods, and to minimize the environmental effects of the drug used in aquaculture.

The low analyte concentrations normally present (ng g^{-1}, ng mL^{-1}), the complexity of matrices and the diverse physico-chemical properties that antibiotics may present make their determination difficult, and highly sensitive, selective methods are necessary for monitoring antibiotics in the aquatic organisms (Hernández et al., 2007). The liquid chromatography is the most widely used method, for quantitative analysis of antibiotics.

The purpose of introduce the chemical method about high-performance liquid chromatography was precise quantitative determination of antibiotics residues in the muscle tissues and hepatopancreas to shrimp. Following this analytical approach, the requirements for veterinary monitoring concerning to antimicrobials residues would be fulfilled. Previously only microbiological methods were used, which gave only orientation results.

In a study performed in our laboratory (unpublished data), an attempt was made to determine the accumulation and elimination of enrofloxacin and its metabolite ciprofloxacin in a crop of *Litopenaeus vannamei* using high-performance liquid chromatography after administering a diet medicated with enrofloxacin at a level of 200 mg kg^{-1} for 14 days. Subsequently, a diet without the antibiotic was administered for 16 days. The study was carried out under controlled laboratory and farm conditions. It was found that the maximum concentrations (C_{max}) reached for enrofloxacin in the muscle and hepatopancreas under laboratory conditions were 0.54±0.26 μg g^{-1} and 3.52±1.9 μg g^{-1}, respectively. For ciprofloxacin, levels of 0.18±0.13 μg g^{-1} and 1.05±0.20 μg g^{-1} were reached in the muscle and hepatopancreas, respectively.

In the farm study, C_{max} enrofloxacin levels of 0.36±0.17 μg g^{-1} and 1.60±0.82 μg g^{-1} were reached in the muscle and hepatopancreas, respectively. For ciprofloxacin, these were 0.03±0.02 and 0.36±0.08 μg g^{-1}, respectively. Once the medicated diet was suspended, enrofloxacin and ciprofloxacin residues in the tissues decreased, requiring four to ten days for the levels of both antibiotics to be undetectable in the muscle and six to fourteen days for elimination from the hepatopancreas.

Under controlled conditions, the greatest accumulations of the antibiotic and its metabolite were reached, showing a reduction of 33% in the muscle and 55% in the hepatopancreas when compared with the levels reached under farm conditions. For ciprofloxacin, the reduction was 66% in both tissues. It is important to relate the accumulations levels reached with the Minimal Inhibitory Concentration of the drug that can treat bacterial infections among shrimp, to determine whether this antibiotic is effective for the control of these diseases.

Oxytetracycline, another of the most widely used antibiotics both in fish and shrimp farms, was studied in relation to its accumulation in shrimp tissues (muscle and hepatopancreas), and this was related to its effect in the inhibition of *Vibrio* bacteria through determination of the Minimal Inhibitory Concentration. In addition, the time of elimination of the antibiotic from shrimp tissues was established.

To perform this study, ponds on a farm of *Litopenaeus vannamei* were employed. The shrimp were treated for 14 days with food that contained a theoretical oxytetracycline concentration of 5000 mg kg^{-1}. Next, an antibiotic-free diet was administered for 16 days. The shrimp were sampled every third day. *Vibrio* bacteria were isolated in the muscle and hepatopancreas of the shrimp, counted and expressed as CFU, and strains belonging to the *Vibrio* genus were identified at the molecular level.

The results obtained show an average C_{max} for oxytetracycline in the shrimp tissues of 31.32±3.44 µg g^{-1} in the muscle and 274.81±62.35 µg g^{-1} in the hepatopancreas. The accumulation levels reached in some of the tissue samples in the treatment stage were greater than the Minimal Inhibitory Concentration determined for oxytetracycline, in the range of 0.75-100 µg mL^{-1}. *Vibrio parahaemolyticus* was identified in 48% of the strains analyzed in the crop system, and no strain was positive for toxigenic *Vibrio cholerae* O1.

The withdrawal times necessary for the oxytetracycline residues to be eliminated were ten days for the hepatopancreas and sixteen days for muscle. Under controlled conditions in the laboratory, a diet was administered to *Litopenaeus vannamei* shrimp that contained 5000 mg Kg^{-1} of oxytetracycline for 14 days. The accumulation levels (C_{max}) for oxytetracycline were 33.54±11.19 µg g^{-1} in the muscle (Fig. 1), 194.37±16.11 µg g^{-1} in the hepatopancreas (Fig. 2), and 18.79±5.87 µg mL^{-1} in the hemolymph. The elimination time for oxytetracycline were six to ten days for all tissues.

The results obtained in the application of these antibiotics demonstrates the importance of applying appropriate therapies with antibiotics, seeking greater effectiveness for the control of bacterial infections, and highlighting the importance of respecting the withdrawal times for each antibiotic, with the purpose of eliminating the residual presence of these compounds from the edible tissues and from the cultivation system in general (to decrease the development of antibiotic resistance in the bacteria).

5. Factors to consider when using antibiotics in aquaculture

Antibiotics must not be used as a preventative measure, since bacteria very rapidly develop resistance to them, leading to their ineffectiveness. Chemical agents should only be applied if there is an appropriate diagnosis of the situation and always under previously established control protocols. To evaluate the impact of antibiotic administration, the usage standards, the hydrology of the area, and the physical-chemical properties of the water should be known (Páez-Osuna et al., 2003; Nakata et al., 2005). Best practices in aquaculture management should be prioritized to avoid the entrance of pathogens into the shrimp cultivation systems, and antibiotics should only be administered as a last resort (Chávez-Sánchez & Higuera-Ciapara, 2003).

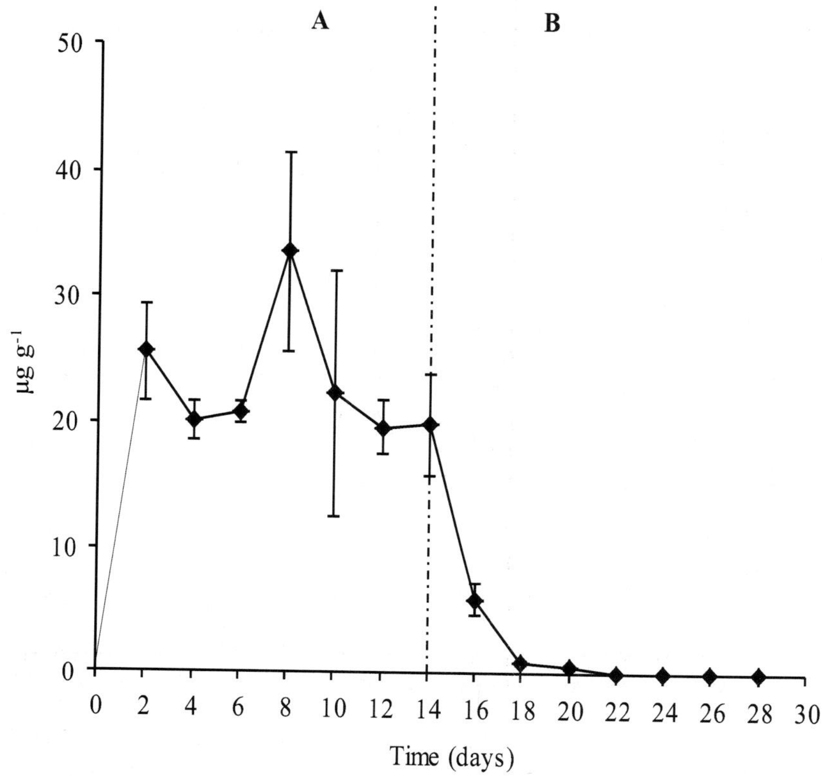

Fig. 1. Muscle oxytetracycline levels -time profiles in the shrimp *Litopenaeus vannamei* after an oral OTC dosage through a medicated feed. The vertical bars represent the Standard Error (n=3). A: treatment period, B: withdrawal period.

The most frequent administration route for antibiotics in shrimp is oral, in which the antibiotic is incorporated in the feed with subsequent exposure to the extremely aggressive aquatic environment. For this reason, it is important that the antibiotic be contained within a pellet to maintain its stability and protect it from factors such as leaching and binding to trivalent and divalent cations (Cabello, 2004).

It must be certain that the shrimp will eat the food when the antibiotic therapy is applied, because the disease will otherwise not be treated, the environment will be contaminated, and the emergence of bacterial resistant strains will be favored. The consumption of food by the farmed organisms may decrease during the molting period, due to environmental factors, or factors related to the infections, reducing the quantity of the antibiotic ingested (Cuzon et al., 2004).

The water temperature of the farm ponds is a critical point to consider, because parameters such as the maximum concentration, distribution volume, and rate of elimination of the

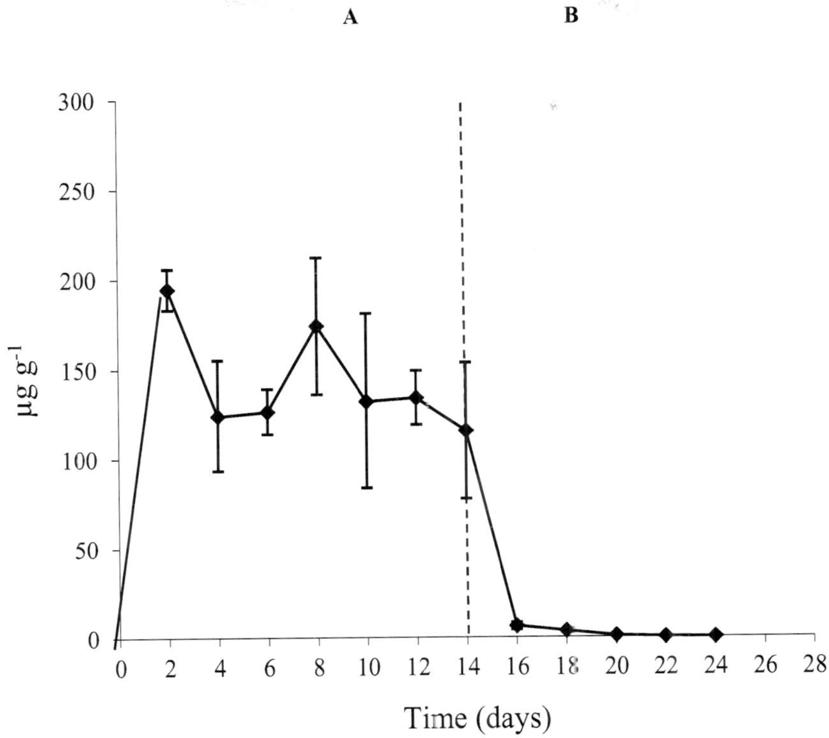

Fig. 2. Hepatopancreas oxytetracycline levels –time profiles in the shrimp *Litopenaeus vannamei* after an oral OTC dosage through a medicated feed. The vertical bars represent the Standard Error (n=3). A: treatment period, B: withdrawal period.

antibiotic may be affected. The pH, oxygenation, salinity, stage of disease, climatic changes, and presence of natural food in the ponds are other factors that affect antibiotic therapies among aquatic organisms (Chávez & Montoya, 2004; Montoya, 2002).

The use of pharmacological agents, antibiotics, and other chemical agents should be considered as methods of last resort in shrimp farming and aquaculture in general. None of the antibiotics is approved for use in shrimp in the United States. The medicated feed is used in an extralabel manner only for treatment of minor species as defined in the Code of Federal Regulations (21 CFR 514.1(d)(1)(ii). In an aquatic species, the extralabel use of medications added to feed in limited to products approved for use in other aquatic species (FDA, 2001).

6. The development of bacterial resistance

Drug resistance in when a formerly effective drug dose is no longer effective. This can be a natural resistance or an acquired resistance. Resistance arises mainly by natural selection, the replication of a naturally resistant strain after the drugs has killed all of the susceptible strains. Since mutagenic drugs generally are not used, resistance by drug-induced mutation seldom occurs. Drug resistance also can develop from gene transfer or gene amplification (Albert, 1985).

The most worrying effect of the use of antibiotics in aquaculture production and its relationship with human health is the generation of resistant bacteria strains and the transfer of this resistance from the aquatic environment to land, where strains that are highly immune to antibiotics may originate that are capable of causing disease among humans. The transfer of resistance may occur through mechanisms as simple as the consumption of seafood products that contain bacteria that are resistant to various antibiotics (Gräslund & Bengtsson, 2001).

It has been demonstrated that the use and abuse of antibiotics has given rise to multiple resistance among microbial populations associated with shrimp production. Various studies have shown that antibiotics persist in the sediment and aquatic environment for several months after their administration (Matyar et al., 2008), and that these may affect native bacterial community in detrimental to the ecosystem since this community plays key roles in biogeochemical processes. Some antimicrobials can inhibit important microbial processes as denitrification or primary production by cyanobacteria (Garcia-Armisen et al., 2011).

The capacity of the microorganisms to reduce sulfates may also be reduced (Páez-Osuna et al., 2003), affecting the quality of the sediment and the environment (Ma et al., 2006), thereby promoting the proliferation of resistant bacterial strains or pathogens (Capone et al., 1996; Hektoen et al., 1995; Tendencia & De la Peña, 2002;), which may place the viability of shrimp crops at risk.

It is estimated that between 15 and 40% of the administered medicated diet is not ingested by the organisms and remains in the substrates. Another part of the medication is not absorbed during its passage through the intestinal tract of the organism and returns to the environment in fecal matter. The amount of antibiotic transferred to the environment varies from 1% (chloramphenicol) to 90% (oxytetracycline) (Capone et al., 1996). Hektoen et al. (1995) reported that approximately 70-90% of the antibiotic used in the therapy of farmed organisms ends up in the environment and sediment, and a high percentage exhibits antibacterial activity. It has been reported that residues of oxolinic acid and oxytetracycline are very persistent under certain conditions, with half-lives exceeding 100 days (Samuelsen et al., 1992).

Three mechanisms of resistance to tetracycline have been described: (1) decreased intracellular accumulation due to either impaired influx or increased efflux by an active transport protein pump; (2) ribosome protection due to production of protein that interfere with tetracycline binding to the ribosome; and (3) enzymatic inactivation of tetracyclines. The most important of these is production of an efflux pump. The pump protein is encoded on a plasmid and may be transmitted by transduction or by conjugation. Because these plasmids commonly encode resistance genes for other drugs, eg, aminoglycosides, sulfonamides and chloramphenicol, tetracycline resistance is marker for resistance to multiple drugs (Tenover, 2006).

Some studies have demonstrated that the concentrations of oxytetracycline in the sediment after therapy may range from 0.4 to 495 $\mu g\ g^{-1}$. Therapeutic dosages of oxytetracycline in fish may cause sub-lethal effects, including alteration of the levels of immunoglobulin in the serum and suppression of the phagocytic response and macrophages (Cyaguari et al., 2009).

International regulations regarding the use of antibiotics in aquaculture have established a list of prohibited products (Stolker & Brinkman, 2005). Shrimp with traces of these products are subject to measures against their importation. The strongest restrictions are on the use of chloramphenicol, dimetridazole, furazolidone, nitrofurazone, other nitrofurans, and fluoroquinolones, and these antibiotics should not be used at any stage of the production process (Defoirdt, et al., 2007; Tittlemier et al., 2007).

Epidemiological and molecular assays have indicated that genes mediating resistance might be transmitted from aquatic bacteria to bacteria capable of producing infections among humans and terrestrial animals. This demonstrates that the aquatic and terrestrial compartments lack borders with respect to the flow of resistance genes and that the resistance phenomenon is global, because the use of antibiotics in an environment will have, over time, repercussions in other, apparently distant, ecosystems (Cabello, 2002; Rhodes et al., 2000). To decrease the contamination of the environment and bacterial resistance, appropriate aquaculture production practices must be carried out, and biosecurity measures must be applied to reduce outbreaks of disease and the propagation of pathogenic agents (Kemper, 2008).

Global efforts are needed to promote more judicious use of antibiotics in aquaculture and the new strategies to control phatogenic bacteria are needed to make the industry more sustainable. However, it is not always economically feasible to culture the organism in the most optimal conditions, so there will always be a risk to infection and a need for effective biocontrol techniques (Defoirdt, et al., 2007).

It is important to highlight that the application of highly sensitive analytical methodologies is indispensably in measuring the concentrations of antibiotics and their metabolites with certainty in the distinct tissues of aquaculture products. This would help in establishing regulations that protect the environment, generating products that are safe for human consumption, and allowing the growth of aquaculture.

A practical use of the pharmacokinetic data is the possibility to design dosage regimens in which levels of a specific drug can be maintained above the Minimum Inhibitory Concentration and below toxic effects by means of repeated dosages. However, this method requires information on Minimal Inhibitory Concentration established for bacterial pathogens of interest. Although there are reports on available Minimal Inhibitory Concentration for bacterial strains potentially pathogens to shrimp species, they reveal a wide range of values. Takahashi et al., (1985) reported that Minimal Inhibitory Concentration of oxytetracycline against 49 strains of *Vibrio sp.*, range from 0.1 to 12.5 µg mL^{-1}. Monhey et al., (1992) found the Minimal Inhibitory Concentration to be in the range of 2.0 µg mL^{-1} or less for *Vibrio* isolated mainly from American shrimp. Furthermore, Roque et al., (2001) in their study of 144 isolated of *Vibrio* reported a Minimal Inhibitory Concentration of 304.0 µg mL^{-1} for oxytetracycline with a range from 0.26 to 1064 µg mL^{-1}. Given the wide range of Minimal Inhibitory Concentration it is recommended to isolate local bacterial strains and evaluate their Minimal Inhibitory Concentration. Additionally, it still requires performing such studies in natural farming conditions (Gómez-Jimenez et al., 2008).

7. Conclusion

Scientific studies have been conclusive with respect to the health risk that the massive and unlimited use of antibiotics in aquaculture represents. When health certifications are implemented for aquaculture products demanded by domestic and foreign markets, the control over the use of these compounds needs to be increased, together with other aspects of primary importance such as food safety, protection of the environment, and the health of farmed organisms. These aspects should be considered and resolved through the implementation of Best Management Practices.

8. Acknowledgment

The authors wish to express their gratitude to PhD. Evelia Acedo Félix and Q.B. Rosalva Pérez Morales of the Molecular Microbiology Laboratory, for their technical support and contribution to help facilitate the completion of the project. Furthermore, we acknowledge the kind donation of the shrimp from the farm "La Borbolla", specialty engineers Roberto Federico Aguayo Valenzuela and César Patiño Patiño, and the technical assistance of Biol. Adolfo Pérez Álvarez and Tech. Juan Carlos Gastélum Domínguez.

9. References

Albert, A. (1985). Drug resistance and synergism; Chemotherapy. In: *The organic chemistry* of *drug design and drug action,*. Richard B. Silverman (Ed), pp. 149-150, Academic Press, ISBN 0-12-643730-0, San Diego, CA.

Banerjee, S.; Devaraja, T.; Shariff, M. & Yusoff, F. (2007). Comparison of four antibiotics with indigenous marine *Bacillus spp*. in controlling pathogenic bacteria from shrimp and Artemia. *Journal of Fish Diseases*, vol.30, pp.383–389.

Bostock, J., McAndrew, B., Richards, R., Jauncey, K., Telfer, T., Lorenzen, K., Little, D., Ross, L., Handisyde, N., Gatward, I. & Corner R. (2010). Aquaculture:global status and trends. *Philosophical Transactions of the Royal Society B*, vol.365, pp.2897-2912.

Cabello, F. C. (2004). Antibióticos y Acuicultura en Chile: consecuencias para la salud humana y animal. *Academia de Medicina, Instituto de Chile,* vol.132, pp.1001-1006.

Capone, G. D.; Weston, P. D.; Miller, V. & Shoemaker, C. (1996). Antibacterial residues in marine sediments and invertebrates following chemotherapy in aquaculture. *Aquaculture,* vol. 145, pp.55-75.

Chambers, H. F. (2004). Chloramphenicol, Tetracyclines, Macrolides, Clindamycin and Streptogramins, In: *Basic and clinical pharmacology,* Katzung B. G. (ed), pp. 754-763, MacGraw-Hill, ISBN 0-07-141092-9, New York.

Chávez-Sánchez, M. C. & Higuera-Ciapara, I. (2003). Manual de Buenas Prácticas de Producción Acuícola de Camarón para la Inocuidad Alimentaria. Servicio Nacional de Sanidad Inocuidad y Calidad Agroalimentaria (SENASICA), ISBN 968-5384-04-5, Mazatlán, Sinaloa, México.

Chávez, S. M. C. & Montoya, R. L. Medidas de bioseguridad para evitar la introducción y dispersión de enfermedades virales en granjas camaronícolas, VII *Simposium Internacional de Nutrición Acuícola,* Hermosillo, Sonora. México, Noviembre, 2004.

Cuzon, G.; Lawrence, A.; Gaxiola, G.; Rosas C. & Guillaume, J. (2004). Nutrition of *Litopenaeus vannamei* reared in tanks or in ponds. *Aquaculture,* vol. 235, pp.513-551.

Defoirdt, T.; Boon, N.; Sorgeloos, P.; Verstraete, W. & Bossier, P. (2007). Alternatives to antibiotics to control bacterial infections: luminescent vibriosis in aquaculture as an example. *TRENDS in Biotechnology*, vol.25, pp.472-479.

FDA. (2001). United States FDA. Aquaculture drugs (a chemical hazard). In: *Fish and fishery products hazards and controls guidance.* 3rd ed. Washington, DC: US FDA, Centre for Food Safety and Applied Nutrition, Office od Seafood, pp.127-144.

Garcia-Armisen, T., Vercammen, K., Passerat, J., Triest, D., Servais, P. & Cornelis, P. (2011). Antimicrobial resistance of heterotrophic bacteria in sewage-contaminated rivers. *Water Research*, vol.45, pp.788-796.

Gräslum, S. & Bengtsson, B. (2001). Chemical and biological products used in south-east Asian shrimp farming and their potential impact on the environment a review. *The Science of the Total Environment*, vol. 280. pp. 93-131.

Gómez-Gil, B.; Roque, A. & Guerra, F. A. (2001). Enfermedades infecciosas más comunes en la camaronicultura en México y el impacto del uso de antimicrobianos. In: *Camaronicultura y Medio Ambiente. México: Unidad Académica Mazatlán.* Páez-Osuna F. (ed). Instituto de Ciencias del Mar y Limnología. Universidad Nacional Autónoma de México. Mazatlán, Sin. pp. 315-346.

Gómez-Jimenez, S.; Espinosa-Plascencia, A.; Valenzuela-Villa, F. & Bermúdez-Almada, M.C. (2008). Oxytetracycline (OTC) accumulation and elimination in hemolymph, muscle and hepatopáncreas of white shrimp *Litopenaeus vannamei* following and OTC-Feed therapeutic treatment. *Aquaculture,* vol. 274, pp.24-29.

Hektoen, H.; Berge, J.A.; Hormazabal, V. & Yndeslad, M. (1995). Persistence of antibacterial agents in marine sediments. *Aquaculture,* vol.133, pp.175-184.

Hernández, F.; Sancho, V. J.; Ibáñez, M. & Guerrero, C. (2007). Antibiotic residue determination in environmental waters by LC-MS. *Trends in Analytical Chemistry,* vol.26, No.6, pp.466-485.

Holmström, K.; Gräslund, S.; Wahlström, A.; Poungshompoo, S.; Bengtsson, B. E. & Kautsky, N. (2003). Antibiotic use in shrimp farming and implications for environmental impacts and human health. *International Journal of Food Science and Technology,* vol. 38, pp.255-266.

Intorre, L.; Cecchini, S. & Bertini, S. (2000). Pharmacokinetics of enrofloxacin in the seabass (*Dicentrarchus labrax*). *Aquaculture,* vol.182, pp.49–59.

Isidori, M.; Lavorgna, M.; Nardelli, A.; Pascarella L. & Parrella, A. (2005). Toxic and genotoxic evaluation of six antibiotics on non-target organisms. *The Science of the Total Environment,* vol.346, pp.87-98.

Jara, M. A. (2007). Tetraciclinas: Un modelo de resistencia antimicrobiana. *Avances en Ciencias Veterinarias,* vol.22, pp.49-55.

Kemper, N. (2008). Veterinary antibiotics in the aquatic and terrestrial environment. *Ecological Indicators,* vol.8, pp.1-13.

King, D.E.; Malone, R. & Lilley, S. H. (2000). New classification and update on the quinolone antibiotics. *American Family Physician,* vol.61, pp.2741-2748.

Le, X.T. ; Munekage, Y. & Kato, S. (2005). Antibiotic resistance in bacteria from shrimp farming in mangrove areas. *Science of the Total Environment,* vol.349, pp.95-105.

Li, C. & Che, J. (2008). The immune response of white shrimp *Litopenaeus vannamei* and its susceptibility to *Vibrio alginolyticus* under low and high pH stress. *Fish & Shellfish Immunology,* vol.25, pp.701-709.

Lightner, D. V. (1996). *A Handbook of pathology and diagnostic procedures of diseases of cultured penaeid shrimp,* World Aquaculture Society. Department of Veterinary Science, University of Arizona, Tucson, Az.

Ma, D.; Hu, Y.; Wang, J.; Ye, S. & Li, A. (2006). Effects of antibacterials use in aquaculture on biogeochemical processes in marine sediment. *The Science of the Total Environment,* vol. 367, No.1, pp.273-277.

Matyar, F., Kaya, A. & Dinçer, S. (2008). Antibacterial agents and heavy metal resistance in Gram-negative bacteria isolated from seawater, shrimp and sediment in Iskenderun Bay, Turkey. *Science of the Total Environment,* vol. 407, pp.279-285.

Monhey, L.L., Bell, T.A. & Lightner, D.V. (1992). Shrimp antimicrobial testing, I. In vitro susceptibility of thirteen Gram-negative bacteria to twelve antimicrobials. *Journal of Aquatic Animal Health,* vol.4, pp.257-261.

Montoya, V. N. (2002). Residuos de antibióticos en camarones: límites residuales y detección de anfenicoles, In: *Centro Nacional de Acuicultura e Investigaciones Marinas "Edgar ArellanoM." (CENAIM),* Julio, 2011, www.cenaim.espol.edu.ec/publicaciones/quincenal/bquinc54.pdf. Boletin 54.

Morales, C. M. S. (Ed). (2004). *Enfermedades del camarón,* Editorial Trillas, ISBN-968-24-7112-5, México, D. F.

Nakata, H.; Kannan, K.; Jones, D. P. & Giesy, P. J. (2005). Determination of fluoroquinolones antibiotics in wastewater effluents by liquid chromatography-mass spectrometry and fluorescente detection. *Chemosphere,* vol.58, pp.759-766.

Oliphant, C.M.; Pharma, D.; Gary, M. & Green, M. D. (2002). Quinolones: a comprehensive review. *American Family Physician,* vol.65, pp.455-464.

Páez-Osuna, F.; Gracia, A.; Flores-Verdugo, F.; Lyle-Fritch, L. P.; Alonso-Rodriguez, R.; Roque, A. & Ruiz-Fernandez, A. C. (2003). Shrimp aquaculture development and the environment in the Gulf of California ecoregion. *Marine Pollution Bulletin,* vol.46, No.7, pp.806-815.

Prescott, J. F.; Baggot, J. D. & Walter, D. R. (2000). Tetracyclines and Glycylcyclines, *In: Antimicrobial Teraphy in Veterinary Medicine.* Giguére, S.; Prescott, J.F.; Baggot, J.D.; Walker, R.D.; Dowling, P.M. (Ed), pp.231, Blackwell Publishing, ISBN-0813806569, Iowa State.

Reed, L. A.; Siewicki, T. C. & Shah, A. C. (2006). The biopharmaceutics and oral bioavailability of two forms of oxytetracycline to the white shrimp, *Litopenaeus setiferus. Aquaculture,* vol.258, pp.42-54.

Rhodes, G.; Huys, G.; Swings, J.; Mc Gann, P.; Hiney M. & Smith, P. (2000). Distribution of oxytetracycline resistance plasmids between aeromonads in hospital and aquaculture environments: Implication of Tn1721 in dissemination of the tetracycline resistance determinant Tet A. *Applied Environmental Microbiology,* vol.66, pp.3883-3890.

Roque, A.; Molina, A. A.; Bolán, M. C. & Gómez, G. B. (2001). *In vitro* susceptibility to 15 antibiotics of vibrios isolated from penaeid shrimps in Northwestern Mexico International *Journal of Antimicrobial Agents,* vol.17, pp.383–387.

Roybal, J. E.; Walker, C. C.; Pfenning, P.A.; Turnipseed, B. S.; Storey, M. J.; Gonzalez, A. S. & Hurlbut, J.A. (2002). Concurrent determination of four fluoroquinolonas in catfish, shrimp, and salmon by liquid chromatography with fluorescence detection. *Journal of Association Official Analytical Chemistry International.* vol.85, No.6, pp.1293-1301.

Sáez, M. A.; Díaz-Rosales, P.; Chabrillón, M.; Smidt, H.; Arijo, S.; León-Rubio, J. M.; Alarcón, F. J.; Balebona, M. C.; Moriñigo, M. A.; Cara, J. B. & Moyano, F. J.(2009). Effect of dietary administration of probiotics on growth and intestine functionality of juvenile Senegalese sole (*Solea senegalensis,* Kaup 1858). *Aquaculture Nutrition,* vol.15, pp.177-185.

Samuelsen, O.B. (1992). The fate of antibiotics/chemotherapeutants in marine aquaculture sediments. In: *Chemotherapy in Aquaculture: From Theory to Reality.* Michel, C. & Alderman, D.J. (Ed), pp. 87-95, Office International des Epizooties, Paris, France.

Schneider, J. M. ; Vazquez-Moreno, L.; Bermudez-Almada, M.C.; Barraza-Guajardo R. & Ortega-Nieblas, M. (2005). Multiresidue determination of fluoroquinolones in shrimp by liquid chromatography-fluorescence-mass spectrometry. *Journal of AOAC International,* vol. 88, No.4, pp.1160-1166.

Soto-Rodríguez, S.; Armenta, M. & Gomez-Gil, B. (2006). Effects of enrofloxacin and florfenicol on survival and bacterial population in an experimental infection with luminescent *Vibrio campbellii* in shrimp larvae of *Litopenaeus vannamei. Aquaculture,* vol.255, pp.48-54.

Stolker, A. A. M. & Brinkman, U.A. Th. (2005). Analytical strategies for residues analysis of veterinary drugs and growth-promoting agents in food-producing animals-a review. *Journal of Chromatography A,* vol.1067, pp.15-53.

Subasinghe, R.; Soto, D. & Jia, J. (2009). Global aquaculture and its role in sustainable development. *Reviews in Aquaculture,* vol.1, pp.2-9.

Takahashi, Y., Itami, T., Nakagawa, A., Nishimura, H. & Abe, T. (1985) Therapeutic effects of oxytetracycline trial tablets against vibriosis in cultured kuruma prawns Penaeus japonicus Bate. *Bulletin of the Japanese Society for the Science of Fish/Nissuishi* vol.51, No.10, pp.1639-1643.

Tittlemier, S. A.; Van de Riet, J.; Burns, G.; Potter, R.; Murphy, C.; Rourke, W.; Pearce, H.; & Dufresne, G. (2007). Analysis of veterinary drug residues in fish and shrimp composites collected during the Canadian Total Diet Study, 1993-2004. *Food Additives and Contaminants,* vol.24, pp.14-20.

Tendencia, E. A. & De la Peña, L. D. (2002). Level and percentage recovery of resistance to oxytetracycline and oxolinic acid of bacteria from shrimp ponds. *Aquaculture,* vol.213, pp.1-13.

Tenover, F. C. (2006). Mechanisms of antimicrobial resistance in bacteria. *The American Journal of Medicine.* vol.119, No.6, Suppl. 1, pp.S3-S10

Tu, T. H.; Silvestre, F.; Bernard, A. ; Douny, C. ; Phuong, T.N ; Tao, T.C. ; Maghuin-Rogiste, G. & Kestemont, P. (2008). Oxidative stress response of black tiger shrimp (*Penaeus monodon*) to enrofloxacin and to culture system. *Aquaculture,* vol.285, pp.244-248.

Uyuguari, M.; Key, P.; Moore, J.; Jackson, K. & Scott, G. (2009). Acute effects of the antibiotic oxytetracycline on the bacterial community of the grass shrimp, *Palaemonetes Pugio. Environmental Toxicology and Chemistry.* vol.28, No.12, pp.2715:2724.

Vandenberghe, J.; Verdonck, L.; Robles-Arozarena, R.; Rivera, G.; Bolland, A.; Balladares, M.; Gomez-Gil, B.; Calderon, J.; Sorgeloos, P. & Swings, J. (1999). Vibrios associated with *Litopenaeus vannamei* larvae, postlarvae, broodstock, and hatchery probionts. *Applied and Environmental Microbiology,* vol.65, No.6, pp.2592-2597.

Vincent, G. A. & Lotz, J. M. (2007). Advances in reseach of Necrotizing Hepatopancreatitis Bacterium (NHPB) affecting penaeid shrimp aquaculture. *Reviews Fisheries Science,* vol.15, pp.63-73.

Walter, J. P. & Winton, R. J. (2010). Emerging viral diseases of fish and shrimp. *Veterinary Research,* vol.51, pp.51-75.

Wang, Q.; Liu, Q. & Li, J. (2004). Tissue distribution and elimination of oxytetracycline in perch *Lateolabrax japonicus* and black seabream (*Sparus macrocephalus*) following oral administration. *Aquaculture,* vol.237, pp.31–40.

Wen, F.; Shuai, Z.; Kai, Z. & Si-cheng, L. (2007). Pharmacokinetics and tissue distribution of enrofloxacin and its metabolite ciprofloxacin in *Scylla serrata* following oral gavage at two salinities. *Aquaculture,* vol.272, pp.180–187.

Williams, R. R.; Bell, T.A. & Lightner, D. V. (2002). Shrimp antimicrobial testing: II. Toxicity testing and safety determinations for twelve antimicrobials with *penaeus* shrimp larvae. *Journal Aquatic Animal Health,* vol.4, pp.262–270.

Xu, W.; Xiaobin, Z.; Xinting, W.; Liping, D. & Gan, Z. (2006). Residues of enrofloxacin, furazolidone and their metabolites in Nile tilapia (*Oreochromis niloticus*). *Aquaculture,* vol.254, pp.1–8.

Yanong, P. R. & Curtis, W. E. (2005). Pharmacokinetic studies of Florfenicol in Koi Carp and threespot gourami *Trichogaster trichopterus* after oral and intramuscular treatment. *Journal of Aquatic Health,* vol.17, pp.129-137.

8

Probiotics in Aquaculture – Benefits to the Health, Technological Applications and Safety

Xuxia Zhou[1] and Yanbo Wang[2,*]
[1]College of Biological and Environmental Engineering, Zhejiang University of Technology, Hangzhou
[2]Food Quality and Safety Department, Zhejiang Gongshang University, Hangzhou
China

1. Introduction

Aquaculture is the fastest growing food-producing sector in the world at an average rate of 8.9% per year since 1970, compared with only 1.2% for capture fisheries and 2.8% for terrestrial farmed meat production systems over the same period (Subasinghe, 2005). Although aquatic food production through aquaculture is the fastest growing sector and vaccines are being developed and marketed in aquaculture, the disease is still a major problem in the aquaculture farming industry (Bondad-Reantaso *et al.*, 2005). During the last decades, chemical additives and veterinary medicines, especially antimicrobial agents, to prevent and control disease have been also applied in aquaculture (Wang and Xu, 2004; Cabello, 2006; Lupin, 2009). However, the risks associated with the transmission of resistant bacteria from aquaculture environments to humans, and the introduction in the human environment of nonpathogenic bacteria, containing antimicrobial resistance genes, and the subsequent transfer of such genes to human pathogens existed according to FAO (2005). Previous studies also show the aquatic bacteria can develop resistance genes as a consequence of exposure to antimicrobial agents (Smith *et al.*, 1994; Kim *et al.*, 2004; Sørum, 2006). Therefore, the need for alternative techniques is increasing and the contribution of probiotics may be considerable.

The use of probiotics in aquaculture is now widely accepted with an increasing demand for environment friendly aquaculture (Ringø and Gatesoupe, 1998; Gatesoupe, 1999; Sharma and Bhukhar, 2000; Irianto and Austin, 2002; Wang and Xu, 2006; Vine *et al.*, 2006; Wang, 2007; Denev *et al.*, 2009; Qi *et al.*, 2009). Nowadays, a number of preparations of probiotics are commercially available and have been introduced to fish, shellfish and molluscan farming as feed additives, or are incorporated in pond water (Moriarty, 1998; Wang *et al.*, 2005; Prado *et al.*, 2010). According to the claims of the producers, these products are effective in supporting the health of aquatic animals and are also safe. However, there are doubts with regard to the general concept of probiotics and to these claims on the other hand. Indeed, the current explanations and principles are still not enough to describe what

* Corresponding Author

probiotics actually are, where they come from, and what they can do (Wang *et al.*, 2008). Thus, there is clearly a need in increasing our knowledge of aquacultural animals and of effective preparation, technological applications and safety evaluation of probiotics. This chapter provided a summary of the status and challenges of probiotics application in aquaculture. In this chapter, the benefits to the health, technological application and safety evaluation were discussed. In addition, the probiotics information in aquaculture obtained from authentic and highly regarded sources was contained and listed.

2. Probiotics and gut microbiota

Three general modes of probiotics actions have been classified and presented by Oelschlaeger (2010) as follow: (1) Probiotics might be able to modulate the host's gut defences including the innate as well as the acquired immune system and this mode of action is most likely important for the prevention and therapy of infectious diseases but also for the treatment of inflammation of the digestive tract or parts thereof. (2) Probiotics can also have a direct effect on other microorganisms, commensal and/or pathogenic ones and this principle is in many cases of importance for the prevention and therapy of infections and restoration of the microbial equilibrium in the gut. (3) Finally, probiotic effects may be based on actions affecting microbial products, host products and food ingredients and such actions may result in inactivation of toxins and detoxification of host and food components in the gut. According to above summary, all three modes of probiotics actions are all likelihood associated with gut and/or gut microbiota. Therefore, it has become apparent that we are in fact dealing with another "organ", the so called "microbiotic canal" with the increased knowledge of the specific activity of the gut microbiota (Wolf, 2006). In general, the gut microbiota remain relatively stable throughout life once established although they can be influenced by several factors such as mode of delivery, hygiene and the use of antibiotics.

The gut microbiota with the epithelium and mucosal immune system orchestrate a network of immunological and nonimmunological defenses, providing both protection against pathogens and tolerance to commensal bacteria and harmless antigens (Sanz and Palma, 2009). The important role of commensal bacteria in development of optimally functioning mucosal immune system was demonstrated in germ-free animals (Tlaskalová-Hogenová, 2004). Therefore, the imbalance of gut microbiota has been linked to several diseases including inflammatory bowel diseases, periodontal disease, rheumatoid arthritis, atherosclerosis and allergy. So probiotics, that is, microbial strains that have beneficial effects on the host, are thought to benefit this intestinal ecosystem (Julio and Marie-Joséé, 2011). In addition, some probiotics strains also induce the secretion of multiple antimicrobial materials by intestinal Paneth cells through cell-autonomous MyD88-dependent toll-like receptor activation (Vaishnava et al., 2008) and regulate the alterations of permeability related with infections, stress, and inflammatory conditions (Lutgendorff *et al.*, 2008). There is evidence that probiotics produce a protective effect on the gut microbiota and the beneficial effects of probiotics on several microbial disorders have been well reviewed (Gismondo *et al.*, 1999).

As for the aquatic animals such as fish and shrimp, the colonization of the gastrointestinal tract starts immediately after hatching and is completed within a few hours to modulate expression of genes in the digestive tract, thus creating a favorable habitat for them and preventing invasion by other bacteria introduced later into the ecosystem (Balcázar *et al.*,

2006). This is attributed to competitive exclusion mechanisms and improved immune system development and maturation. Intake of probiotics has been demonstrated to modify the composition of the microbiota, and therefore assist in returning a disturbed microbiota (by antibiotics or other risk factors) to its normal beneficial composition (Gómez and Balcázar, 2008). As for the mechanisms during this physiological process, the production of antimicrobial substances, competition for nutrients or adhesion receptors, inhibition of virulence gene expression and enhancement of the immune response are all included (Irianto and Austin, 2002; Nikoskelainen, *et al.*, 2003; Vine *et al.*, 2004; Kim and Austin, 2006; Balcázar, *et al.*, 2007). However, the exact mechanism by which these probiotics do this is not known. Advances in the understanding of the mechanisms between gut microbiota and probiotics and how the immune system of aquatic animals generally responds to gut microbiota would be of great help to identify the molecular targets of probiotics and the biomarkers of their effects, and to provide sounder evidences on their benefits on physiologic conditions and immune-mediated disorders.

3. Probiotics effects in aquaculture: Benefits to the health

When looking at probiotics intended for an aquatic usage it is important to consider certain influencing factors that are fundamentally different from terrestrial based probiotics (Kesarcodi-Watson *et al.*, 2008). Indeed, aquatic animals are quite different from the land animals and a consequence of the specificity of aquatic microbiota is that the most efficient probiotics for aquaculture may be different from those of terrestrial species (Gatesoupe, 1999). A fairly constant habitat of resident microbiota in the gastrointestinal tract of terrestrial livestock is important, whereas most microbiota is transient in aquatic animals (Moriarty, 1990). Shift in intestinal microflora of Atlantic halibut (*Hippoglossus hippoglossus*) larvae during first feeding was studied and the results showed the transition from a prevailing *Flavobacterium* spp. intestinal flora to an *Aeromonas* spp./*Vibrio* spp. dominant flora occurred when first feeding commenced (Bergh et al., 1994). It indicated that the gut microbiota of aquatic animals may change rapidly with the intrusion of microflora from water, live food and artificial diet. In addition, aquatic animal and microorganisms share the same ecosystem in the aquatic environment and it suggested that the interaction between the microbiota, including probiotics, and the host is not limited to the intestinal tract. Therefore, the definition of a probiotic for aquatic environments needs to be modified, which allows a broader application of the term "probiotic". A probiotic is then defined by Verschuere *et al.* (2000) as a live microbial adjunct which has a beneficial effect on the host by modifying the host-associated or ambient microbial community, by ensuring improved use of the feed or enhancing its nutritional value, by enhancing the host response towards disease, or by improving the quality of its ambient environment.

Most probiotics used in aquaculture belong to the lactic acid bacteria, of the genus *Bacillus*, to the photosynthetic bacteria or to the yeast, although other genera or species have also been mentioned (Fig. 1). Many studies have reported promising results using a single beneficial bacterial strain as probiotic in the culture of many aquatic species (Gatesoupe, 1991; Noh *et al.*, 1994; Bogut *et al.*, 1998; Carnevali, *et al.*, 2006; Díaz-Rosales *et al.*, 2009; Li *et al.*, 2009; Zhou *et al.*, 2009; Tovar-Ramírez *et al.*, 2010; Wang and Gu, 2010; Zhou *et al.*, 2010; Wang, 2011). It is important to consider the possibility of using different species, as suggested by Noh *et al.* (1994) and Bogut *et al.* (1998). The effect of probiotics, photosynthetic

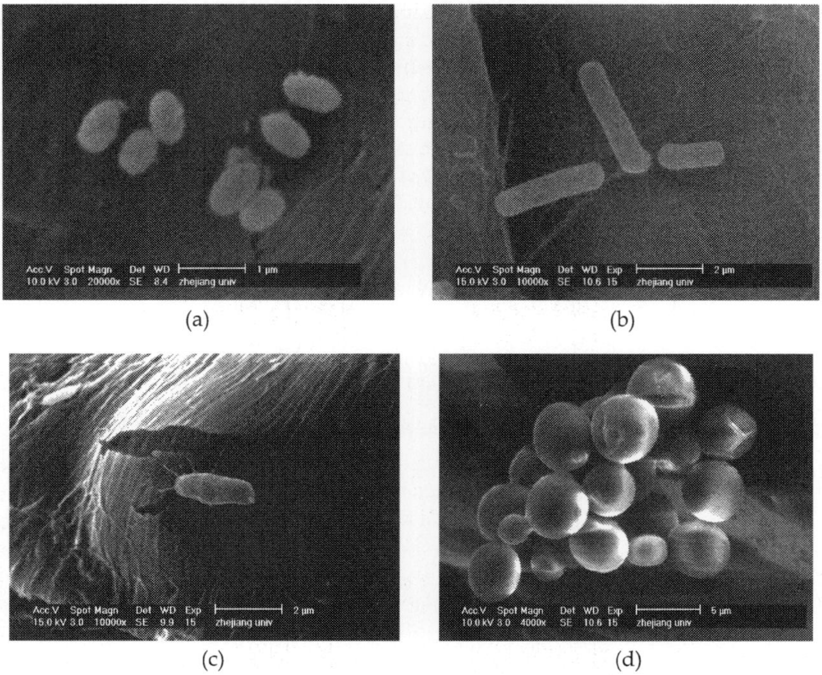

Fig. 1. The configurations of putative probiotics strains isolated and stored in our laboratory using scanning electron microscope (Philips XL30ESEM, Netherlands). a, *Lactococcus lactis*; b, *Bacillus coagulans*; c, *Rhodopseudomonas palustris*; d, *Saccaromyces cerevisiae*.

bacteria (*Rhodobacter sphaeroides*) and *Bacillus* sp. (*B. coagulans*), on growth performance and digestive enzyme activity of the shrimp, *Penaeus vannamei*, was investigated and the results showed that the effects were related with supplementation concentrations of probiotics and thus use of a 10 g/kg (wet weight) supplement of probiotics in shrimp diet was recommended to stimulate productive performance (Wang, 2007). A mixture of *Bacillus* probiotic bacteria (*Bacillus subtilis, Bacillus licheniformis* and *Bacillus pumilus*) was also evaluated in the gilthead sea bream (*Sparus aurata*) larviculture focusing on their effects on survival, growth and general welfare (Avella *et al.*, 2010). The data generated in this study show the benefit of the administration of *Bacillus* probiotic mixture in terms of stress response and growth and provide scientific and technical support for the implementation of sustainable development of sea bream aquaculture. Similar results were also observed in olive flounder supplemented with *Lactobacillus plantarum, Lactobacillus acidophilus, Lactobacillus sakei, Bacillus subtilis,* and *Saccharomyces cerevisiae* as individual and mixed enriched diet (Harikrishnan *et al.*, 2011a). Lactobacil probiotics individually or mixed with Sporolac enriched diet were used to enhance the immune status, thereby improving the disease resistance in lymphocystis disease virus infected olive flounder (*Paralichthys olivaceus*) and the results showed that the better innate immune response and disease resistance were found in groups supplemented with mixed probiotics (Harikrishnan *et al.*,

2010). However, feeding experiments conducted on 600 *O. niloticus* using the diets containing single or mixed isolated probiotic bacteria show the different results in survival rates and the highest with fish fed diets supplemented with *B. pumilus* was observed, followed by a mixture of probiotics (*B. firmus*, *B. pumilus* and *C. freundii* in equal numbers), and then *C. freundii* (Aly *et al.*, 2008). It indicates that the beneficial effects of probiotics fed aquatic animals are associated with probiotic strains, isolation species, culture animals and water quality. Altogether, the data reported above may well explain the current trend to prefer alternative probiotics for the application in aquaculture.

Additionally, a large number of studies have combined probiotics with prebiotics, a selectively fermented ingredient that allows specific changes both in the composition and/or activity in the gastrointestinal microflora that confers benefits upon host well-being and health (Gibson *et al.*, 2004). Thus the synbiotics, as a combination of probiotics and prebiotics, have been studied to expect the synergistic effects. Nowadays, there are several recognized functional prebiotic oligosaccharides such as fructooligosaccharides (FOS), mannan oligosaccharides (MOS), insulin, ß-glucan, and xylo-oligosaccharides (XOS) in use around the world. The effect of dietary application of a commercial probiotic (*Bacillus* spp.) and MOS, used singularly and combined, on the survival, growth performance and feed cost-benefit of European lobster (*Homarus gammarus*) larval was assessed and the results in this study strongly suggest that the dietary combination of Bacillus spp. And MOS is cost effective when used to promote survival and provides the added benefits of improved growth performance, compared to their individual supplementation (Daniels *et al.*, 2010). Similar results have been reported on shrimp, *Litopenaeus vannamei*, and the disease resistance was also improve by enhancing immunity, as well as presumably modulating microflora in the shrimp's gut (Li *et al.*, 2009). It suggested that the combined application of probiotics and prebiotics is an interesting prospect for replacement of growth-promoting chemotherapeutics in the aquaculture industry and could be a useful tool in the rearing of certain aquatic animals. Recently, herbs and probiotics are combined in diet and treated as one of the promising alternative tools to supplement and supplant antibiotics, chemicals or vaccines (Sahu *et al.*, 2008; Nayak, 2010). According to Harikrishnan *et al.* (2011b), administration of probiotics (*Lactobacillus sakei* BK19) and herb (*Scutellaria baicalensis*) can effectively minimize the mortality and restore the altered heamatological parameters and enhancing the innate immunity in *O. fasciatus* against *Edwardsiella tarda*, which indicate a promising role to prevent diseases and disease outbreaks in aquaculture. Similar results were also determined in olive flounder, *Paralichthys olivaceus*, against *Streptococcus parauberis* and the enhanced growth, blood biochemical constituents, and nonspecific immunity were observed in the groups treated with probiotics and herbals mixture supplementation diet (Harikrishnan *et al.*, 2011c). Further investigations on the interaction between probiotics and other functional additives at molecular level are warranted in aquaculture.

4. Manufacture and safety evaluation of probiotics

The continuing expansion of interest in probiotic bacteria has led to an increase in manufactured functional foods and feeds containing these bacteria. Given the natural and/or intestinal origin of these microorganisms, the challenges these putative probiotics face in order to be in a highly viable state throughout processing, manufacture, and storage are enormous. Environmental stresses such as temperature, acid, exposure and osmotic pressure, oxygen have important effects on probiotics survival and activity both in product and animal gut. However, like all bacteria, probiotic bacteria retain a broad arsenal of molecular mechanisms to combat the often lethal environmental stresses encountered during processing and following ingestion and therefore the comprehensive appreciation of these mechanisms should inevitably lead to the design and manufacture of probiotic cultures, which retain greater viability through to the target site in the intestine (Corcoran *et al.*, 2008). Environmental stress responses in *Lactobacillus*, which have been investigated mainly by proteomics approaches, are reviewed by De Angelis and Gobbetti (2004) and the physiological and molecular mechanisms of responses to heat, cold, acid, osmotic, oxygen, high pressure and starvation stresses are described. As for the proteomics approaches, the technique primarily bases on two-dimensional gel electrophoresis (2-DE) (Kellner, 2000). The intensity of an individual spot indicates how much the cell has produced of that actual protein and thus it has facilitated the rapid characterization of thousands of proteins in a single polyacrylamide gel for the molecular mechanism studies of probiotics. Such studies associated with the cellular processes and metabolism mechanisms available to probiotic bacteria to facilitate survival in various stressful conditions can lead to production of designer probiotic strains with enhanced viability in feed systems and efficacy following ingestion for aquatic animals. Additionally, several other factors including the physiologic state of the probiotics, the chemical composition of the product and possible interactions of the probiotics with the starter cultures must be considered to ensure the abilities of probiotics in aquaculture.

Although the probiotic species such as *Lactobacillus acidophilus* have been safely used for a long time, the safety aspects have always to be considered and possible adverse effects should continuously be evaluated as illustrated by literature (Salminen *et al.*, 1998). However, a growing number of diseases that appeared with the worldwide development of aquaculture may be assigned to distinct bacteria belonging to the genera *Streptococcus*, *Lactococcus*, *Vagococcus* and *Carnobacterium*, but, in most cases, the clear mechanisms have not been found (Ringø and Gatesoupe, 1998). In addition, safety considerations regarding antimicrobial resistance neglected for a long time are now taken into account for the development and marketing of probiotics (Courvalin, 2006). The question whether genetic exchange may occur between probiotics and gut microflora or pathogens is raised because the genes can be transferred between microorganisms. As a result, the antibiotic multi-resistance existent of probiotics shows the possible insecurity caused by the possibility of resistance genes transfer from probiotic strains to bacterial pathogens or from aquatic commensals to probiotics. According to O'Brien et al. (1999), it is important to differentiate between intrinsic resistance and that mediated by special genetic elements when evaluating the antibiotic resistance profiles among different species and strains. Indeed, safety is the state of being certain that adverse effects will not be caused by an agent under defined conditions. Therefore, feeding of novel probiotics to healthy aquatic animals is not only concerned with efficacy but safety even though lactobacilli and bifidobacteria are generally regarded as safe. With the development of molecular biology and other advanced modern techniques, the critical, tailored approaches such as cell culture to safety evaluation of probiotics can ensure that healthy benefits are accessible to aquatic animals. The epithelial cells of tilapia (*Oreochromis nilotica*) were isolated and primarily cultured as the cells model to evaluate the probiotic, *Rhodopseudomonas palustris*, through the morphologic characters, cells viability, livability and permeability (Wang and

Xu, 2007). This study shows cell culture is one of the promising approaches to safety evaluation of probiotic in the future.

5. Future probiotics for aquaculture

The important role of the gut flora in the maintenance of health and in the prevention of disease is well recognized (Holzapfel and Schillinger, 2002). Use of probiotics is likely to be the most natural and safe means for improving gut flora balance to prevent bacterial pathogens by competing for essential nutrients or attachment sites (Chukeatirote, 2003). As for aquatic animals gut flora, the continuous interaction with the environment, the body system and intrinsic microorganisms is very complex. Although the explosion in recent years of publications dealing with probiotic organisms has been increased, central and vital information is still needed and therefore more advanced methods should be developed to assess the changes in the composition of the gut flora and their mutual interaction with the metabolism of aquatic animals. Currently, probiotics may serve to partially replace the presently reduced or even prohibited application of nutritive antibiotics or chemotherapeutics in animal nutrition and in fulfillment of health claims in man and animals (Reuter, 2001). According to Kesarcodi-Watson *et al.* (2008), a probiotic for the new, effective and safe products in aquaculture must possess certain properties as follow: (1) the probiotic should not be harmful to the host it is desired for, (2) it should be accepted by the host, e.g. through ingestion and potential colonization and replication within the host, (3) it should reach the location where the effect is required to take place, (4) it should actually work in vivo as opposed to in vitro findings, and (5) it should preferably not contain virulence resistance genes or antibiotic resistance genes. These properties should be considered during the manufacture process and safety evaluation of novel probiotics. Then the future will provide targeted probiotic bacteria accord with above properties for specific use with carefully controlled studies on clearly defined selected strains. In addition, an increasing demand for alternative to antibiotics products applied in aquaculture indicates a bright future for probiotics and a number of better commercial probiotics will be available, particular directed at larval culture.

6. Acknowledgments

This study is supported by the National Natural Science Foundation of China (No. 30901044 and 31072221) and Zhejiang Provincial Natural Science Foundation of China (No. R3110345). Special appreciation goes to Mr Junhui Wen, Mrs. Yaqing Zhu and Mr. Gentu Wu for their works which lead to success of this review.

7. References

Aly, S.M., Abd-El-Rahman, A.M., John, G., Mohamed, M.F. 2008. Characterization of some bacteria isolated from *Oreochromis niloticus* and their potential use as probiotics. Aquaculture. 277: 1-6.

Avella, M.A., Gioacchini, G., Decamp, O., Makridis, P., Bracciatelli, C., Carnevali, O. 2010. Application of multi-species of *Bacillus* in sea bream larviculture. Aquaculture. 305: 12-19.

Balcázar, J.L., Blas, I., Ruiz-Zarzuela, I., Cunningham, D., Vendrell, D., Muzquiz, J.L. 2006. The role of probiotics in aquaculture. Vet Microbiol. 114: 173-186.

Balcázar, J.L., Vendrell, D., Blas, I., Ruiz-Zarzuela, I., Gironés, O., Muzquiz, J.L. 2007. In vitro competitive adhesion and production of antagonistic compounds by lactic acid bacteria against fish pathogens. Vet Microbiol. 122: 373-380.

Bergh, O., Naas, K.E., Harboe, T. 1994. Shift in intestinal microflora of Atlantic halibut (*Hippoglossus hippoglossus*) larvae during first feeding. Canadian Journal of Fisheries and Aquatic Science. 51: 1899-1903.

Bogut, I., Milakovic, Z., Bukvic, Z., Brkic, S., Zimmer, R. 1998. Influence of probiotic *Streptococcus faecium* M74 on growth and content of intestinal microflora in carp *Cyprinus carpio*. Czech J. Anim. Sci. 43: 231-235.

Bondad-Reantaso, M., Subasinghe, R.P., Arthur, J.R., Ogawa, K., Chinabut, S., Adlard, R., Tan Z., Shariff, M. 2005. Disease and health management in Asian aquaculture. Vet. Parasitology. 132: 249-272.

Cabello, F.C. 2006. Heavy use of prophylactic antibiotics in aquaculture: A growing problem for human and animal health and for the environment. Environmental Microbiology. 8: 1137-1144.

Carnevali, O., Vivo, L., Sulpizio, R., Gioacchini, G., Olivotto, L., Silvi, S., Cresci, A. 2006. Growth improvement by probiotic in European sea bass juveniles (*Dicentrarchus labrax*, L.), with particular attention to IGF-1, myostatin and cortisol gene expression. Aquaculture. 258: 430-438.

Chukeatirote, E. 2003. Potential use of probiotics. Songklanakarin J. Sci. Technol. 25: 275-282.

Corcoran, B.M., Stanton, C., Fitzgerald, G., Ross, R.P. 2008. Life under stress: the probiotic stress response and how it may be manipulated. Curr Pharm Des. 14: 1382-1399.

Courvalin, P. 2006. Antibiotic resistance: he pros and cons of probiotics. Digestive and Liver Disease. 38: S261-S265.

Daniels, C.L., Merrifield, D.L., Boothroyd, D.P., Davies, S.J., Factor, J.R., Arnold, K.E. 2010. Effect of dietary *Bacillus* spp. and mannan oligosaccharides (MOS) on European lobster (*Homarus gammarus* L.) larvae growth performance, gut morphology and gut microbiota. Aquaculture. 304: 49-57.

De Angelis, M., Gobbetti, M. 2004. Environmental stress responses in Lactobacillus: a review. Proteomics. 4: 106-122.

Denev, S., Staykov, Y., Moutafchieva, R., Beev, G. 2009. Microbial ecology of the gastrointestinal tract of fish and the potential application of probiotics and prebiotics in finfish aquaculture. International Aquatic Research. 1: 1-29.

Díaz-Rosales, P., Arijo, S., Chabrillón, M., Alarcón, F.J., Tapia-Paniagua, S.T., Martínez - Manzanares, E., Balebona, M.C., Moriñigo M.A. 2009. Effects of two closely related probiotics on respiratory burst activity of Senegalese sole (*Solea senegalensis*, Kaup) phagocytes, and protection against *Photobacterium damselae* subsp. *Piscicida*. Aquaculture. 293: 16-21.

FAO. 2005. Responsible Use of Antibiotics in Aquaculture (Ed. Serrano, P.H.), FAO Fisheries Technical Paper. 469, FAO, Rome, Italy, pp 98.

Gatesoupe, F.J. 1991. The effect of three strains of lactic bacteria on the production rate of rotifers, *Brachionus plicatilis*, and their dietary value for larval turbot, *Scophthalmus maximus*. Aquaculture. 96: 335-342.

Gatesoupe, F.J. 1999. The use of probiotics in aquaculture. Aquaculture. 180: 147-165.

Gibson, G.R., Probert, H.M., Van Loo, J., Rastall, R.A., Roberfroid, M.B. 2004. Dietary modulation of the human colonic microbiota: Updating the concept of prebiotics. Nutrition Research Reviews. 17: 259-275.

Gismondo, M.R., Drago, L., Lombardi, A. 1999. Review of probiotics available to modify gastrointestinal flora. International Journal of Antimicrobial Agents. 12: 287-292.

Gómez, G.D., Balcázar, J.L. 2008. A review on the interactions between gut microbiota and innate immunity of fish. FEMS Immunol Med Microbiol. 52: 145-154.

Harikrishnan, R., Balasundaramb, C., Heo, M.S. 2010. Effect of probiotics enriched diet on *Paralichthys olivaceus* infected with lymphocystis disease virus (LCDV). Fish & Shellfish Immunology. 29: 868-874.

Harikrishnan, R., Kim, M.C., Kim, J.S., Balasundaram, C., Heo, M.S. 2011c. Probiotics and herbal mixtures enhance the growth, blood constituents, and nonspecific immune response in *Paralichthys olivaceus* against *Streptococcus parauberis*. Fish Shellfish Immunol. 31: 310-317.

Harikrishnan, R., Kim, M.C., Kim, J.S., Balasundaramb, C., Heo, M.S. 2011b. Protective effect of herbal and probiotics enriched diet on haematological and immunity status of *Oplegnathus fasciatus* (Temminck & Schlegel) against *Edwardsiella tarda*. Fish & Shellfish Immunology. 30: 886-893.

Harikrishnan, R., Kim, M.C., Kim, J.S., Balasundaramb, C., Heo, M.S. 2011a. Immunomodulatory effect of probiotics enriched diets on *Uronema marinum* infected olive flounder. Fish & Shellfish Immunology. 30: 964-971.

Holzapfel, W.H., Schillinger, U. 2002. Introduction to pre- and probiotics. Food Research International. 35: 109-116.

Irianto, A., Austin, B. 2002. Use of probiotics to control furunculosis in rainbow trout *Oncorhynchus mykiss* (Walbaum). J Fish Dis. 25: 333-342.

Julio, A., Marie-Joséé, B. 2011. Proteomics, human gut microbiota and probiotics. Expert Review of Proteomics. 8: 279-288.

Kellner, R. 2000. Proteomics: concepts and perspectives. Fresenius J Anal Chem. 366: 517-524.

Kesarcodi-Watson, A., Kaspar, H., Lategan, M.J., Gibson, L. 2008. Probiotics in aquaculture: the need, principles and mechanisms of action and screening processes. Aquaculture. 274: 1-14.

Kim, D.H., Austin, B. 2006. Innate immune responses in rainbow trout (*Oncorhynchus mykiss*, Walbaum) induced by probiotics. Fish Shellfish Immunol. 21: 513-524.

Kim, S., Nonaka, L., Suzuki, S. 2004. Occurrence of tetracycline resistance genes tet (M) and tet (S) in bacteria from marine aquaculture sites. FEMS Microbiol. Lett. 237: 147-156.

Li, J., Tan, B., Mai, K. 2009. Dietary probiotic *Bacillus* OJ and isomaltooligosaccharides influence the intestine microbial populations, immune responses and resistance to white spot syndrome virus in shrimp (*Litopenaeus vannamei*). Aquaculture. 291: 35-40.

Lupin, H.M. 2009. Human health aspects of drug and chemical use in aquaculture. Options Méditerranéennes A. 86: 95-103.

Lutgendorff, F., Akkermans, L.M., Söderholm, J.D. 2008. The role of microbiota and probiotics in stress-induced gastro-intestinal damage. Curr Mol Med. 8: 282-298.

Moriarty, D.J.W. 1990. Interactions of microorganisms and aquatic animals, particularly the nutritional role of the gut flora. In: Lésel, R. (Ed.), Microbiology in Poecilotherms. Elsevier, Amsterdam, pp. 217-222.

Moriarty, D.J.W. 1998. Control of luminous *Vibrio* species in penaeid aquaculture ponds. Aquaculture. 164: 351-358.

Nayak, S.K. 2010. Probiotics and immunity: a fish perspective. Fish Shellfish Immunol. 29: 2-14.

Nikoskelainen, S., Ouwehand, A.C., Bylund, G., Salminen, S., Lilius, E.M. 2003. Immune enhancement in rainbow trout (*Oncorhynchus mykiss*) by potential probiotic bacteria (*Lactobacillus rhamnosus*). Fish Shellfish Immunol. 15: 443-452.

Noh, S.H., Han, K., Won, T.H., Choi, Y.J. 1994. Effect of antibiotics, enzyme, yeast culture and probiotics on the growth performance of Israeli carp. Korean J. Anim. Sci. 36: 480-486.

O'Brien, J., Crittenden, R., Ouwehand, A.C., Salminen, S. 1999. Safety evaluation of probiotics. Trends in Food Science & Technology. 10: 418-424.

Oelschlarger, T.A. 2010. Mechanisms of probiotic actions - a review. International Journal of Medical Microbiology. 300: 57-62.

Prado, S., Romalde, J.L., Barja, J.L. 2010. Review of probiotics for use in bivalve hatcheries. Veterinary Microbiology. 145: 187-197.

Qi, Z., Zhang, X.H., Boon, N., Bossier, P. 2009. Probiotics in aquaculture of China – current state, problems and prospect. Aquaculture. 290: 15-21.

Reuter, G. 2001. Probiotics--possibilities and limitations of their application in food, animal feed, and in pharmaceutical preparations for men and animals. Berl Munch Tierarztl Wochenschr. 114: 410-419.

Ringø, E., Gatesoupe, F.J. 1998. Lactic acid bacteria in fish: a review. Aquaculture. 160: 177-203.

Sahu, M.K., Swarnakumar, N.S., Sivakumar, K., Thangaradjou, T., Kannan, L. 2008. Probiotics in aquaculture: importance and future perspectives. Indian Journal of Microbiology. 48: 299-308.

Salminen, S., Wright, A., Morelli, L., Marteau, P., Brassart, D., Vosf, W.M., et al. 1998. Demonstration of safety of probiotics - a review. International Journal of Food Microbiology. 44: 93-106.

Sanz, Y., Palma, G.D. 2009. Gut Microbiota and probiotics in modulation of epithelium and gut-associated lymphoid tissue function. International Reviews of Immunology. 28: 397-413.

Sharma, O.P., Bhukhar, S.K.S. 2000. Effect of Aquazyn-TM-1000, a probiotic on the water quality and growth of *Cyprinus carpio* var. communis (L.). Indian J. Fish. 47: 209-213.

Smith, P., Hiney, M.P., Samuelsen, O.B. 1994. Bacterial resistance to antimicrobial agents used in fish farming: a critical evaluation of method and meaning. Annu. Rev. Fish Dis. 4: 273-313.

Sørum, H. 2006. Antimicrobial drug resistance in fish pathogens. In: Aarestrup, F.M. (Ed.), Antimicrobial Resistance in Bacteria of Animal origin. ASM Press, Washington DC, pp. 213-238.

Subasinghe, R.P. 2005. Epidemiological approach to aquatic animal health management: opportunities and challenges for developing countries to increase aquatic production through aquaculture. Prev. Vet. Med. 67: 117-124.

Tlaskalová-Hogenová, H., Stepánková, R., Hudcovic, T., Tucková, L., Cukrowska, B., Lodinová - Zádníková, R., et al. 2004. Commensal bacteria (normal microflora), mucosal immunity and chronic inflammatory and autoimmune diseases. Immunol Lett. 93: 97-108.

Tovar-Ramírez, D., Mazurais, D., Gatesoupe, J.F., Quazuguel, P., Cahu, C.L., Zambonino-Infante, J.L. 2010. Dietary probiotic live yeast modulates antioxidant enzyme activities and gene expression of sea bass (*Dicentrarchus labrax*) larvae. Aquaculture. 300: 142-147.

Vaishnava, S., Behrendt, C.L., Ismail, A.S., Eckmann, L., Hooper, L.V. 2008. Paneth cells directly sense gut commensals and maintain homeostasis at the intestinal host-microbial interface. Proc Natl Acad Sci USA. 105: 20858-20863.

Verschuere, L., Rombaut, G., Sorgeloos, P., Verstraete, W. 2000. Probiotic bacteria as biological control agents in aquaculture. Microbiology and Molecular Biology Review. 64: 655-671.

Vine, N.G., Leukes, W.D., Kaiser, H. 2004. In vitro growth characteristics of five candidate aquaculture probiotics and two fish pathogens grown in fish intestinal mucus. FEMS Microbiol Lett. 231: 145-152.

Vine, N.G., Leukes, W.D., Kaiser, H. 2006. Probiotics in marine larviculture. FEMS Microbiol. Rev. 30: 404-427.

Wang, Y. 2007. Effect of probiotics on growth performance and digestive enzyme activity of the shrimp *Penaeus vannamei*. Aquaculture. 269: 259-264.

Wang, Y. 2011. Use of probiotics *Bacillus coagulans, Rhodopseudomonas palustris* and *Lactobacillus acidophilus* as growth promoters in grass carp (*Ctenopharyngodon idella*) fingerlings. Aquaculture Nutrition. 17: 372-378.

Wang, Y., Gu, Q. 2010. Effect of probiotics on white shrimp (*Penaeus vannamei*) growth performance and immune response. Marine Biology Research. 6: 327-332.

Wang, Y., Li, J., Lin, J. 2008. Probiotics in aquaculture: Challenges and outlook. Aquaculture. 281: 1-4.

Wang, Y., Xu, Z. 2004. Probiotics treatment as method of biocontrol in aquaculture. Feed Research. 12: 42-45.

Wang, Y., Xu, Z. 2006. Effect of probiotics for common carp (*Cyprinus carpio*) based on growth performance and digestive enzyme activities. Animal Feed Science and Technology. 127: 283-292.

Wang, Y., Xu, Z. 2007. Safety evaluation of probiotic PSB0201 (*Rhodopseudomonas palustris*) using primary culture epithelial cells isolated from tilapia (*Oreochromis nilotica*) intestine. Journal of Agricultural Biotechnology. 15: 233-236.

Wang, Y., Xu, Z., Xia, M. 2005. The effectiveness of commercial probiotics in northern white shrimp (*Penaeus vannamei* L.) ponds. Fisheries Science. 71: 1034-1039.

Wolf, G. 2006. Gut microbiota: a factor in energy regulation. Nutrition Review. 64: 47-50.

Zhou, X., Tian, Z., Wang, Y., Li W. 2010. Effect of treatment with probiotics as water additives on tilapia (*Oreochromis niloticus*) growth performance and immune response. Fish Physiol Biochem. 36: 501-509.

Zhou, X., Wang, Y., Li, W. 2009. Effect of probiotic on larvae shrimp (*Penaeus vannamei*) based on water quality, survival rate and digestive enzyme activities. Aquaculture. 287: 349-353.

9

Probiotics in Aquaculture of Kuwait – Current State and Prospect

Ahmed Al-marzouk and Azad I. Saheb
Mariculture and Fisheries Department (MFD)
Kuwait Institute for Scientific Research (KISR), Salmiya
Kuwait

1. Introduction

The blue-fin porgy, *Sparidentex hasta,* known as sobaity in Kuwait, is a commercially valuable food fish greatly preferred in Kuwait and other Arabian Gulf countries. This species has been cultured at the Kuwait Institute for Scientific Research (KISR) since 1979 (Hussain et al., 1981). From 1982 to1986, research efforts were carried out with the intension of developing a commercially valid culture technology for sobaity (Teng et al., 1984). The research efforts were completed, and the results were assessed to formulate a culture technology recommended for commercial application (Teng et al., 1999). However, a number of bacterial diseases have been reported that cause severe losses in sobaity larvae. Outbreaks of vibriosis caused by Gram-negative bacteria *Vibrio* spp., is the most serious bacterial disease of both grouper and sobaity (Rasheed, 1989a). They were identified as *V. anguillarum*, *V. ordalli* and *V. carchariae* and *V. damsela*. In addition, *V. harveyi* was associated with mortalities in hamoor and sobaity (Saeed, 1995). So far, conventional approaches, such as the use of antimicrobial drugs, have had limited success in the prevention or cure of aquatic animal disease. Furthermore, there is a growing concern about the use and, particularly, the abuse of antimicrobial drugs in aquaculture. The practice of using antibiotics indiscriminately for the treatment of diseases in aquaculture could result in the accumulation of residues and the development of resistant strains of bacteria (Uma, 1999). This leads to the search for new, more effective antibiotics thus increasing the consumption of antibiotics in aquaculture. Vaccination can not prevent the development of the disease in young and small fish (Ellis, 1999; Magnadottir et al. 2006). Conventional vaccination is, accordingly, not of value before this time, and the larvae are wholly reliant on the non-specific immune parameters. Thus, an alternative methods are to be evolved to maintain a healthy microbial environment in fish rearing tanks. One such method that is gaining recent acceptance within the aquaculture industry is the use of probiotics bacteria to control potential bacterial pathogens (Wang et al., 2008; Decamp et al., 2010). Thus, the application of probiotics may provide a potential alternative method to protect fish from infectious diseases and improve the survival of cultured marine fish (Irianto and Austin, 2003). The aim of the present study was to evaluate the effect of three autochthonous probiotics, coded as SHBP, 4SQ and 5L8 and a standard isolate *Lactobacillus divergens* (ATCC, 35677) on growth performance of rotifers and the effect of dietary administration of SHBP, *L. divergens* and a combination (SHBP& *L. divergens*) on the survival rate of sobaity larvae.

2. Screening, isolation and *in vitro* antagonism test of autochthonous probiotics

Performing an *in vitro* antagonism test is considered to be an important step in screening potential probiotics, in which pathogenic bacteria are exposed to the selective probiotics in liquid (Gildberg, et al., 1995, 1997) or solid (Austin et al., 1992; Dopazo, et al., 1988; Westerdahl et al., 1990) medium. The preselection of candidate probiotics based on these *in vitro* antagonism tests has usually led to the finding of effective probiotics (Verschuere et al., 2000). Bacterial isolates were obtained from the mid and hind gut of cultured yellow-fin porgy *Acanthopagrus latus* (shaem), wild silver pomfret *Pampus argenteus* (zobaidy), wild orange-spotted grouper *Epinephelus coioides* (hamoor), wild tigertooth croaker, *Otolithes argenteus* (newaiby), cultured blue- fin porgy *Sparidentex hasta* (sobaity), *Lactobacillus* sp., (Alken-Clear FIO-1006, Alken-Murray Corp, USA) and a *Lactobacillus divergens* (ATCC, 35677). Autochthonous probiotic, Gram-positive bacteria isolated from cultured shaem was coded as SHPB, form wild zobaidy, coded as 4SQI, from cultured sobaity coded as S24, form wild newaiby coded as 5L82 and from wild hamoor coded as 5M99b. Probiotic bacteria were cultured in brain heart infusion broth (BHIB, Oxoid, Basingstoke, UK) with 2% NaCl. After cultivation, bacteria were harvested by centrifugation (2000 rpm for 10 min), washed twice and resuspended in phosphate-buffered saline (PBS). The *in vitro* antagonism of the isolated autochthonous probiotics against *Vibrio alginolyticus* (locally isolated from diseased cultured sobaity), *V. anguillarum* (ATCC 43310), *V. harveyi* (locally isolated from diseased cultured mullet *Liza klunzingeri*), *V. parahaemolyticus* (ATCC, 27159), *V. vulnificus* (ATCC, 33149) *and Streptococcus agalactiae* (locally isolated from diseased zobaidy) was investigated. The inhibitory activity was assessed by three antagonism tests, the first test is the well-diffusion test (Perez et al., 1990) and the second test is the double-layer method (Dopazo et al., 1988). The third test was a modification to the double-layer method. The new modified technique applied was named as "filter paper disc method", which aimed to obtain precise and accurate inhibition zones created by the probiotics bacteria against the pathogenic bacteria. During the current research the third method was selected. Five sterile filter paper disks were placed on the BHIA, and a drop of the probiotic culture (2 μL) was placed on the sterile filter papers, and incubated overnight at 30°C. The overnight cultures of the pathogenic bacteria were prepared at 1:10 dilution using sterile phosphate buffered saline. All the probiotic cells were killed by exposing the plate to each diluted pathogenic bacteria were overlayed. The cultures were incubated overnight at 30 °C and the zone of inhibition was measured. The results showed that all the methods were suitable to assess the effect of probiotic bacteria on pathogenic bacteria. However, the modified overlay method seems to be more effective for the selection of probiotics. This method showed consistent results on the zone of inhibition compared to the other two methods that some times produced doubtful false negative results mainly due to the swarming nature of growth in the case of *V. alginolyticus*. In addition, the administration of suitable concentration of probiotics and allowing growth and production of antimicrobial compounds before the addition of *Vibrio* spp produced a reliable inhibition results. The probiotics coded as SHPB and 4SQI showed a significant zone of inhibition against all the pathogenic bacteria (Plate1). The other probiotics showed some effect, but they were unable to inhibit the growth of all the pathogenic bacteria.

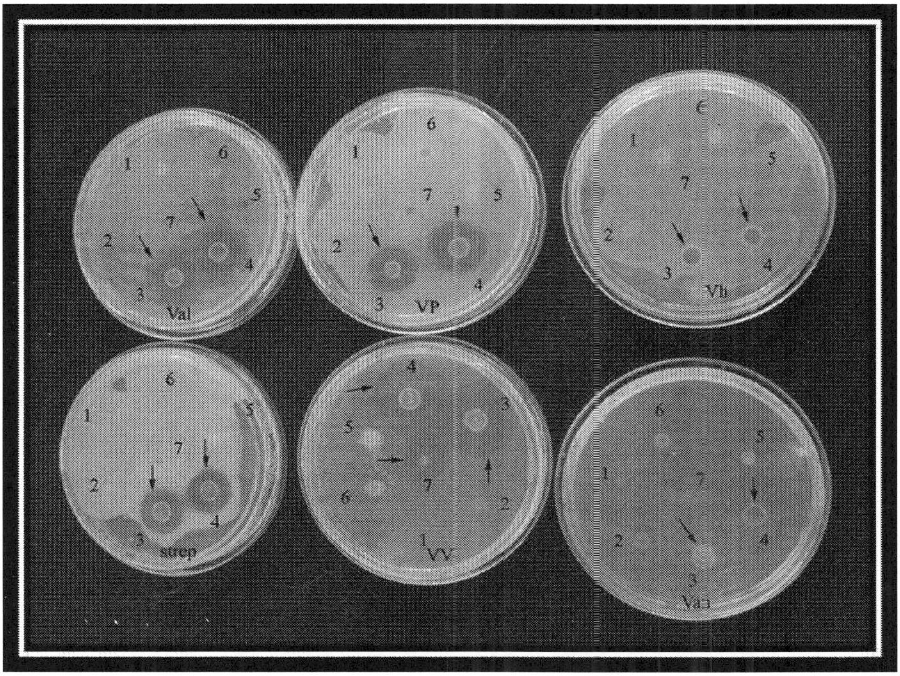

1: Lactobacillus sp; 2: Lactobacillus divergens; 3: 4SQI; 4: SHPB; 5: 5L82; 6: M99b; 7:S24.
Val: V. alginolyticus; VP: V. parahaemolyticus; Vh: V. harveyi; Strep: Streptococcus agalactiae; VV: V. vulnificus; Van: V. anguillarum.

Plate 1. Antagonism test of seven putative probiotics against six pathogenic bacteria by the modified double-layer method.

3. Competitive exclusion of vibrio co-cultured with autochthonous probiotics (SHPB)

Competitive exclusion of potential pathogenic bacteria effectively reduces or eliminates the need for antibiotic prophylaxis in intensive larviculture systems (Garriques & Arevalo, 1995). It has been reported that bacterial strains associated with intestinal and skin mucus of adult marine turbot *Scophthalmus maximus* and dab *Limanda limanda*, suppressed the growth of the fish pathogen *V. anguillarum* (Olsson et al., 1992). Thus the manipulation of microbial constitutes is a viable tool to reduce or eliminate the incidence of opportunist pathogens (Balcazar et al., 2006). In this study, co-culture of *Vibrio* sp. and SHPB was plated on a BHI agar plate. A 24 h BHI broth culture of *Vibrio* spp., and SHPB was used. The suspensions of individual bacteria after harvesting and adjusting the cell density to 10^4 / mL were used. Both the suspensions were spread (100 µL each) on a BHI agar plate and incubated for observing the colony formation. The colonies were observed, under the microscope, at 0, 3 and 6 h of incubation at 30°C for competitive exclusion or invasive growth. The SHPB showed a distinctive competitive exclusion against *Vibrio* spp., (plate 2).

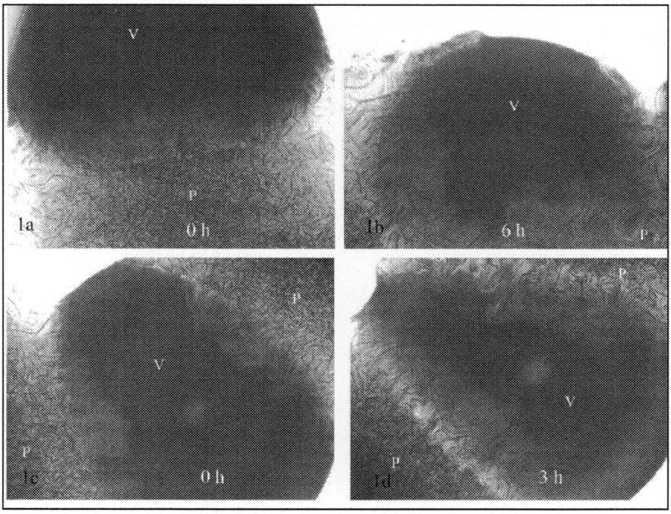

Plate 2. Competitive exclusion of SHPB (P) against one pathogenic Vibrio (V) After 6 h (1a &1b) and after 3 h (1c & 1d). 20 x.

4. Effect of autochthonous probiotics on rotifer proliferation

Several studies related to marine fish larviculture have attempted to find suitable probiotics that has a positive effect on the live food and improves their dietary value (Benetti et al., 2004; Robertson et al., 2000; Gomez-Gil et al., 2000; Ringo and Birkbeck, 1999; Harzevilli et al., 1998; Gatesoupe, 1991a). Douillet (2000) reported that rotifers cultured with an *Altermonas* strain or blend of strains resulted in a consistent enhancement of the rotifers culture. In this study, rotifers *Brachionus plicatilis* were enriched for 24 h with a high quality mixture of algae (Nanochloropsis, Tetraselmis and Isochrysis) along with the available commercial enrichments (Super-Selco and DHA protein selco). The rotifer density was 22 cell/ mL. A one week experiment was carried out to evaluate the proliferation of the rotifers treated with probiotics SHPB, 4SQI, 5L82 and *L. divergens*. Daily 1 mL of the probiotic and 1 mL of the algal suspension was added to each flask, which corresponds to 10^{10} CFU/ mL. The control flasks received 2 mL algae daily. The rotifer count was determined for seven days to assess the effect of the probiotics. An increase in rotifers proliferation and reproducibility within different treatments was obtained, as described by the differences in the coefficient of determination for four probiotics treatments and the control (Fig. 1). The probiotic SHPB and *L.divergens* showed better enhancement on the cell population with time, while 4SQI and 5L8 showed lower cell proliferation response compared to the control. The SHPB gave the best compared to the control and other probiotics, mainly during the first three days. All the probiotics tested showed better effect on the rotifer counts compared to the control. The *Vibrio* load in the rotifer population without probiotic treatment was significantly higher than that of the probiotic treated sets during the first three days. However, the vibrio population was significantly lower in SHPB treated rotifer all through the week. The sobaity larvae usually start feeding on day two or three post hatch. Thus, the

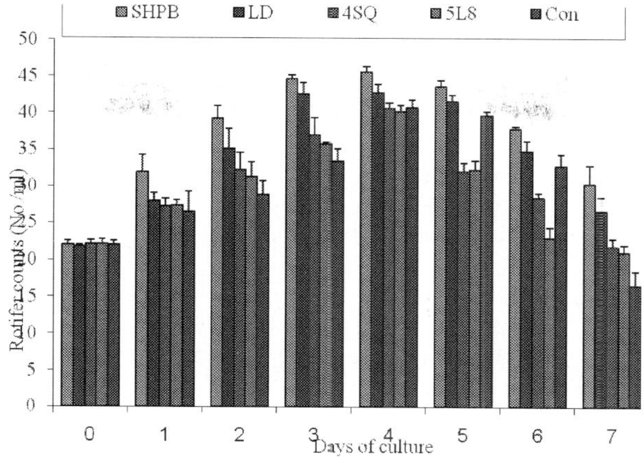

Fig. 1. The effect of SHPB (*B. halotolerance*), LD (*L. divergens*), 4SQI, 5L82 on the growth performance of rotifers compared with the control (Con).

rotifers are initially added on these days with a starting density of 5 rotifer/ mL. The probiotics were added daily, this could explain the significant effect mainly for the SHPB on rotifers proliferation all through out the experiment. Repetitive addition of the probiotics can significantly enhance the rotifers performance and survival. The result of this study is in agreement with the results obtained by Planas et al., (2006). They reported that continuous additions of probiotic (*Rosebacter* strain) are necessary to maintain a minimum level of it in the rotifer and the culture water. In addition, exposing the larvae to sufficient probiotic concentrations will increase the chance of the probiotics being ingested by the larvae. Thus, this could lead to the advantage of improving the survival rate of the fish larvae.

5. Effect of SHPB and *L. divergens* used alone and in combination (SHPB & *L. divergens*) on the survival of sobaity larvae

Owing to the problem of antibiotic resistance and subsequent reluctance of using antibiotics, the use of probiotics in larviculture is becoming increasingly popular. During the early stages of development, manipulation of the larval digestive system seems possible through the addition of probiotics, either through the culture water, or via the live food (Vine et al., 2006). Geovanny et al. (2007) reported that the use of probiotics can increase the survival rate and control the high incidences of larval diseases. Thus, by shifting the bacterial flora in live feed organisms to probiotic species, this can assist the fish larvae to minimize the pathogenic bacteria from the feed, and the fry will benefit from the probiotic bacteria. Several bacterial probiotics were used in the larval culture of aquatic organisms. Kozasa (1986) reported that the spores of *Bacillus toyoi* increased the growth rate of yellowtail and reduced the mortality of Japanese eel that were infected by *Edwardsiella* sp. The Gram-negative *Vibrio pelagius* decreased the mortality of the turbot larvae *Scophthalmus maximus* challenged with *Aeromonas caviae* (Ringo and Vadstein, 1998). Gatesoupe (1991b) showed that *Bacillus toyoi* and *Bacillus* sp spores increased the growth rate of larval turbot introduced via the rotifer *Brachionus plicatilis*. Pirarat et al., (2006), reported that supplementation of

L. rhamnosus significantly reduced cumulative mortality due to *E. tarda*, confirming the protective effect of a probiotic bacterium against this pathogenic bacterium. Suzer et al., 2008 showed significant increase in the survival rate of *Sparus auratus* larvae fed with *Lactobacillus* spp. via live food and water. In this study, the effect of single and combined administration of SHPB and *L. divergens* on the survival rate of sobaity larvae was evaluated. The larvae were reared in 1m^3 circular fiberglass tanks with stocking density of 60 larvae/ L. The enriched rotifers were added to the larval tanks and the rotifer density was maintained at 5/ mL. The survival rate of the larvae was determined at 28 days post hatch. The results showed that all the treatment with probiotics significantly enhanced the survival rate compared to the control. The survival in *L. divergens* fed larvae was the highest (11.7%) whereas it was 9.2% in the mixed probiotics, 8.9% in SHPB and 6.3% in the control (Fig 2). Balcazar et al., 2007, demonstrated that the administration of a mixture of bacterial strains (*Bacillus* sp. and *Vibrio* sp.) positively influenced the growth and survival of white shrimp juvenile. Salinas et al., 2008, showed that the combined probiotics, *L. delbruekii* and *B. subtilis* enhanced the cellular innate immune system of gilthead seabream. Suzer et al., 2008, showed that *Sparus aurata* larvae fed with commercial *Lactobacillus* spp via live food increases survival rate and specific growth rate. To our knowledge there have been no studies on any probiotic on the blue- finned *Sparidentex hasta* (sobaity). So, in our study, since all the tested probiotics showed significant survival rate compared to the control, the dietary administration of both singly SHPB and *L.divergens* and in combination (SHPB and *L. divergens*) were used and they showed significant increased in the survival rate of sobaity larvae and seems to be a promising probiotic candidate. However, feeding sobaity larvae with a combination of them needs further dose adjustment to achieve the best survival rate and possible beneficial interaction between both bacteria in sobaity gut microenvironment, which may make the use of a mixture of different bacterial strains more interesting than using a single bacterium.

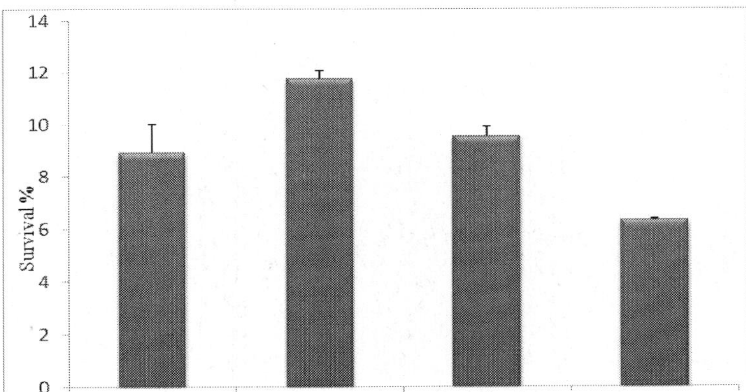

Fig. 2. The survival rate of sobaity larvae fed with SHPB (*B. halotolerance*), LD (*L. divergens*) and their combination (Mix) compared with a control (Con) group fed with un-supplemented diet.

6. Pathogen challenge test for probiotic-fed sobaity larvae

Increased resistance to the pathogen by probiotics has been extensively reported. Growth inhibition against pathogens by *Carnobacterium* was reported (Joborn et al., 1997). The tolerance of rainbow trout *Oncorhynchus mykiss* to furunculosis was enhanced when fed with a diet including the probiotic *L. rhamnosus* (Nikoskelainen et al., 2001). In Atlantic cod *Gadus morhua,* tolerance to *V. anguillarum* increased by feeding with lactic acid bacteria *Carnobacterium divergens* supplemented in the diet (Gildberg et al., 1998). Robertson *et al* ., 2000) reported that Atlantic salmon *Salmo salar* and rainbow trout *O. mykiss,* fed with *Carnobacterium* spp. supplemented in the diet were more tolerant to disease. Chiu et al., 2010, showed that *S. cerevisiae* colonized the intestines of the grouper *E. coioides* fed *S. cerevisiae*-supplemented diets improved and increased the resistance to challenge by *Streptococcus* sp. and a grouper iridovirus. In our study, after feeding the larvae with probiotic-enriched rotifers (SHBP/ *L. divergens* / mixture of SHBP and *L. divergens*) for 28 days, the larvae challenged through immersion against virulent *Vibrio harveyi*. The bath suspension contained a cell density of 10^7 cells m^{-1} and the duration of challenge was 30 min. Fish larvae were transferred to 50 L aquarium tanks (50 fish/ tank) after the challenge, mortalities were observed and recorded for one week and the squash preparations of washed freshly dead larvae were plated on TCBS agar to record specific mortalities. The results showed that fish larvae fed with SHPB and combination probiotics showed clear disease resistance as indicated by distinctive 85.3% and 58% survival rate for SHPB and mixed treatments respectively compared to *L.divergens*-fed fish 54% and control 50%. Pieters et al., 2008, reported that challenge with *A. bestiarum,* the probiotics GC2 and BA211 led to 76% and 38% survival, respectively, in contrast to 22% survival for the controls. In the current study, it is apparent that SHPB and the mixed probiotics (SHPB & *L. divergens*) fulfilled the major requirements of being an effective probiotics by enhancing the survival rate of sobaity larvae after challenge with virulent *V. harveyi*. However, the survival of larvae fed a combination of probiotics was not as good as that of SHPB and here, probably the dose structuring needs to be optimized for furthering the effects of a combination of probiotics (Fig 3).

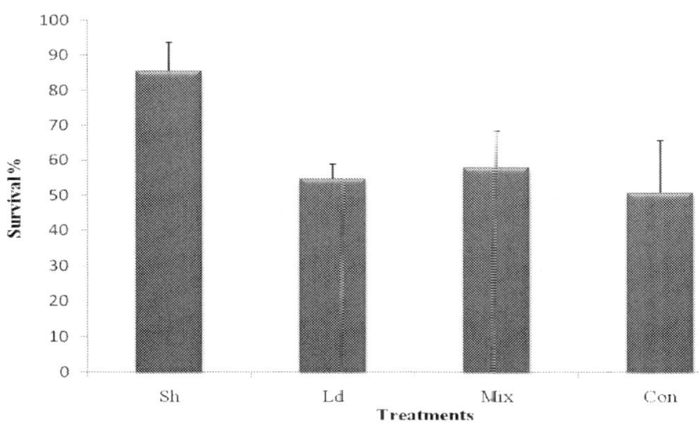

Fig. 3. The survival rate of sobaity larvae challenged against *V. harveyi* after feeding with SHPB (*B. halotolerance*), LD (*L. divergens*) and their combination (Mix) compared with a control (Con) group fed with un-supplemented diet.

7. Molecular characterization of SHPB probiotic and bacteriocin-like compound

The SHPB was characterized using the PCR and 16s rDNA gene amplification (Al-Marzouk et al., 2009). The identification of SHBP probiotic confirmed as *Bacillus halotolerance*. The modes of action of probiotics include the inhibition of a pathogen through the production of bacteriocin-like compounds, competition for attachment sites, competition for nutrients (particularly iron in marine microbes), alteration of enzymatic activity of pathogens, immunostimulatory functions, and nutritional benefits such as improving feed digestibility and feed utilization (Kesarcodi-Watson et al., 2008; Fuller, 1989). Thus, an understanding of the mechanisms probiotics use to compete with pathogens is important when designing a protocol for their selection. Bacteriocins are antibacterial proteins produced by bacteria to kill or inhibit the growth of other bacteria (Cleveland et al., 2001). They are ribosomally synthesized unlike antibiotics, which are synthesized by other mechanisms (Brock & Madigan, 1997). In this study, *B. halotolerance* (SHBP) cultures of different age (12h, 24h, 36h and 48h) were used for detecting the possible role of bacteriocin in the antibacterial activity. The mode of action of *B. halotolerance* was confirmed through its ability to produce bacteriocin-like compound, which is considered as a significant criterion of the defense system displayed by it. It produced an amplicon of approximately 1500 bp and for the bacteriocin gene a 1000 bp amplicon (Plate 3). Cultures of different age, however, showed interesting amplification pattern with clear amplification of the bacteriocin-like gene in 24-h culture and a very mild amplification in 12-h culture. This state was tested by treating the compound (probiotic bacteria free BHI broth) with different pH and temperatures. Later, the treated broth was used in antibacterial assay. The persistence of antibacterial activity of the

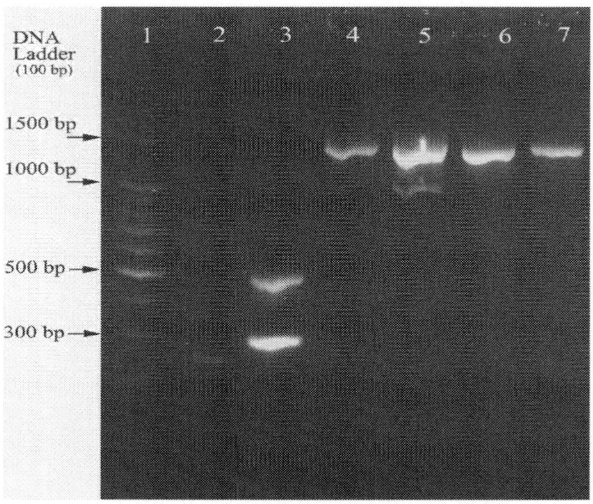

Plate 3. Polymerase Chain Reaction (PCR) amplification of the bacteria specific 16s rDNA (approximately 1500 bp, brighter band) and bacteriocin gene (approximately 1000 bp). Lanes: 1: 100 bp DNA ladder; 2: Negative sample; 3: Positive amplicon (kit); 4 to 7– probiotic bacterial cultures (12h, 24h, 36 h and 48h) of *B. halotolerance*.

treated broth confirmed that it a bacteriocin- like compound of the probiotic that was responsible for the antagonism. One of the most well known bacteriocins is nisin, which is a ribosomally synthesized antimicrobial peptide produced by certain strains of *Lactococcus lactis* which has been proved to act against human multidrug resistant pathogens such *Streptococcus pneumoniae, Staphylococcus aureus, Staphylococcus epidermidis, Enterococcus faecalis* and others (Gatesoupe, 2008). Further research will be required to specify the exact type of bacteriocin produced by the probiotic *B. halotolerance*.

8. Persistence of probiotics in the fish gut

Adhesion and colonization are important for selection and use of probiotic strains (Bussarin & Rakshit, 2006; Vine et al., 2004; Olsson et al., 1992) and because it is considered a pre-requisite for colonization (Beachey, 1981). Adhesion of probiotic bacteria to the intestinal mucosa has been shown to enhance their antagonistic activity against pathogens (Coconnier et al., 2003). In this study, the ability of SHPB, *L. divergens* and their combination to attach to fish intestinal mucus were examined. Colonization of probiotics fed through diets was monitored at 15 and 30 d during the probiotic feeding and at one month after the probiotic-enriched feed was withdrawn. Colonization of the SHPB was evident even at 15 d (Plate 4). The intense localization on the brush boarders of the hindgut intestinal epithelium noticed at 15 and 30 d (Plates 4 & 5) during feeding, suggests that the SHPB is more likely to

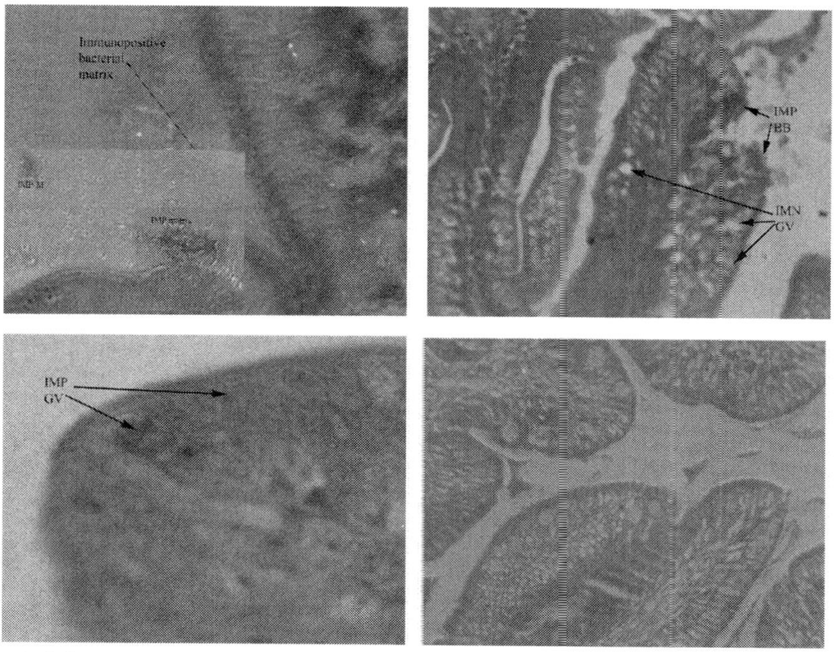

Plate 4. Sobaity gut fed with SHPB, L. divergens & mixed (SHPB & L. divergens) and control diets. Arrows: IMP: Immunopositive; GV: Gut vacuole; BB: brush boarder. Left (40 x) and right. 10 x.

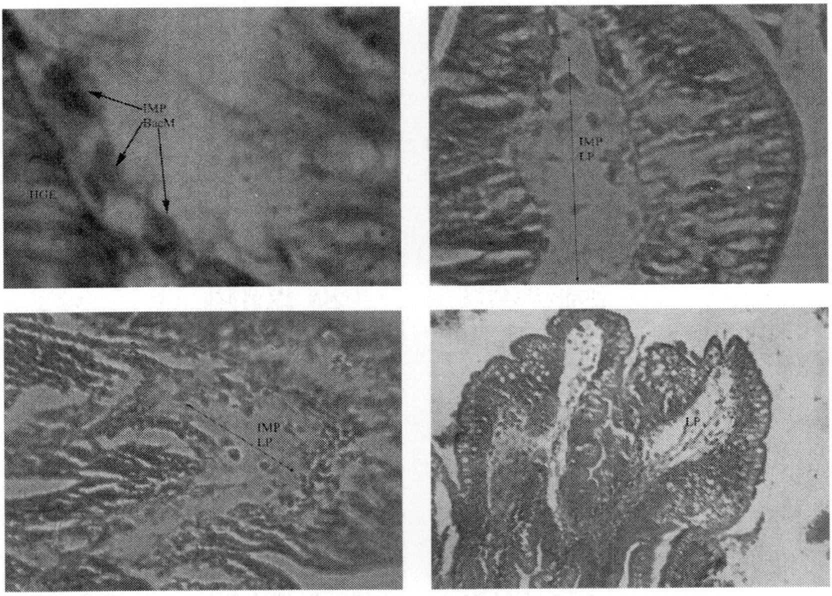

Arrows: HE: hindgut epithelium; LP= lamina propria of the hindgut.

Plate 5. Gut of sobaity sampled after 30 days of continuous feeding with SHPB and L. divergens-(top right, 20 x), mixed probiotics (bottom left, 40x) and control (bottom right, 10x) diets. Note immunopositive SHPB colonization on the hindgut epithelium (top left). No immunopositive bluish-purple reaction in the gut of control fish (bottom right, 10x).

colonize in the hind gut of sobaity than the *L. divergens*, which was processed as an antigen and did not remain in the gut long enough (almost no localization at one month post withdrawal). Clear and significant immunopositive localization (Plate 4) was noticed in gut of both probiotic SHPB and mixed (SHPB and *L. divergens*) probiotic-fed sobaity even after withdrawal of probiotic feeding. This indicates the higher persistence of the SHPB than that of *L. divergens*. The process of colonization is characterized by attraction of bacteria to the mucosal surface, followed by association within the mucous gel or attachment to epithelial cells. Adhesion and colonization of the mucosal surfaces are possible protective mechanisms against pathogens through competition for binding sites and nutrients, or immune modulation (Balcazar et al., 2006). However it needs to be accepted that the efficiency of a selected probiotic *in vitro* may significantly change when administered to the host because it is influenced by more complex factors such as the selective ingestion and the death in the intestinal tract (Vine et al., 2006) caused by the failure of the probiotic to maintain its *in vitro* physiology under circumstances of a more complex microbial interactions and/or nutritional environment. In general, there is a sense of the lack of correlation between in vitro and in vivo experiments in the latest reviews on probiotic use in aquaculture (Balcazar et al., 2006; Vine et al., 2006). The main claimed mechanisms are: competitive exclusion, digestion enhancement, immune response enhancement, water quality improvement and antiviral effects.

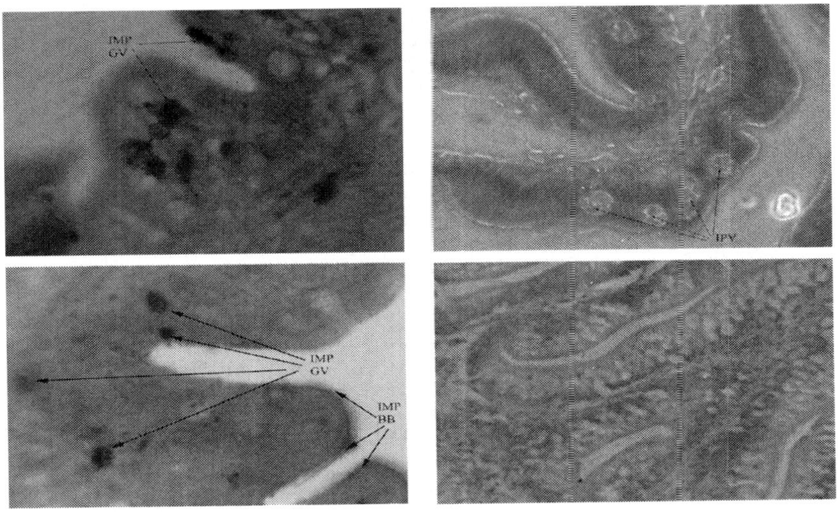

No immunopositive bluish-purple reaction in the gut of control fish (bottom right).
Arrows: IPV: immunopositive vacuoles.

Plate 6. After 30 days of withdrawal of feeding with SHPB-(top left, 40x), L.divergens-(top right, 40x), mixed probiotics (bottom left 40x)) and control (bottom right (40 x)) diets.

9. Bacterial count in gastrointestinal tract in probiotics-fed sobaity

No *Vibrio* sp., was detected in the vibrio selective medium (TCBS), from the fish gut fed from SHPB and mixed probiotics (SHPB and *L. divergens*). The highest bacterial count was counted in the control and *L.divergens* treatment (Table 1). The main bacterial colonies that were detected in the brain heart infusion agar (BHIA) from fish fed with SHBP.

Treatment	Media	45d	60d	75d	105d
SHPB	BHIA	6.30×10^4	$[illegible].00 \times 10^3$	0.30×10^2	0.80×10^3
	TCBS	-	-	-	-
LD	BHIA	3.20×10^3	4.00×10^2	1.70×10^3	0.90×10^4
	TCBS	-	-	1.00×10^2	15.00×0^4
Mix	BHIA	3.00×10^3	6.60×10^3	6.20×10^3	0.30×10^4
	TCBS	-	-	-	-
Con	BHIA	1.30×10^4	6.50×10^2	4.10×10^3	2.10×10^3
	TCBS	-	-	0.30×10^3	1.00×10^3

Table 1. Bacterial Count from the Gut of Sobaity Fed with SHPB (*B. halotolerance*), LD (*L. divergens*) and their Combination (Mix) Compared with Control (Con).

10. Pathogen challenge test for probiotic-fed sobaity fry

Vendrell et al., 2008 showed that Probiotic supplementation of *Leuconostoc mesenteroides* and *Lactobacillus plantarum* reduced fish mortality significantly from 78% in the control group to 46-54% in the probiotic groups after challenged *with Lactococcus garvieae*. In this study, sobaity fry fed with SHPB and mixed probiotics showed clear disease resistance as indicated by distinctive 100% survival rate compared to *L.divergens* fed fish and control (Table 2). Based on the ability of the SHPB and mixed probiotics to attach to fish gut, the growth of the pathogen in the digestive tract might be suppressed by the candidate probiotics presence.

Treatments	Survival Rate (%)
SHBP	100
LD	16.6
Mix	100
Con	66

Table 2. Survival Rate of Control (Con) and Probiotic-Fed (SHPB, LD and Mix) Sobaity Fry after Challenge with *V. anguillarum*

11. Immunological assays

The non-specific immune system can be stimulated by probiotics. Taoka et al., 2006 indicated that probiotics supplied in the rearing water and the diet of fish enhanced the stress tolerance and the non-specific immune system of Japanese flounder, providing them a higher resistance against stress conditions and pathogens. It has been demonstrated that oral administration of *Clostridium butyricum* bacteria to rainbow trout enhanced the resistance of fish to vibriosis, by increasing the phagocytic activity of leucocytes (Sakai et al., 1995). Balcazar (2003) demonstrated that the administration of a mixture of bacterial strains (*Bacillus* sp. & *Vibrio* sp.) positively influenced the growth and survival of juveniles of white shrimp and presented a protective effect against the pathogens *Vibrio harveyi* and white spot syndrome virus. This protection was due to a stimulation of the immune system, by increasing phagocytosis and antibacterial activity. In this study, custom-production and characterization of rabbit polyvalent antibodies against *B. halotolerance* and *L. divergens* were carried out using sodium dodecyl sulfate polyacrylamide gel electrophoresis (SDS-PAGE) and Western blot analysis, and in addition, to evaluate both the humoral and cellular innate immunity responses of sobaity fed for 75 d with the above mentioned probiotics. These included the phagocytic activity; lysozyme activity, serum immunoglobulin and alternative complement (ACH50). Different immunological parameters, mainly serum and mucus lysozyme, phagocyte activity and complement were enhanced in sobaity fry fed with *B. halotolerance* and the mixed probiotics for 75 and 105 days.

11.1 Lysozyme activity

Lysozyme has an important role in non-specific immune defense system and it contained in the mucus on the fish body surface, and in plasma and liver. Lysozyme has an antibiotic ability and is released by leukocytes. It can damage bacterial cell walls, especially of Gram-positive and some Gram-negative bacteria (Grinde, 1989). Lysozyme activity varies between species of fish, genetic strains and different pathogens. Several reports showed the effect of probiotics on lysozyme activity. A significant increase in lysozyme activity was observed in Nile tilapia, *O. niloticus*, fed diets containing *S. cerevisiae* for 21 days (El-Boshy et al., 2010). Gatesoupe (2008) reported that feeding with Gram-positive and Gram negative potential probiotics caused increase in the cellular parameters such as macrophages and enhanced lysozyme activity. Kim & Austin (2006) recorded high gut mucosal lysozyme activity in fish fed with *Caronbacterium divergens* B 33 and *Carnobacterium maltaromaticum* B26. Taoka et al., 2006, also reported an enhanced lysozyme level in tilapia fed with live and dead probiotics and recorded a high survival rate when challenged with *Edwardsiella tarda.* Panigrahi et al., 2004, showed significantly higher lysozyme activity in rainbow trout fed with *Lactobacilllus rhamnosus.* Apart from serum lysozyme content, probiotics can also enhance the lysozyme level in skin mucosa of fish (Song et al., 2006). Lysozyme in fish can be measured either by the turbidimetric method or the agarose plate assay. Each method was developed based on the amount of lysis of the gram positive bacteria *Micrococcus lysodeikticus.* In this study, lysozyme activity in serum was determined according to the method of Demers and Bayne (1997) based on the lysis of the lysozyme- sensitive Gram positive bacterium, *Micrococcus lysodeikticus* (Sigma, St. Louis). With the help of a computer application software, Delta Soft 3 (Biometalics Inc., New Jersey, USA) the equivalent unit of activity of the sample as compared to the standard were determined and expressed as μg ml^{-1} serum. The turbidimetric assay results can be reported by two methods; relative lysozyme activity measured in Units/min, and by comparison to a standard curve generated using purified. Our turbidimetric assay results were reported by relative lysozyme activity which measured in Units/min, and by comparison to a standard curve generated using purified hen egg white lysozyme. The results showed that sobaity fry fed with three types of probiotics has significant higher lysozyme activity (Table 3).

Treatments	ACH50 (log_2) titers		Lysozyme (units/min)			
			Serum		Mucus	
	75d	105d	75d	105d	75d	105d
SHPB	6.67 ± 0.58	5.67 ± 0.58	18.63 ± 4.50	17.00 ± 6.61	21.25 ± 4.53	26.38 ± 7.37
LD	3.67 ± 0.58	4.67 ± 0.58	15.50 ± 4.34	15.38 ± 5.66	18.63 ± 5.15	20.63 ± 4.60
Mix	4.67 ± 0.58	6.67 ± 0.58	20.13 ± 4.26	17.13 ± 7.24	22.38 ± 4.69	20.38 ± 2.26
Con	2.33 ± 0.58	3.67 ± 0.58	11.63 ± 3.81	11.75 ± 2.38	14.13 ± 3.98	12.38 ± 3.50

ACH50 Tukey's HSD=1.35 for treatments; for time HSD= 1.0. ; serum lysozyme HSD= 5.74 for treatments; for time HSD= 4.344; mucus lysozyme HSD= 4.56 for treatments; for time HSD= 3.45. Values less than HSD values considered to be not significant

Table 3. Alternative Complement (ACH50) Titers ($\log_2$), Serum and Mucus Lysozyme Content of Probiotic-Fed Sobaity (75 d) = After Probiotic feeding & (105 d) = After 30 d of Withdrawal of Probiotic feed

11.2 Alternative complement activity (ACH50)

The alternative pathway of **complement activity** has emerged as a powerful nonspecific defense mechanism for protecting fish against a wide range of potentially invasive organisms, such as bacteria, **fungi, viruses** and **parasites (Muller-Eberhard, 1988).** Chiu et al., 2010, showed that *Epinephelus coioides* fed *S. cerevisiae*-supplemented diets exhibited significant increases in both **serum lysozyme** and **alternative complement pathway** activities (ACH50). The same result was also observed in hybrid **tilapia** fed diets containing DVAQUA® diet **(Zohu** et al., 2009). Dietary administration of the probiotic, *Lactobacillus plantarum*, enhanced the growth, lysozyme, phagocytic activity, alternative complement activity (ACH50) and disease resistance of the grouper *E. coioides* (Son et al., 2009).

Pirarat et al., 2006 reported that *Lactobacillus rhamnosus* probiotic enhanced the ACH50 of tilapia *Orechromis niloticus* and protect the fish from acute septicemic death by *E. tarda*. Balcazar et al., 2007, showed that in comparison to untreated control fish, the alternative complement activity and phagocytic activity of head kidney leukocytes in serum were significantly greater in all probiotic groups (*Lactococcus lactis* ssp. lactis CLFP 100, *Leuconostoc mesenteroides* CLFP 196, and *Lactobacillus sakei* CLFP 202) at the end of the second week of feeding. Our results showed that ACH50 activity was significantly higher in the probiotic-supplemented groups than in the control group after 75 d feeding and 105 d post feeding. After the feeding, the ACH50 activity gradually decreased in the control group, whereas it remained high in both SHPB and the mixed probiotic groups throughout the test period (Table 3). These results indicate that probiotics incorporated in the feed of sobaity resulted in increased serum complement activity and hence enhanced the immunity of the fish against any virulent pathogens.

11.3 Phagocytic assay

Phagocytic activity is responsible for early activation of the inflammatory response before antibody production and plays an important role in antibacterial defenses (Nayak, 2010).

Several studies showed the effect of different probiotics on phagocytic activity in several fish species. Probiotics such as *Lactobacillus rhamnosus, L. lactis* and *L. acidophilus* can effectively trigger the pahgocytic cells in fish species. These probiotics used either in viable or inactivated form were found to have the ability to stimulate phagocytic activity in different fish species (Brunt et al., 2007; Brunt &Austin, 2005; Irianto & Austin, 2003). Black tiger shrimp fed with a probiotic diet had greater tolerance to *V. harveyi*, and phagocytic activity in hemolymph was activated (Rengpipat et al., 2000; 1998). Sun et al., (2010) reported that phagocytic activity and phagocytic index of grouper fed probiotic *Bacillus pumilus* and *Bacillus clausii* were significantly higher than those of fish fed the control diet for 60 days. Pieters et al., 2008, showed that the analysis of innate immune responses revealed that probiotic GC2, *Aeromonas sobria* promoted higher phagocytic activity in rainbow trout against *Aeromonas bestiarum* after the probiotic was administered orally (10^8 cells per g feed for 14 d). Patricia et al., 2006, reported that the phagocytic ability of gilthead seabream, *Sparus aurata* fed with a mixture of two inactivated bacteria from the Vibrionaceae family was significantly higher than in the control after three weeks feeding. Pirarat et al., 2006, reported that *Lactobacillus rhamnosus* probiotic increases the phagocytic activity of tilapia *Orechromis niloticus* and protect the fish from acute septicemic death by *Edwardsiella tarda*. In this study, the phagocytic ability of head kidney leucocytes was significantly increased after 75 d feeding with SHPB and *L.divergens* supplemented diet. No statistically significant difference was detected in the phagocytic ability of fish fed with a mixture of SHPB and *L.divergens*. The activity fell in a time-dependent manner after 105 d for all the treatments (Table 4). After the completion of 75 d,

Treatments	Phagocytic Index (PI) Number of Fluorescent Beads per Field of Observation	
	FP-PI	EP-PI
SHPB	77.00 ± 13.23	56.00 ± 7.94
LD	65.67 ± 15.67	43.83 ± 5.53
Mix	58.83 ± 20.97	37.17 ± 9.72
Con	43.33 ± 5.13	29.00 ± 2.00

FP= End of probiotic feeding period (75 d) and EP = End of experiment period (105 d).
Treatments Tukey's HSD=22.28; Time Tukey's HSD=16.50. Values less than HSD values considered to be not significant

Table 4. Phagocytic Index of Fresh Head-Kidney Tissue Imprints from Probiotic-Fed (SHPB, LD and Mix) and Control (Con) Sobaity.

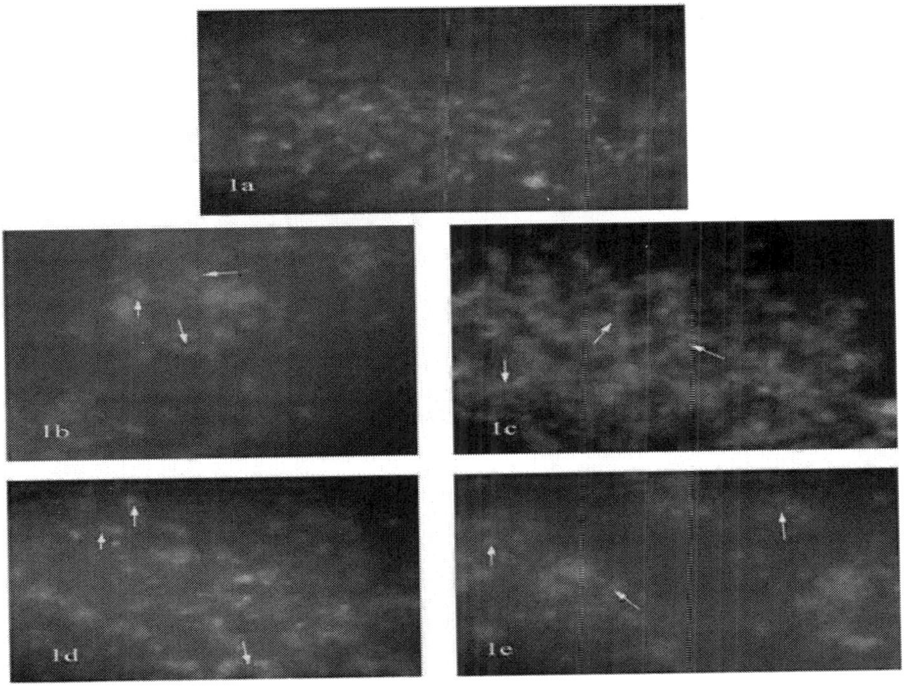

1a – Negative control,
1b – head-kidney imprint from control fish,
1c - head-kidney imprint from SHPB-fed fish, 1d - head-kidney imprint from *L.divergens*-fed fish and 1e - head-kidney imprint from fish fed with a 1:1 mix of SHPB and L.d.

Plate 7. Phagocytic assay for sobaity sampled soon after the completion of 75 d of probiotic feeding showing fresh head-kidney tissue imprints after incubation with fluorescent latex beads. Photographed at 20x magnification.

the fresh head-kidney tissue imprints after incubation with fluorescent latex beads showed clear phagocytic ability in both SHPB and *L.divergens* supplemented diet compared to the mixed treatment and control (Plate 3). Thus, phagocytic activity played an important role in sobaity innate immunity as indicated by the phagocytic assay from fresh head kidney tissue imprints and phagocytic index. Our finding indicate that phagocytosis against bacterium can be enhanced by incorporating autochthonous probiotic alone or mixing it with commercial probiont such as *L. divergens*.

12. Conclusion

A protocol for the isolation and selection of candidate probiotic bacteria based on several selective criteria was accomplished. These criteria include *in vitro* antagonism ability of probiotics against pathogenic bacteria, adhesion ability in the intestinal epithelium to compete for attachment sites on the gut wall, improvement of fish survival, disease resistance and immune responses. Results point out a significant effect of *Bacillus halotolerance* (SHBP) on rotifers proliferation compared to other autochthonous (4SQIb, 5L82) and a commercial (*L. divergens*) probiotics. Results showed that *B. halotolerance* **and** *L. divergens* **had a significant positive effect on the survival mainly during day 12 and 22, compared to the control. Dietary administration of combined** *B. halotolerance* **and** *L. divergens* **and single SHPB and** *L. divergens* **showed significant survival rate compared to the control.** Also, the survival rate of larvae fed with these probiotics was improved after challenging with *V. harveyi*, indicating the positive effect of the probiotics used. Molecular characterization confirmed the identification of SHBP as *Bacillus halotolerance*. It also confirmed the ability of *B. halotolerance* to produce bacteriocin-like compound, which is considered as a significant criterion of the defense systems displayed by it.

The probiotics persistence in the gut indicated that they are able to establish themselves in the digestive tract. Also, different immunological parameters, mainly serum and mucus lysozyme, phagocytic activity and complement activity were enhanced in sobaity fry fed with *B. halotolerance* and the mixed probiotics for 75 and 105 d.

These finding can be applied to other potential cultured fish in Kuwait through future research. In addition the most potential autochthonous probiotics will be evaluated in large scale fish production.

13. Acknowledgment

Our sincere thanks are due to Kuwait Foundation for the Advancement of Sciences (KAFAS), for funding all the capital equipment and operating items used in this study.

14. References

Al-Marzouk, A.; Saheb, A.; Al-Abdul-Elah, K.; Abu-Rezq, T. & Al-Gharabally, H. (2009). Probiotics: Alternative Methods for Disease Control in sobaity larvae Production., Kuwait Institute for Scientific Research, Report No. KISR 9666. Kuwait.

Austin, B.; Baudet, E. & Stobie, M. (1992). Inhibition of Bacterial fish Pathogens by *Tetraselmis suecica*. *Journal of Fish Diseases*, Vol.15, pp.55-61

Balcazar, J.L. (2003). Evaluation of Probiotic Bacterial Strains in *Litopenaeus vannamei*. Final Report, National Center for Marine and Aquaculture Research, Guayaquil, Ecuador.

Balcazar J.L.; Blas, I, D.; Ruiz-Zarzuela, I.; Cunningham, D.; Vendrell, D. & Muzquiz, J.L (2006). The Role of Probiotics in Aquaculture. *Veterinary Microbiology*, Vol.114, pp. 173–186.

Balcazar J. L.; Blas, I. D.; Ruiz-Zarzuela, I.; Vendrell, D.; O. Gironés, O. & Muzquiz, J. L. (2007). Enhancement of the Immune Response and Protection Induced by Probiotic Lactic Acid Bacteria Against Furunculosis in Rainbow Trout, *Oncorhynchus mykiss. Immunology and Medical Microbiology*, Vol. 23, pp. 132-138

Beachey, E. H. (1981). Bacterial Adherence: Adhesion-receptor Interactions Mediating the Attachment of Bacteria to Mucosal Surfaces. *Journal of Infectious Diseases*. Vol. 143, pp. 325-345.

Benetti, D. D.; O'Hanlon, B.; Ayvazian, J.; Orhun, M.R.; Rivera, J.A.; Rice, P.H. & Douillet, P.A. (2004). From Egg to Market: The Development of A new Sustainable Offshore Aquaculture Industry in Florida (US and the Caribbean). In: *Abstracts of Aquaculture 2004, Aquaculture: An Ecologically Sustainable and Profitable Venture.* Hawaii, USA: p. 708.

Brock. T.D & Madigan, M.T (1997). Biology of Microorganisms, 968 pp, Prentice Hall Inc., New Jersey.

Brunt, J. & Austin, B. (2005). Use of A probiotic to Control Lactococcosis and Streptococcosis in Rainbow trout, *Oncorhynchus mykiss* (Walbaum). *Journal of Fish Diseases*. Vol. 28, pp.693-701.

Brunt, J.; Newaj-Fyzul, A.; Austin, B. (2007). The Development of Probiotics for the Control of Multiple Bacterial Diseases of Rainbow trout, *Oncorhynchus mykiss* (Walbaum). *Journal of Fish Diseases*, Vol. 30, pp.57-39.

Bussarin, K. & Rakshit, S.K. (2006). Microbial and Processing Criteria for Productionof Probiotics: A Review. *Food Technology and Biotechnology, Vol. 44, No.3*, pp. 371–379.

Chiu, C. H.; Cheng, C. H.; Gua, W. R., Guu, Y. Y.& Cheng, W. (2010). Dietary Administration of the Probiotic, *Saccharomyces cerevisiae* P13, Enhanced the Growth, Innate Immune Responses, and Disease Resistance of the Grouper, *Epinephelus coioides. Fish and Shell Fish Immunogy*, Vol. 29, No. 6, pp. 1053-1059.

Cleveland, J.; Montville, T. J.; Nes, I. F. & Chikindas, M. L. (2001). Bacteriocins: Safe, Natural Antimicrobials for Food Preservation. *International Journal of Food Microbiology*, Vol. 71, pp. 1-20.

Coconnier, M. H.; Bernet, M. F.; Kemeis, S; Chauviere, G.; Fourmait, J. & Servin, A. L. (2003). Inhibition of Adhesion of Enteroinvasive Pathogens to Human Intestinal Caco-2 Cells by *Lactobacillus acidophilus* Strain LB Decreases Bacterial Invasion. *FEMS Microbiology Letters*, Vol. 110, pp. 229-306.

Decamp, O.; Moriatry, D.J. & Lavens, P. (2010). Probiotics for Shrimp Larviculture: Review of Field Data from Asia and Latin America. *Aquaculture Research*, Vol.39, No.4, pp. 334-338.

Demers, N. E. & Bayne, C. J. (1997). The Immediate Effect of Stress on Hormones and Plasma Lysozyme in Rainbow trout. *Developmental and Comparative Immunology*, Vol.21, pp. 363-373.

Douillet, P. A. (2000). Bacterial Additives that Consistently Enhance Rotifer Growth under Synxenic Culture Conditions and Use of Single and Multiple Bacterial Probiotics. *Aquaculture*, Vol.182, pp. 241-248.

Dopazo, C. P., Lemos, M. L.; Lodeiros, C.; Bolinches, J.; J. L. Barja, & Toranzo, A. E. (1988). Inhibitory Activity of Antibiotic-Producing Marine Bacteria Against Fish Pathogens. *Journal of Applied Bacteriology*, Vol. 65, pp. 97-101

El-Boshy, M. E.; El-Ashram, A. M.; AbdelHamid, F. M.& Gadalla, H.A. (2010). Immunomodulatory Effect of Dietary *Saccharomyces cerevisiae*, ß-glucan and Laminaran in Mercuric Chloride Treated Nile tilapia (*Oreochromis niloticus*) and Experimentally Infected with *Aeromonas hydrophila. Fish and Shellfish Immunology*, Vol.28, pp. 802–808.

Ellis, A. E. (1999). Immunity to Bacteria in Fish. *Fish and Shellfish Immunology*, Vol.9, pp.291-308.

Fuller, R. (1989). A review: Pobiotics in Man and Animals. *Journal of Applied Bacteriology*,Vol. 66, pp: 365–37.

Garriques, D. & Arevalo, G. (1995). An Evaluation of the Production and Use of a Live Bacterial Isolate to Manipulate the Microbial Flora in the Commercial Production of *Penaeus vannamei* Postlarvae in Ecuador. , pp. 53–59.Hopkins, J.S. (Ed.), *In: Proceedings of the Special Session on Shrimp Farming*, Aquaculture'95. Baton Rouge,World Aquaculture Society

Gatesoupe, F. J. (1991a). *Bacillus* sp. Spores as Food Additive for the Rotifer *Brachionus plicatilis*: Improvement of their Bacterial Environment and their Dietary Value for Larval Turbot. Kaushik,S (Ed.), In: *Proceeding of the 4th International Symposium on Fish Nutrition and Feeding*. pp. 561-568. Paris, France.

Gatesoupe, F.J.(1991b). The Effect of Three Strains of Lactic Acid Bacteria on the Production Rate of Rotifers, *Brachinous plicatilis* and Their Dietary Value for Larval Turbot *Scophthalmus maximus*. *Aquaculture*, Vol. 96, pp. 335-342.

Gatesoupe, F.J. (2008). Updating the Importance of Lactic Acid Bacteria in Fish Farming: Natural Occurrence and Probiotic Treatments. *Journal of Molecular Microbiology and Biotechnology*, Vol. 14, No.(1-3), pp. 107-14.

Geovanny, G. R., LuisB. J. & Shen, M.A. (2007). Probiotics as Control Agents in Aquaculture. *Journal of Ocean University of China, Vol.* 6, No.1, pp.76-79.

Gildberg, A.; Johansen, A & Bogwald, J. (1995). Growth and Survival of Atlantic Salmon (*Salmo salar*) Fry Given Diets Supplemented with Fish Protein Hydrolysate and Lactic Acid Bacteria During Challenge Trial with *Aeromonas salmonicida. Aquaculture*, Vol. 138, pp 23-34.

Gildberg, A.; Mikkelsen. H,; Sandaker. E & Ringø, E. (1997). Probiotic Effect of Lactic Acid Bacteria in the Feed on Growth and Survival of Fry of Atlantic Cod (*Gadus morhua*). *Hydrobiologia*, Vol. 352, pp 279-285.

Gildberg A. & Mikkelsen H. (1998). Effects of Supplementing the Feed to Atlantic Cod (*Gadus morhua*) Fry with Lactic Acid Bacteria and Immuno-Stimulating Peptides

During Challenge Trial with *Vibrio anguillarum*. *Aquaculture*, Vol. 167, pp. 103–113.

Gomez-Gil, B.; Roque, A. & Turnbull, J.F. (2000). The Use and Selection of Probiotic Bacteria for Use in the Culture of Larval Aquatic Organisms. Aquaculture, Vol.191, pp. 259–270.

Grinde, B. (1989). Lysozyme from Rainbow Trout *Salmo gairdneri* Richardson as an Anti bacterial Agent Against Fish Pathogens. *Journal of Fish Diseases*; Vol. 12, pp. 207–210.

Hussain, N. A., Akatsu, S. & El-Zahr, C. (1981). Spawning, Egg and Early Larval Development and Growth of *Acanthopagrus cuvieri*. *Aquaculture*, Vol. 22, pp.125-136.

Harzevili, A. R. S.; Vanduffel,H.; Dhert, P.; Swing, J. & Sorgeloos, P. (1998). Use of Potential Probiotic *Lactococcus lactis* AR21 Strain for the Enhancement of Growth in the Rotifer *Brachinous plicatilis* (Muller). *Aquaculture Research, Vol.* 29, pp. 411-417.

Irianto, A. & Austin, B. (2003). Use of Dead Probiotic Cells to Control Furunculosis in Rainbow Trout, *Onchorhynchus mykiss* (Walbaum). *Journal of Fish Diseases*. Vol.26, pp. 59-62.

Joborn, A..; Olsson, J.C.; Westerdahl, A.; Conway, P.L. & Kjelleberg, S. (1997) Colonization in the Fish Intestinal Tract and Production of Inhibitory Substances in Intestinal Mucus and Faecal extracts by *Carnobacterium* sp. strain K1. Journal of Fish Diseases, Vol.20, pp. 383–392.

Kesarcodi-Watson A, Kaspar H, Lategan MJ, Gibson L (2008). Probiotics in Aquaculture: The Need, Principles and Mechanisms of Action and Screening Process. *Aquaculture*, Vol. 274, pp. 1-14.

Kim, D. H. & Austin, B. (2006). Innate Immune Responses in Rainbow trout *Oncorhynchus mykiss* Induced by Probiotics. *Fish and Shellfish Immunology*, Vol.21, pp. 513-524.

Kozasa, M. (1986). Toyocerin (*Bacillus toyoi)* as Growth Promoter for Animal Feeding. *Microbiologie-Aliments-Nutrition*, Vol.4, pp. 121-135.

Magnadottir, B., B. K. Gudmundsdottir, B. K.; Lange, S.; Steinarsson, A.; Oddgeirsson,M.; Bowden, M.; Bricknell, I.; Daimo, R. A. & Gudmundsdottir, S. (2006). Immunostimulation of Larvae and Juveniles of Cod, *Gadus morhua* L. *Journal of Fish Diseases*, Vol. 29, pp.47-155.

Müller-Eberhard, H. J. (1988). Molecular Organization and Function of the Complement System. *Annual Review of Biochemistry*, Vol. 57, pp. 321–347.

Nayak, S.K. (2010). Probiotics and Immunity: A fish Perspective. *Fish and Shellfish Immunology*, Vol.29, pp. 2-14.

Nikoskelainen S.; Ouwehand, A.; Salminen, S. & Bylund, G. (2001). Protection of Rainbow Trout (*Oncorynchus myliss*) from Frunculosis by *Lactobacillus rhamnosus*. *Aquaculture*, Vol. 198, pp 229–236.

Olsson, J. C., Westerdahl, A., Conway, P. L., Kjelleberg, S. (1992). Intestinal Colonization Potential of Turbot (*Scophthalmus maximus*) and Dab (*Limanda limanda*) -Associated

Bacteria with Inhibitory Effects against *Vibrio anguillarum. Applied and Environmental Microbiology,* Vol. 58, No.2, pp. 551-556

Panigrahi, A.; Kiron, V.; Puangkacw, J.; Satoh, S. & Sugita, H. (2004). Immune Responses in Rainbow trout *Oncorhynchus mykiss* Induced by A potential Probiotic Bacteria *Lactobacillus rhamnosus* JCM 1136. *Veterniary Immunology and Immunolpathology,* Vol.102: 379-388.

Patricia, D.R., Irene, S.; Alejandro, R.; Alberto, C.; Mariana, C.; Balebona, C.; Morinigo, M.; Esteban, M. & Meseguer, J. (2006). Gilthead Seabream *Sparus aurata* L. Innate Immune Response After Dietary Administration of Heat-Inactivated Potential Probiotics. *Fish and Shellfish Immunology,* Vol.5, pp. 1-11.

Perez, C.; Paul, M. & Bazerque, P. (1990). An Antibiotic Assay by the Agar Well Diffusion Method. *ACTA Biologica Medical Experimentalis.* Vol. 15, pp. 113-115.

Planas M.; Perez-Lorenzo, M.; Hjelm, M.; Gram, L.; Fiksdal, I. U.; O. Bergh, O. & Pintado, J. (2006). Probiotic Effect In Vivo of *Roseobacter* Strain 27-4 Against *Vibrio* (*Listonella*) *anguillarum* Infections in Turbot (*Scophthalmus maximus* L.) Larvae. *Aquaculture,* Vol. 255, pp. 323-333

Pieters, N.; Brunt, J.; Austin, B. & Lyndon, A. R. (2008). Efficacy of in-Feed Probiotics Against *Aeromonas bestiarum* and *Ichthyophthirius multifiliis* Skin Infections in Rainbow trout (*Oncorhynchus mykiss,* Walbaum). *Applied Microbiology,* Vol. 105, No.3, pp. 723-732

Pirarat, N., Obayashi, T.; Katagiri, T.; Maita, M.; & Endo, M. (2006). Protective Effects and Mechanisms of A probiotic Bacterium *Lactobacillus rhamnosus* Against Experimental *Edwardsiella tarda* Infection in Tilapia *Oreochromis niloticus. Veterniary Immunology and Immunolpathology,* Vol. 113, pp. 339-347

Rasheed,V. (1989). Vibriosis Outbreaks Among Cultured Seabream *Acanthopagrus cuvieri* broodstock in Kuwait. *Aquaculture,* Vol.76, pp.189-197.

Ringo, E. & Vadstein, O. (1998). Colonization of *Vibrio pelagius* and *Aeromonas caviae* in Early Developing Turbot *Scophthalmus maximus* Larvae. *Journal of Applied Microbiology,* Vol.84, pp. 227-233.

Ringo, E. & Birkbeck, T. H. (1999). Intestinal Microflora of Fish Larvae and Fry. *Aquaculture Research,* Vol.30, pp.73-93.

Rengpipat S, Phianphak W, Piyatiratitivorakul S, Menasveta P. (1998). Effects of a Probiotic Bacterium on Black Tiger Shrimp *Penaeus monodon* Survival and Growth. *Aquaculture,* Vol. 167, pp 301–313.,

Rengpipat, S.; Rukpratanporn, S.; Piyatiratitivorakul, S.; Menasaveta, P. (2000). Immunity Enhancement in Black Tiger Shrimp (*Penaeus monodon*) by A probiont Bacterium (*Bacillus* S11). *Aquaculture,* Vol. 191, pp 71–288.

Robertson, P.; O'Dowd, C.; Burrells, C.; Williams, P.; Austin, B. (2000). Use of *Carnobacterium* sp. as A probiotic for Atlantic Salmon (*Salmo salar* L.) and Rainbow trout (*Oncorynchus mykiss,* Walbaum). *Aquaculture,* Vol. 185, pp. 235–243.

Saeed, M. O. (1995). Association of *Vibrio harveyi* with Mortalities in Cultured Marine Fish in Kuwait. *Aquaculture,* Vol. 136, pp. 21-29.

Son, V. M.; Chang, C.C.; Wu, M.C.; Gu, Y.K, Chiu, C.H & Cheng, W. (2009). Dietary Administration of the Probiotic, *Lactobacillus plantarum,* Enhanced the Growth,

Innate Immune Responses, and Disease Resistance of the Grouper *Epinephelus coioides*. *Fish and ShellFish Immunology*, Vol. 26, No. 5, pp. 691-698.

Song, Z.; Wu, T.; Cai, L.; Zhang, L.; Zheng, X. (2006). Effects of Dietary Supplementation with *Clostridium butyricumon* the Growth Performance and Humoral Immune Response in *Miichthys miiuy*. *Journal of Zhejiang University Science*, Vol.7, No.7, pp. 596-602.

Sun, Y.Z.; Yang, H.L.; Ma, R.L. & Lin, W.Y. (2010). Probiotic Applications of Two Dominant gut Bacillus Strains with Antagonistic Activity Improved the Growth Performance and Immune Responses of Grouper *Epinephelus coioides*. *Fish and Shellfish Biology*, Vol. 29, No. 5, pp. 803-809.

Suzer, C.; Coban, D.; Kamaci, H. O.; Saka, S.; Firat, K.; Otgucuoglu, O. & Kucuksari, H. (2008). *Lactobacillus* spp. Bacteria as Probiotics in Gilthead Seabream *Sparus aurata* Larvae: Effects on Growth Performance and Digestive Enzyme activities. *Aquaculture*, Vol. 280, pp. 140-145.

Taoka, Y.; Maeda, H.; Jo,J.Y.; Jeon, M.J.; Bai, S.C.; Lee, W. J.; Yuge, K. & Koshio, S. (2006).Growth, Stress Tolerance and Non-Specific Immune Response of Japanese Flounder *Paralichthys olivaceus* to Probiotics in A closed Recalculating System. *Fisheries Science*, Vol. 72, pp: 310–321

Teng, S. K.; James, C.M.; Tareen, I. U. & Dakour, S. (1984). Fish Culture Project, Phase 1. Kuwait Institute for Scientific Research, Report No. KISR 1238, Kuwait.

Teng, S.K; Al-Zahar, C.; Al-Abdul-Elah, K. & S. AlMatar.(1999). Pilot-Scale Spawning and Fry Production of Blue-Fin Porgy, *Sparidentex hasta* in Kuwait. *Aquaculture*, Vol. 178, pp.27-41.

Uma, A. (1999). Feed Probiotics for Sustainable Aquaculture. *Seafood Export Journal*, Vol.30, No.3, pp.9-11.

Vendrell, D.; Balcazar, J.L.; Blas, I.D.; Ruiz-Zzarzuela, I.; Girones, O. & Muzquiz, J. L. (2008). Protection of Rainbow Trout (*Oncorhynchus mykiss*) from Lactococcosis by Probiotic Bacteria. *Comparative Immunology, Microbiology & Infectious Diseases*, Vol. 31, No.4, pp. 337-345

Verschuere. L., Geert Rombaut, G.; Sorgeloos, P.& Willy Verstraete. W. (2000). Probiotic Bacteria as Biological Control Agents in Aquaculture. *Microbiology and Molecular Biology Reviews*, Vol.64, No.4, pp. 655-671.

Vine, N. G., Leukes, W.D.; Kaiser, H.; Daya, S.; Baxter,J. & Hecht, T. (2004). Competition for Attachment of Aquaculture Candidate Probiotic and Pathogenic Bacteria on Fish Intestinal Mucus. *Journal of Fish Diseases*, Vol.27, No. 6, pp. 319-326.

Vine, N. G.; Leukes, W. D. & Kaiser, H. (2006). Probiotics in Marine Larviculture. *FEMS Microbiology Reviews*, Vol. 30, pp. 404–427

Wang, Y. B.; Li, J. R. & Lin, J. (2008). Probiotics in Aquaculture: Challenge and Outlook. *Aquaculture*, Vol. 281, pp. 1-4.

Westerdahl, A.; Olsson, C.; Kjelleberg, S. & Conway, P. (1991). Isolation and Characterization of Turbot (*Scophthalmus maximus*)-Associated Bacteria with Inhibitory Effects Against *Vibrio anguillarum*. *Applied and Environmental Microbiology*, Vol. 57, pp.2223-2228

Zhou, S. He.Z.; Liu, Y.; Shi, P.; Yao, B. &E. Ringo, E. (2009). Effects of Dietary *Saccharomyces cerevisiae* Fermentation Product (DVAQUA®) on Growth Performance, Intestinal Autochthonous Bacterial Community and Non-Specific Immunity of Hybrid Tilapia (*Oreochromis niloticus* ♀ × *Oreochromis aureus* ♂) cultured in cages. *Aquaculture,* Vol. 294, pp. 99–107

Part 4

Applied Topics of Cellular and Molecular Biology

10

Use of Microarray Technology to Improve DNA Vaccines in Fish Aquaculture – The Rhabdoviral Model

P. Encinas[1], E. Gomez-Casado[1], A. Estepa[2] and J.M. Coll[1]
[1]*Instituto Nacional Investigaciones Agrarias, Departamento Biotecnología, INIA, Madrid*
[2]*Universidad Miguel Hernández, IBMC Elche*
Spain

1. Introduction

Fish rhabdoviral infections, specially those caused by novirhabdoviruses, can be tackled with commertial DNA vaccines such as the one against infectious haematopoietic necrosis virus (IHNV) (Salonius et al., 2007). Nevertheless, rhabdoviral diseases continue to pose a considerable threat to aquaculture because a number of practical problems regarding vaccination remain unsolved (for instance, mass delivery methods for small fish and requirements for safer vectors). Furthermore, some fish rhabdoviruses appear to be spreading to wild-type species.

Theoretically, efficient DNA vaccines could be used for any fish pathogen, such as other viruses (nodaviruses and orthomyxoviruses, for instance), bacteria and parasites. However, in practice, many fish DNA vaccines do not perform satisfactorily in most other pathogens than novirhabdoviruses. Therefore, fish novirhabdoviral vaccines are suitable models in which to study why similar DNA vaccines have not been successfully developed for other viruses (Gomez-Casado et al., 2011; Kurath, 2008). Those studies include the use of microarrays.

To date, effective vaccines against fish rhabdoviruses have been achieved simply by using their glycoprotein G gene (Einer-Jensen et al., 2009; Kurath, 2008; Kurath et al., 2007; Lorenzen, 2000; Lorenzen et al., 2009; Lorenzen & LaPatra, 2005). The glycoprotein G of rhabdoviruses is a widely studied antigen in fish (Bearzotti et al., 1995; LaPatra et al., 1994; McAllister et al., 1974; Vestergaard-Jorgensen, 1972; Winton et al., 1988) and its crystal structure has recently been elucidated in a similar mammalian rhabdovirus (Roche et al., 2006; Roche et al., 2007).

Most of our present knowledge about the factors that affect DNA vaccination efficacy in fish (vaccine dosage, delivery route, water quality, host species/size, time to challenge, severity of challenge, viral strain, etc) derives from work on fish rhabdoviral models (Kurath, 2008). Thus, the first fish DNA vaccines against IHNV were reported in 1996 (Anderson et al., 1996a; Anderson et al., 1996b) and against viral haemorrhagic septicemia virus (VHSV) in 1998 (Lorenzen et al., 1998). In 2005, Vical-Aqua Health Ltd. of Canada (Novartis APEX-

IHN) received authorization to commertialize an IHNV DNA vaccine (Salonius et al., 2007). In support of their licensing, millions of salmon were vaccinated in British Columbia in 2004 and 2005. However, there are no reports on the efficacy of this vaccine against natural viral challenges (Kurath, 2008; Salonius et al., 2007).

When injected intramuscularly in each fish, plasmid-based G glycoprotein-coding rhabdoviral vaccines induce long-term (months) specific immunity, preceded by an early (4-8 days) non-specific protective response (Kim et al., 2000; Lorenzen, 2000; Lorenzen et al., 2002). Non-specific short-term protective immunity results from the induction of interferon-*mx* and related genes, while specific long-term protection may have this effect as a result of the induction of G glycoprotein gene-specific antibody or cellular responses (Kurath et al., 2007). However, most changes in gene expression that occur with resistance mechanisms in short-term and long-term immunity are not fully understood (Goetz & MacKenzie, 2008). Furthermore, more basic knowledge on mucosal immunity is required to move rhabdoviral DNA vaccines from the laboratory into the field, as existing vaccines still require either intramuscular injection in individual fish or stronger (adjuvanted) immune responses to facilitate mass delivery methods, such as those using oral- (delasHeras et al., 2010; Tian et al., 2008) or ultrasound-aided (Fernandez-Alonso et al., 2001) immunization. Studies using microarrays could greatly contribute to furthering this basic knowledge (Secombes, 2008).

Theoretically, for best performance an optimal vaccination should mimic viral infection steps such as entry and replication. For instance, since the entry of rhabdoviruses would be first detected by cellular membrane toll-like receptors (TLRs) through the G glycoprotein and their later cytoplasmic replication by endosomal TLRs through dsRNA intermediates, the question arises as to whether DNA vaccines should include not only the G glycoprotein gene but also dsRNA intermediates (ie.: RNA hairpins). Again, new data obtained from microarrays could shed some light on these possibilities.

As established by quantitative RT-qPCR before the advent of microarrays, 4 to 8 days after DNA vaccination by intramuscular injection, gene expression by fish haematopoietic organs showed an increase in interferon-inducible *mx* (Acosta et al., 2005; Boudinot et al., 1998; McLauchlan et al., 2003; Purcell et al., 2004; Robertsen, 2008; Tafalla et al., 2007), virally-induced genes (Vig) (Boudinot et al., 1999; Boudinot et al., 2001) and *mhc* and *tcr* genes (Takano et al., 2004).

In this context, the recent availability of fish microarrays (Martin et al., 2008), which allow the expression profiling of thousands of genes simultaneously, has provided new opportunities to further study fish immunological responses in several rhabdovirus/fish models.

Expressed sequence tag (EST)-based microarrays of the Japanese flounder, trout, salmon and zebrafish have been used in gene-discovery efforts. These studies included infections with IHNV (MacKenzie et al., 2008; Purcell et al., 2006a), VHSV(Byon et al., 2005; Byon et al., 2006; Encinas et al., 2010) and hirame rhabdovirus (HRV)(Yasuike et al., 2007) (Table 1). However, no studies have reported on the largest microarrays that have recently become available, such as the ~ 32 K cDNA of salmonids (von Schalburg et al., 2008) and the ~ 37 K 60-mer oligos of trout (Salem et al., 2008), most probably due to the complexity of the interpretation of the data.

Fish name	Genus species	Size, ~K	Type	Name	References
Flatfish					
Japanese flounder	*Paralichthys olivaceus*	1	cDNA	HRV-infected Leukocytes	(Aoki et al., 1999) (Nam et al., 2000)
European flounder	*Platichthys flesus*	3	cDNA	GENIPOL	(Diab et al., 2008)
Atlantic halibut	*Hipppoglossus hyppoglossus*	7	oligo	---	(Douglas et al., 2008)
Turbot	*Scophthalmus maximus*	3	oligo	Aeromonas-infected organs	(Pardo et al., 2009)
Salmonids					
Atlantic salmon	*Salmo salar*	16	cDNA	TRAITS-SGP	(Salem et al., 2008)
Atlantic salmon	*salmonids*	32	cDNA	GRASP	(von Schalburg et al., 2008) (Koop et al., 2008)
Rainbow trout	*Oncorhynchus mykiss*	37	oligo 60mer	RTGI	(Salem et al., 2008)
Other species					
Zebrafish	*Danio rerio*	40	oligo 60mer	----	Agilent (commertial)

GENIPOL, http://www.pleuronectes.ca. GRASP, http://web.uvic.ca/grasp/microarray/array.html. TIGR, http://biocomp.dfci.harward.edu/tgi/cgi-bin/tgi/gimain.pl?gudb=salmon. TRAITS-SGP, http://www.traitsdb.stir.ac.uk. RTGI, http://compbio.dfci.harvard.edu/tgi/cgi-bin/tgi/gimain.pl?gudb=r_trout. Zebrafish, http://www.ensembl.org/Danio_rerio/Info/Index and Sanger zebrafish project http://www.sanger.ac.uk/Projects/D_rerio/.

Table 1. Summary of rhabdoviral-sensitive fish species with microarrays in different stages of development.

In this review we focus on the data published on the use of microarrays for the identification of rhabdoviral-induced genes with properties that make them candidate adjuvants for the improvement of fish DNA vaccines.

2. Vertebrate viral infections, vaccination and adjuvants

Pathogen-associated molecular patterns (PAMPs) are sensed in higher vertebrates by pattern recognition receptors (PRRs). There are several PRR classes (retinoic acid-inducible gene-like helicases, nucleotide-binding oligomerization domain-like receptor, peptide recognition proteins, etc). The most studied PRRs belong to the family of toll like receptors (TLRs) (Manicassamy & Pulendran, 2009) When expressed at the cell (TLRs numbers 1,2,4,5,6,10,11) and at the endosomal (3,7,8,9) membranes, TLRs detect PAMPs outside and inside the cells, respectively. Most natural infections start through mucosal surfaces that contain dendritic cells (DCs) specialized in sensing PAMPs through their cell-specific TLRs-enriched membranes (Iwasaki, 2007a; Iwasaki, 2007b; Thompson & Iwasaki, 2008). After recognition of their corresponding PAMP, TLRs generates TLR-mediated signals, these resulting in a complex signalling network whose integration by the host determines the final immune response (Manicassamy & Pulendran, 2009).

Since the most effective vaccinations are obtained after infections with live or attenuated pathogens, several PAMPs from a unique pathogen (such as external glycoproteins and internally synthesized dsRNA/glycoproteins in rhabdoviruses) simultaneously stimulate several TLRs. In contrast, dead recombinant protein subunits and antigenic genes contain fewer PAMPs than live/attenuated pathogens. Nevertheless, single PAMPs have also been used to immunize against live pathogens, mostly with the help of adjuvants to replace the missing PAMPs. Therefore, the purpose of vaccine adjuvants is to increase the immune responses of otherwise weak individual PAMPs.

Most adjuvants in mammals are believed to target professional antigen-presenting cells, such as tissue DCs (De Gregorio et al., 2009; Lambrecht et al., 2009). The expression patterns of pro-inflammatory genes such as cytokines, chemokines, MHC and co-estimulatory molecules are altered in adjuvant-targeted DCs (Figure 1). Subsequently, maturing DCs migrate to lymph nodes and activate naive CD4+ (helper) and CD8+ (cytotoxic) T cells to produce antigen-specific antibodies, cytotoxic cells, antimicrobial peptides and regulatory cytokines (Craig et al., 2009; Longhi et al., 2009; Manicassamy & Pulendran, 2009; Secombes, 2008). DCs also process PAMPs into peptides for presentation onto major histocompatibility (*mhc*) molecules to T cell receptors (*tcr*). Thus DCs are crucial for both adjuvant effects and innate/adaptive immune responses (Figure 1).

Although most PAMP-derived vaccine adjuvants act through TLRs on mammalian DCs (Figure 1), other internal adjuvants, such as *hmgb1* released from lysed cells, exert their action through cell damage molecules (Lambrecht et al., 2009). Artifitial TLR-independent adjuvants, such as those derived from particulate compounds administered together with mammalian vaccines (mineral salts, liposomes, microparticles, saponins, and emulsions) either increase antigen persistence or uptake by DCs. Traditionally, vaccine adjuvants have been empirically identified as enhancers of antibody responses to a co-administered antigen. However, new adjuvant candidates have also been found among molecules of the signalling cascades of DC activation. According to a recent review (Secombes, 2008), the molecules with potential capacity to act as fish vaccine adjuvants might be found among: i) cytokine/chemokine molecules; ii) co-stimulatory cluster differentiation (*cd*) antigen receptors; and iii) blocking molecules, which might inhibit negative regulators. Microarray analysis of rhabdoviral fish immunizations have identified some of these molecules, as it will be reviewed here.

3. Microarrays in the study of the flatfish/HRV/VHSV models

Traditional sequencing, annotation and estimation of frequencies of each rhabdovirally-induced transcript in flatfish, is one of the strategies designed to identify genes transcribed after rhabdoviral infections (pathogen-induced gene approach)(Aoki et al., 2011). Thus, the first attempts to identify HRV-induced genes were made in the Japanese flounder *Paralichthys olivaceus* by sequencing 300-596 expressed sequence tag (EST) clones from leukocytes 2-5 days after infection. The frequencies of each EST were estimated within a short 1 to 10 range (Aoki et al., 1999; Nam et al., 2000).

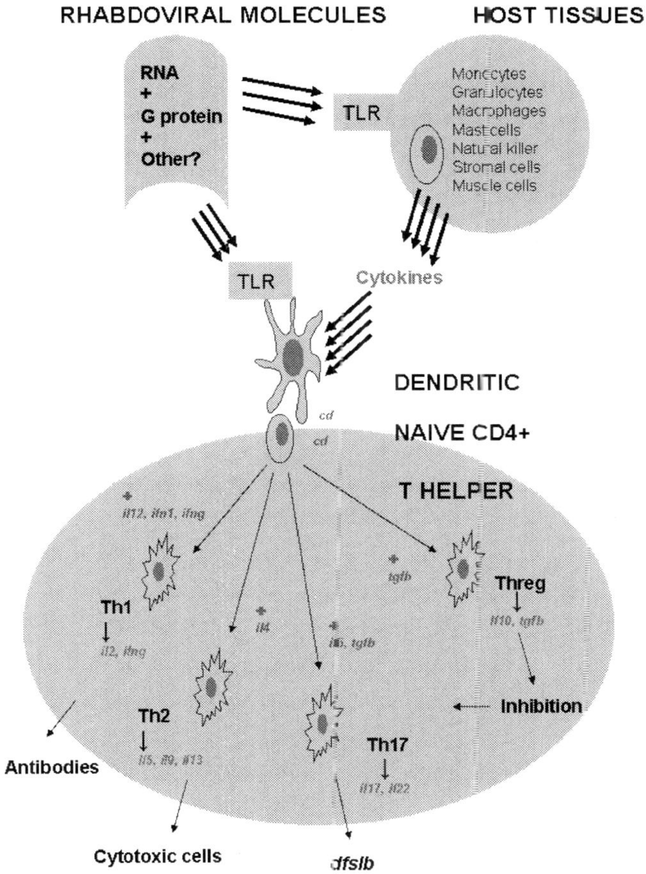

Fig. 1. Scheme of possible mechanisms of adjuvanticity. Modified from several published schemes (De Gregorio et al., 2009; Manicassamy & Pulendran, 2009; Secombes, 2008). External incoming or internally synthesized rhabdoviral molecules (dsRNA, glycoprotein G, other viral proteins, etc) activate dendritic cells (DCs). These recognize rhabdoviral molecules and are activated either through toll-like receptors (TLRs) or cytokines produced by other cells (monocytes, granulocytes, macrophages, mast cells, natural killer cells, stromal cells, muscle cells, etc). Each combination of rhabdoviral molecules induces simultaneous stimulation of DCs to induce the expression of secreted cytokines and costimulatory membrane *cd*s. The induced membrane *cds,* together with other signals (blue), induce differentiation of *cd4*+ cells to T helper cells (Th1, Th2, Th17 and/or Threg). Each differentiated Th cell produces a series of cytokines (red), which are required to make antibodies, cytotoxic lymphocytes, antimicrobial peptides and molecules involved in the regulation of other Ths. Theoretically, any of the up- or down-regulatory molecules that increase defensive responses could be candidate molecular adjuvants for vaccines.

The new mass-sequencing technologies, such as those offered by the Genome Sequencer FLX (454 Life Sciences, Branford, CT, USA), Illumina Solexa (Illumina Inc., San Diego, CA, USA) or ABI SOLiD (Applied Biosystems, Foster City, CA, USA), could improve the pathogen-induced transcript frequency-estimation strategy. Thus the mass-sequencing technologies produce millions of sequences per run, facilitating significant statistical data for the quantitation of each sequence frequency. However correct annotation of such a mass of new sequences continues to be a problem (Goetz & MacKenzie, 2008). For instance, of 58 million cDNA sequences of ~ 100 bp from largemouth bass, only 31391 unique sequences could be annotated (Garcia-Reyero et al., 2008). Although, the recent production of longer sequence sizes (200–400 bp), will facilitate their annotation, comparison of transcripts from many samples by this ultra-high-throughput sequencing technology is still not economically feasible. Massive sequencing could be used as a first approach, while a more focused microarray developed with selected genes could then be used for quantification of larger numbers of samples (Goetz & MacKenzie, 2008).

Japanese flounder EST-derived cDNA microarrays were applied to *in vitro* kidney cell cultures 3-6 h after HRV infection (Kurobe et al., 2005). The number of expressed transcripts changed in 20.8 % of the genes after HRV infection. The 91 immune-related genes of the microarray were preliminarily categorized into 8 clusters on the basis of their known pattern of gene expression. After 3 h of HRV infection, several genes included in the chemotaxis, apoptosis, cell growth and antigen-presenting clusters were increased while the expression of some genes, including *mx,* decreased. Among the genes of unknown function that changed after HRV infection, 13 showed a similar response profile to that of the genes of known function mentioned above. This observation may be indicative of their association.

Improved versions of Japanese flounder EST-derived cDNA microarrays (779 spots containing 228 immune-related genes) were used for *in vivo* differential gene expression after intramuscular injection of DNA vaccines containing the G gene of VHSV (Byon et al., 2006) and/or HRV (Yasuike et al., 2007). The differential expression of their transcripts was studied in kidney tissue 1, 3, 7 and 21 days after vaccination. The greatest number of differentially expressed genes (Figure 2) was observed 3 days after injection (91.4 % were increased, of which 31 % were known genes). Genes with increased expression/transcription include those related to the non-specific immune responses, such as *tnf, il1r, ccr,* and *mx,* transcription factors, and even a few genes associated with the late specific antibody response, such as *cd20*. Many interferon-inducible genes including *mx* and interferon regulatory genes were the most strongly induced genes 3 and 7 days after injection. The expression of a number of unknown genes was also increased(Aoki et al., 2011). Among these, the *LB3*(8) gene increased a maximum of 56-fold 3 days after infection and then remained increased during one week (Byon et al., 2005).

Later versions of the Japanese flounder EST-derived cDNA microarrays of up to 1187 unique flounder ESTs (691 identified genes) were then used to compare the injection of recombinant G protein (non-protective) with the *G* gene (protective) (Byon et al., 2006). A number of IFN-related genes (including the unknown *LB3*(8)) and *mx* increased 7 days after injection, thereby confirming the observations made in previous studies using reverse transcriptase-quantitative polymerase chain reaction (RT-qPCR) (Acosta et al., 2005; Robertsen, 2008). Further studies included differential gene expression in kidneys from Japanese flounder injected with the *HRV G* gene (protective) in comparison with the *N* gene

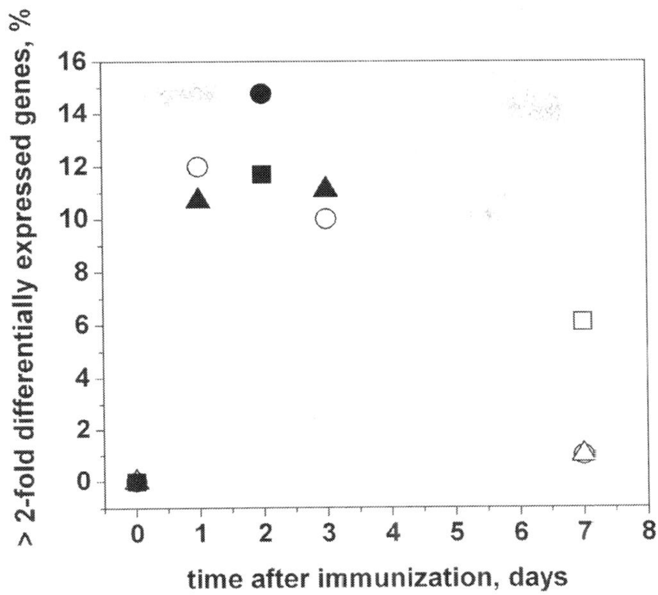

Fig. 2. Number of differentially expressed genes after rhabdoviral immunization. Genes with increase expression were defined as those genes with more than 2-fold increase in expression. ip, intraperitoneal injection. im, intramuscular injection. imm, immersion. Flounder, Japanese flounder, *Paralichthys olivaceus*. Trout, rainbow trout, *Oncorhynchus mykiss*. Zebrafish, *Danio rerio*. *, number of unique sequences or features and type of microarray (cDNA or oligo DNA). HRV, hirame rhabdovirus. IHNV, infectious haematopoietic necrosis virus. VHSV, viral haemorragic septicemia virus. ▲, infection-by-injection of trout with the IHNV G gene and expression on head kidney with $p<0.01$ (MacKenzie et al., 2008).●, Infection-by-immersion of zebrafish with VHSV and expression on fins (Encinas et al., 2010). ■, VHSV infection-by-immersion of zebrafish and expression in internal organs (head kidney, liver and spleeen) (Encinas et al., 2010). □, Injection of trout with the IHNV G gene and expression in muscle tissue with $p>0.01$ (Purcell et al., 2006b). Δ, injection of Japanese flounder with the HRV G gene and expression in head kidney (Yasuike et al., 2007). ○, Injection of Japanese flounder with the VHSV G gene and expression in head kidney (Byon et al., 2006).

(non protective). Results confirmed that the IFN-inducible genes, *LB3*(8) and *mx*, were also increased 7 days after vaccination but only when the *G* gene was used (Yasuike et al., 2007). Furthermore, it was shown that the *LB3*(8) gene has an homologous domain to that of a mammal IFN-inducible protein. Thus, this gene is an example of how new genes involved in rhabdoviral immunization can be discovered by the microarray approach.

However, in all the series of experiments on flounder commented above, only transcripts from pooled organs from 3-5 fish were compared. Biological replicates were not reported and therefore statistical biological variation could not be estimated. Furthermore, the number of genes in the microarrays were relatively small and their collected data has not been deposited in any known banks to allow for independent or comparative analysis. Nevertheless, two main conclusions can be drawn from these experiments. Firstly, the largest number of differentially expressed kidney genes after fish rhabdoviral (VHSV or HRV) immunization-by-injection occurs 2-3 days after vaccination (Figure 2) and, secondly, IFN-induced gene responses are stimulated after 3-7 days (Byon et al., 2005; Byon et al., 2006; Kurobe et al., 2005).

In these earliest experimentations, microarrays based on cDNAs (100-500-mer) rather than oligos (60-70-mer) were used. Because one of the greatest concerns with cDNA arrays is cross-hybridization between similar genes or between repeated elements of different genes as a result of the pseudotetraploidy of many fish, the use of oligo microarrays would increase specificity (von Schalburg et al., 2008). However, in contrast to cDNA microarrays, oligo microarrays have a poorer performance when used for other related species. The current tendency appears to favour the use of the former. Thus, by using oligo microarrays, the printing layouts, total number of sequences and number of sequence replicates can be modified to meet any formats. Furthermore, oligo microarrays do not required maintenance of collections of bacterial clones coding for cDNAs. In addition, oligonucleotides can be selected and used in a range of various formats suitable for each experimental design. Improved sensitivity, increased dynamic range, lower variance and fewer outliers have also been demonstrated when using oligo rather than cDNA microarrays. Correlation between cDNA and oligo microarray results has been demonstrated, although some discrepancies have also been reported (Salem et al., 2008). High density oligo microarrays have been developed in other fish such as salmonids (von Schalburg et al., 2008), rainbow trout (Dios et al., 2008; Salem et al., 2008) and zebrafish (Cameron et al., 2005) (Table 1).

4. Microarrays in the study of the salmonid/IHNV/VHSV models

Large-scale genomic projects for salmon have been initiated by groups in Canada, the USA, the UK, Norway and France. As a result there are many physical and genetic maps, large collections of ESTs and a growing number of genomic sequences and derived microarrays. Thus three projects have developed salmonid microarrays. The first salmonid 16K cDNA microaarray appeared in 2004. This array was developed by the Genomic Research on Atlantic Salmon Project (GRASP)(von Schalburg et al., 2005a) and led to the most recent 32K cDNA (von Schalburg et al., 2008) and the first 5K oligo DNA of 70-mer (Koop et al., 2008) microarrays. The high sequence similarity (~ 86 %) between salmonids (9 genera and 68 species) indicates that cDNA microarrays may be suitable for studies involving any member of this fish family. Transcriptome Analysis of Important Traits of Salmon (TRAITS) and the Norwegian Salmon Genome Project (SGP) also developed a 16K cDNA microarray (http://www.abdn.sfirc/salmon) based on two independent collections of their bacterial clones kept in ARK, Genomics Facility at Roslin Institute, UK and at SGP Genetics Laboratory at the University of Oslo, respectively. The TRAITS-SGP cDNA array was obtained from ESTs from 15 tissues (pathogen-induced libraries, trait-specific substractive EST, starvation-induced libraries, diet-response libraries, smoltification-response libraries and well-known genes). This array was conceived as a preliminary tool to develop an oligo microarray for routine health monitoring of Atlantic salmon. The first results found some artefactual expression patterns caused by cross-hybridization of similar transcripts and underlined the greater relevance of biological over technical replicates (Taggart et al., 2008).

By using all the tentative consensus sequences available at the Rainbow Trout Gene Index (RTGI) data base, a 37K oligo microarray was constructed (Salem et al., 2008), which is available at Agilent (design number 16271, deposited on the GEO with the GPL6018 number). The new rainbow trout (*Oncorhynchus mykiss*) high-density oligonucleotide microarray was developed using 37394 specific 60-mer oligonucleotide probes assembled from 244984 ESTs from 12 tissues (http://compbio.dfci.harvard.edu/tgi/cgi-bin/tgi/gimain.pl?gudb¼r_trout). The specificity of each probe was checked for possible non-specific mRNA cross-hybridization by comparing all individual probes with all rainbow trout transcriptome sequences. Approximately 91 % of the sequences used for this microarray matched a previously annotated sequence in the GenBank.

Few attempts have been made to use these microarrays to study the rhabdoviral immunization of salmonids. In homozygous trout, the 16 K cDNA GRASP microarray was used to profile 7-day muscle transcripts after intramuscular injection of the IHNV G gene (Purcell et al., 2006a; von Schalburg et al., 2005b). After immunization, *irf3*, *mx*, *vig1*, and *vig8* transcripts were increased (Purcell et al., 2006a). Genes associated with antigen-presenting cells, lymphocytes, leukocytes, inflammation, antigen presentation, and interferon pathways were also augmented. The increased levels of transcripts associated with type I IFN pathways in systemic organs (gill, spleen and kidney) were corroborated by RT-qPCR. These observations confirmed that, when intramuscularly injected, the host-expressed viral G gene induces a systemic non-specific type 1 IFN innate immune response.

Using a 1.8K cDNA salmonid microarray, comparison of infection-by-injection with IHNV and attenuated IHNV in rainbow trout after 1 and 3 days showed an IHNV-dependent change in differential transcription in kidney towards adaptive immunity genes (MacKenzie et al., 2008). Thus, the rapid spread of the IHNV infection inhibited *tnfa*, *mhc1*, and several other gene markers while favouring *mhc2* and *ig* responses. The molecular mechanism for the development of late (months) specific cytotoxic T or B cell-mediated humoral responses has not been addressed by means of microarrays (Kurath, 2008; Kurath et al., 2006).

More recently, trout families with low (32% survival following challenge) and high susceptibility to VHSV (18% survival following challenge) were infected with VHSV by bath exposure and transcriptional data from internal organs were analyzed with the 16K GRASP microarray from day 3 post-challenge (Jorgensen et al., 2011). In total, 939 genes were differentially expressed between infected and non-infected fish. The genes increased in infected fish belonged to the following categories: stress and defence response, NFkappaB signal transduction, response to non-self, antigen processing and presentation, and proteasome complexes. Most were also increased among the 642 differentially expressed genes in the low-susceptibility trout family but not among the 556 differentially expressed genes in the high-susceptibility family. These results suggest that the innate immune system of internal organs plays a crucial role in eliciting an effective immune response to VHSV infection in rainbow trout (Jorgensen et al., 2011).

5. Microarrays in the study of the VHSV/zebrafish model

The zebrafish *Danio rerio* is one of the most suitable models in which to carry out microarray studies because, compared to other fish, its genome sequence is one of the most advanced. Furthermore, ~ 40 K annotated quantitative polymerase chain reaction (qPCR) arrays and annotated oligo microarrays are available. In addition, large-scale experimentation with zebrafish is easier than with other fish models and zebrafish are susceptible to several viruses, most of these belonging to the fish rhabdoviral family (Sullivan & Kim, 2008). Of these, VHSV (Novoa et al., 2006) was chosen in a recent study using microarrays (Encinas et al., 2010) over IHNV (LaPatra et al., 2000), snake-head rhabdovirus (SHRV)(Phelan et al., 2005) and spring viremia of carp (SVC)(Sanders et al., 2003), because only in the VHSV/zebrafish model have infection-by-immersion (the natural route of infection) and successful vaccination been described (Novoa et al., 2006).

Damage and epithelial cell death immediately after VHSV infection in the surface portals of entry of these viruses, such as the fins (Harmache et al., 2006), should alert surrounding cells to promote epithelial cell division to replace dead cells, recruit inflammatory cells to the infection site, and send signals to internal immune organs. However, viral-induced signals to inhibit the most relevant host responses have also been detected. Detection of natural early responses may contribute to identifying vaccine adjuvants. Thus, the expression of the 636 immune-related transcripts that were increased after VHSV infection, as estimated by hybridization to oligo microarrays (confirmed by RT-qPCR arrays), was higher in fins than in organs. In contrast, the number of decreased transcripts was higher in organs than in fins (Figure 3). Therefore, an upregulated response of immune-related genes was greatest in fin tissues, while a downregulated response was most detected in the internal organ responses. The latter might be targets of viral inhibitory signals early after infection (Encinas et al., 2010). These results showed that 2 days after infection-by-immersion, VHSV had not yet caused an strong response from zebrafish internal organs, which contrasts with reports in other fish at later times after infection-by-injection (such as *ifn1, mx, il1b, tnfa,* etc) (Acosta et al., 2006; Samuel, 2001; Tafalla et al., 2007; Tafalla et al., 2005) or infection-by-immersion (Jorgensen et al., 2011; Zhang et al., 2009).

The zebrafish are refractory to rhabdoviral infection-by-immersion at high temperatures or without acclimatation to low temperatures with IHNV (LaPatra et al., 2000), VHSV (Novoa et al., 2006) or SVC (Sanders et al., 2003). Therefore, a temperature-dependent response mechanism(s) that inhibits rhabdoviral infection and spread may occur. While these preliminary findings shed some light on the earliest effects of VHSV infection at the molecular level, some of the new immune-related genes identified might be suitable candidate adjuvants for fish vaccines (Rajcani et al., 2005; Secombes, 2008).

6. Comparative microarray study of fish/rhabdoviral models

To best detect innate immune responses, early times after rhabdoviral infection should be studied. Thus, according to the data obtained from flatfish, salmonid and zebrafish studies, the maximal number of >2-fold differentially expressed genes in microarrays was detected 2-3 days after rhabdoviral infection (Aoki et al., 1999; Byon et al., 2005; Byon et al., 2006; Kurobe et al., 2005; MacKenzie et al., 2008; Nam et al., 2000; Purcell et al., 2006a; von Schalburg et al., 2005b; Yasuike et al., 2007) (Figure 2).

Table 2 shows a list of some of the differentially transcribed immune-related genes detected using microarrays after rhabdoviral immunization, independently of immunization mechanism, fish species, rhabdoviruses and organs. Among the gene list, *ifn* and *irf*-related genes were expected to be present; however, their presence was scarce. As with many other

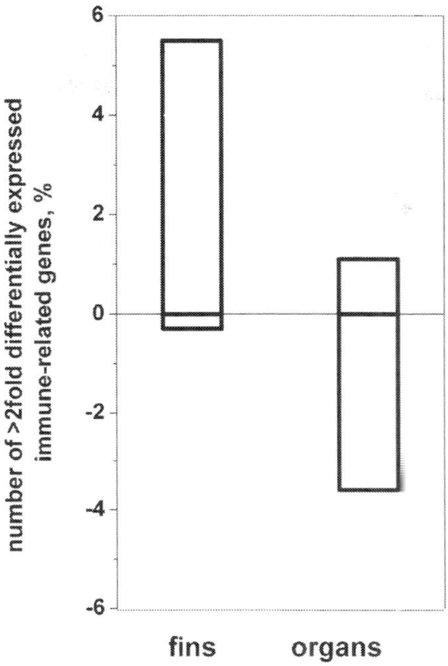

Fig. 3. Comparison of the number of differentially expressed genes 2 days after VHSV infection in zebrafish fins and organs (modified from Encinas et al., 2010). Groups of 10 zebrafish (*Danio rerio*) were infected with VHSV. To perform the hybridization to oligo microarrays, 636 selected immune-related sequences from the 44K microarray (Appligene) were used to analyze mRNA levels. Normalization was made with the ribosomal phosphoprotein p0 (*rplp0*) gene. Folds were calculated by the following formula: mRNA levels in VHSV-infected zebrafish / mRNA levels in non-infected zebrafish. The total number of genes increasing (+) or decreasing (-) expression > 2-fold and $p<0.05$ was represented as the percentage of the total number of immune-related genes assayed (n = 4). Percentages were calculated by the formula, 100 x the number of differentially expressed genes / total number of immune-related genes assayed (Encinas et al., 2010).

viruses and host species, an increase in *ifn1* expression is one of the first responses to the injection of any DNA vaccine (Acosta et al., 2006; Samuel, 2001) and to rhabdoviral infections (Samuel, 2001; Theofilopoulos et al., 2005).

Transcripts encoding several forms of *il17* were detected as differentially expressed only in one of the studies using microarrays (Encinas et al., 2010). *Il17* is produced by T helper 17 (Th17) cells (Figure 1) and acts together with *il22* on epithelial cells (Trifari et al., 2009) and other types of skin cells to trigger *il1b* and *tnf*. These responses induce neutrophil/macrophage recruitment in epithelial surfaces (Qiu et al., 2009), stimulate keratinocytes (Nograles et al., 2008) and increase the production of antimicrobial peptides such as hepcidin (*hamp1*) and defensin ß like-2 (*defbl2*). Both *hamp1* and *defbl2* were found

Gene classes	Genes	References
Interleukins	*il17*	(Encinas et al., 2010)
&	*il1r*	(Kurobe et al., 2005)
Receptors	*il8r*	(Nam et al., 2000)
	il1b	(Encinas et al., 2010)
Interferon -	*irf1*	(Kurobe et al., 2005)
related	*mx*	(Yasuike et al., 2007)
molecules	*isg15, 56*	(Yasuike et al., 2007)
	iip56	(Byon et al., 2006)
	iip54	(Aoki et al., 1999)
	ifn3	(Encinas et al., 2010)
Major Histocompatibility	*mhc1*	(Aoki et al., 1999; Byon et al., 2006; Encinas et al., 2010; MacKenzie et al., 2008)
Complex	*mhc2*	(Aoki et al., 1999; Byon et al., 2006; Kurobe et al., 2005; MacKenzie et al., 2008; Nam et al., 2000)
Antimicrobial	*hamp1*	(Aoki et al., 2011; Encinas et al., 2010)
peptides	*defbl2*	(Encinas et al., 2010)
Chemokines & receptors	*ccr*	(Byon et al., 2006; Kurobe et al., 2005)
Complement	*c3*	(Byon et al., 2006; Encinas et al., 2010)
components	*cfb/c2b, crpp, c3b, bfb, cfhp, clu, c6, c8a, c8g, c9, c1q*	(Encinas et al., 2010)
	c3ar	(MacKenzie et al., 2008)
High Mobility proteins	*hmgb*	(Aoki et al., 1999; Encinas et al., 2010; MacKenzie et al., 2008; Nam et al., 2000)
G proteins	*gnb*	(Byon et al., 2006; Encinas et al., 2010)
TNF-related molecules	*tnf*	(Byon et al., 2006; Encinas et al., 2010; Kurobe et al., 2005)
	tnfr	(MacKenzie et al., 2008)
	tnfr1	(Nam et al., 2000)
	tnfr2-traf	(Kurobe et al., 2005; Nam et al., 2000)
Toll-like	*tlr2*	(Kurobe et al., 2005)
receptors	*tlr5*	(Encinas et al., 2010)
	tlr7	(Encinas et al., 2010)
	tlr9	(Encinas et al., 2010)
Immunoglobulin	*igh*	(Byon et al., 2006; Encinas et al., 2010)
chain domains	*ighz*	(Encinas et al., 2010)
	igl	(Aoki et al., 1999; MacKenzie et al., 2008)
	sid4	(Encinas et al., 2010)

Table 2. Some of the fish immune-related genes differentially transcribed in microarray studies after rhabdoviral immunization. Independently of fish species, rhabdovirus, organ and time after immunization, annotated genes with a differential expression >2 fold were searched in the original papers and some of the most common were ordered on the basis of gene classes and listed in the table.

to be differentially expressed in some studies (Table 2) (Liang et al., 2006; Yu & Gaffen, 2008).

Differentially upregulated transcripts of *il12* were detected in a zebrafish fin study (Encinas et al., 2010). In that context, *il12* is crucial because it has been widely described as a vaccine adjuvant in mammals (Bliss et al., 1996; Chong et al., 2007; Hirao et al., 2008; Stevceva et al., 2006), specifically increasing protective mucosal immunity (Arulanandam et al., 1999; Wright et al., 2008) to viral infection (Hancock et al., 2000; Jacobson et al., 2006; Skeen et al., 1996; Zheng et al., 2005). However, *il12* has not been tested in fish.

Co-stimulatory cell membrane cluster differentiation antigens (*cd*) molecules responsible for the antigen-presenting cell interactions with T cells were not differentially expressed, except those belonging to *mhc1* and *mhc2* molecules (Table 2). Nevertheless, the use of *cds* as vaccine adjuvants has been described for the *cd154* gene in zebrafish (Gong et al., 2009); the reasoning being that co-expression of *cds* with antigen in the same cell might accelerate specific immune responses. Thus, specific antibody responses obtained using *cd154* and the pMCV1.4 plasmid coding for the G gene of VHSV (Ruiz et al. 2008) were increased 3-4-fold with respect to the plasmid alone (Gong et al., 2009).

Transcripts of *c3* and *c3a* were differentially expressed in salmonid and flatfish internal organs after immunization-by-injection while many more complement components (*cfb/c2b, crpp, bfb, cfhp, clu* and *c6* and *c8a*) were found in zebrafish fins after infection-by-immersion (Encinas et al., 2010). The use of *c3* derivatives (*c3a, c3d, c4a* and *c5a*) (Green et al., 2002; Ross et al., 2001; Sunyer et al., 2005; Villiers et al., 1999a; Villiers et al., 1999b) as vaccine adjuvants has been reported in mammals but not in fish. A possible relationship between *c3* trout genetic polymorphism and VHSV resistance (not confirmed by genetic evidence) (Slierendrecht et al., 1993; Slierendrecht et al., 1995; Slierendrecht et al., 1996) may require further physiological studies.

TLRs, immunoglobulin chains, TNF -related molecules, and high mobility proteins were also found amongst the differentially expressed genes in several studies using microarrays and thus might deserve some consideration as adjuvant candidates (Table 2).

7. Future research

Microarray and mass sequencing technologies have opened up new avenues to analyze gene expression profiles. The data obtained by these technologies might facilitate the discovery of new immune-related genes (immunogenomics), clarify the molecular mechanisms of immunity and identify new candidates for vaccine adjuvants. Nevertheless, some problems remain in the application of these technologies to the amelioration of fish rhabdoviral vaccines. For instance there is a need to improve comparison of the data obtained from different models, to complete present gene annotations, to confirm transcriptional data with protein data and to develop mathematical models to facilitate interpretation of the abundant data.

In addition, the number of immune-related genes on zebrafish, trout, salmon, human and mice microarrays (Table 3), shows that more fish immune-related genes might have to be included in future microarray designs. The number of immune-related genes are still much lower in other cultured fish species (turbot, sea bream, sea bass, etc).

	Zebrafish		Trout	Salmonids		Human	Mouse
Immune-related	AFYz	AGIz	AGIt	GRAs	TRAs	AFYh U133v2.0	AFYm 430v2.0
key words	**14K**	**45K**	**37K**	**36K**	**36K**	**47K**	**39K**
interferon	19	53	62	92	42	103	91
chemokine	14	36	29	56	10	104	100
interleukin	8	57	62	49	40	187	142
cytokine	13	43	28	49	13	87	68
defensin	1	3	2	0	1	29	32
antiviral	0	2	1	0	0	9	5
LPS	0	0	0	0	2	2	7
histocompatibility	13	18	13	59	3	76	74
MHC	2	16	70	375	433	10	4
viral	40	73	29	16	8	171	96
Mx	0	1	0	0	1	0	0
complement	28	79	168	88	53	129	102
immunoglobulin	14	53	116	54	46	294	156
Toll	3	22	5	17	12	39	31
TNF	15	22	7	13	3	7	38
macrophage	5	11	22	25	25	34	36
lymphocyte	2	15	22	14	8	49	52
neutrophil	2	4	0	4	7	11	7
leukocyte	6	15	12	18	4	50	21
cytotoxic	5	5	14	3	4	30	22
natural killer	3	0	3	2	5	19	8
T cell	13	64	63	59	14	88	112
B cell	20	42	29	36	11	102	97
dendritic	0	3	15	0	3	0	0
TOTAL	**226**	**637**	**772**	**1029**	**748**	**1630**	**1301**

Table 3. Estimation of the numbers of immune-related genes in fish microarrays compared to human and mouse. Microarrays vary in the number of probes per gene, and gene nomenclatures. Many fish genes might be duplicated variants (due to pseudotetraploid genomes or transposon variations) and arrays may use different genes and/or cDNA or oligos per gene. All these facts make comparison of microarray platform gene contents difficult. The use of immune-related key words to preliminarily compare the relative abundance of the genes might serve for a first estimation. The future should bring about the use of a common languages such as gene abbreviations following the HUGO Gene Nomenclature Committee for human orthologues (http://www.genenames.org) and/or UniGene entries (http://www.ncbi.nlm.nih.gov/unigene). Genes should be also grouped by functional categories such as by using gene ontology (GO annotation for the immune system http://www.geneontology.org/GO.immunology.shtml) or the clusters of orthologous genes (COG, http://www.ncbi.nlm.nih.gov/COG).

AFYz (AFFYMETRIX, zebrafish)	14K, ~ 15 oligos 25-mer/gen (Santa Clara, CA,USA).
AGIz (AGILENT, zebrafish vs2)	45K, 1 oligo 60-mer/gen (Palo Alto,CA, USA)
AGIt (AGILENT, trout)	37K, 1 oligo 60-mer/gen
GRAs (GRASP, salmonids)	36K, 1 cDNA/gen
TRAs (TRAITS-SGP, salmonids)	36K, 1 cDNA/gen
AFYh (AFFYMETRIX, human)	47K, U133plus v2.0 ~ 11 oligos 25-mer /gen
AFYm (AFFYMETRIX, mouse)	39K, 430 v2.0 11 oligos 25-mer/gen

The use of microarray technology stems from the availability of genome, mRNA and EST (http://www.ncbi.nlm.nih.gov/dbEST) sequences to build representative annotated (gene-identified sequences) microarrays. For most commertial fish species, there is a lack of information on the annotated genome or known mRNA sequences and thus most microarrays used for these species mostly apply EST sequences. However, correct and complete annotation continues to be a bottleneck.

At present, it is quite difficult to compare data from distinct microarrays, even between salmonid microarrays such as GRASP, TRAITS-SGP and RTGI. We consider that reanalysis of the data deposited in data banks, for instance by the NCBI Gene Expression Omnibus (GEO) (http://www.ncbi.nlm.nih.gov/geo), to identify similar or identical genes should be undertaken.

For the most studied fish species, such as zebrafish, trout and salmon, identified mRNA sequences present in current GenBanks can alternatively be used to design focused microarrays enriched in some gene classes. For instance, the zebrafish and trout microarrays that are currently available omit a number of immune-related genes, the number of which vary depending on each microarray. A possible alternative to this problem in fish species for which abundant annotated mRNA sequences are held in GenBanks is to search for keyword-selected sequences to build up the corresponding microarrays.

One example of using some of the most obvious keywords corresponding to immune-related genes for the trout *O. mykiss* and the zebrafish *D. rerio* is shown in Table 4. Thus, this table shows that the number of some immune-related genes extracted from GenBank data are 2-24-fold higher than their corresponding numbers in the microarrays of trout and zebrafish commertially available. Furthermore, trout and zebrafish 2-4K microarrays designed on unique sequences selected from GenBank immune-related genes (using ~50 keywords in GenBanks) are enriched an average of 2-3-fold in immune-related sequences with respect to the more general 37-44K commertial alternatives (data not shown). The design of smaller, focused (ie: immune-related) microarrays based on existing GenBank sequences could contribute to making the experiments less expensive and their results easier to interpret.

Although the advent of mass sequencing technologies might soon change this scenario, correct annotation will still require a considerable research effort for most fish species.

Care must be taken not to over interpret differential transcript gene expression. Thus, some rhabdoviral-induced changes might involve protein cleavages (complement *c3*) and/or protein post-translation modifications (*hmgb1*) in which transcriptional control may not be essential. Although in most reports some of the microarray results were confirmed using RT-qPCR, true confirmation would require estimation of its corresponding protein levels by parallel proteomic studies. Thus, although the differential expression of some fin proteins

Gene names	Zebrafish		Trout	
	*Available microarrays	GenBank	**Available microarrays	GenBank
Interleukin	88	355	47	97
Chemokine	93	367	24	84
Interferon	89	299	48	107
Toll	41	107	8	37
Immunoglobulin	96	2207	98	1234
MHC	13	320	56	411
Vig	0	0	6	14

* vs 3 of 44K oligo microarray of zebrafish (Agilent's ID 26437)
** 37K oligo microarray of trout (Agilents ID 16271) (Salem et al., 2008)
MHC, major histocompatibility complex
Vig, VHSV important genes

Table 4. Comparison of some immune-related genes found in commertially available microarrays with those obtained from GenBank sequences. GenBank at http://www.ncbi.nlm.nih.goc/nuccore

(transferrin, hemopexin, annexin, ATP binding, alpha actin, and kinesin) show a parallel variation with their transcript levels, in most of them, the changes in the differential expression of proteins do not correlate with their corresponding transcript changes (Encinas et al., 2010). This observation suggests that regulation of their expression is not at the transcriptional level, at least in that study. Although correlation of gene and protein expression has been found in some plants (Gallardo et al., 2007; Joosen et al., 2007), most studies found no correlation, including a recent report on individual *E. coli* cells (Taniguchi et al., 2010). Correlation values comparing gene/protein expression levels in several systems are consistently very low (Hack, 2004), suggesting that mRNA levels are poor indicators of the expression of their corresponding protein. Therefore, the study of mRNA levels is justified only when protein levels cannot be detected by the proteomic approach because of their low concentrations or short lives.

Finally, mathematical modelling of microarray data may shed light on gene changes and be useful for testing new hypotheses. From the first symposium held on 2003 (Petrovsky et al., 2003), some progress has been reported on the use of mathematical modelling for early response genes (Lawrence et al., 2007), whole immune responses (Ahmed & Hashish, 2006; Kalita et al., 2006), immunity to infectious diseases, including microarray data (Morel et al., 2006) and future perspectives (Li et al., 2009; Ta'asan & Gandlin, 2009). Mathematical modelling is expected to develop further since there are few other alternatives available to interpret the massive amount of information generated by microarrays.

8. Conclusions

Novirhabdoviroses are among the few fish viral diseases for which efficacious DNA vaccines are available; however, they continue to affect aquacultured fish worldwide.

Despite DNA vacines being commertial in Canada, the actual method of delivery by fish-to-fish intramuscular injection and safety concerns are the major bottle necks to wide acceptance of DNA vaccination. In addition, a complete understanding of the molecular events induced after rhabdoviral fish infection and immunization may contribute to improving DNA vaccines not only for rhabdoviroses but also for other fish infections for which there are no current remedies.

Knowledge about infection, vaccination and adjuvant mechanisms in mammal models, together with high throughput genomic techniques, such as hybridization to microarrays (cDNA or oligo, wide or focused) and new massive sequencing technologies (largely unexplored in fish), offer the opportunity to gather a considerable amount of new transcriptional data in fish models.

Indeed, microarrays have already been used to quantify fish gene expression as well as to discover new genes involved in defense in several fish rhabdovirus models, such as flatfish, salmonid (salmon and trout) and zebrafish.

Genes that show increased transcription after infection (hypothetically signalling internal organs to react against the viral invasion) and also genes whose transcription is inhibited (possibly due to viral shut-off of critical host defences) might help researchers in their quest to identify new adjuvant candidates for fish vaccines.

9. Acknowledgments

This work was supported by CICYT projects AGL08-03519-CO4-ACU, AGL2011-28921-CO3-02 and INGENIO 2010 CONSOLIDER 2007-00002 awarded by the *Ministerio de Ciencia e Innovación* of Spain.

10. References

Acosta F., Collet B., Lorenzen N. & Ellis A. E. (2006). Expression of the glycoprotein of viral haemorrhagic septicaemia virus (VHSV) on the surface of the fish cell line RTG-P1 induces type 1 interferon expression in neighbouring cells. *Fish & Shellfish Immunology*, Vol. 21, No.2, (n.d.), pp. 272-278

Acosta F., Petrie A., Lockhart K., Lorenzen N. & Ellis A. E. (2005). Kinetics of Mx expression in rainbow trout (Oncorhynchus mykiss) and Atlantic salmon (Salmo salar L.) parr in response to VHS-DNA vaccination. *Fish & Shellfish Immunology*, Vol. 18, No. 1, (n.d.), pp. 81-89

Ahmed E. & Hashish A. H. (2006). On modelling the immune system as a complex system. *Theory in Biosciences*, Vol. 124, No. 3-4, (n.d.), pp. 413-418

Anderson E. D., Mourich D. V., Fahrenkrug S. C., LaPatra S. C., Shepherd J. & Leong J. C. (1996a). Genetic immunization of rainbow trout (Oncorhynchus mykiss) against infectious hematopoietic necrosis virus. *Molecular Marine Biology and Biotechnology*, Vol. 5, No. 1, (n.d.), pp. 114-122

Anderson E. D., Mourich D. V. & Leong J. C. (1996b). Gene expression in rainbow trout (Onchorynchus mykiss) following intramuscular injection of DNA. *Molecular Marine Biology and Biotechnology*, Vol. 5, No. 1, (n.d.), pp. 105-113

Aoki T., Hirono I., Kondo H., Hikima J. & Jung T. S. (2011). Microarray technology is an effective tool for identifying genes related to the aquacultural improvement of

Japanese flounder, Paralichthys olivaceus. *Comarative Biochemical Physiology Part D Genomics Proteomics*, Vol. 6, No. 1, (n.d.), pp. 39-43

Aoki T., Nam B. H., Hirono I. I. & Yamamoto E. (1999). Sequences of 596 cDNA Clones (565,977 bp) of Japanese Flounder (Paralichthys olivaceus) Leukocytes Infected with Hirame Rhabdovirus. *Marine Biotechnology (NY)*, Vol. 1, No. 3, (n.d.), pp. 477-0488

Arulanandam B. P., O'Toole M. & Metzger D. W. (1999). Intranasal interleukin-12 is a powerful adjuvant for protective mucosal immunity. *Journal of Infectious Disease*, Vol. 180, No. 4, (n.d.), pp. 940-949

Bearzotti M., Monnier A. F., Vende P., Grosclaude J., DeKinkelin P. & Benmansour A. (1995). The glycoprotein of viral hemorrhagic septicemia virus (VHSV): antigenicity and role in virulence. *Veterinary Research*, Vol. 26, No. 3, (n.d.), pp. 413-422

Bliss J., Maylor R., Stokes K., Murray K. S., Ketchum M. A. & Wolf S. F. (1996). Interleukin-12 as vaccine adjuvant. Characteristics of primary, recall, and long-term responses. *Annual New York Academy Sciences*, Vol. 795, No.1, (n.d.), pp. 26-35

Boudinot P., Blanco M., de Kinkelin P. & Benmansour A. (1998). Combined DNA immunization with the glycoprotein gene of viral hemorrhagic septicemia virus and infectious hematopoietic necrosis virus induces double-specific protective immunity and nonspecific response in rainbow trout. *Virology*, Vol. 249, No. 2, (n.d.), pp. 297-306

Boudinot P., Massin P., Blanco M., Riffault S. & Benmansour A. (1999). vig-1, a new fish gene induced by the rhabdovirus glycoprotein, has a virus-induced homologue in humans and shares conserved motifs with the MoaA family. *Journal of Virology*, Vol. 73, No. 3, (n.d.), pp. 1846-1852

Boudinot P., Salhi S., Blanco M. & Benmansour A. (2001). Viral haemorrhagic septicaemia virus induces vig-2, a new interferon-responsive gene in rainbow trout. *Fish & Shellfish Immunology*, Vol. 11, No. 5, (n.d.), pp. 383-397

Byon J. Y., Ohira T., Hirono I. & Aoki T. (2005). Use of a cDNA microarray to study immunity against viral hemorrhagic septicemia (VHS) in Japanese flounder (Paralichthys olivaceus) following DNA vaccination. *Fish & Shellfish Immunology*, Vol. 18, No. 2, (n.d.), pp. 135-147

Byon J. Y., Ohira T., Hirono I. & Aoki T. (2006). Comparative immune responses in Japanese flounder, Paralichthys olivaceus after vaccination with viral hemorrhagic septicemia virus (VHSV) recombinant glycoprotein and DNA vaccine using a microarray analysis. *Vaccine*, Vol. 24, No. 8, (n.d.), pp. 921-930

Cameron D. A., Gentile K. L., Middleton F. A. & Yurco P. (2005). Gene expression profiles of intact and regenerating zebrafish retina. *Molecular Vision*, Vol. 11, No. 8, (n.d.), pp. 775-791

Chong S. Y., Egan M. A., Kutzler M. A., Megati S., Masood A., Roopchard V., Garcia-Hand D., Montefiori D. C., Quiroz J., Rosati M., Schadeck E. B., Boyer J. D., Pavlakis G. N., Weiner D. B., Sidhu M., Eldridge J. H. & Israel Z. R. (2007). Comparative ability of plasmid IL-12 and IL-15 to enhance cellular and humoral immune responses elicited by a SIVgag plasmid DNA vaccine and alter disease progression following SHIV(89.6P) challenge in rhesus macaques. *Vaccine*, Vol. 25, No. 26, (n.d.), pp. 4967-4982

Craig D. H., Shiratsuchi H. & Basson M. D. (2009). Increased extracellular pressure provides a novel adjuvant stimulus for enhancement of conventional dendritic cell maturation strategies. *Biochemical Biophysical Research Communications*, Vol. 387, No. 1, (n.d.), pp. 174-179

De Gregorio E., D'Oro U. & Wack A. (2009). Immunology of TLR-independent vaccine adjuvants. *Current Opinion on Immunology*, Vol. 21, No. 3, (n.d.), pp. 339-345

delasHeras A. I., Saint-Jean S. R. & Pérez-Prieto S. I. (2010). Immunogenic and protective effects of an oral DNA vaccine against infectious pancreatic necrosis virus in fish. *Fish & Shellfish Immunology*, Vol. 28, No. 4, (n.d.), pp. 562-570

Diab A. M., Williams T. D., Sabine V. S., Chipman J. K. & George S. G. (2008). The GENIPOL European flounder Platichthys flesus L. toxicogenomics microarray: application for investigation of the response to furunculosis vaccination. *Journal of Fish Biology*, Vol. 72, No. 21, (n.d.), pp. 2154-2169

Dios S., Novoa B., Buonocore F., Scapigliati G. & Figueras A (2008). Genomic Resources for Immunology and Disease of Salmonid and Non-Salmonid Fish. *Reviews in Fisheries Science*, Vol. 16, No.1, (n.d.), pp. 119-132

Douglas S. E., Knickle L. C., Williams J., Flight R. M. & Reith M. E. (2008). A first generation Atlantic halibut Hippoglossus hippoglossus (L.) microarray: application to developmental studies. *Journal of Fish Biology*, Vol. 72, No. 21, (n.d.), pp. 2391-2406

Einer-Jensen K., Delgado L., Lorenzen E., Bovo G., Evensen O., Lapatra S. & Lorenzen N. (2009). Dual DNA vaccination of rainbow trout (Oncorhynchus mykiss) against two different rhabdoviruses, VHSV and IHNV, induces specific divalent protection. *Vaccine*, Vol. 27, No. 8, (n.d.), pp. 1248-1253

Encinas P., Rodriguez-Milla M. A., Novoa B., Estepa A., Figueras A. & Coll J. M. (2010). Zebrafish fin immune responses during high mortality infections with viral haemorrhagic septicemia rhabdovirus. A proteomic and transcriptomic approach. *BMC Genomics*, Vol. 11, No. 3, (n.d.), pp. 518-534

Fernandez-Alonso M., Rocha A. & Coll J. M. (2001). DNA vaccination by immersion and ultrasound to trout viral haemorrhagic septicaemia virus. *Vaccine*, Vol. 19, No. 11, (n.d.), pp. 3067-3075

Gallardo K., Firnhaber C., Zuber H., Hericher D., Belghazi M., Henry C., Kuster H. & Thompson R. (2007). A combined proteome and transcriptome analysis of developing Medicago truncatula seeds: evidence for metabolic specialization of maternal and filial tissues. *Molecular Cell Proteomics*, Vol. 6, No. 12, (n.d.), pp. 2165-2179

Garcia-Reyero N., Griffitt R. J., Liu L., Kroll K. J., Farmerie W. G., Barber D. S. & Denslow N. D. (2008). Construction of a robust microarray from a non-model species largemouth bass, Micropterus salmoides (Lacepede), using pyrosequencing technology. *Journal of Fish Biology*, Vol. 72, No. 11 (n.d.), pp. 2354-2376

Goetz F. W. & MacKenzie S. (2008). Functional genomics with microarrays in fish biology and fisheries. *Fish and Fisheries*, Vol. 9, No. 3, (n.d.), pp. 378-395

Gomez-Casado E., Estepa A. & Coll J. M. (2011). A comparative review on European-farmed finfish RNA viruses and their vaccines. *Vaccine*, Vol. 29, No. 15, (n.d.), pp. 2657-2671

Gong Y. F., Xiang L. X. & Shao J. Z. (2009). CD154-CD40 interactions are essential for thymus-dependent antibody production in zebrafish: insights into the origin of costimulatory pathway in helper T cell-regulated adaptive immunity in early vertebrates. *Journal of Immunology*, Vol. 182, No. 12, (n.d.), pp. 7749-7762

Green T. D., Newton B. R., Rota P. A., Xu Y., Robinson H. L. & Ross T. M. (2002). C3d enhancement of neutralizing antibodies to measles hemaglutinin. *Vaccine*, Vol. 20, No. 1, (n.d.), pp. 242-248

Hack C. J. (2004). Integrated transcriptome and proteome data: the challenges ahead. *Brief Functional Genomic Proteomic*, Vol. 3, No. 3, (n.d.), pp. 212-219

Hancock G. E., Heers K. M. & Smith J. D. (2000). QS-21 synergizes with recombinant interleukin-12 to create a potent adjuvant formulation for the fusion protein of respiratory syncytial virus. *Viral Immunology*, Vol. 13, No. 4, (n.d.), pp. 503-509

Harmache A., Leberre M., Droineau S., Giovannini M. & Bremont M. (2006). Bioluminescence Imaging of Live Infected Salmonids Reveals that the Fin Bases Are the Major Portal of Entry for Novirhabdovirus. *Journal of Virology*, Vol. 103, No. 8, (n.d.), pp. 3655-3659

Hirao L. A., Wu L., Khan A. S., Hokey D. A., Yan J., Dai A., Betts M. R., Draghia-Akli R. & Weiner D. B. (2008). Combined effects of IL-12 and electroporation enhances the potency of DNA vaccination in macaques. *Vaccine*, Vol. 26, No. 25, (n.d.), pp. 3112-3120

Iwasaki A. (2007a). Division of labor by dendritic cells. *Cell*, Vol. 128, No. 3, (n.d.), pp. 435-436

Iwasaki A. (2007b). Mucosal dendritic cells. *Annual Review of Immunology*, Vol. 25, No. 2, (n.d.), pp. 381-418

Jacobson M. A., Sinclair E., Bredt B., Agrillo L., Black D., Epling C. L., Carvidi A., Ho T., Bains R., Girling V. & Adler S. P. (2006). Safety and immunogenicity of Towne cytomegalovirus vaccine with or without adjuvant recombinant interleukin-12. *Vaccine*, Vol. 24, No. 25, (n.d.), pp. 5311-5319

Joosen R., Cordewener J., Supena E. D., Vorst O., Lammers M., Maliepaard C., Zeilmaker T., Miki B., America T., Custers J. & Boutilier K. (2007). Combined transcriptome and proteome analysis identifies pathways and markers associated with the establishment of rapeseed microspore-derived embryo development. *Plant Physiology*, Vol. 144, No. 1, (n.d.), pp. 155-172

Jorgensen H. B., Sorensen P., Cooper G. A., Lorenzen E., Lorenzen N., Hansen M. H., Koop B. F. & Henryon M. (2011). General and family-specific gene expression responses to viral hemorrhagic septicaemia virus infection in rainbow trout (Oncorhynchus mykiss). *Molecular Immunology*, Vol. 48, No. 8, (n.d.), pp. 1046-1058

Kalita J. K., Chandrashekar K., Hans R. & Selvam P. (2006). Computational modelling and simulation of the immune system. *International Journal of Bioinformatics and Research Applications*, Vol. 2, No. 1, (n.d.), pp. 63-88

Kim C. H., Johnson M. C., Drennan J. D., Simon B. E., Thomann E. & Leong J. A. (2000). DNA vaccines encoding viral glycoproteins induce nonspecific immunity and Mx protein synthesis in fish. *Journal of Virology*, Vol. 74, No. 15, (n.d.), pp. 7048-7054

Koop B. F., von Schalburg K. R., Leong J., Walker N., Lieph R., Cooper G. A., Robb A., Beetz-Sargent M., Holt R. A., Moore R., Brahmbhatt S., Rosner J., Rexroad C. E., 3rd, McGowan C. R. & Davidson W. S. (2008). A salmonid EST genomic study: genes, duplications, phylogeny and microarrays. *BMC Genomics*, Vol. 9, No. 5, (n.d.), pp. 545-555

Kurath G. (2008). Biotechnology and DNA vaccines for aquatic animals. *Reviews in Science and Technology*, Vol. 27, No. 1, (n.d.), pp. 175-196

Kurath G., Garver K. A., Corbeil S., Elliott D. G., Anderson E. D. & LaPatra S. E. (2006). Protective immunity and lack of histopathological damage two years after DNA vaccination against infectious hematopoietic necrosis virus in trout. *Vaccine*, Vol. 24, No. 3, (n.d.), pp. 345-354

Kurath G., Purcell M. K. & Garver K. A. (2007). Fish rhabdovirus models for understanding host response to DNA vaccines. *CAB reviews*, Vol. 2, No. 1, (n.d.), pp. 1-12

Kurobe T., Yasuike M., Kimura T., Hirono I. & Aoki T. (2005). Expression profiling of immune-related genes from Japanese flounder Paralichthys olivaceus kidney cells using cDNA microarrays. *Developmental Comparative Immunology*, Vol. 29, No. 6, (n.d.), pp. 515-523

Lambrecht B. N., Kool M., Willart M. A. & Hammad H. (2009). Mechanism of action of clinically approved adjuvants. *Current Opinion on Immunology*, Vol. 21, No. 1, (n.d.), pp. 23-29

LaPatra S. E., Barone L., Jones G. R. & Zon L. I. (2000). Effects on infectious hematopoietic necrosis virus and infectious necrosis virus infection on hematopoietic precursosrs of the zebrafish. *Blood Cell Molecular Diseases*, Vol. 26, No. 3, (n.d.), pp. 445-452

LaPatra S. E., Lauda K. A. & Jones G. R. (1994). Antigenic variants of infectious hematopoietic necrosis virus and implications for vaccine development. *Diseases Aquatic Organisms*, Vol. 20, No. 1, (n.d.), pp. 119-126

Lawrence S., Tang Y., Frank M. B., Dozmorov I., Jiang K., Chen Y., Cadwell C., Turner S., Centola M. & Jarvis J. N. (2007). A dynamic model of gene expression in monocytes reveals differences in immediate/early response genes between adult and neonatal cells. *Journal of Inflammation (London)*, Vol. 4, No. 1, (n.d.), pp. 4-14

Li X., Wang Z., Lu T. & Che X. (2009). Modelling immune system: principles, models, analysis and perspectives. *Journal of Bionic Enginneering*, Vol. 6, No. 1, (n.d.), pp. 77-85

Liang S. C., Tan X. Y., Luxenberg D. P., Karim R., Dunussi-Joannopoulos K., Collins M. & Fouser L. A. (2006). Interleukin (IL)-22 and IL-17 are coexpressed by Th17 cells and cooperatively enhance expression of antimicrobial peptides. *Journal of Experimental Medicine*, Vol. 203, No. 10, (n.d.), pp. 2271-2279

Longhi M. P., Trumpfheller C., Idoyaga J., Caskey M., Matos I., Kluger C., Salazar A. M., Colonna M. & Steinman R. M. (2009). Dendritic cells require a systemic type I interferon response to mature and induce CD4+ Th1 immunity with poly IC as adjuvant. *Journal of Experimental Medicine*, Vol. 206, No. 7, (n.d.), pp. 1589-1602

Lorenzen E., Einer-Jensen K., Rasmussen J. S., Kjaer T. E., Collet B., Secombes C. J. & Lorenzen N. (2009). The protective mechanisms induced by a fish rhabdovirus DNA vaccine depend on temperature. *Vaccine*, Vol. 27, No. 15. (n.d.), pp. 3870–3880

Lorenzen E., Einer-Jensen, K., Martinussen, T., LaPatra, S. E. & Lorenzen, N. (2000). DNA vaccination of rainbow trout against viral hemorrhagic septicemia virus: a dose-response and time-course study. *Journal of Aquatic Animal Health*, Vol. 12, No. 1, (n.d.), pp. 167-180

Lorenzen N. & LaPatra S. E. (2005). DNA vaccines for aquacultured fish. *Reviews in Science and Technology of the Office of International Epizooties*, Vol. 24, No. 1, (n.d.), pp. 201-213

Lorenzen N., Lorenzen E., Einer-Jensen K., Heppell J., Wu T. & Davis H. (1998). Protective immunity to VHS in rainbow trout (Oncorhynchus mykiss, Walbaum) following DNA vaccination. *Fish & Shellfish Immunology*, Vol. 8, No. 4, (n.d.), pp. 261-270

Lorenzen N., Lorenzen E., Einer-Jensen K. & LaPatra S. E. (2002). Immunity induced shortly after DNA vaccination of rainbow trout against rhabdoviruses protects against heterologous virus but not against bacterial pathogens. *Developmental Comparative Immunology*, Vol. 26, No. 2, (n.d.), pp. 173-179

MacKenzie S., Balasch J. C., Novoa B., Ribas L., Roher N., Krasnov A. & Figueras A. (2008). Comparative analysis of the acute response of the trout, O. mykiss, head kidney to in vivo challenge with virulent and attenuated infectious hematopoietic necrosis virus and LPS-induced inflammation. *BMC Genomics*, Vol. 9, No. 1, (n.d.), pp. 141-151

Manicassamy S. & Pulendran B. (2009). Modulation of adaptive immunity with Toll-like receptors. *Seminars in Immunology*, Vol. 21, No. 4, (n.d.), pp. 185-193

Martin S. A. M., Collet B., Mackenzie S., Evensen O. & Secombes C. J. (2008). Genomic tools for examining immune gene function in salmonid fish. *Reviews on Fisheries Science*, Vol. 16, No. 1, (n.d.), pp. 112-118

McAllister P. E., Fryer J. L. & Pilcher K. S. (1974). An antigenic comparison between infectious haematopoietic necrosis virus (OSV strain) and the virus of haemorrhagic septicaemia of rainbow trout (Salmo gairdneri) (Denmark strain) by cross neutralization. *Journal of Wildlife Diseases*, Vol. 10, No. 1, (n.d.), pp. 101-103

McLauchlan P. E., Collet B., Ingerslev E., Secombes C. J., Lorenzen N. & Ellis A. E. (2003). DNA vaccination against viral haemorrhagic septicaemia (VHS) in rainbow trout: size, dose, route of injection and duration of protection-early protection correlates with Mx expression. *Fish & Shellfish Immunology*, Vol. 15, No. 1, (n.d.), pp. 39-50

Morel P. A., Ta'asan S., Morel B. F., Kirschner D. E. & Flynn J. L. (2006). New insights into mathematical modeling of the immune system. *Immunology Research*, Vol. 36, No. 1-3, (n.d.), pp. 157-165

Nam B. H., Yamamoto E., Hirono I. & Aoki T. (2000). A survey of expressed genes in the leukocytes of Japanese flounder, Paralichthys olivaceus, infected with Hirame rhabdovirus. *Developmental Comparative Immunology*, Vol. 24, No. 1, (n.d.), pp. 13-24

Nograles K. E., Zaba L. C., Guttman-Yassky E., Fuentes-Duculan J., Suarez-Farinas M., Cardinale I., Khatcherian A., Gonzalez J., Pierson K. C., White T. R., Pensabene C., Coats I., Novitskaya I., Lowes M. A. & Krueger J. G. (2008). Th17 cytokines interleukin (IL)-17 and IL-22 modulate distinct inflammatory and keratinocyte-response pathways. *British Journal of Dermatology*, Vol. 159, No. 5, (n.d.), pp. 1092-1102

Novoa B., Romero A., Mulero V., Rodriguez I., Fernandez I. & Figueras A. (2006). Zebrafish (Danio rerio) as a model for the study of vaccination against viral haemorrhagic septicemia virus (VHSV). *Vaccine*, Vol. 24, No. 31-32, (n.d.), pp. 5806-5816

Pardo B. G., Fernandez C., Millan A., Bouza C., Vazquez-Lopez A., Vera M., Alvarez-Dios J. A., Calaza M., Gomez-Tato A., Vazquez M., Cabaleiro S., Magarinos B., Lemos M. L., Leiro J. M. & Martinez P. (2009). Expressed sequence tags (ESTs) from immune tissues of turbot (Scophthalmus maximus) challenged with pathogens. *BMC Veterinary Research*, Vol. 4, No. 1, (n.d.), pp. 37-47

Petrovsky N., Schonbach C. & Brusic V. (2003). Bioinformatic strategies for better understanding of immune function. *In Silico Biology*, Vol. 3, No. 4, (n.d.), pp. 411-416

Phelan P. E., Pressley M. E., Witten P. E., Mellon M. T., Blake S. & Kim C. H. (2005). Characterization of snakehead rhabdovirus infection in zebrafish (Danio rerio). *Journal of Virology*, Vol. 79, No. 10, (n.d.), pp. 1842-1852

Purcell M. K., Kurath G., Garver K. A., Herwig R. P. & Winton J. R. (2004). Quantitative expression profiling of immune response genes in rainbow trout following infectious haematopoietic necrosis virus (IHNV) infection or DNA vaccination. *Fish & Shellfish Immunology*, Vol. 17, No. 5, (n.d.), pp. 447-462

Purcell M. K., Nichols K. M., Winton J. R., Kurath G., Thorgaard G. H., Wheeler P., Hansen J. D., Herwig R. P. & Park L. K. (2006a). Comprehensive gene expression profiling following DNA vaccination of rainbow trout against infectious hematopoietic necrosis virus. *Molecular Immunology*, Vol. 43, No. 21, (n.d.), pp. 2089-2106

Purcell M. K., Nichols K. M., Winton J. R., Kurath G., Thorgaard G. H., Wheeler P., Hansen J. D., Herwig R. P. & Park L. K. (2006b). Comprehensive gene expression profiling following DNA vaccination of rainbow trout against infectious hematopoietic necrosis virus. *Molecular Immunology*, Vol. 43, No. 13, (n.d.), pp. 2089-2106

Qiu Z., Dillen C., Hu J., Verbeke H., Struyf S., Van Damme J. & Opdenakker G. (2009). Interleukin-17 regulates chemokine and gelatinase B expression in fibroblasts to recruit both neutrophils and monocytes. *Immunobiology*, Vol. 214, No. 9-10, (n.d.), pp. 835-842

Rajcani J., Mosko T. & Rezuchova I. (2005). Current developments in viral DNA vaccines: shall they solve the unsolved? *Reviews in Medical Virology*, Vol. 15, No. 5, (n.d.), pp. 303-325

Robertsen B. (2008). Expression of interferon and interferon-induced genes in salmonids in response to virus infection, interferon-inducing compounds and vaccination. *Fish & Shellfish Immunology*, Vol. 25, No. 4, (n.d.), pp. 351-357

Roche S., Bressanelli S., Rey F. A. & Gaudin Y. (2006). Crystal structure of the low-pH form of the vesicular stomatitis virus glycoprotein G. *Science*, Vol. 313, No. 5784, (n.d.), pp. 187-191

Roche S., Rey F. A., Gaudin Y. & Bressanelli S. (2007). Structure of the prefusion form of the vesicular stomatitis virus glycoprotein G. *Science*, Vol. 315, No. 5813, (n.d.), pp. 843-848

Ross T. M., Xu Y., Green T. D., Montefiori D. C. & Robinson H. L. (2001). Enhanced avidity maturation of antibody to human immunodeficiency virus envelope: DNA vaccination with gp120-C3d fusion proteins. *AIDS Research on Human Retroviruses*, Vol. 17, No. 9, (n.d.), pp. 829-835

Ruiz S., Tafalla C., Cuesta A., Estepa A. & Coll J. M. (2008). In vitro search for alternative promoters to the human immediate early-cytomegalovirus (IE-CMV) to express the G gene of viral haemorrhagic septicemia virus (VHSV) in fish epithelial cells. *Vaccine*, Vol. 26, No. 10, (n.d.), pp. 6620-6629

Salem M., Kenney P. B., Rexroad C. E. & Yao J. (2008). Development of a 37 k high-density oligonucleotide microarray: a new tool for functional genome research in rainbow trout. *Journal of Fish Biology*, Vol. 72, No. 12, (n.d.), pp. 2187-2206

Salonius K., Simard N., Harland R. & Ulmer J. B. (2007). The road to licensure of a DNA vaccine. *Current Opinion in Investigational Drugs*, Vol. 8, No. 6, (n.d.), pp. 635-641

Samuel C. E. (2001). Antiviral Actions of Interferons. *Clinical Microbiological Reviews*, Vol. 14, No. 4, (n.d.), pp. 778-809

Sanders G. E., Batts W. N. & Winton J. R. (2003). Susceptibility of zebrafish (Danio rerio) to a model pathogen, spring viremia of carp virus. *Comparative Medicine*, Vol. 53, No. 5, (n.d.), pp. 514-521

Secombes C. (2008). Will advances in fish immunology change vaccination strategies? *Fish & Shellfish Immunology*, Vol. 25, No. 4, (n.d.), pp. 409-416

Skeen M. J., Miller M. A. & Ziegler H. K. (1996). Interleukin-12 as an adjuvant in the generation of protective immunity to an intracellular pathogen. *Annual New York Academy Sciences*, Vol. 795, No. 4, (n.d.), pp. 416-419

Slierendrecht W. J., Jensen L. B., Horlyck V. & Koch C. (1993). Genetic-Polymorphism of Complement Component C3 in Rainbow-Trout (Oncorhynchus-Mykiss) and Resistance to Viral Hemorrhagic Septicemia. *Fish & Shellfish Immunology*, Vol. 3, No. 1, (n.d.), pp. 199-206

Slierendrecht W. J., Olesen N. J., Lorenzen N., Gottschau A. & Koch C. (1995). Alloforms of complement component C3 and resistance to viral haemorrhagic septicaemia in rainbow trout (Oncorhynchus mykiss). *Aquaculture*, Vol. 137, No. 1, (n.d.), pp. 278-279

Slierendrecht W. J., Olesen N. J., Lorenzen N., Jorgensen P. E. V., Gottschau A. & Koch C. (1996). Genetic alloforms of rainbow trout (Oncorhynchus mykiss) complement component C3 and resistance to viral haemorrhagic septicaemia under experimental conditions. *Fish & Shellfish Immunology*, Vol. 6, No. 2, (n.d.), pp. 235-237

Stevceva L., Moniuszko M. & Ferrari M. G. (2006). Utilizing IL-12, IL-15 and IL-7 as mucosal vaccine adjuvants. *Letters in Drug Design & Discovery*, Vol. 3, No. 4, (n.d.), pp. 586-592

Sullivan C. & Kim C. H. (2008). Zebrafish as a model for infectious disease and immune function. *Fish & Shellfish Immunology*, Vol. 25, No. 2, (n.d.), pp. 341-350

Sunyer J. O., Boshra H. & Li J. (2005). Evolution of anaphylatoxins, their diversity and novel roles in innate immunity: insights from the study of fish complement. *Veterinary Immunology and Immunopathology*, Vol. 108, No. 1-2, (n.d.), pp. 77-89

Ta'asan S. & Gandlin R. (2009). BioLogic: A mathematical modeling framework for immunologists. *Methods in Molecular Biology*, Vol. 500, No. 3, (n.d.), pp. 429-443

Tafalla C., Chico V., Perez L., Coll J. M. & Estepa A. (2007). In vitro and in vivo differential expression of rainbow trout (Oncorhynchus mykiss) Mx isoforms in response to viral haemorrhagic septicaemia virus (VHSV) G gene, poly I:C and VHSV. *Fish & Shellfish Immunology*, Vol. 23, No. 1, (n.d.), pp. 210-221

Tafalla C., Coll J. & Secombes C. J. (2005). Expression of genes related to the early immune response in rainbow trout (Oncorhynchus mykiss) after viral haemorrhagic septicemia virus (VHSV) infection. *Developmental Comparative Immunology*, Vol. 29, No. 7, (n.d.), pp. 615-626

Taggart J. B., Bron J. E., Martin S. A. M., Seear P. J., Hoyheim B., Talbot R., Carmichael S. N., Villeneuve L. A. N., Sweeney G. E., Houlihan D. F., Secombes C. J., Tocher D. R. & Teale A. J. (2008). A description of the origins, design and performance of the

TRAITS-SGP Atlantic salmon Salmo salar L. cDNA microarray. *Journal of Fish Biology*, Vol. 72, No. 15, (n.d.), pp. 2071-2094

Takano T., Iwahori A., Hirono I. & Aoki T. (2004). Development of a DNA vaccine against hirame rhabdovirus and analysis of the expression of immune-related genes after vaccination. *Fish & Shellfish Immunology*, Vol. 17, No. 3, (n.d.), pp. 367-374

Taniguchi Y., Choi P. J., Li G. W., Chen H., Babu M., Hearn J., Emili A. & Xie X. S. (2010). Quantifying E. coli proteome and transcriptome with single-molecule sensitivity in single cells. *Science*, Vol. 329, No. 5991, (n.d.), pp. 533-538

Theofilopoulos A. N., Baccala R., Beutler B. & Kono D. H. (2005). Type I interferons (alpha/beta) in immunity and autoimmunity. *Annual Review of Immunology*, Vol. 23, No. 2, (n.d.), pp. 307-336

Thompson J. M. & Iwasaki A. (2008). Toll-like receptors regulation of viral infection and disease. *Advances in Drug Delivery Reviews*, Vol. 60, No. 7, (n.d.), pp. 786-794

Tian J. Y., Sun X. Q. & Chen X. G. (2008). Formation and oral administration of alginate microspheres loaded with pDNA coding for lymphocystis disease virus (LCDV) to Japanese flounder. *Fish & Shellfish Immunology*, Vol. 24, No. 5, (n.d.), pp. 592-599

Trifari S., Kaplan C. D., Tran E. H., Crellin N. K. & Spits H. (2009). Identification of a human helper T cell population that has abundant production of interleukin 22 and is distinct from T(H)-17, T(H)1 and T(H)2 cells. *Nature Immunology*, Vol. 10, No. 8, (n.d.), pp. 864-871

Vestergaard-Jorgensen P. E. (1972). Egtved virus: antigenic variation in 76 virus isolates examined in neutralization test and by means of the fluorescent antibody technique. *Symposiums of the Zoological Society*, Vol. 30, No. 3, (n.d.), pp. 333-339

Villiers M. B., Villiers C. L., Laharie A. M. & Marche P. N. (1999a). Amplification of the antibody response by C3b complexed to antigen through an ester link. *Journal Immunology*, Vol. 162, No. 6, (n.d.), pp. 3647-3652

Villiers M. B., Villiers C. L., Laharie A. M. & Marche P. N. (1999b). Different stimulating effects of complement C3b and complete Freund's adjuvant on antibody response. *Immunopharmacology*, Vol. 42, No. 1-3, (n.d.), pp. 151-157

von Schalburg K. R., Cooper G. A., Leong J., Robb A., Lieph R., Rise M. L., Davidson W. S. & Koop B. F. (2008). Expansion of the genomics research on Atlantic salmon Salmo salar L. project (GRASP) microarray tools. *Journal of Fish Biology*, Vol. 72, No. 10, (n.d.), pp. 2051-2070

von Schalburg K. R., Rise M. L., Brown G. D., Davidson W. S. & Koop B. F. (2005a). A comprehensive survey of the genes involved in maturation and development of the rainbow trout ovary. *Biological Reproduction*, Vol. 72, No. 6, (n.d.), pp. 687-699

von Schalburg K. R., Rise M. L., Cooper G. A., Brown G. D., Gibbs A. R., Nelson C. C., Davidson W. S. & Koop B. F. (2005b). Fish and chips: various methodologies demonstrate utility of a 16,006-gene salmonid microarray. *BMC Genomics*, Vol. 6, No. 1, (n.d.), pp. 126-136

Winton J. R., Arakawa C. K., Lannan C. N. & Fryer J. L. (1988). Neutralizing monoclonal antibodies recognize antigenic variants among isolates of infectious hematopoietic necrosis. *Diseases Aquatic Organisms*. Vol. 4, No. 1, (n.d.), pp. 199-204

Wright A. K., Briles D. E., Metzger D. W. & Gordon S. B. (2008). Prospects for use of interleukin-12 as a mucosal adjuvant for vaccination of humans to protect against respiratory pneumococcal infection. *Vaccine*, Vol. 26, No. 38, (n.d.), pp. 4893-4903

Yasuike M., Kondo H., Hirono I. & Aoki T. (2007). Difference in Japanese flounder, Paralichthys olivaceus gene expression profile following hirame rhabdovirus (HIRRV) G and N protein DNA vaccination. *Fish & Shellfish Immunology*, Vol. 23, No. 4, (n.d.), pp. 531-541

Yu J. J. & Gaffen S. L. (2008). Interleukin-17: a novel inflammatory cytokine that bridges innate and adaptive immunity. *Frontiers of Bioscience*, Vol. 13, No. 1, (n.d.), pp. 170-177

Zhang Z., Swain T., Bogwald J., Dalmo R. A. & Kumari J. (2009). Bath immunostimulation of rainbow trout (Oncorhynchus mykiss) fry induces enhancement of inflammatory cytokine transcripts, while repeated bath induce no changes. *Fish & Shellfish Immunology*, Vol. 26, No. 5, (n.d.), pp. 677-684

Zheng L. Y., Mou L., Lin S., Lu R. M. & Luo E. J. (2005). Enhancing DNA vaccine potency against hantavirus by co-administration of interleukin-12 expression vector as a genetic adjuvant. *Clinical Medicine Journal (England)*, Vol. 118, No. 4, (n.d.), pp. 313-319

11

Fighting Virus and Parasites with Fish Cytotoxic Cells

M. Ángeles Esteban, José Meseguer and Alberto Cuesta
*Fish Innate Immune System Group,
Department of Cell Biology and Histology,
Faculty of Biology, University of Murcia, Murcia
Spain*

1. Introduction

Cell-mediated cytotoxic (CMC) activity is the main cellular immunological response to kill tumor cells, virus-infected cells and parasites (Groscurth, 1989). In mammalian species this is carried out by several leucocyte populations depending on the non-specific/innate and specific/adaptive immune response. Among the last ones, the CMC activity is carried out by cytotoxic T lymphocytes (CTLs), expressing the co-receptor CD8, after repeated antigen contact and restricted to major histocompatibility complex (MHC) I. Among the innate cytotoxic cells, acting without previous neither sensitization nor MHC I restriction, the most important are the natural killer (NK) cells, which consist on large granular lymphocytes (markers: $CD16/56^{+}CD8^{-}$). However, other cell types such as the lymphokine-activated killer cells (LAK), adherent lymphokine-activated killer cells (ALAK), antibody-dependent cytotoxic cells (ADCC), macrophages, neutrophils and acidophils are also responsible for innate CMC activity (Groscurth, 1989). This CMC activity has been described in all the vertebrate animals with substantial differences. In the case of fish, the first vertebrate group showing both innate and adaptive immune system, they are not an exception. However, deeper studies are needed to clearly understand the appearance and evolution of the fish cytotoxic cells and their activity from an evolutionary point of view. Furthermore, the great potential of aquaculture industry and lack of commercial antiviral and antiparasitic vaccines for fish make necessary to increase the knowledge on the CMC activity of fish.

2. Cell-mediated cytotoxic activity in fish

In all the fish studied so far, different populations of leucocytes from head-kidney (the main haematopoietic tissue in fish), peripheral blood, spleen, thymus, peritoneal exudates or gut display variable cytotoxic activity. The fish innate CMC, not restricted to the MHC, is mainly carried out by the named non-specific cytotoxic cells (NCC), which show great differences at morphological and functional levels between fish species (Carlson et al., 1985; Evans et al., 1984a-d, 1987; Graves et al., 1984). The adaptive cytotoxic activity is restricted to the MHC, shows memory and is formed by CTLs (Fischer et al., 2006; Nakanishi et al., 2002, in press; Somamoto et al., 2000; Verlhac et al., 1990). Most of the data from fish CMC come from the activity against xenogeneic tumor cells but recently the interest to evaluate their potential against viral infections and the generation of proper tools is increasing. One of the main problems associated with the study of the fish immune system, and the CMC in particular, is the lack of proper tools. Most of the studies are based on morpho-functional data but the lack of commercial antibodies is a serious task to definitely identify the leucocyte-types involved. Furthermore, most data obtained in mammalian CMC come from very few species such as human, rat and mouse which show some differences at molecular and cellular levels. However, in fish, the number of evaluated species is larger, including nurse shark (*Ginglymostoma cirratum*), rainbow trout (*Oncorhynchus mykiss*), common carp (*Cyprinus carpio*), tilapia (*Oreochromis niloticus* or *Tilapia mossambica*), channel catfish (*Ictalurus punctatus*), bicolour damselfish (*Stegastes partitus*), Atlantic salmon (*Salmo salar*), Japanese flounder (*Paralichthys olivaceus*), orange-spotted grouper (*Epinephelus coioides*), zebrafish (*Danio rerio*), European sea bass (*Dicentrarchus labrax*) or gilthead seabream (*Sparus aurata*), what greatly increases the variability and difficult the interpretation and correlation between species. However, most of the knowledge comes from the cytotoxic activity against xenogeneic or allogeneic cells and there is few information regarding the role and importance in combating fish virus and parasites.

2.1 Fish innate cytotoxic cells

First evidences showed that head-kidney leucocytes from several freshwater fish (common carp; crucian carp, *Carassius cuvieri*; grass carp, *Ctenopharyngodon idella*; pond loach, *Misgurnus anguillicandatus*; and northern snakehead, *Channa argus*) were cytotoxic towards mammalian cell lines (Hinuma et al., 1980). Afterwards, a series of reports on channel catfish widely described the morphology, biochemical and physical requirements, killing mechanisms, etc. of these leucocytes (Carlson et al., 1985; Evans et al., 1984a-d, 1987; Graves et al., 1984). They showed for the first time that head-kidney catfish have small non-adherent, non-phagocytic and agranular cells displaying the cytotoxic activity, which were catalogued as monocyte-like but also resembled to lymphocytes (Evans et al., 1988). Obviously, they showed different morphological features than the mammalian NK cells, but very similar functional properties. These leucocytes were called non-specific cytotoxic cells (NCC) and are considered phylogenetical precursors of the mammalian NK cells. However, studies since then, including more fish species, have shown that there are many different leucocyte-types displaying the innate CMC activity but sharing the NK cell functions. Thus, the term of fish NCC population should be renamed as NCC activity more than a subpopulation since it is not a discrete and concrete leucocyte type. After that, fish NCCs have been characterized as single or heterogeneous population of leucocytes (Table 1) including lymphocytes, monocyte-macrophages and/or granulocytes (neutrophils and/or acidophils) (Bielek, 1988, 1991; Cammarata et al., 2000; Cuesta et al., 1999; Greenlee et al., 1991; Kurata et al., 1995; McKinney et al., 1986; Meseguer et al., 1994, 1996; Mulero et al., 1994; Ordás et al., 2011; Pettey & McKinney, 1983; Sasaki et al., 2002; Seeley & Weeks-Perkins, 1993).

Though very different in terms of origin and morphology, fish NCCs share the cytotoxic activity and showed the same mechanism as the mammalian NK cells (Groscurth, 1989; Lancki, 1998; Roitt et al., 1996): target recognition and binding, activation and delivery of the lethal hit and finally the target death. In the first step, some membrane molecules have been identified playing a role in the fish CMC. Vimentin-like proteins were identified in the

Fish species	Tissues	Effector cells	References
Ictalurus punctatus	HK, Sp, PBL	Small agranular, non-adherent leucocytes (NCC)	Evans et al., 1984c Evans et al., 1988 Graves et al., 1984
Oncorhynchus mykiss	HK, Th, PBL, Sp	Small agranular monocuclear leucocytes RTS11 cell line	Greenlee et al., 1991 Hayden & Laux, 1985 Moody et al., 1985 Ordás et al., 2011
Salmo salar	HK, PBL, Sp	Small agranular monocuclear leucocytes	Moody et al., 1985
Ginglymostoma cirratum	PBL	Macrophages	McKinney et al., 1986 Pettey & McKinney, 1983
Notemigonus crysoleucas	HK, PBL, Sp	ND	Moody et al., 1985
Stegastes partitus	HK, Sp	ND	McKinney & Schmale, 1994a
Oreochromis sp.	HK, PBL, Sp, PE	Lymphocytes Monocyte-macrophages Granulocytes	Faisal et al., 1989 Jaso-Friedmann & Evans, 1999
Fundulus heteroclitus	HK, Sp	ND	Faisal et al., 1991
Opsanus tau	HK, PBL, Sp, PE	Lymphocytes?	Seeley & Weeks-Perkins, 1993
Cyprinus carpio	HK, Sp, PBL, Th	Lymphocytes Monocyte-macrophages Neutrophils	Bielek 1988, 1991 Kurata et al., 1995
Sparus aurata	HK, PBL, Sp, PE, Th	Lymphocytes Monocyte-macrophages Acidophils	Cuesta et al., 1999 Meseguer et al., 1994, 1996 Mulero et al., 1994
Diecentrarchus labrax	HK, PBL, Sp, PE, Th	Lymphocytes Monocyte-macrophages Neutrophils	Cammarata et al., 2000 Meseguer et al., 1994, 1996 Mulero et al., 1994
Danio rerio	PE	ND	Moss et al., 2009

HK, head-kidney; PBL, peripheral blood leucocytes; Th, thymus; Sp, spleen; PE, peritoneal exudate; ND, not determined.

Table 1. Characteristics of representative fish NCCs.

catfish NCCs and inferred to be important in the recognition and binding to the target cells (Jaso-Friedmann et al., 1993). However, the best characterization of this first step was achieved by the finding of the non-specific cytotoxic cell receptor protein-1 (NCCRP-1) by the generation and selection of a monoclonal antibody (5C6) that completely blocked catfish NCC activity (Evans et al., 1988; Jaso-Friedmann et al., 1988, 2001). This receptor showed important features: 1) the 5C6 antibody recognizes the NCCs of most studied fish and even the mammalian NK and LAK cells, demonstrating its conservation (Cuesta et al., 2005a; Evans et al., 1988; Jaso-Friedmann & Evans, 1999; McKinney & Schmale, 1997); 2) the NCC activity is blocked by the 5C6 antibody (Evans et al., 1988; Iwanowicz et al., 2004; Jaso-Friedmann et al., 1988, 2001); 3) the NCCRP-1 is a 32-34 kDa protein found in the membrane of NCCs and binds to a 42 and 46 kDa from the tumor targets and protozoan that they kill, respectively (Evans et al., 1996; Jaso-Friedmann et al., 1997a, 1997b, 2001; Lester et al., 1994); and 4) it is a type-III membrane protein and its activation led to tyrosine and serine phosphorylation and uses the Jak-STAT signalling pathway (Evans et al., 1999; Jaso-Friedmann et al., 1995, 2001). After binding to the target cell, mammalian NKs and fish NCCs share the same killing mechanisms including granule-dependent (release of perforin and granzymes) and granule-independent (Fas/FasL system) (Cuesta et al., 2003a; Hogan et al., 1999; Jaso-Friedmann et al., 2000; Shen et al., 2002). The release of perforin and granzyme contained in the granules is calcium-dependent and the NCC activity is greatly inhibited or completely abrogated by Ca^{2+}-chelators demonstrating their involvement in the NCC-mediated cytotoxic activity (Carlson et al., 1985; Hogan et al., 1999). In the last decade, fish perforin (Athanasopoulou et al., 2009; Hwang et al., 2004; Toda et al., 2011a) and granzyme (Huang et al., 2010; Praveen et al., 2004, 2006; Wernersson et al., 2006) sequences have been obtained but their gene expression or function has been scarcely related to the innate cytotoxic activity (Ordás et al., 2011; Praveen et al., 2006). The granule-independent killing mechanism has also been identified in fish NCCs by the use of commercial antibodies or functional studies (Ca-chelators) leading to the identification of the Fas/FasL system in fish NCCs (Bishop et al., 2002; Cuesta et al., 2003a; Evans et al., 200, 2001; Jaso-Friedmann et al., 2000; Kaur et al., 2004; Long et al., 2004). After the delivery of the lethal hit, the killing of the target cells occurs by two conserved pathways: necrosis and apoptosis (Cuesta et al., 1999; Meseguer et al., 1994, 1996; Mulero et al., 1994). At the end of the cytotoxic reaction, while NK cells are able to recycle, inactivate or dye (Leibson, 1997) the very few data available in fish NCCs demonstrate that they are unable to recycle and dye by apoptosis after encounter the target cells and kill them (Bishop et al., 2000; Evans et al., 1984a). Finally, it is important to note that in most studies the ratios between fish NCCs and targets is usually higher than when using mammalian NK cells, a fact demonstrated by the very low fish NCC kinetic parameters (V_{max} or K_M) observed (Cuesta et al., 2002a; Evans et al., 1984a). Further characterization of the fish NCCs at molecular and cellular levels will help to elucidate their role in the immune response against virus-infected cells and parasites and the mechanisms involved.

Apart from fish NCCs, other innate cytotoxic cells resembling the mammalian NK cells have been discovered. In catfish peripheral blood leucocytes (PBL), two populations of NK-like cells have been identified: one able to kill allogeneic, but not autologous, cells and the other able to kill virus-infected catfish cells (Hogan et al., 1996, 1999; Shen et al., 2002, 2004; Stuge et al., 1997, 2000; Yoshida et al., 1995). These NK-like cells were able to proliferate after weak alloantigen stimulation and presence of specific growth factors giving to clonal NK-like cells, what has greatly allowed further characterization. First, they morphologically resembled the mammalian NK cells and resulted in large granular lymphocytes, similarly to those previously identified in carp (Bielek, 1988, 1991; Shen et al., 2002, 2004). Second, they were negative for 5C6 antibody and this NCC-marker failed to block the NK-like cells-mediated cytotoxic activity (Shen et al., 2002, 2004). Moreover, they express neither T (T cell receptor -TCR- α, β or γ chains) nor B (immunoglobulin -Ig- chains) lymphocyte markers

(Shen et al., 2002, 2004). Clonal catfish NK cells induced apoptosis in their target cells by means of the calcium-dependent perforin/granzyme-mediated secretory lytic pathway since Ca-chelators completely abolished their cytotoxic activity (Hogan et al., 1999). Moreover, an antibody against leucocyte-function-associated antigen (LFA)-1, which is an adhesion molecule, inhibited the clonal catfish NK-like cell activity (Yoshida et al., 1995). Finally, clonal catfish NK-like cells bound to IgM through an FcμR and exerted an ADCC activity (Shen et al., 2002, 2003, 2004), which has been related to the presence of a similar antibody receptor (CD16) in mammalian NK cells.

Unfortunately, very little is still known about the fish innate receptors involved in the proper recognition of target cells. In mammals, it is widely known the presence of activating and inhibitory NK receptors that mediate the recognition and differentiate between self, normal and altered cells (Bakker et al., 2000). In humans, they belong to the killer immunoglobulin (KIR) or C-type lectin membrane (NKG2/CD94) receptors with either activating (ITAM) or inhibitory (ITIM) intracellular motifs. In fish, orthologs to human KIR and NKG2/CD94 gene receptors have been identified and named novel immune-type receptor (NITR) and KLR, respectively (Litman et al., 2001; Yoder 2004). Functional characterization of these receptors will help to elucidate the innate cytotoxic populations in fish, their regulation and role in disease.

2.2 Specific cytotoxic cells or CTLs

First evidences of the existence of specific cytotoxic cells in fish come from *in vivo* studies of allograft rejection (skin and scales), graft-versus-host reaction or delayed hypersensitivity reaction (DTH) (Manning & Nakanishi 1996; Nakanishi et al., 2002, in press). These experiments showed a great infiltration of lymphocytes and macrophages to the graft site, thymectomy greatly reduced these responses and the second exposure greatly reduced the time of response and increased the fish survival. All together clearly demonstrated the necessity of repeated sensitization and suggested the role of T lymphocytes. Afterwards, with the use of specific antibodies, it has been clearly demonstrated that the infiltrated lymphocytes were T-type, and very recently that were $CD4^+$ (T helper) and $CD8^+$ (T cytotoxic or CTL) (Abelli et al., 1999; Nakanishi et al., in press). However, these aspects are not reviewed here in depth since these concepts do not apply to aquaculture industry.

In vitro studies with PBLs from channel catfish, rainbow trout and ginbuna crucian carp (*Carassius auratus langsdorfii*) have been used as models for fish CTL characterization and have also demonstrated the restriction to the MHC class I (Fischer et al., 2006; Manning & Nakanishi 1996; Nakanishi et al., 2002, in press; Shen et al., 2002; Somamoto et al., 2002). These have demonstrated that immunized fish are able to kill hapten-modified autologous cells, allogeneic cell lines and allogeneic erythrocytes (Fischer et al., 1998; Nakanishi et al., 2002; Verlhac et al., 1990). In the case of channel catfish, the use of mixed leucocyte reactions (MLR) from PBLs and cloning of the cytotoxic effectors resulted in five types of clones (Stuge et al., 1997, 2000). The first type of clones (I) was related to catfish CTLs since they showed the following characteristics: exerted specific cytotoxic activity to the allogeneic cells used for immunization, expressed TCRαβ genes but not the Ig, were large granular lymphocytes and killed their targets by the calcium-dependent perforin/granzyme-mediated secretory lytic pathway (Shen et al., 2002, Stuge et al., 1997,

2000). Unfortunately, it is not known whether they express the CD8αβ co-receptor that will definitely demonstrate that they are CTLs. This approach also produced other type of clones: group II) clones consisting on TCRαβ$^+$ and CD4$^+$ lymphocytes showing non-specific cytotoxic cells and killing the targets by both the perforin/granzyme and Fas/FasL system pathways (Edholm et al., 2007; Stuge et al., 2000; Zhou et al., 2001); group III) alloantigen-specific TCRαβ$^+$ non-cytotoxic cells presumed to be T helper lymphocytes; group IV) TCRαβ$^-$ non-specific cytotoxic cells defined as NK-like cells and described above (Shen et al., 2002, 2004); and group V) TCRαβ$^-$ alloantigen-specific cytotoxic cells presumed to be γδT cells (Zhou et al., 2001). In the model using rainbow trout PBLs, the presence and function of CTLs has been documented thanks to the use of clonal trout effectors and MHC I-matching RTG-2 cell line targets, both sharing the same allele (Dijkstra et al., 2003). Sensitized-rainbow trout showed that only sorted IgM negative (sIgM$^-$) PBLs were able to kill the targets in a specific manner (Fischer et al., 2003, 2006). These data suggested the involvement of trout CTLs that was further evidenced by the up-regulation of TCRα and CD8α genes in these sIgM$^-$ cells after allogeneic cell immunization. The generation of monoclonal antibodies for rainbow trout CD8α has allowed further characterization of this population (Takizawa et al., 2011). Sorted trout CD8α$^+$ lymphocytes showed great expression of perforin or NK-lysin genes (related to the cytotoxic activity, either specific or not), as well as their up-regulation upon stimulation with the T-lymphocyte-mitogen PHA-L. However, further studies are still needed to clearly identify them as the trout specific cytotoxic cells or effectors since tissue distribution and gene expression pattern in CD8α$^+$ cells show some contradictory results and deserve deeper analysis. In the last model, the use of clonal ginbuna crucian carp has been very productive. They firstly proved the existence of specific cytotoxic response against syngeneic virus-infected cells (Somamoto et al., 2000, 2002, 2006). Afterwards, they have purified CD8α$^+$, CD4$^+$, IgM$^+$ and CD8α$^-$CD4$^-$IgM$^-$ leucocytes by means of house-produced antibodies and found that only the CD8α$^+$ population was able to kill the allotargets in a specific manner, what definitely demonstrates the specific cytotoxic activity of fish CTLs (Toda et al., 2009). Moreover, they have also showed that these CTLs mediate the target cell killing by the perforin-mediated pathway since perforin and granzyme B inhibitors abolished almost completely the cytotoxic activity (Toda et al., 2011a, 2011b).

Further studies in other fish species have documented the presence of TCR and CD8 genes indicating presence of CTLs, but functional characterization of the CTL-mediated CMC activity is still lacking. Thus, CD8 genes, alpha or beta chains, have also been sequenced in fugu (*Takifugu rubripes*) (Suetake et al., 2007), Atlantic salmon (*Salmo salar*) (Moore et al., 2005), European sea bass (Buonocore et al., 2006), gilthead seabream (Randelli et al., 2006), Atlantic halibut (*Hippoglossus hippoglossus*) (Patel et al., 2008), common carp (Sun et al., 2007) or orange-spotted grouper (Xu et al., 2011). Unfortunately, CD8α gene might not be the definite CTL marker. In mammals, CTLs are characterized by the presence of the CD8αβ while the expression of the homodimer CD8αα is detected in NK cells, γδT cells and intestinal intraepithelial lymphocytes (Bonneville & Lang, 2002). Thus, unexpected data obtained in the functional characterization of CD8α$^+$-purified lymphocytes could reside in the potential purification of other cells different to CTLs with non-specific activity. However, further studies are needed to clearly ascertain the CTL presence, distribution and role in these fish species, some of them with aquaculture interest.

3. Cytotoxic response against fish tumors

Fish tumors are quite rare in the wild. However, aquaculture management, intensive culture conditions and environmental contamination may increase the incidence of fish tumors. Although some aspects, such as tumour structure and nature, metastasis or lethal effects have been studied, little information exists concerning the involvement of the immune system in protection against tumours (Campbell et al., 2001; McKinney & Schmale, 1994a, 1994b, 1997; Romano & Marozzi, 2004; Schmale et al., 1994, 2004; Thompson & Kostiala, 1990; Vicha & Schmale, 1994). Thus, most of the information regarding fish cytotoxic activity comes from the use of hapten-modified autologous cells or xenogeneic/allogeneic cell lines (Evans et al., 1984a-d, 1987; Fischer et al., 2006; Graves et al., 1984; Manning & Nakanishi, 1996; Nakanishi et al., 2002; Shen et al., 2002; Verlhac et al., 1990). So far, fish immune response against tumors has been slightly evaluated. In the bicolour damselfish naturally suffering of neurofibromatosis (DNF) (caused by a retrovirus), study of the immune response has provided information with respect to CMC activity, morphology and distribution, degranulation of eosinophilic granular cells (EGCs) and lymphocyte proliferation (Vicha & Schmale, 1994; McKinney and Schmale, 1994a, 1994b; Campbell et al. 2001; Schmale et al. 2004). Most of the cytotoxic activity of damselfish leucocytes against DNF-derived target cell lines resided in the spleen whilst in the head-kidney it was quite low. Interestingly, specificity suggested that this activity was likely carried out by CTLs in the spleen and by NCCs in the pronephros (McKinney & Schmale, 1994a). Later, they demonstrated that the 5C6$^-$ lymphocytes showed all the cytotoxic activity against the retrovirus-infected DNF tumor cell lines, suggesting the presence and role of damselfish CTLs, whilst the 5C6$^+$ leucocytes were only able to kill xenogeneic erythrocytes (McKinney & Schmale, 1997). Unfortunately, deeper characterization of this CMC model has been abandoned.

The use of zebrafish as a model for human cancer would also help to understand the fish immune response against tumors, and concretely the role played by cytotoxic cells. As mentioned above, zebrafish showed NCCs in the peritoneal cavity that were positive for the 5C6 antibody and exerted cytotoxic activity against xenogeneic tumor cells (Moss et al., 2009). Moreover, the complete genome sequence allow to identify major molecules involved in the cytotoxic activity such as NCCRP-1, TCR, CD8, perforin, granzymes, Fas/FasL, etc. The easy generation of transgenic zebrafish and mutants would also be a very valuable tool to study the fish CMC activity against tumors. Further studies should focus on the leucocyte infiltration to the tumor site and identification of the potential molecules involved in the activity of the cytotoxic cells.

4. Cytotoxic response against parasites

Fish parasites represent a serious problem in the aquaculture since there are no available vaccines or effective treatments. Whilst some aspects of the fish immune response against parasites have been studied very little is known about the role of the cell-mediated cytotoxic activity (Buchmann et al., 2001; Jones, 2001) First study evaluated the NCC activity in catfish parasitized with *Ichthyophthirius multifiliis* (Graves et al., 1985a). They found that moribund *Ichthyophthirius multifiliis*-infected fish showed decreased NCC activity in the head-kidney against xenogeneic cells when compared to control specimens. Strikingly, this activity was increased in the PBLs of the same fish as consequence of an activation of the

NCC killing capacity and affinity (Graves et a., 1985a). A second study determined that catfish NCC were able to bind and kill 50-60% of *Tetrahymena pyriformis* after 10 h of co-incubation (Graves et al., 1985b). Furthermore, NCC binding to xenogeneic tumor cells and *Ichthyophthirius multifiliis* or *Tetrahymena pyriformis* parasites shared the same antigen, that in the case of parasites, consist on a 46 kDa (Evans et al., 1998a, 1996; Graves et al., 1985a; Jaso-Friedmann et al., 1997b; Lester et al., 1994). In another study, gilthead seabream specimens were parasitized with the enteric *Enteromyxum leei* parasite (Cuesta et al., 2006). This parasitation increased head-kidney NCC activity against tumor cells indicating that parasitized fish posses enhanced cytotoxic cells activity. Moreover, parasite-exposed fish either parasitized or not, showed increased NCC activity. However, no other study has evaluated the role of the cell-mediated cytotoxicity against fish parasites and deserves further evaluation due to the interest for aquaculture industry.

5. Cytotoxic response against viral infections

Viral diseases are responsible for most of the economic losses suffered in modern aquaculture since they produce high levels of mortality and no effective antiviral treatments are available. Moreover, fish farming practices such as growth under very high densities, introduction of species in new areas, continuous transport between hatcheries, nurseries and growing plants are increasing the spread of pathogens and the number of susceptible and reservoir species. However, while most available information focuses on the mechanisms involved in pathogen susceptible fish immune responses, further knowledge is also important in pathogen-reservoir fish systems. Among the major immune mechanisms to kill virus, the interferon (IFN) and the CMC are the most important, but most efforts have only focused on the IFN pathway (Ellis, 2001; Robertsen, 2006). Regarding the CMC activity against virus, this can be mediated by innate or specific cytotoxic cells (Table 2). Regarding the innate CMC activity against virus-infected cells, first studies demonstrated that salmonid kidney, spleen and PBL leucocytes were able to kill infectious pancreatic necrosis virus (IPNV)-infected cells much more than to non-infected cells (Moody et al., 1985; Yoshinaga et al., 1994), and similarly in catfish against channel catfish virus (CCV)-infected cells (Hogan et al., 1996), demonstrating the antiviral activity of fish NCC and NK-like cells, respectively. In the orange-spotted grouper, $CD8^+$ PBLs also showed non-specific cytotoxic activity against nodavirus (nervous necrosis virus or NNV)-infected cells suggesting a role for NK-like or $\gamma\delta$T cells (Chang et al., 2011). Fish exposure to virus also increases the fish innate cytotoxic activity. Thus, gilthead seabream injected with viral hemorrhagic septicemia virus (VHSV), which did not replicate at the assayed conditions, increased the NCC activity, demonstrating the importance of studying the antiviral immune response in reservoir fish species (Esteban et al., 2008). Moreover, NNV-infection increased the NCC activity of head-kidney leucocytes from 1 to 15 days post-injection in both gilthead seabream and European sea bass (unpublished data). Recently, we have also demonstrated that trout RTS11 (monocyte-macrophage cell line) cells exposed to VHSV increased their cytotoxic activity against xenogeneic tumor cells and up-regulated the NKEF (natural killer enhancing factor), granzyme and perforin gene expression whilst trout head-kidney leucocyte infection with the VHSV increased the innate cytotoxic activity but failed to significantly change the expression of these genes (Ordás et al., 2011).

Fish	CMC activity	References
Channel catfish	NK-like activity against CCV-infected cells	Hogan et al., 1996
Atlantic salmon	CMC activity against IPNV-infected cells	Moody et al., 1985
Rainbow trout	CMC activity against IPNV-infected cells	Moody et al., 1985 Yoshinaga et al., 1994
	VHSV infection induced innate CMC activity, up-regulated NKEF, CD8α, perforin and granzyme genes	Cuesta & Tafalla, 2009 Utke et al., 2007 Unpublished data
	VHSV infection elicited specific CMC activity, up-regulated CD8 α gene	Fischer et al., 2006 Utke et al., 2007
	VHSV DNA vaccine elicited specific CMC activity	Utke et al., 2008
	VHSV and IPNV DNA vaccines up-regulated NKEF, perforin and granzyme genes	Cuesta & Tafalla, 2009 Cuesta et al., 2010 Unpublished data
	VHSV infection of RTS11 cells increased the CMC activity, up-regulated NKEF, granzyme and perforin genes	Ordás et al., 2011
Ginbuna crucian carp	IPNV or EVA infection elicited specific CMC activity	Somamoto et al., 2000
	CHNV infection elicited specific CMC activity, up-regulated TCRβ and CD8α genes	Somamoto et al., 2002 Somamoto et al., 2006
	Generation of *in vitro* virus-specific CTLs and up-regulation of TCRβ and CD8 α genes	Somamoto et al., 2009
	Anal immunization with CHNV-infected cells elicited specific CMC activity	Sato & Okamoto, 2010
Common carp	SVCV infection up-regulated, granzyme A/K or CD8α genes	Forlenza et al., 2008 Huang et al., 2010
Gilthead seabream	NCC activity induced by VHSV injection	Esteban et al., 2008
	NCC activity induced by NNV infection	Unpublished data
Sea bass	NNV infection no affected TCRβ and CD8α genes	Scapigliati et al., 2010
Atlantic halibut	NNV infection no affected CD8α and CD8β genes	Patel et al., 2008
Orange-spotted grouper	CMC activity against NNV- or RSIV-infected cells	Chang et al., 2011
	NNV infection elicited specific CMC activity, increased CD8α$^+$ cells and CD8α gene	Chang et al., 2011
Japanese flounder	VHSV infection up-regulated CD8 gene	Byon et al., 2005
	VHSV DNA vaccine up-regulated CD8 gene	Byon et al., 2006

CMC, cell-mediated cytotoxicity; CCV, channel catfish virus; IPNV, infectious pancreatic necrosis virus; VHSV, viral hemorrhagic septicaemia virus; EVA, eel virus from America; CHNV, crucian carp haematopoietic virus; RSIV, red seabream iridovirus; SVCV, spring viremia carp virus; NNV, nervous necrosis virus; CTL, cytotoxic T lymphocytes; TCR, T cell receptor; NKEF, natural killer enhancing factor.

Table 2. Major studies evaluating the fish CMC activity against virus.

Viral infections also elicited the specific immune response by inducing antibody production and CTL activity (Table 2) (Nakanishi et al., in press). First studies demonstrated that isogeneic ginbuna crucian carp elicited CTL activity against virus. Thus, ginbuna crucian carps immunized with hematopoietic necrosis virus (CHNV) specifically killed CHNV-infected syngeneic cells in a viral antigen and MHC I-restricted manner (Somamoto et al., 2000, 2002), increased the TCRβ and CD8α gene expression (Somamoto et al., 2006) and helped to establish virus-dependent CTL clones *in vitro* (Somamoto et al., 2009). In rainbow trout, infection with VHSV greatly elicited specific- and MHC I-matched cytotoxic cells but a non-specific and MHC I-mismatched cytotoxic activity was also found (Fischer et al., 2006; Utke et al., 2007). Surprisingly, they found that specific CMC activity mediated by CTLs was produced much earlier than the innate activity, in sharp contrast to all the information at this respect. Strikingly, the NKEF gene expression followed the same time-profile than the CTL activity but in the case of CD8α was opposite, adding more controversy to these data (Utke et al., 2007). Furthermore, trout vaccination with VHSV DNA vaccines also elicited CMC activity against MHC I-matched infected cells, suggesting a role for CTLs (Utke et al., 2008). However, they also found a bit lower CMC activity against non-matching-infected cells or cells infected with a different virus, suggesting a role for NCCs or even the ADCC activity since these fish showed high antibody levels, but this has not been confirmed. In other studies, VHSV infection increased the trout NKEF and CD8α gene *in vivo* but failed to modulate the NKEF, perforin and granzyme genes *in vitro* (Cuesta & Tafalla, 2009; Ordás et al., 2011). VHSV and IPNV DNA vaccination also up-regulated the trout CD8α, perforin and granzyme gene expression (Cuesta et al., 2010; unpublished data), giving more consistency to the involvement of CMC activity against viral infections and its activation by DNA vaccines. Moreover, oral vaccination with inactivated CHNV elicited specific CMC activity that resulted viral antigen-specific and restricted to the MHC I (Sato & Okamoto, 2010). In the orange-spotted grouper, nodavirus infection also elicited a CTL response when viral antigens were properly presented by MHC I receptors, as well as increased the CD8α expression at gene and CTL surfaces (Chang et al., 2011). This study represents the first one demonstrating the CTL role against viral infection in marine fish species with great interest for aquaculture industry. Further studies would help to understand the CMC activity against viral infections and to design and probe viral vaccines.

6. Modulation of the fish cytotoxic activity

Fish CMC activity regulation has been widely evaluated and mostly focused on NCC modulation. Fish NCC activity has been shown to be modulated by several chemicals, cytokines, environmental contaminants, stress factors, immunostimulants, etc. First studies dealt with the NCC inhibition by blocking the binding to target in order to characterize the role of NCCRP-1, or inhibiting the killing mechanisms in order to evaluate the perforin- or Fas/FasL-mediated lytic pathway (Bishop et al., 2002; Carlson et al., 1985; Evans et al., 1988, 2000; Hogan et al., 1999; Iwanowicz et al., 2004; Jaso-Friedmann et al., 1988, 2001; Kaur et al., 2004; Shen et al., 2002). Further studies demonstrated that catfish NCC activity is increased by leucocyte treatment with ionophore A23187, A23187 plus phorbol myristate acetate (PMA) or vanadate but no with PMA alone or poly I:C (a mimic for viral infections) (Evans et al., 1984b, 1990, 1998b). Moreover, serum from stressed fish contained cytokine-like factors able to increase the tilapia NCC activity suggesting a role for FasL (Jaso-Friedmann et al., 2000; Ruiz et al., 2001). Fish NCC activity is also increased by bacterial infections: *Edwardsiella ictaluri* in channel catfish (Evans et al., 1998b), *Aeromonas salmonicida* in brook trout (*Salvelinus fontinalis*) (Dautremepuits et al., 2006) or *Streptococcus iniae* in tilapia (Taylor et al., 2001). In our lab, we have been investigating the immunostimulatory role of many substances and conditions in the gilthead seabream, one of the most important farmed species in the marine aquaculture. This has allowed us to get a lot of information about the regulation of the seabream immune response, and concretely the NCC activity. Thus, we have shown *in vitro* and/or *in vivo* modulation of seabream NCC activity by vitamins C (Cuesta et al., 2002b), E (Cuesta et al., 2001), A (Cuesta et al., 2003b) and D3 (Cerezuela et al., 2009), chitin (Cuesta et al., 2003c; Esteban et al., 2000, 2001), levamisole (Cuesta et al., 2002c), lactoferrin (Esteban et al., 2005), melatonin (Cuesta et al., 2008a), propolis (Cuesta et al., 2005b), inulin (Cerezuela et al., 2008), unmethylated oligodeoxynucleotides (ODNs) containing cytosine-phosphodiester-guanosine (CpG) motifs (Cuesta et al., 2008b, 2008c), probiotic bacteria (Díaz-Rosales et al., 2006; Salinas et al., 2005, 2006, 2008), yeast (Cuesta et al., 2007; Ortuño et al., 2002; Reyes-Becerril et al., 2008; Rodríguez et al., 2003), fungi (Rodríguez et al., 2004), virus (Esteban et al., 2008), environmental contaminants (p,p´-DDE and lindane) (Cuesta et al., 2008d) or stress factors (air exposure, crowding and anaesthetics) (Cuesta et al., 2003d). In general, we have demonstrated great NCC increments after these treatments. Moreover, we have also observed that NCC activity reached the greatest activation, compared to other innate cellular immune responses such as phagocytosis or respiratory burst activity, and did at shorter treatment times and lower dosages. Unfortunately, most of this information has been obtained evaluating the NCC activity against xenogeneic tumor cells and whether this is correlated to the *in vivo* activity against viral infections deserves further investigation. In this sense, few recent studies have correlated the stimulatory role of immunostimulants with an increased viral disease resistance. Thus, probiotic-supplemented diets resulted in reduced mortality of Japanese flounder specimens exposed to lymphocystis disease virus (LCDV) (Harikrishnan et al., 2010) whilst feeding of shrimp with immunostimulant herbs reduced their mortality upon viral disease (Citarasu et al., 2006). Further characterization of the beneficial immunostimulants against viral diseases is needed to control the virus spreading and lethal effects.

Apart from the direct activation of fish cytotoxic activity, the expression of some CMC-related genes (NCCRP-1, CD8, perforin, granzyme, etc.) is also modulated (Table 2), suggesting an increase in the CMC activity. First, the NCCRP-1 gene expression was altered after bacterial infection (Reyes-Becerril et al., 2011; Sakata et al., 2005), administration of immunostimulants (Cuesta et al., 2008b, 2008d; Lazado et al., ; Reyes-Becerril et al., 2008), exposure to contaminants (Cuesta et al., 2008d) or bacterial vaccination (Caipang et al., 2008), depending on the fish species, tissue, time and dose of exposure, and suggests a parallel effect of fish NCC activity. Perforin gene expression is usually up-regulated after immunization of ginbuna crucian carp with tumor cells (Toda et al. 2011a), after PHA-L (*Phaseolus vulgaris* leucoagglutinin) stimulation of trout CD8α^+ cells (Takizawa et al., 2011), after VHSV infection of RTS11 cell line (Ordás et al., 2011) and after viral infection or DNA vaccination in rainbow trout (unpublished data), whilst down-regulated after cadmium exposure (Auslander et al., 2008). In a similar fashion, granzyme genes are up-regulated by bacterial vaccination (Caipang et al., 2008), viral infections (VHSV in RTS11 cell line and SVCV in carp) (Huang et al., 2010; Ordás et al., 2011) and viral infection or DNA vaccination (unpublished data). The transcript level of CD8 gene is related to the CTL presence, abundance and activity. Thus, fish CD8 transcripts are up-regulated by viral and bacterial infections, viral DNA vaccines, scale grafts, poly I:C or mitogens (Byon et al., 2005 2006; Cuesta et al., 2010; Cuesta & Tafalla, 2009; Forlenza et al., 2008; Overturf & LaPatra, 2006; Somamoto et al., 2005, 2006; Utke et al., 2007; Xu et al., 2011). In some studies, these CD8 gene levels have been correlated with increased CTL activity. Finally, other genes related to the cytotoxic activity have received less attention. In this category, the natural killer enhancing factor (NKEF), which increase the cytotoxic activity in humans but its role is unknown in fish, is up-regulated by viral infections and DNA vaccines (Cuesta & Tafalla, 2009; Ordás et al., 2011; Utke et al., 2007) while granulysin, which is secreted together to granzymes and lyses target cells, gene is up-regulated in $CD8^{+}$ lymphocytes by mitogen stimulation (Takizawa et al., 2011). Further studies are needed to clearly state the gene expression with either innate or specific cytotoxic activity in fish. Future development of more molecular tools will help to elucidate this fascinating and complex immune response.

7. Future directions

As summarized above, fish posses a wide range of cytotoxic cells with killing activity against tumor cells, virus-infected cells and parasites. Further studies in the future should identify, describe and characterize the cytotoxic cells and mechanisms in the most cultured fish species and those susceptible to be farmed in the future. Another issue is the generation of molecular tools to evaluate the fish CMC and clearly identify the function of NCCs, NK-like and CTLs as well as assay models such as clonal fish, cytotoxic cell clones or MHC I-paired effector and targets (virally infected or not). These tools will also help to design powerful and safe vaccines against problematic virus and parasites for fish aquaculture. Finally, these studies have also to be applied to marine fish, which culture is continuously increasing because of the human demand and high economic value.

8. Glossary

ADCC	Antibody-dependent cytotoxic cells
ALAK	Lymphokine-activated killer cells
CCV	Channel catfish virus
CD4+	T helper lymphocyte
CD8+	T cytotoxic lymphocyte or CTL
CHNV	Crucian carp haematopoietic virus
CMC	Cell-mediated cytotoxicity
CpG	Cytosine-phosphodiester-guanosine
CTLs	Cytotoxic T lymphocytes
DNA	Deoxyribonucleic acid
DNF	Damselfish neurofibromatosis
DTH	Delayed hypersensitivity reaction
EGCs	Eosinophilic granular cells
EVA	Eel virus from America
HK	Head-kidney
IgM	Immunoglobulin M
IPNV	Infectious pancreatic necrosis virus

ITAM	Activating intracellular motifs
ITIM	Inhibitory intracellular motifs
Jak	Janus kinase
KIR	Killer immunoglobulin
LAK	Lymphokine-activated killer cells
LCDV	Lymphocystis disease virus
LFA-1	Leucocyte-function-associated antigen-1
MHC	Major histocompatibility complex
MLR	Mixed leucocyte reaction
NCC	Non-specific cytotoxic cells
NCCRP-1	non-specific cytotoxic cell receptor protein-1
NITR	Novel immune-type receptor
NK	Natural killer
NKEF	Natural killer enhancing factor
NKG2/CD94	C-type lectin membrane receptors
NNV	Nervous necrosis virus
ODNs	Unmethylated oligodeoxynucleotides
PBL	Peripheral blood leucocytes
PE	Peritoneal exudate
PHA-L	Phaseolus vulgaris leucoagglutinin
PMA	Phorbol myristate acetate
RSIV	Red seabream iridovirus
RTG-2	Rainbow trout gonad cell line
Sp	Spleen
STAT	Signal Transducer and Activator of Transcription
SVCV	Spring viremia carp virus
TCR	T cell receptor
Th	Thymus
VHSV	Viral hemorrhagic septicaemia virus

9. Acknowledgments

This work has been funded by national (AGL2008-05119-C02-01 and AGL2010-20801-C02-02) and regional projects (04538/GERM/06). A. Cuesta thanks to the *Ministerio de Ciencia e Innovación* for the *Ramón y Cajal* contract.

10. References

Abelli, L.; Baldassini, M.R.; Mastrolia, L. & Scapigliati, G. (1999). Immunodetection of lymphocyte subpopulations involved in allograft rejection in a teleost, *Dicentrarchus labrax* (L.). *Cellular Immunology*, Vol.191, No.2, pp. 152-160, ISSN 0008-8749.

Athanasopoulou, S.; Marioli, D.; Mikrou, A.; Papanastasiou, A.D. & Zarkadis, I.K. (2009). Cloning and characterization of the trout perforin. *Fish & Shellfish Immunology*, Vol.26, No.6, pp. 908-912, ISSN 1095-9947.

Auslander, M.; Yudkovski, Y.; Chalifa-Caspi, V.; Herut, B.; Ophir, R.; Reinhardt, R.; Neumann, P.M. & Tom, M. (2008). Pollution-affected fish hepatic transcriptome and its expression patterns on exposure to cadmium. *Marine Biotechnology (NY)*, Vol.10, No.3, pp. 250-261, ISSN 1436-2228.

Bakker, A.B.; Wu, J.; Phillips, J.H. & Lanier, L.L. (2000). NK cell activation: distinct stimulatory pathways counterbalancing inhibitory signals. *Human Immunology*, Vol.61, No.1, pp. 18-27, ISSN 0198-8859.

Bielek, E. (1988). Ultrastructural analysis of leucocyte interaction with tumour targets in a teleost, *Cyprinus carpio* L. *Developmental and Comparative Immunology*, Vol.12, No.4, pp. 809-821, ISSN 0145-305X.

Bielek, E. (1991). Morphological differentiation on presumptive "natural killer" cells in a teleost (*Cyprinus carpio* L.). *Fischerei-Forschung Rostock*, Vol.29, pp. 58-60.

Bishop, G.R.; Jaso-Friedmann, L. & Evans, D.L. (2000). Activation-induced programmed cell death of nonspecific cytotoxic cells and inhibition by apoptosis regulatory factors. *Cellular Immunology*, Vol.199, No.2, pp. 126-137, ISSN 0008-8749.

Bishop, G.R.; Taylor, S.; Jaso-Friedmann, L. & Evans, D.L. (2002). Mechanisms of nonspecific cytotoxic cell regulation of apoptosis: cytokine-like activity of Fas ligand. *Fish & Shellfish Immunology*, Vol.13, No.1, pp. 47-67, ISSN 1050-4648.

Bonneville, M. & Lang, F. (2002). CD8: from coreceptor to comodulator. *Nat Immunol*, Vol.3, No.1, pp. 12-14, ISSN 1529-2908.

Buchmann, K.; Sigh, J.; Nielsen, C.V. & Dalgaard, M. (2001). Host responses against the fish parasitizing ciliate *Ichthyophthirius multifiliis*. *Vet Parasitol*, Vol.100, No.1-2, pp. 105-116, ISSN 0304-4017.

Buonocore, F.; Randelli, E.; Bird, S.; Secombes, C.J.; Costantini, S.; Facchiano, A.; Mazzini, M. & Scapigliati, G. (2006). The CD8alpha from sea bass (*Dicentrarchus labrax* L.): Cloning, expression and 3D modelling. *Fish & Shellfish Immunology*, Vol.20, No.4, pp. 637-646, ISSN 1050-4648.

Byon, J.Y.; Ohira, T.; Hirono, I. & Aoki, T. (2005). Use of a cDNA microarray to study immunity against viral hemorrhagic septicemia (VHS) in Japanese flounder (*Paralichthys olivaceus*) following DNA vaccination. *Fish & Shellfish Immunology*, Vol.18, No.2, pp. 135-147, ISSN 1050-4648.

Byon, J.Y.; Ohira, T.; Hirono, I. & Aoki, T. (2006). Comparative immune responses in Japanese flounder, Paralichthys olivaceus after vaccination with viral hemorrhagic septicemia virus (VHSV) recombinant glycoprotein and DNA vaccine using a microarray analysis. *Vaccine*, Vol.24, No.7, pp. 921-930, ISSN 0264-410X.

Caipang, C.M.; Hynes, N.; Puangkaew, J.; Brinchmann, M.F. & Kiron, V. (2008). Intraperitoneal vaccination of Atlantic cod, *Gadus morhua* with heat-killed *Listonella anguillarum* enhances serum antibacterial activity and expression of immune response genes. *Fish & Shellfish Immunology*, Vol.24, No.3, pp. 314-322, ISSN 1050-4648.

Cammarata, M.; Vazzana, M.; Cervello, M.; Arizza, V. & Parrinello, N. (2000). Spontaneous cytotoxic activity of eosinophilic granule cells separated from the normal peritoneal cavity of *Dicentrarchus labrax*. *Fish & Shellfish Immunology*, Vol.10, No.2, pp. 143-154, ISSN 1050-4648.

Campbell, C.E.; Gibbs, P.D. & Schmale, M.C. (2001). Progression of infection and tumor development in damselfish. *Marine Biotechnology (NY)*, Vol.3, No.Supplement 1, pp. S107-114, ISSN 1436-2228.

Carlson, R.L.; Evans, D.L. & Graves, S.S. (1985). Nonspecific cytotoxic cells in fish (*Ictalurus punctatus*). V. Metabolic requirements of lysis. *Developmental and Comparative Immunology*, Vol.9, No.2, pp. 271-280, ISSN 0145-305X.

Cerezuela, R.; Cuesta, A.; Meseguer, J. & Esteban, M.A. (2008). Effects of inulin on gilthead seabream (*Sparus aurata* L.) innate immune parameters. *Fish & Shellfish Immunology*, Vol.24, No.5, pp. 663-668, ISSN 1050-4648.

Cerezuela, R.; Cuesta, A.; Meseguer, J. & Esteban, M.A. (2009). Effects of dietary vitamin D3 administration on innate immune parameters of seabream (*Sparus aurata* L.). *Fish & Shellfish Immunology*, Vol.26, No.2, pp. 243-248, ISSN 1095-9947.

Citarasu, T.; Sivaram, V.; Immanuel, G.; Rout, N. & Murugan, V. (2006). Influence of selected Indian immunostimulant herbs against white spot syndrome virus (WSSV) infection in black tiger shrimp, *Penaeus monodon* with reference to haematological, biochemical and immunological changes. *Fish & Shellfish Immunology*, Vol.21, No.4, pp. 372-384, ISSN 1050-4648.

Chang, Y.T.; Kai, Y.H.; Chi, S.C. & Song, Y.L. (2011). Cytotoxic CD8alpha+ leucocytes have heterogeneous features in antigen recognition and class I MHC restriction in grouper. *Fish & Shellfish Immunology*, Vol.30, No.6, pp. 1283-1293, ISSN 1095-9947.

Cuesta, A.; Esteban, M.A. & Meseguer, J. (1999). Natural cytotoxic activity of gilthead seabream (*Sparus aurata* L.) leucocytes. Assessment by flow cytometry and microscopy. *Veterinary Immunology and Immunopathology*, Vol.71, No.3-4, pp. 161-171, ISSN 0165-2427.

Cuesta, A.; Esteban, M.A.; Ortuño, J. & Meseguer, J. (2001). Vitamin E increases natural cytotoxic activity in seabream (Sparus aurata L.). *Fish & Shellfish Immunology*, Vol.11, No.4, pp. 293-302, ISSN 1050-4648.

Cuesta, A.; Esteban, M.A. & Meseguer, J. (2002a). Natural cytotoxic cells of gilthead seabream: maximum percentage of lysis. *Fish & Shellfish Immunology*, Vol.12, No.2, pp. 111-118, ISSN 1050-4648.

Cuesta, A.; Esteban, M.A. & Meseguer, J. (2002b). Natural cytotoxic activity in seabream (*Sparus aurata* L.) and its modulation by vitamin C. *Fish & Shellfish Immunology*, Vol.13, No.2, pp. 97-109, ISSN 1050-4648.

Cuesta, A.; Esteban, M.A. & Meseguer, J. (2002c). Levamisole is a potent enhancer of gilthead seabream natural cytotoxic activity. *Veterinary Immunology and Immunopathology*, Vol.89, No.3-4, pp. 169-174, ISSN 0165-2427.

Cuesta, A.; Esteban, M.A. & Meseguer, J. (2003a). Identification of a FasL-like molecule in leucocytes of the teleost fish gilthead seabream (*Sparus aurata* L.) *Developmental and Comparative Immunology*, Vol.27, No.1, pp. 21-27, ISSN 0145-305X.

Cuesta, A.; Esteban, M.A. & Meseguer, J. (2003b). Tumouricidal activity of gilthead seabream (*Sparus aurata* L.) natural cytotoxic cells: the role played *in vitro* and *in vivo* by retinol acetate. *Fish & Shellfish Immunology*, Vol.14, No.2, pp. 133-144, ISSN 1050-4648.

Cuesta, A.; Esteban, M.A. & Meseguer, J. (2003c). *In vitro* effect of chitin particles on the innate cellular immune system of gilthead seabream (*Sparus aurata* L.). *Fish & Shellfish Immunology*, Vol.15, No.1, pp. 1-11, ISSN 1050-4648.

Cuesta, A.; Esteban, M.A. & Meseguer, J. (2003d). Effects of different stressor agents on gilthead seabream natural cytotoxic activity. *Fish & Shellfish Immunology*, Vol.15, No.5, pp. 433-441, ISSN 1050-4648.

Cuesta, A.; Esteban, M.A. & Meseguer, J. (2005a). Molecular characterization of the nonspecific cytotoxic cell receptor (NCCRP-1) demonstrates gilthead seabream NCC heterogeneity. *Developmental and Comparative Immunology*, Vol.29, No.7, pp. 637-650, ISSN 0145-305X.

Cuesta, A.; Rodríguez, A.; Esteban, M.A. & Meseguer, J. (2005b). *In vivo* effects of propolis, a honeybee product, on gilthead seabream innate immune responses. *Fish & Shellfish Immunology*, Vol.18, No.1, pp. 71-80, ISSN 1050-4648.

Cuesta, A.; Salinas, I.; Rodríguez, A.; Muñoz, P.; Sitjà-Bobadilla, A.; Álvarez-Pellitero, P.; Meseguer, J. & Esteban, M.A. (2006). Cell-mediated cytotoxicity is the main innate immune mechanism involved in the cellular defence of gilthead seabream (Teleostei: Sparidae) against *Enteromyxum leei* (Myxozoa). *Parasite Immunology*, Vol.28, No.12, pp. 657-665, ISSN 0141-9838.

Cuesta, A.; Rodríguez, A.; Salinas, I.; Meseguer, J. & Esteban, M.A. (2007). Early local and systemic innate immune responses in the teleost gilthead seabream after intraperitoneal injection of whole yeast cells. *Fish & Shellfish Immunology*, Vol.22, No.3, pp. 242-251, ISSN 1050-4648.

Cuesta, A.; Cerezuela, R.; Esteban, M.A. & Meseguer, J. (2008a). *In vivo* actions of melatonin on the innate immune parameters in the teleost fish gilthead seabream. *J Pineal Res*, Vol.45, No.1, pp. 70-78, ISSN 1600-079X.

Cuesta, A.; Esteban, M.A. & Meseguer, J. (2008b). The expression profile of TLR9 mRNA and CpG ODNs immunostimulatory actions in the teleost gilthead seabream points to a major role of lymphocytes. *Cellular and Molecular Life Sciences*, Vol.65, No.13, pp. 2091-2104, ISSN 1420-682X.

Cuesta, A.; Salinas, I.; Esteban, M.A. & Meseguer, J. (2008c). Unmethylated CpG motifs mimicking bacterial DNA triggers the local and systemic innate immune parameters and expression of immune-relevant genes in gilthead seabream. *Fish & Shellfish Immunology*, Vol.25, No.5, pp. 617-624, ISSN 1050-4648.

Cuesta, A.; Meseguer, J. & Esteban, M.A. (2008d). Effects of the organochlorines p,p'-DDE and lindane on gilthead seabream leucocyte immune parameters and gene expression. *Fish & Shellfish Immunology*, Vol.25, No.5, pp. 682-688, ISSN 1050-4648.

Cuesta, A. & Tafalla, C. (2009). Transcription of immune genes upon challenge with viral hemorrhagic septicemia virus (VHSV) in DNA vaccinated rainbow trout (*Oncorhynchus mykiss*). *Vaccine*, Vol.27, No.2, pp. 280-289, ISSN 0264-410X.

Cuesta, A.; Chaves-Pozo, E.; de Las Heras, A.I.; Saint-Jean, S.R.; Pérez-Prieto, S. & Tafalla, C. (2010). An active DNA vaccine against infectious pancreatic necrosis virus (IPNV) with a different mode of action than fish rhabdovirus DNA vaccines. *Vaccine*, Vol.28, No.19, pp. 3291-3300, ISSN 1873-2518.

Dautremepuits, C.; Fortier, M.; Croisetiere, S.; Belhumeur, P. & Fournier, M. (2006). Modulation of juvenile brook trout (*Salvelinus fontinalis*) cellular immune system after *Aeromonas salmonicida* challenge. *Veterinary Immunology and Immunopathology*, Vol.110, No.1-2, pp. 27-36, ISSN 0165-2427.

Díaz-Rosales, P.; Salinas, I.; Rodríguez, A.; Cuesta, A.; Chabrillón, M.; Balebona, M.C.; Moriñigo, M.A.; Esteban, M.A. & Meseguer, J. (2006). Gilthead seabream (*Sparus aurata* L.) innate immune response after dietary administration of heat-inactivated potential probiotics. *Fish & Shellfish Immunology*, Vol.20, No.4, pp. 482-492, ISSN 1050-4648.

Dijkstra, J.M.; Kollner, B.; Aoyagi, K.; Sawamoto, Y.; Kuroda, A.; Ototake, M.; Nakanishi, T. & Fischer, U. (2003). The rainbow trout classical MHC class I molecule Onmy-UBA*501 is expressed in similar cell types as mammalian classical MHC class I molecules. *Fish & Shellfish Immunology*, Vol.14, No.1, pp. 1-23, ISSN 1050-4648.

Edholm, E.S.; Stafford, J.L.; Quiniou, S.M.; Waldbieser, G.; Miller, N.W.; Bengten, E. & Wilson, M. (2007). Channel catfish, *Ictalurus punctatus*, CD4-like molecules. *Developmental and Comparative Immunology*, Vol.31, No.2, pp. 172-187, ISSN 0145-305X.

Ellis, A.E. (2001). Innate host defense mechanisms of fish against viruses and bacteria. *Developmental and Comparative Immunology*, Vol.25, No.8-9, pp. 827-839, ISSN 0145-305X.

Esteban, M.A.; Mulero, V.; Cuesta, A.; Ortuño, J. & Meseguer, J. (2000). Effects of injecting chitin particles on the innate immune response of gilthead seabream (*Sparus aurata* L.). *Fish & Shellfish Immunology*, Vol.10, No.6, pp. 543-554, ISSN 1050-4648.

Esteban, M.A.; Cuesta, A.; Ortuño, J. & Meseguer, J. (2001). Immunomodulatory effects of dietary intake of chitin on gilthead seabream (*Sparus aurata* L.) innate immune system. *Fish & Shellfish Immunology*, Vol.11, No.4, pp. 303-315, ISSN 1050-4648.

Esteban, M.A.; Rodríguez, A.; Cuesta, A. & Meseguer, J. (2005). Effects of lactoferrin on non-specific immune responses of gilthead seabream (*Sparus auratus* L.). *Fish & Shellfish Immunology*, Vol.18, No.2, pp. 109-124, ISSN 1050-4648.

Esteban, M.A.; Meseguer, J.; Tafalla, C. & Cuesta, A. (2008) NK-like and oxidative burst activities are the main early cellular innate immune responses activated after virus inoculation in reservoir fish. *Fish & Shellfish Immunology*, Vol.25, No.4, pp. 433-438, ISSN 1050-4648.

Evans, D.L.; Carlson, R.L.; Graves, S.S. & Hogan, K.T. (1984a). Nonspecific cytotoxic cells in fish (*Ictalurus punctatus*). IV. Target cell binding and recycling capacity. *Developmental and Comparative Immunology*, Vol.8, No.4, pp. 823-833, ISSN 0145-305X.

Evans, D.L.; Graves, S.S.; Blazer, V.S.; Dawe, D.L. & Gratzek, J.B. (1984b). Immunoregulation of fish nonspecific cytotoxic cell activity by retinolacetate but not poly I:C. *Comparative Immunology, Microbiology & Infectious Diseases*, Vol.7, No.2, pp. 91-100, ISSN 0147-9571.

Evans, D.L.; Graves, S.S.; Cobb, D. & Dawe, D.L. (1984c). Nonspecific cytotoxic cells in fish (*Ictalurus punctatus*). II. Parameters of target cell lysis and specificity. *Developmental and Comparative Immunology*, Vol.8, No.2, pp. 303-312, ISSN 0145-305X.

Evans, D.L.; Hogan, K.T.; Graves, S.S.; Carlson, R.L., Jr.; Floyd, E. & Dawe, D.L. (1984d). Nonspecific cytotoxic cells in fish (*Ictalurus punctatus*). III. Biophysical and biochemical properties affecting cytolysis. *Developmental and Comparative Immunology*, Vol.8, No.3, pp. 599-610, ISSN 0145-305X.

Evans, D.L.; Smith, E.E., Jr. & Brown, F.E. (1987). Nonspecific cytotoxic cells in fish (*Ictalurus punctatus*). VI. Flow cytometric analysis. *Developmental and Comparative Immunology*, Vol.11, No.1, pp. 95-104, ISSN 0145-305X.

Evans, D.L.; Jaso-Friedmann, L.; Smith, E.E., Jr.; St John, A.; Koren, H.S. & Harris, D.T. (1988). Identification of a putative antigen receptor on fish nonspecific cytotoxic cells with monoclonal antibodies. *Journal of Immunology*, Vol.141, No.1, pp. 324-332, ISSN 0022-1767.

Evans, D.L.; Harris, D.T. & Jaso-Friedmann, L. (1990). Effects of phorbol esters and calcium ionophore on nonspecific cytotoxic cells. *Developmental and Comparative Immunology*, Vol.14, No.2, pp. 223-230, ISSN 0145-305X.

Evans, D.L.; Leary, J.H., 3rd; Weisman, Z.; Warren, J. & Jaso-Friedmann, L. (1996). Mapping of the epitope recognized by non-specific cytotoxic cells: determination of the fine specificity using synthetic peptides. *Scandinavian Journal of Immunology*, Vol.43, No.5, pp. 556-565, ISSN 0300-9475.

Evans, D.L.; Leary, J.H., Jr.; Nadella, P. & Jaso-Friedmann, L. (1998a). Evidence for antigen recognition by nonspecific cytotoxic cells: initiation of 3H-thymidine uptake following stimulation by a protozoan parasite and homologous cognate synthetic peptide. *Developmental and Comparative Immunology*, Vol.22, No.2, pp. 161-172, ISSN 0145-305X.

Evans, D.L.; Shotts, E.B. & Jaso-Friedmann, L. (1998b). *In vivo* modulation of innate resistance to *Edwardsiella ictaluri* with a phosphatase inhibitor. *Diseases of Aquatic Organisms*, Vol.33, No.1, pp. 19-24, ISSN 0177-5103.

Evans, D.L.; Leary, J.H., 3rd & Jaso-Friedmann, L. (1999). An antigen receptor (NCCRP-1) on nonspecific cytotoxic cells is a phosphoprotein associated with the JAK-STAT activation pathway. *Cellular Signaling*, Vol.11, No.4, pp. 287-292, ISSN 0898-6568.

Evans, D.L.; Taylor, S.L.; Leary, J.H., 3rd; Bishop, G.R.; Eldar, A. & Jaso-Friedmann, L. (2000). *In vivo* activation of tilapia nonspecific cytotoxic cells by *Streptococcus iniae* and amplification with apoptosis regulatory factor(s). *Fish & Shellfish Immunology*, Vol.10, No.5, pp. 419-434, ISSN 1050-4648.

Evans, D.L.; Leary, J.H., 3rd & Jaso-Friedmann, L. (2001). Nonspecific cytotoxic cells and innate immunity: regulation by programmed cell death. *Developmental and Comparative Immunology*, Vol.25, No.8-9, pp. 791-805, ISSN 0145-305X.

Faisal, M.; Ahmed, I.I.; Peters, G. & Cooper, E.L. (1989). Natural cytotoxicity of tilapia leukocytes. *Diseases of Aquatic Organisms*, Vol.7, pp. 17-22, ISSN 0177-5103.

Faisal, M.; Weeks, B.A.; Vogelbein, W.K. & Huggett, R.J. (1991). Evidence of aberration of the natural cytotoxic cell activity in *Fundulus heteroclitus* (Pisces: Cyprinodontidae) from the Elizabeth River, Virginia. *Veterinary Immunology and Immunopathology*, Vol.29, No.3-4, pp. 339-351, ISSN 0165-2427.

Fischer, U.; Ototake, M. & Nakanishi, T. (1998). *In vitro* cell-mediated cytotoxicity against allogeneic erythrocytes in ginbuna crucian carp and goldfish using a non-radioactive assay. *Developmental and Comparative Immunology*, Vol.22, No.2, pp. 195-206, ISSN 0145-305X.

Fischer, U.; Utke, K.; Ototake, M.; Dijkstra, J.M. & Kollner, B. (2003). Adaptive cell-mediated cytotoxicity against allogeneic targets by CD8-positive lymphocytes of rainbow trout (*Oncorhynchus mykiss*). *Developmental and Comparative Immunology*, Vol.27, No.4, pp. 323-337, ISSN 0145-305X.

Fischer, U.; Utke, K.; Somamoto, T.; Kollner, B.; Ototake, M. & Nakanishi, T. (2006). Cytotoxic activities of fish leucocytes. *Fish & Shellfish Immunology*, Vol.20, No.2, pp. 209-226, ISSN 1050-4648.

Forlenza, M.; de Carvalho Dias, J.D.; Vesely, T.; Pokorova, D.; Savelkoul, H.F. & Wiegertjes, G.F. (2008). Transcription of signal-3 cytokines, IL-12 and IFN alpha beta, coincides with the timing of CD8 alpha beta up-regulation during viral infection of common carp (*Cyprinus carpio* L). *Molecular Immunolog*, Vol.45, No.6, pp. 1531-1547, ISSN 0161-5890.

Graves, S.S.; Evans, D.L.; Cobb, D. & Dawe, D L. (1984). Nonspecific cytotoxic cells in fish (*Ictalurus punctatus*). I. Optimum requirements for target cell lysis. *Developmental and Comparative Immunology*, Vol.8, No.2, pp. 293-302, ISSN 0145-305X.

Graves, S.S.; Evans, D.L. & Dawe, D.L. (1985a). Antiprotozoan activity of nonspecific cytotoxic cells (NCC) from the channel catfish (*Ictalurus punctatus*). *Journal of Immunology*, Vol.134, No.1, pp. 78-85, ISSN 0022-1767.

Graves, S.S.; Evans, D.L. & Dawe, D.L. (1985b). Mobilization and activation of nonspecific cytotoxic cells (NCC) in the channel catfish (*Ictalurus punctatus*) infected with *Ichthyophthirius multifiliis*. *Comparative Immunology, Microbiology & Infectious Diseases*, Vol.8, No.1, pp. 43-51, ISSN 0147-9571.

Greenlee, A.R.; Brown, R.A. & Ristow, S.S. (1991). Nonspecific cytotoxic cells of rainbow trout (*Oncorhynchus mykiss*) kill YAC-1 targets by both necrotic and apoptic mechanisms. *Developmental and Comparative Immunology*, Vol.15, No.3, pp. 153-164, ISSN 0145-305X.

Groscurth, P. (1989). Cytotoxic effector cells of the immune system. *Anatomy and Embryology (Berl)*, Vol.180, No.2, pp. 109-119, ISSN 0340-2061.

Harikrishnan, R.; Balasundaram, C. & Heo, M.S. (2010). Effect of probiotics enriched diet on *Paralichthys olivaceus* infected with lymphocystis disease virus (LCDV). *Fish & Shellfish Immunology*, Vol.29, No.5, pp. 868-874, ISSN 1050-4648.

Hayden, B.J. & Laux, D.C. (1985). Cell-mediated lysis of murine target cells by nonimmune salmonid lymphoid preparations. *Developmental and Comparative Immunology*, Vol.9, No.4, pp. 627-639, ISSN 0145-305X.

Hinuma, S.; Abo, T.; Kumagai, K. & Hata, M. (1980). The potent activity of fresh water fish kidney cells in cell-killing. I. Characterization and species-distribution of cytotoxicity. *Developmental and Comparative Immunology*, Vol.4, No.4, pp. 653-666, ISSN 0145-305X.

Hogan, R.J.; Stuge, T.B.; Clem, L.W.; Miller, N.W. & Chinchar, V.G. (1996). Anti-viral cytotoxic cells in the channel catfish (*Ictalurus punctatus*) *Developmental and Comparative Immunology*, Vol.20, No.2, pp. 115-127, ISSN 0145-305X.

Hogan, R.J.; Taylor, W.R.; Cuchens, M.A.; Naftel, J.P.; Clem, L.W.; Miller, N.W. & Chinchar, V.G. (1999). Induction of target cell apoptosis by channel catfish cytotoxic cells. *Cellular Immunology*, Vol.195, No.2, pp. 110-118, ISSN 0008-8749.

Huang, R.; Zhong, S.; Liu, H.; Kong, R.; Wang, Y.; Hu, W. & Guo, Q. (2010). Identification and characterization of common carp (*Cyprinus carpio* L.) granzyme A/K, a cytotoxic cell granule-associated serine protease. *Fish & Shellfish Immunology*, Vol.29, No.3, pp. 388-398, ISSN 1050-4648.

Hwang, J.Y.; Ohira, T.; Hirono, I. & Aoki, T. (2004). A pore-forming protein, perforin, from a non-mammalian organism, Japanese flounder, *Paralichthys olivaceus*. *Immunogenetics*, Vol.56, No.5, pp. 360-367, ISSN 0093-7711.

Iwanowicz, L.R.; Densmore, C.L. & Ottinger, C.A. (2004). Calcein AM release-based cytotoxic cell assay for fish leucocytes. *Fish & Shellfish Immunology*, Vol.16, No.2, pp. 127-137, ISSN 1050-4648.

Jaso-Friedmann, L.; Evans, D.L.; Grant, C.C.; St John, A.; Harris, D.T. & Koren, H.S. (1988). Characterization by monoclonal antibodies of a target cell antigen complex recognized by nonspecific cytotoxic cells. *Journal of Immunology*, Vol.141, No.8, pp. 2861-2868, ISSN 0022-1767.

Jaso-Friedmann, L.; Leary, J.H., 3rd & Evans, D.L. (1993). Nonspecific cytotoxic cells in fish: antigenic cross-reactivity of a function associated molecule with the intermediate filament vimentin. *Cellular Immunology*, Vol.148, No.1, pp. 208-217, ISSN 0008-8749.

Jaso-Friedmann, L.; Leary, J.H., 3rd & Evans, D.L. (1995). Monoclonal antibody binding to a receptor on nonspecific cytotoxic cells (NCC) increases the expression of proto-oncogene kinases and protein kinase C. *Cellular Signalling*, Vol.7, No.5, pp. 463-470, ISSN 0898-6568.

Jaso-Friedmann, L.; Leary, J.H., 3rd & Evans, D.L. (1997a). NCCRP-1: a novel receptor protein sequenced from teleost nonspecific cytotoxic cells. *Molecular Immunology*, Vol.34, No.12-13, pp. 955-965, ISSN 0161-5890.

Jaso-Friedmann, L.; Leary, J.H., 3rd; Warren, J.; McGraw, R.A. & Evans, D.L. (1997b). Molecular characterization of a protozoan parasite target antigen recognized by nonspecific cytotoxic cells. *Cellular Immunology*, Vol.176, No.2, pp. 93-102, ISSN 0008-8749.

Jaso-Friedmann, L. & Evans, D.L. (1999). Mechanisms of cellular cytotoxic innate resistance in tilapia (*Oreochromis nilotica*). *Developmental and Comparative Immunology*, Vol.23, No.1, pp. 27-35, ISSN 0145-305X.

Jaso-Friedmann, L.; Leary, J.H., 3rd & Evans, D.L. (2000). Role of nonspecific cytotoxic cells in the induction of programmed cell death of pathogenic protozoans: participation of the Fas ligand-Fas receptor system. *Experimental Parasitology*, Vol.96, No.2, pp. 75-88, ISSN 0014-4894.

Jaso-Friedmann, L.; Leary, J.H., 3rd & Evans, D.L. (2001). The non-specific cytotoxic cell receptor (NCCRP-1): molecular organization and signaling properties. *Developmental and Comparative Immunology*, Vol.25, No.8-9, pp. 701-711, ISSN 0145-305X.

Jones, S.R. (2001). The occurrence and mechanisms of innate immunity against parasites in fish. *Developmental and Comparative Immunology*, Vol.25, No.8-9, pp. 841-852, ISSN 0145-305X.

Kaur, H.; Jaso-Friedmann, L. & Evans, D.L. (2004). Single base oligodeoxyguanosine upregulates Fas ligand release by nonspecific cytotoxic cells. *Developmental and Comparative Immunology*, Vol.28, No.6, pp. 571-579, ISSN 0145-305X.

Kurata, O.; Okamoto, N. & Ikeda, Y. (1995). Neutrophilic granulocytes in carp, *Cyprinus carpio*, possess a spontaneous cytotoxic activity. *Developmental and Comparative Immunology*, Vol.19, No.4, pp. 315-325, ISSN 0145-305X.

Lancki, D.W. (1998). Cell-mediated lysis. In: *Encyclopedia of Immunology*, P.J. Delves & I.M. Roitt (Eds.), Academic Press, ISBN 0122267656, UK.

Lazado, C.C.; Caipang, C.M.; Gallage, S.; Brinchmann, M.F. & Kiron, V. (2010). Expression profiles of genes associated with immune response and oxidative stress in Atlantic cod, *Gadus morhua* head kidney leukocytes modulated by live and heat-inactivated intestinal bacteria. *Comparative Biochemistry and Physiology Part A: Molecular & Integrative Physiology*, Vol.155, No.3, pp. 249-255, ISSN 1095-6433.

Leibson, P.J. (1997). Signal transduction during natural killer cell activation: inside the mind of a killer. *Immunity*, Vol.6, No.6, pp. 655-661, ISSN 1074-7613.

Lester, J.P., 3rd; Evans, D.L.; Leary, J.H., 3rd; Fowler, S.C. & Jaso-Friedmann, L. (1994). Identification of a target cell antigen recognized by nonspecific cytotoxic cells using an anti-idiotype monoclonal antibody. *Developmental and Comparative Immunology*, Vol.18, No.3, pp. 219-229, ISSN 0145-305X.

Litman, G.W.; Hawke, N.A. & Yoder, J.A. (2001). Novel immune-type receptor genes. *Immunological Reviews*, Vol.181, pp. 250-259, ISSN 0105-2896.

Long, S.; Wilson, M.; Bengten, E.; Clem, L.W.; Miller, N.W. & Chinchar, V.G. (2004). Identification and characterization of a FasL-like protein and cDNAs encoding the channel catfish death-inducing signaling complex. *Immunogenetics*, Vol.56, No.7, pp. 518-530, ISSN 0093-7711.

Manning, M.J. & Nakanishi, T. (1996). The specific immune system: cellular defenses. In: *The Fish Immune System. Organism, Pathogen, and Environment*, I. Iwama & T. Nakanishi, (Eds.), 160-205, Academic Press, ISBN 978-0123504395, San Diego, CA, USA.

McKinney, E.C.; Haynes, L. & Droese, A.L. (1986). Macrophage-like effector of spontaneous cytotoxicity from the shark. *Developmental and Comparative Immunology*, Vol.10, No.4, pp. 497-508, ISSN 0145-305X.

McKinney, E.C. & Schmale, M.C. (1994a). Damselfish with neurofibromatosis exhibit cytotoxicity toward tumor targets. *Developmental and Comparative Immunology*, Vol.18, No.4, pp. 305-313, ISSN 0145-305X.

McKinney, E.C. & Schmale, M.C. (1994b). Immune dysfunction in experimental versus naturally occurring neurofibromatosis in damselfish. *Anticancer Research*, Vol.14, No.1A, pp. 201-204, ISSN 0250-7005.

McKinney, E.C. & Schmale, M.C. (1997). Damselfish with neurofibromatosis exhibit cytotoxicity towards retrovirus infected cells. *Developmental and Comparative Immunology*, Vol.21, No.3, pp. 287-298, ISSN 0145-305X.

Meseguer, J.; Esteban, M.A.; López-Ruiz, A. & Bielek, E. (1994). Ultrastructure of nonspecific cytotoxic cells in teleosts. I. Effector-target cell binding in a marine and a freshwater species (seabream: *Sparus aurata* L., and carp: *Cyprinus carpio* L.). *Anatomical Record*, Vol.239, No.4, pp. 468-474, ISSN 0003-276X.

Meseguer, J.; Esteban, M.A. & Mulero, V. (1996). Nonspecific cell-mediated cytotoxicity in the seawater teleosts (*Sparus aurata* and *Dicentrarchus labrax*): ultrastructural study of target cell death mechanisms. *Anatomical Record*, Vol.244, No.4, pp. 499-505, ISSN 0003-276X.

Moody, C.E.; Serreze, D.V. & Reno, P.W. (1985). Non-specific cytotoxic activity of teleost leukocytes. *Developmental and Comparative Immunology*, Vol.9, No.1, pp. 51-64, ISSN 0145-305X.

Moore, L.J.; Somamoto, T.; Lie, K.K.; Dijkstra, J.M. & Hordvik, I. (2005). Characterisation of salmon and trout CD8alpha and CD8beta. *Molecular Immunology*, Vol.42, No.10, pp. 1225-1234, ISSN 0161-5890.

Moss, L.D.; Monette, M.M.; Jaso-Friedmann, L.; Leary, J.H., 3rd; Dougan, S.T.; Krunkosky, T. & Evans, D.L. (2009). Identification of phagocytic cells, NK-like cytotoxic cell activity and the production of cellular exudates in the coelomic cavity of adult zebrafish. *Developmental and Comparative Immunology*, Vol.33, No.10, pp. 1077-1087, ISSN 0145-305X.

Mulero, V.; Esteban, M.A.; Muñoz, J. & Meseguer, J. (1994). Non-specific cytotoxic response against tumor target cells mediated by leucocytes from seawater teleosts, *Sparus aurata* and *Dicentrarchus labrax*: an ultrastructural study. *Archives of Histology and Cytology*, Vol.57, No.4, pp. 351-358, ISSN 0914-9465.

Nakanishi, T.; Fischer, U.; Dijkstra, J.M.; Hasegawa, S.; Somamoto, T.; Okamoto, N. & Ototake, M. (2002). Cytotoxic T cell function in fish. *Developmental and Comparative Immunology*, Vol.26, No.2, pp. 131-139, ISSN 0145-305X.

Nakanishi, T.; Toda, H.; Shibasaki, Y. & Somamoto, T. (in press). Cytotoxic T cells in teleost fish. *Developmental and Comparative Immunology*, doi:10.1016/j.dci.2011.03.033. ISSN 0145-305X.

Ordás, M.C.; Cuesta, A.; Mercado, L.; Bols, N.C. & Tafalla, C. (2011). Viral hemorrhagic septicaemia virus (VHSV) up-regulates the cytotoxic activity and the perforin/granzyme pathway in the rainbow trout RTS11 cell line. *Fish & Shellfish Immunology*, Vol.31, No.2, pp. 252-259, ISSN 1050-4648.

Ortuño, J.; Cuesta, A.; Rodríguez, A.; Esteban, M.A. & Meseguer, J. (2002). Oral administration of yeast, Saccharomyces cerevisiae, enhances the cellular innate immune response of gilthead seabream (*Sparus aurata* L.). *Veterianry Immunology and Immunopathology*, Vol.85, No.1-2, pp. 41-50, ISSN 0165-2427.

Overturf, K. & LaPatra, S. (2006). Quantitative expression (Walbaum) of immunological factors in rainbow trout, *Oncorhynchus mykiss* (Walbaum), after infection with either *Flavobacterium psychrophilum, Aeromonas salmonicida,* or infectious haematopoietic necrosis virus. *Journal of Fish Diseases*, Vol.29, No.4, pp. 215-224, ISSN 0140-7775.

Patel, S.; Overgard, A.C. & Nerland, A.H. (2008). CD8alpha and CD8beta in Atlantic halibut, *Hippoglossus hippoglossus*: cloning, characterization and gene expression during viral and bacterial infection. *Fish & Shellfish Immunology*, Vol.25, No.5, pp. 570-580, ISSN 1050-4648.

Pettey, C.L. & McKinney, E.C. (1983). Temperature and cellular regulation of spontaneous cytotoxicity in the shark. *European Journal of Immunology*, Vol.13, No.2, pp. 133-138, ISSN 0014-2980.

Praveen, K.; Evans, D.L. & Jaso-Friedmann, L. (2004). Evidence for the existence of granzyme-like serine proteases in teleost cytotoxic cells. *Journal of Molecular Evolution*, Vol.58, No.4, pp. 449-459, ISSN 0022-2844.

Praveen, K.; Leary, J.H., 3rd; Evans, D.L. & Jaso-Friedmann, L. (2006). Molecular characterization and expression of a granzyme of an ectothermic vertebrate with chymase-like activity expressed in the cytotoxic cells of Nile tilapia (*Oreochromis niloticus*). *Immunogenetics*, Vol.58, No.1, pp. 41-55, ISSN 0093-7711.

Randelli, E.; Foglietta, A.; Mazzini, M.; Scapigliati, G. & Buonocore, F. (2006). Cloning and expression analysis of the co-receptor CD8a in sea bream (*Sparus aurata* L.). *Aquaculture*, Vol.256, pp. 631-637, ISSN 0044-8486.

Reyes-Becerril, M.; Salinas, I.; Cuesta, A.; Meseguer, J.; Tovar-Ramírez, D.; Ascencio-Valle, F. & Esteban, M.A. (2008). Oral delivery of live yeast *Debaryomyces hansenii* modulates the main innate immune parameters and the expression of immune-relevant genes in the gilthead seabream (*Sparus aurata* L.). *Fish & Shellfish Immunology*, Vol.25, No.6, pp. 731-739, ISSN 1050-4648.

Reyes-Becerril, M.; López-Medina, T.; Ascencio-Valle, F. & Esteban, M.A. (2011). Immune response of gilthead seabream (*Sparus aurata*) following experimental infection with *Aeromonas hydrophila. Fish & Shellfish Immunology*, pp. ISSN 1050-4648.

Robertsen, B. (2006). The interferon system of teleost fish. *Fish & Shellfish Immunology*, Vol.20, No.2, pp. 172-191, ISSN 1050-4648.

Rodríguez, A.; Cuesta, A.; Ortuño, J.; Esteban, M.A. & Meseguer, J. (2003). Immunostimulant properties of a cell wall-modified whole *Saccharomyces cerevisiae* strain administered by diet to seabream (*Sparus aurata* L.). *Veterianry Immunology and Immunopathology*, Vol.96, No.3-4, pp. 183-192, ISSN 0165-2427.

Rodríguez, A.; Cuesta, A.; Esteban, M.A. & Meseguer, J. (2004). The effect of dietary administration of the fungus *Mucor circinelloides* on non-specific immune responses of gilthead seabream. *Fish & Shellfish Immunology*, Vol.16, No.2, pp. 241-249, ISSN 1050-4648.

Roitt, I.M.; Brostoff, J. & Male, D.K. (1996). *Immunology* (4th edition), Times Mirror International Publishers Limited, ISBN 978-0723421788, London, UK.

Romano, L.A. & Marozzi, V.A. (2004). Epithelio-reticular cell thymoma in carp, *Cyprinus carpio* L: an ultrastructural study. *Journal of Fish Diseases*, Vol.27, No.6, pp. 369-373, ISSN 0140-7775.

Ruiz, J.; Leary, J.H., 3rd & Jaso-Friedmann, L. (2001). Phosphorylation-induced activation of tilapia nonspecific cytotoxic cells by serum cytokines. *Diseases of Aquatic Organisms*, Vol.46, No.2, pp. 129-137, ISSN 0177-5103.

Sakata, H.; Savan, R.; Sogabe, R.; Kono, T.; Taniguchi, K.; Gunimaladevi, I.; Tassakka, A.C. & Sakai, M. (2005). Cloning and analysis of non-specific cytotoxic cell receptor (NCCRP)-1 from common carp *Cyprinus carpio* L. *Comparative Biochemistry and Physiology Part C: Toxicology & Pharmacology*, Vol.140, No.3-4, pp. 287-294, ISSN 1532-0456.

Salinas, I.; Cuesta, A.; Esteban, M.A. & Meseguer, J. (2005). Dietary administration of *Lactobacillus delbrueckii* and *Bacillus subtilis*, single or combined, on gilthead seabream cellular innate immune responses. *Fish & Shellfish Immunology*, Vol.19, No.1, pp. 67-77, ISSN 1050-4648.

Salinas, I.; Díaz-Rosales, P.; Cuesta, A.; Meseguer, J.; Chabrillón, M.; Moriñigo, M.A. & Esteban, M.A. (2006). Effect of heat-inactivated fish and non-fish derived probiotics on the innate immune parameters of a teleost fish (*Sparus aurata* L.). *Veterianry Immunology and Immunopathology*, Vol.111, No.3-4, pp. 279-286, ISSN 0165-2427.

Salinas, I.; Abelli, L.; Bertoni, F.; Picchietti, S.; Roque, A.; Furones, D.; Cuesta, A.; Meseguer, J. & Esteban, M.A. (2008). Monospecies and multispecies probiotic formulations produce different systemic and local immunostimulatory effects in the gilthead seabream (*Sparus aurata* L.). *Fish & Shellfish Immunology*, Vol.25, No.1-2, pp. 114-123, ISSN 1050-4648.

Sasaki, Y.; Maita, M. & Okamoto, N. (2002). Rainbow trout neutrophils are responsible for non-specific cytotoxicity. *Fish & Shellfish Immunology*, Vol.12, No.3, pp. 243-252, ISSN 1050-4648.

Sato, A. & Okamoto, N. (2010). Induction of virus-specific cell-mediated cytotoxic responses of isogeneic ginbuna crucian carp, after oral immunization with inactivated virus. *Fish & Shellfish Immunology*, Vol.29, No.3, pp. 414-421, ISSN 1050-4648.

Scapigliati, G.; Buonocore, F.; Randelli, E.; Casani, D.; Meloni, S.; Zarletti, G.; Tiberi, M.; Pietretti, D.; Boschi, I.; Manchado, M.; Martin-Antonio, B.; Jimenez-Cantizano, R.; Bovo, G.; Borghesan, F.; Lorenzen, N.; Einer-Jensen, K.; Adams, S.; Thompson, K.; Alonso, C.; Bejar, J.; Cano, I.; Borrego, J.J. & Alvarez, M.C. (2010). Cellular and molecular immune responses of the sea bass (*Dicentrarchus labrax*) experimentally infected with betanodavirus. *Fish & Shellfish Immunology*, Vol.28, No.2, pp. 303-311, ISSN 1050-4648.

Schmale, M.C.; Gill, K.A.; Cacal, S.M. & Baribeau, S.D. (1994). Characterization of Schwann cells from normal nerves and from neurofibromas in the bicolour damselfish. *Journal of Neurocytology*, Vol.23, No.11, pp. 668-681, ISSN 0300-4864.

Schmale, M.C.; Vicha, D. & Cacal, S.M. (2004). Degranulation of eosinophilic granule cells in neurofibromas and gastrointestinal tract in the bicolor damselfish. *Fish & Shellfish Immunology*, Vol.17, No.1, pp. 53-63, ISSN 1050-4648.

Seeley, K.R. & Weeks-Perkins, B.A. (1993). Preliminary characterizarion of the non-specific cytotoxic cells of the oyster toadfish (*Opsanus tau* L.). *Fish & Shellfish Immunology*, Vol.3, No.2, pp. 131-141, ISSN 1050-4648

Shen, L.; Stuge, T.B.; Zhou, H.; Khayat, M.; Barker, K.S.; Quiniou, S.M.; Wilson, M.; Bengten, E.; Chinchar, V.G.; Clem, L.W. & Miller, N.W. (2002). Channel catfish cytotoxic cells: a mini-review. *Developmental and Comparative Immunology*, Vol.26, No.2, pp. 141-149, ISSN 0145-305X.

Shen, L.; Stuge, T.B.; Evenhuis, J.P.; Bengten, E.; Wilson, M.; Chinchar, V.G.; Clem, L.W. & Miller, N.W. (2003). Channel catfish NK-like cells are armed with IgM via a putative FcmicroR. *Developmental and Comparative Immunology*, Vol.27, No.8, pp. 699-714, ISSN 0145-305X.

Shen, L.; Stuge, T.B.; Bengten, E.; Wilson, M.; Chinchar, V.G.; Naftel, J.P.; Bernanke, J.M.; Clem, L.W. & Miller, N.W. (2004). Identification and characterization of clonal NK-like cells from channel catfish (*Ictalurus punctatus*). *Developmental and Comparative Immunology*, Vol.28, No.2, pp. 139-152, ISSN 0145-305X.

Somamoto, T.; Nakanishi, T. & Okamoto, N. (2000). Specific cell-mediated cytotoxicity against a virus-infected syngeneic cell line in isogeneic ginbuna crucian carp. *Developmental and Comparative Immunology*, Vol.24, No.6-7, pp. 633-640, ISSN 0145-305X.

Somamoto, T.; Nakanishi, T. & Okamoto, N. (2002). Role of specific cell-mediated cytotoxicity in protecting fish from viral infections. *Virology*, Vol.297, No.1, pp. 120-127, ISSN 0042-6822.

Somamoto, T.; Yoshiura, Y.; Nakanishi, T. & Ototake, M. (2005). Molecular cloning and characterization of two types of CD8alpha from ginbuna crucian carp, *Carassius auratus langsdorfii*. *Developmental and Comparative Immunology*, Vol.29, No.8, pp. 693-702, ISSN 0145-305X.

Somamoto, T.; Yoshiura, Y.; Sato, A.; Nakao, M.; Nakanishi, T.; Okamoto, N. & Ototake, M. (2006). Expression profiles of TCRbeta and CD8alpha mRNA correlate with virus-specific cell-mediated cytotoxic activity in ginbuna crucian carp. *Virology*, Vol.348, No.2, pp. 370-377, ISSN 0042-6822.

Somamoto, T.; Okamoto, N.; Nakanishi, T.; Ototake, M. & Nakao, M. (2009). *In vitro* generation of viral-antigen dependent cytotoxic T-cells from ginbuna crucian carp, *Carassius auratus langsdorfii*. *Virology*, Vol.389, No.1-2, pp. 26-33, ISSN 0042-6822.

Stuge, T.B.; Yoshida, S.H.; Chinchar, V.G.; Miller, N.W. & Clem, L.W. (1997). Cytotoxic activity generated from channel catfish peripheral blood leukocytes in mixed leukocyte cultures. *Cellular Immunology*, Vol.177, No.2, pp. 154-161, ISSN 0008-8749.

Stuge, T.B.; Wilson, M.R.; Zhou, H.; Barker, K.S.; Bengten, E.; Chinchar, G.; Miller, N.W. & Clem, L.W. (2000). Development and analysis of various clonal alloantigen-dependent cytotoxic cell lines from channel catfish. *Journal of Immunology*, Vol.164, No.6, pp. 2971-2977, ISSN 0022-1767.

Suetake, H.; Araki, K.; Akatsu, K.; Somamoto, T.; Dijkstra, J.M.; Yoshiura, Y.; Kikuchi, K. & Suzuki, Y. (2007). Genomic organization and expression of CD8alpha and CD8beta genes in fugu *Takifugu rubripes*. *Fish & Shellfish Immunology*, Vol.23, No.5, pp. 1107-1118, ISSN 1050-4648.

Sun, X.F.; Shang, N.; Hu, W.; Wang, Y.P. & Guo, Q.L. (2007). Molecular cloning and characterization of carp (*Cyprinus carpio* L.) CD8beta and CD4-like genes. *Fish & Shellfish Immunology*, Vol.23, No.6, pp. 1242-1255, ISSN 1050-4648.

Takizawa, F.; Dijkstra, J.M.; Kotterba, P.; Korytar, T.; Kock, H.; Kollner, B.; Jaureguiberry, B.; Nakanishi, T. & Fischer, U. (2011). The expression of CD8alpha discriminates distinct T cell subsets in teleost fish. *Developmental and Comparative Immunology*, Vol.35, No.7, pp. 752-763, ISSN 0145-305X.

Taylor, S.L.; Jaso-Friedmann, L.; Allison, A.B.; Eldar, A. & Evans, D.L. (2001). *Streptococcus iniae* inhibition of apoptosis of nonspecific cytotoxic cells: a mechanism of activation of innate immunity in teleosts. *Diseases of Aquatic Organisms*, Vol.46, No.1, pp. 15-21, ISSN 0177-5103.

Thompson, J.S. & Kostiala, A.A. (1990). Immunological and ultrastructural characterization of true histiocytic lymphoma in the northern pike, *Esox lucius* L. *Cancer Research*, Vol.50, No.17 Suppl, pp. 5668S-5670S, ISSN 0008-5472.

Toda, H.; Shibasaki, Y.; Koike, T.; Ohtani, M.; Takizawa, F.; Ototake, M.; Moritomo, T. & Nakanishi, T. (2009). Alloantigen-specific killing is mediated by CD8-positive T cells in fish. *Developmental and Comparative Immunology*, Vol.33, No.4, pp. 646-652, ISSN 0145-305X.

Toda, H.; Araki, K.; Moritomo, T. & Nakanishi, T. (2011a). Perforin-dependent cytotoxic mechanism in killing by CD8 positive T cells in ginbuna crucian carp, *Carassius auratus langsdorfii*. *Developmental and Comparative Immunology*, Vol.35, No.1, pp. 88-93, ISSN 0145-305X.

Toda, H.; Yabu, T.; Shiba, H.; Moritomo, T. & Nakanishi, T. (2011b). Evaluating antigen-specific cytotoxicity of CD8+ T cells in fish by granzyme B-like activity. *Veterianry Immunology and Immunopathology*, Vol.141, No.1-2, pp. 168-172, ISSN 0165-2427.

Utke, K.; Bergmann, S.; Lorenzen, N.; Kollner, B.; Ototake, M. & Fischer, U. (2007). Cell-mediated cytotoxicity in rainbow trout, *Oncorhynchus mykiss*, infected with viral haemorrhagic septicaemia virus. *Fish & Shellfish Immunology*, Vol.22, No.3, pp. 182-196, ISSN 1050-4648.

Utke, K.; Kock, H.; Schuetze, H.; Bergmann, S.M.; Lorenzen, N.; Einer-Jensen, K.; Kollner, B.; Dalmo, R.A.; Vesely, T.; Ototake, M. & Fischer, U. (2008). Cell-mediated immune responses in rainbow trout after DNA immunization against the viral hemorrhagic septicemia virus. *Developmental and Comparative Immunology*, Vol.32, No.3, pp. 239-252, ISSN 0145-305X.

Verlhac, V.; Sage, M. & Deschaux, P. (1990). Cytotoxicity of carp (*Cyprinus carpio*) leucocytes induced against TNP-modified autologous spleen cells and influence of acclimatization temperature. *Developmental and Comparative Immunology*, Vol.14, No.4, pp. 475-480, ISSN 0145-305X.

Vicha, D.L. & Schmale, M.C. (1994). Morphology and distribution of eosinophilic granulocytes in damselfish neurofibromatosis, a model of mast cell distribution in neurofibromatosis type 1. *Anticancer Research*, Vol.14, No.3A, pp. 947-952, ISSN 0250-7005.

Wernersson, S.; Reimer, J.M.; Poorafshar, M.; Karlson, U.; Wermenstam, N.; Bengten, E.; Wilson, M.; Pilstrom, L. & Hellman, L. (2006). Granzyme-like sequences in bony fish shed light on the emergence of hematopoietic serine proteases during vertebrate evolution. *Developmental and Comparative Immunology*, Vol.30, No.10, pp. 901-918, ISSN 0145-305X.

Xu, S.W.; Wu, J.Y.; Hu, K.S.; Ping, H.L.; Duan, Z.G. & Zhang, H.F. (2011). Molecular cloning and expression of orange-spotted grouper (*Epinephelus coioides*) CD8alpha and CD8beta genes. *Fish & Shellfish Immunology*, Vol.30, No.2, pp. 600-608, ISSN 1050-4648.

Yoder, J.A. (2004). Investigating the morphology, function and genetics of cytotoxic cells in bony fish. *Comparative Biochemistry and Physiology Part C: Toxicology & Pharmacology*, Vol.138, No.3, pp. 271-280, ISSN 1532-0456.

Yoshida, S.H.; Stuge, T.B.; Miller, N.W. & Clem, L.W. (1995). Phylogeny of lymphocyte heterogeneity: cytotoxic activity of channel catfish peripheral blood leukocytes directed against allogeneic targets. *Developmental and Comparative Immunology*, Vol.19, No.1, pp. 71-77, ISSN 0145-305X.

Yoshinaga, K.; Okamoto, N.; Kurata, O. & Ikeda, Y. (1994). Individual variations of natural killer activity of rainbow trout leukocytes against IPNV virus-infected and uninfected RTG-2 cells. *Fish Pathology*, Vol.29, pp. 1-4, ISSN 0388-788X.

Zhou, H.; Stuge, T.B.; Miller, N.W.; Bengten, E.; Naftel, J.P.; Bernanke, J.M.; Chinchar, V.G.; Clem, L.W. & Wilson, M. (2001). Heterogeneity of channel catfish CTL with respect to target recognition and cytotoxic mechanisms employed. *Journal of Immunology*, Vol.167, No.3, pp. 1325-1332, ISSN 0022-1767.

12

Bacteriocins of Aquatic Microorganisms and Their Potential Applications in the Seafood Industry

Suphan Bakkal, Sandra M. Robinson and Margaret A. Riley
University of Massachusetts Amherst
USA

1. Introduction

Bacteriocins are potent antimicrobial polypeptides and proteins produced by most lineages of Bacteria and, perhaps, by all members of Archaea (O'Connor & Shand, 2002; Riley & Wertz, 2002a, 2002b; Tagg et al., 1976). Although initially the focus of numerous biochemical, evolutionary, and ecological studies, more recently, their potential to serve in human and animal health applications has taken center stage (Gillor et al., 2008). The use of bacteriocins in probiotic applications, as preservatives, and, (most excitingly) as alternatives to classical antibiotics is being broadly explored (Abee et al., 1995; Einarsson & Lauzon, 1995; Gillor & Ghazaryan, 2007; Gillor et al., 2007).

Most bacterial species produce one or more bacteriocins (Cascales et al., 2007). One of the most prolific bacteriocin-producing species is *Pseudomonas aeruginosa*, of which 90% or more of the strains tested produce their own version of bacteriocins, known as pyocins (Govan & Harris, 1985). In contrast, only 15-50% of *Escherichia coli* produce their brand of bacteriocins, known as colicins (Riley & Gordon, 1992). The colicins are exceedingly well characterized proteins, and have been the subject of numerous detailed biochemical, molecular, evolutionary, and ecological analyses (Cascales et al., 2007; Riley et al., 2003; Riley & Gordon, 1999; Riley & Wertz, 2002a, 2002b). Some species of bacteria produce toxins that may exhibit numerous bacteriocin-like features, but have not yet been fully characterized; these toxins are referred to as bacteriocin-like inhibitory substances, or BLIS (Messi et al., 2003; Moro et al., 1997).

In this chapter, we will explore the bacteriocins of aquatic bacteria, particularly those of potential interest in the seafood industry. A short primer of bacteriocin biology is followed by a detailed review of the diversity of bacteriocins described from marine microorganisms. These toxins have received far less attention than bacteriocins produced by terrestrial or human-commensal bacteria, yet they have equivalent potential as antibiotics and even greater promise for use in the creation of probiotic strains for the seafood industry.

2. Bacteriocin basics

Bacteriocins are proteins or short polypeptides, which are generally only toxic to bacteria that are closely related to the producing strain. A typical bacteriocin contains a toxin (bacteriocin) gene, an immunity gene (which confers resistance to the aforementioned toxin), and a lysis gene, which encodes a protein that aids in toxin release from the producing cell (Chavan & Riley, 2007). Bacteriocins work by binding to and killing only cells with surface receptors that are recognized by that specific bacteriocin (Cascales et al., 2007; Chavan & Riley, 2007). In a microbial community, cells can either be bacteriocinogenic (produce bacteriocin), sensitive, or resistant to each bacteriocin. When all three cell-types are present and are competing for limiting resources, the strain interactions mimic the children's game "rock-paper-scissors" (Kerr et al., 2002). The premise of this game is that paper covers rock, scissors cut paper, and rock breaks scissors, creating a cycle of wins and losses with no one matter dominating as long as all three states are present. The same interaction is observed in microbial communities that employ bacteriocins (Table 1). Only a small percentage of bacteriocinogenic cells will be induced to produce and release bacteriocin. Some sensitive cells are immediately killed by the bacteriocin, while others harbor mutations that confer resistance. These resistant cells rapidly displace the producer cells, due to the cost of bacteriocin production. However, the resistant cells grow more slowly than their sensitive counterparts, because resistance mutations often have a negative effect on fitness (Kerr et al., 2002).

Strain	More Fit Than	Less Fit Than
Bacteriocin-producer	Sensitive	Resistant
Sensitive	Resistant	Bacteriocin-producer
Resistant	Bacteriocin-producer	Sensitive

Table 1. Competition for resources results in a "rock-paper-scissors"-like interaction of microorganisms (adapted from Kerr et al., 2002).

In contrast to traditional antibiotics, which are used in human health applications precisely because of their ability to kill a diversity of bacterial pathogens, bacteriocins generally target only members of their producing species and its closest relatives (although numerous exceptions abound)(Riley et al., 2003; Tagg et al., 1976). Riley et al. (2003) mapped the killing spectrum of bacteriocins from seven enteric species (*Escherichia coli*, *Klebsiella pneumoniae*, *Klebsiella oxytoca*, *Enterobacter cloacae*, *Citrobacter freundii*, *Hafnia alvei*, and *Serratia plymuthica*) onto the molecular phylogeny of the same species (Fig. 1). This study showed that bacteriocin producers tend to kill strains belonging to their same species. However, there are some exceptions, such as bacteriocins of *E. coli* that inhibit distantly related *H. alvei* (Fig. 1).

2.1 Bacteriocin naming

Bacteriocins were originally named based on the producer species such as colicins produced by *Escherichia coli*, pyocins of *Pseudomonas aeruginosa* (formerly named *pyocyania*), cloacins of *Enterobacter cloacae*, cerecins of *Bacillus cereus*, and pesticins of *Yersinia pestis* (Reeves, 1965). Fredericq (1957) created the first classification, and thus nomenclature, of bacteriocins focusing on the colicins of *E. coli* (Fredericq, 1957). Fredericq grouped colicins into 17 different types (colicins A, B, C, D, E, F, G, H, I, J, K, V, S1, S2, S3, S4, and S5) based on their receptor specificity. These colicins were then further subtyped (colicin E1, E2, and E3, etc.) based on their immunity patterns. In this scheme, all subtypes were recognized by the same receptor, but they possessed different immunity phenotypes (Fredericq, 1957). Later, the addition of the producer strain's name provided further differentiation of bacteriocins produced by strains of the same species (Daw & Falkiner, 1996). This scheme is still used today.

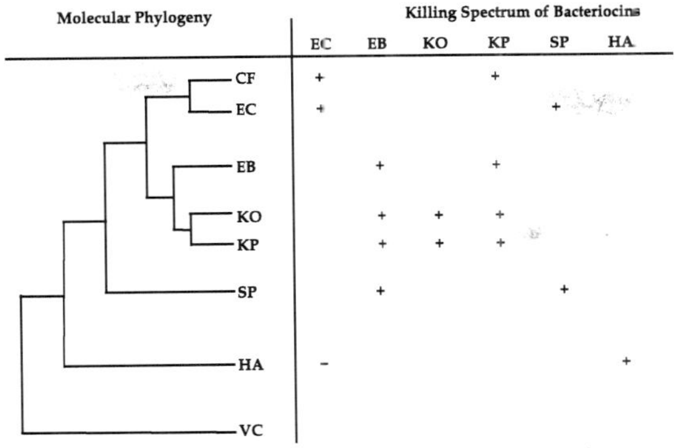

Fig. 1. The breadth of bacteriocin killing in enteric bacteria (adapted from Riley et al., 2003). The bacteriocin killing phenotype of six enteric bacterial species were mapped onto their molecular phylogeny constructed with concatenated sequences of five housekeeping genes and 16s RNA. *Vibrio cholerae* (VC) was used as an outgroup to root the phylogenetic tree. EC, *Escherichia coli*; KP, *Klebsiella pneumoniae*; KO, *Klebsiella oxytoca*; EB, *Enterobacter cloacae*; CF, *Citrobacter freundii*; HA, *Hafnia alvei*; SP, *Serratia plymuthica*.

2.2 Bacteriocin classes

In general, bacteriocins are produced by Bacteria and studied based on the gram designation of their producing species (Gram-negative versus Gram-positive). Additionally, a relatively small number of bacteriocins from Archaeal species have also been characterized. A comprehensive review of bacteriocins from Bacteria and Archaea can be found elsewhere (O'Connor & Shand, 2002; Reeves, 1965; Riley & Gordon, 1999; Riley & Wertz, 2002a, 2002b; Tagg et al., 1976). Below are short descriptions of the bacteriocin classes of Bacteria and Archaea and examples of bacteriocins belonging to each class (Table 2).

	Bacteriocins	Bacteriocin Types /Class	Size (kDa)	Examples	References
Gram-negative Bacteria	Colicins	Pore Formers Nucleases	20-80	Colicins A, B Colicins E2, E3	Cascales et al., 2007 Michel-Briand & Baysse, 2002 Gillor et al., 2004 Reeves, 1965
	Colicin-like	NA	20-80	S-pyocins Klebicins	
	Phage-tail like	NA	> 80	R and F pyocins	
	Microcins	Post-translationally modified Unmodified	< 10	Microcin C7 Microcin B17 Colicin V	
Gram-positive Bacteria	Class I	Type A-positively charged and linear Type B-uncharged or negatively charged globular Type C-synergistic	< 5	Nisin Mersacidin Lacticin 3147	Heng et al., 2007 Drider et al., 2006 Field et al., 2007 Maqueda et al., 2004
	Class II	Class IIa-antilisterial Class IIb-synergistic	< 10	Pediocin PA1 Carnobacteriocin B2	
	Class III	Type IIIa-Bacteriolytic enzymes Type IIIb-Nonlytic peptides	> 10	Lysostaphin Helveticin	
	Class IV	Cyclic peptides	< 10	Enterocin AS-48	
Archaea	Halocins	Microhalocins Protein halocins	< 10 > 10	Halocin A4, C8, G1 Halocin H1, H4	Shand et al., 2007 O'Connor & Shand, 2002 Ellen et al., 2011 Sun et al., 2005
	Sulfolobicin	NA	~20	Sulfolobicin	

Table 2. Bacteriocins of Bacteria and Archaea

2.2.1 Bacteriocins of Gram-negative bacteria

Bacteriocins of Gram-negative bacteria are categorized into four main classes: colicins, colicin-like bacteriocins, phage-tail like bacteriocins, and microcins (Table 2) (Chavan & Riley, 2007). The colicins, produced by *E. coli*, are the most studied bacteriocins (Cascales et al., 2007). Indeed, they have been used as a model system to study bacteriocin structure, function, and evolution (Cascales et al., 2007; Riley & Gordon, 1999; Riley & Wertz, 2002a, 2002b). In general, colicins are protease sensitive, thermosensitive proteins that vary in size from 25 to 90 kDa (Pugsley & Oudega, 1987). There are two major colicin types based on their mode of killing; pore former and nuclease colicins. Pore former colicins (colicins A, B, E1, Ia, Ib, K, E1, 5) kill sensitive strains by forming pores in the cell membrane. Nuclease colicins (Colicins E2, E3, E4, E5, E6, E7, E8, E9) kill by acting as DNases, RNases, or tRNAses (Gillor et al., 2004). Proteinaceous bacteriocins produced by other Gram-negative species are classified as colicin-like due to the presence of similar structural and functional characteristics (Table 2). Like colicins, they can be nucleases (pyocins S1, S2) and pore formers (pyocin S5) (Michel-Briand & Baysse, 2002). Klebicins of *Klebsiella* species, S-pyocins of *Pseudomonas aeruginosa*, and alveicins of *Hafnia alvei* are among the most studied colicin-like bacteriocins.

Phage-tail like bacteriocins are larger structures that resemble the tails of bacteriophages. Some even argue that they are defective phage particles (Bradley, 1967). R and F pyocins of *P. aeruginosa* are some of the most thoroughly studied phage-tail like bacteriocins (Michel-Briand & Baysse, 2002; Nakayama et al., 2000). They are encoded in a large gene cluster, which spans a DNA region greater than 40 kb (Nakayama et al., 2000). There are 44 open reading frames associated with the R2/F2 phenotypes, which include regulatory, lysis, and toxin genes. The R2 and F2 pyocins show sequence similarity to the tail fiber genes of P2 and lambda phages, respectively (Nakayama et al., 2000). Finally, Gram-negative bacteria produce much smaller (<10 kDa) peptide bacteriocins called microcins. Microcins can be divided into two classes: post-translationally modified (microcins B17, C7, J25, and D93) and unmodified microcins (microcins E492, V, L, H47, and 24) (Table 2). Microcins are chromosomally encoded (Gillor et al., 2004).

2.2.2 Bacteriocins of Gram-positive bacteria

Bacteriocins of Gram-positive bacteria are generally divided into four classes based on size, morphology, physical, and chemical properties (Lee & Kim, 2011). Class I bacteriocins are lantibiotics, which are small peptides (<5 kDa)(Field et al., 2007). They are post-translationally modified, incorporating non-traditional amino acids such as dehydroalanine, dehydrobutyrine, methyl-lanthione, and lantionine (Cleveland et al., 2001). This class is subdivided into Type A, B, and C with the distinction being that members of Type A are positively charged, linear peptides whereas those in Type B are either neutrally or negatively charged rigid globular peptides. Members of Type C require synergistic activity of two peptides to be active. This class includes the well-studied bacteriocins nisin and lacticin (McAuliffe et al., 2001).

Class II bacteriocins are small (<10 kDa), heat-stable peptides that are not post-translationally modified (Heng et al., 2007). Class II is also subdivided into two subgroups. Class IIa are pediocin-like or *Listeria*-active peptides, which contain a conserved N-terminal sequences (YGNGVxCxxxxCxV). Class IIb bacteriocins require the synergistic activity of two peptides to be fully active (Nissen-Meyer et al., 1992). Class III bacteriocins are generally large (>10 kDa), heat-labile peptides. They are subdivided into two subtypes. Type IIIa are bacteriolysins, which are bacteriolytic enzymes. Lysostaphin is the most studied bacteriocin in this subtype (Schindler & Schuhardt, 1964). Type IIIb are non-lytic bacteriocins. Helveticin J (37 kDa) produced by *Lactobacillus helveticus* belongs to this type (Joerger & Klaenhammer, 1986). Finally, Class IV bacteriocins have unique structural characteristics. The first and last amino acids of these bacteriocins are covalently bonded, thus they have cyclic structures. Enterocin AS-48 produced by *Enterococcus faecalis* subsp. *liquefaciens* S-48 was the first characterized bacteriocin belonging to this class (Maqueda et al., 2004).

2.2.3 Bacteriocins of Archaea

The Archaea also produce unique types of bacteriocin-like antimicrobial compounds, called archaeocins (Shand & Leyva, 2007). They have been much less studied than the bacteriocins of Bacteria. Thus far, two major types of archaeocins have been identified: halocins of halobacteria and sulfolobicins of *Sulfolobus* species. Halocins can be peptides (< 10 kDa) and/or proteins (>10 kDa) (Shand & Leyva, 2007). According to Torreblanca and Meseguer (1994), halocin production is a universal feature of halobacteria. Halocins are located on megaplasmids (or minichromosomes). Halocins H4 and S8 are located on ~300 kbp and ~200 kbp plasmids, respectively (Cheung et al., 1997; Price & Shand, 2000). Their activity is usually detected at the late exponential to early stationary growth phase (Cheung et al., 1997; Price & Shand, 2000). There is not much known about sulfolobicins. Prangishvili et al. (2000) screened sulfolobicin production from *Sulfolobus islandicus* isolated from volcanic vents throughout Iceland. This study predicted that sulfolobicin activity is membrane-associated and is not released from the cell. Sulfolobicins are also associated with membranous vesicles ranging in size from 90 to 180 nm in diameter (Prangishvili et al., 2000). Like many bacteriocins, they are thermostable and sensitive to protease treatment. Their mode of action is still unknown (Ellen et al., 2011).

2.3 Bacteriocin genetics

Bacteriocins can be encoded on chromosomes, plasmids, and other transposable elements. For example, the colicins of *E. coli* and halocin H4 are plasmid-encoded while the pyocins of *P. aeruginosa* are chromosomal (Chavan & Riley, 2007; Cheung et al., 1997; Michel-Briand & Baysse, 2002). Lacticin 481 has been shown to reside on the transposon Tn*5721* (Dufour et al., 2000) and some bacterial species such as *Serratia marcescens* possess both plasmid and chromosomally encoded bacteriocins (Riley & Wertz, 2002b). Just as we see differences in the function of Gram-negative, Gram-positive, and Archaeal bacteriocins, we can also trace these distinctions to the genetic level.

In general, the full function of bacteriocins produced by Gram-negative bacteria is encoded via three tightly linked genes, the toxin, immunity, and lysis genes (Fig. 2A). However, there are significant differences in the genetics of colicins, colicin-like bacteriocins, phage-tail-like bacteriocins, and microcins (Fig. 2). For example, the colicin gene cluster consists of the three bacteriocin-related genes in close proximity, whereas colicin-like pyocin S3 does not possess a lysis gene (Fig. 2A-B). Two representative phage-tail like bacteriocins, R and F-type pyocins, R2 and F2, are encoded in a large gene cluster that spans more than 40 kb (Nakayama et al., 2000) and includes 44 open reading frames (Nakayama et al., 2000). The open reading frames include regulatory, lysis, R, and F pyocin genes (Fig. 2C).

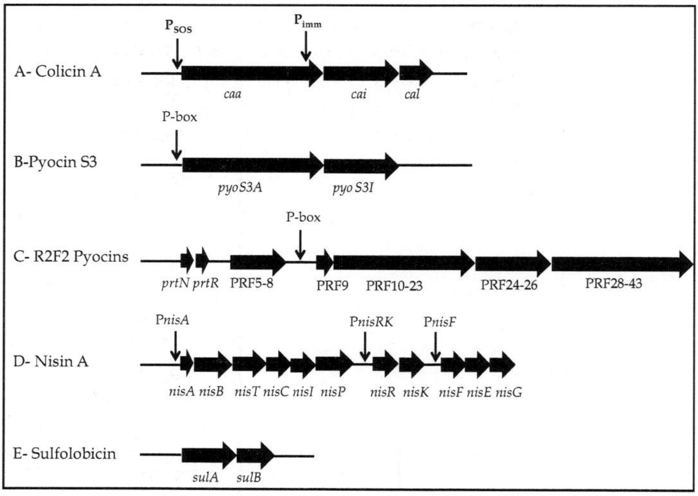

Fig. 2. Genetic organization of bacteriocins
(Horizontal arrows represent genes; not to scale)
A- Genetic organization of colicin gene cluster. P_{SOS} and P_{imm} represent promoter regions of the SOS and immunity genes. Gene *caa* encodes colicin A; *cai* encodes the immunity gene; and *cal* encodes the lysis gene (adapted from Cascales et al., 2007).
B-Genetic organization of colicin-like gene cluster. P-box refers to the binding site for transcriptional regulator (PrtN); *pyoS3A* gene encodes pyocin S3; *pyoS3I* encodes the immunity gene (adapted from Duport et al., 1995).
C- Genetic organization of phage-tail like bacteriocin gene cluster. *prtN* and *prtR* encode transcriptional activator (PrtN) and repressor (PrtR), respectively. PRF 9 and PRF24-26 encodes the lysis genes; PRF10-23 and PRF28-43 encode the R2 and F2 pyocin structural genes, respectively (adapted from Nakayama et al., 2000).
D-Genetic organization of lantibiotic gene cluster. P_{nisA}, P_{nisRK}, and P_{nisF} encode promoter genes for *nis A*, *nisRK*, and *nisFEG*, respectively; *nisA* encodes nisin A precursor; *nisR* and *nisK* encode proteins for nisin biosynthesis; *nisB*, *nisC*, *nisT*, and *nisP* encode proteins for nisin processing and translocation; *nisI*, *nisF*, *nisE*, and *nisG* encode immunity proteins (adapted from Kuipers et al., 1993; Mierau & Kleerebezem, 2005).
E-Genetic organization of archaeocin gene cluster. Sulfolobicin is composed of SulA and SulB proteins, encoded by *sulA* and *sulB* genes (adapted from Ellen et al., 2011).

The Gram-positive bacteriocins are more complicated genetically, with genes that encode post-translational modification of the toxin. The genetic organization varies between the Gram-positive bacteriocin classes as well - such as the requirement of two peptides for the full activation of Class II bacteriocins. An example is provided by the nisin gene cluster (Fig. 2D), which includes 11 genes (nis*ABTCIPRKFEG*) encoding functions such as synthesis of the nisin precursor (*nis*A), regulation of nisin biosynthesis (*nis*RK), the processing and translocation of nisin (*nis*BCTP), and immunity (*nis*IFEG) (Kuipers et al., 1993; Mierau & Kleerebezem, 2005).

The genetic organization of archaeocins is relatively unknown in comparison to other bacteriocins. A recent study showed that sulfolobicin of *Sulfolobus acidocaldarius* is comprised of two proteins, SulA and SulB, which are encoded by *sulA* and *sulB* genes, respectively (Fig. 2E). These two proteins are associated with membrane vesicles in the extracellular medium (Ellen et al., 2011). The gene organization of archaeocins is also unique. Sun et al. (2005) showed that the gene *halC8* encoded both halocin C8 and its immunity protein HalI (Sun et al., 2005). It differs from other bacteriocins in that typically separate genes encode bacteriocin toxin and immunity proteins.

2.4 Bacteriocin biosynthesis

Bacteriocins are often produced under stress conditions, such as nutrient limitation and overpopulation (Riley & Gordon, 1999). The biosynthesis of Gram-negative bacteriocins is regulated by the host; often involving the SOS system. The SOS system is comprised of RecA and LexA proteins. LexA is a transcriptional repressor, which binds to the bacteriocin promoter and prevents its transcription. In stress conditions such as DNA damage and UV exposure expression of the *recA* gene is induced. RecA binds to LexA and therefore prevents the repression of bacteriocin expression (Cascales et al., 2007). The expression of colicin-like S-pyocins and phage-tail like RF pyocins also depends on RecA, except these genes possess a P-box in their promoter region instead of an SOS box (Nakayama et al., 2000; Sano et al., 1993). Further, there is no LexA-dependent repression of bacteriocin expression. Activated RecA cleaves the PtrR transcriptional repressor protein, which leads to expression of the transcriptional activator *ptrN* gene. PtrN binds to the P-box and induces the expression of the pyocin genes (Matsui et al., 1993; Michel-Briand & Baysse, 2002). Alternatively, Gram-positive bacteria possess bacteriocin-specific biosynthesis pathways. For example, nisin regulates its own biosynthesis in cell-density dependent conditions (Eijsink et al., 2002). Finally, there is much less known about archaeocins. Regulation of their biosynthesis is still under investigation.

3. Bacteriocins of marine bacteria

Bacteriocins produced by marine bacteria have generated a great deal of excitement due primarily to their potential to serve as probiotics and antibiotics in the seafood industry (Galvez et al., 2008; García et al., 2010; Pilet & Leroi, 2011). A recent antimicrobial screen of 258 bacterial strains isolated from water and sediment in the Yucatan peninsula revealed that 46 strains belonging to the genera *Aeromonas, Bacillus, Burkholderia, Photobacterium, Pseudomonas, Serratia,* and *Stenotrophomonas* possessed antimicrobial activity. Approximately fifty percent of this antimicrobial activity was due to bacteriocins or BLIS (De la Rosa-Garcia et al., 2007). Further, Wilson et al. (2000) investigated surface-attached bacteria isolated from Sydney Harbor, Australia. He showed that approximately 10% of surface-attached marine bacteria possess antibacterial activity. Proteinase K treatment showed that this inhibitory activity was associated with proteinaceous substances such as bacteriocins or BLIS (Wilson et al., 2010). Given the fact that bacteriocins and BLIS have been characterized in most culturable bacterial species, it is tempting to speculate about the diversity of new substances the marine environment will reveal.

The first bacteriocin isolated from marine microorganisms was detected in *Vibrio harveyi* (formerly *Beneckea harveyi*). McCall and Sizemore (1979) screened a total of 795 *Vibrio* spp. strains isolated from Galveston Island, Texas for bacteriocin production (McCall & Sizemore, 1979). This study revealed that approximately 5% of the *Vibrio* spp. possessed a high molecular weight bacteriocin-like killing agent. Further investigation revealed that the source of the bacteriocin-like killing was a plasmid. It was also determined that the killing range was limited to strains of *B. harveyi*. This bacteriocin was named harveyicin (McCall & Sizemore, 1979).

In 1982, Hoyt and Sizemore investigated the chemical nature of harveyicin. They showed that it is a proteinaceous substance, sensitive to protease, trypsin, and papain treatment, and resistant to pepsin and lipase treatment. Furthermore, it is stable at room temperature and -20 °C for several weeks and several months, respectively. However, the protein did lose killing activity after heat treatment at 55 °C for 4 hours (Hoyt & Sizemore, 1982). Hoyt and Sizemore (1982) also investigated the role of harveyicin in bacterial competition in enteric and planktonic environments at various temperatures (4-39°C), pH's (5-9.5), and salt concentrations (1.75% and 3.5%). They also performed a competition assay on nonluminous bacteriocinogenic (SYLum-) and luminous bacteriocin sensitive strains (SYcured) in equal concentrations (10^6 cells/ml). The ratio of harveyicin producer strain (SYLum-) to the harveyicin sensitive strain (SYcured) peaked in an enteric environment (25 °C, high salinity (3%), and alkaline pH (pH 9.5)). The harveyicin producer strain also outcompeted the harveyicin sensitive strain at an acidic pH (pH 5.0) (Hoyt and Sizemore, 1982). Given the fact that the natural habitat of these species is in the guts of fish (an acidic environment), bacteriocin production may serve as a competitive advantage to the bacteriocin producer strain in this environment.

The identification of harveyicin led to numerous bacteriocin-screening studies in marine bacteria, which focused primarily on biochemical characterization of bacteriocins and bacteriocin-like inhibitory substances (Bagenda et al., 2008; Bhugaloo-Vial et al., 1996; Carraturo et al., 2006; Hosseini et al., 2009; Hoyt & Sizemore, 1982; Longeon et al., 2004; McCall & Sizemore, 1979; Messi et al., 2003; Metivier et al., 1998; Moro et al., 1997; Nilsson et al., 2002; Pinto et al., 2009; Pirzada et al., 2004; Prasad et al., 2005; Rajaram et al., 2010; Selvin et al., 2004; Shehane & Sizemore, 2002; Stoffels et al., 1992; Sugita et al., 1997; Suzuki et al., 2005; Tahiri et al., 2004; Valenzuela et al., 2010; Yamazaki et al., 2005; Zai et al., 2009). The majority of these studies focused on the killing breadth of the producing strains due to their potential for use as antimicrobials and probiotics. However, the authors generally did not further characterize the identified bacteriocins/BLIS, although see Table 3 for rough classifications (Table 3).

The relatively few characterized marine bacteriocins are primarily isolated from *Carnobacterium* species, which are ubiquitous, Gram-positive lactic acid bacteria isolated from marine organisms (such as fish and marine sponges), from cold and temperate environments, as well as from terrestrial environments including Canadian winter soil, permafrost ice, composite piles, and horse manure (Leisner et al., 2007). *C. divergens* and *C. maltaromaticum* (formerly known as *C. piscicola*) are the most-studied species from this genus.

Carnobacterium species can produce bacteriocins at low temperatures and high salt concentrations (Buchanan & Bagi, 1997). Further, the bacteria survive in fish products and have low acidifying capacity (Tahiri et al., 2009). Thus, these species have been the focus of intense research due to their potential as probiotics (Leisner et al., 2007; Rihakova et al., 2009). Piscicocin V1a, V1b, divercin V41, piscicocin CS526, divergicin M35, carnocin U149, and carnobacteriocin B2 are some of the bacteriocins isolated from marine *Carnobacterium* species (Table 3) (Bhugaloo-Vial et al., 1996; Duffes et al., 1999; Metivier et al., 1998; Stoffels et al., 1992; Suzuki et al., 2005; Tahiri et al., 2004; Yamazaki et al., 2005). These bacteriocins share similar characteristics with the Class II bacteriocins of Gram-positive bacteria.

Bacteriocin	Producer Strain	Enzyme Sensitivity	Enzyme Resistance	Molecular Weight	Killing Breadth	Source of Isolation	Reference
BLIS	*Aeromonas hydrophila*	NA	NA	NA	*Staphylococcus aureus*, MRSA	Water sample from a water tank containing alligators	Moro et al., 1997
BLIS	*Aeromonas hydrophila*	NA	NA	NA	*Staphylococcus, Listeria, Streptococcus, Lactobacillus*	Water samples	Messi et al., 2003
Carnocin U149	*Carnobacterium* sp.	NA	NA	4.5-5 kDa	*Lactobacillus, Lactococcus, Pediococcus, Carnobacterium*	Fish	Stoffels et al., 1992
Divergicin M35	*Carnobacterium divergens* M35	α-chymotrypsin, Proteinase K, Pronase E	Trypsin	~4.5 kDa	*Listeria, Carnobacterium*	Frozen smoked mussel	Tahiri et al., 2004
Divercin V41	*Carnobacterium divergens* V41	Pronase E	Catalase	4.5 kDa	*Listeria, Carnobacterium, Enterococcus*	Fish viscera	Métivier et al., 1998; Duffes et al., 1999
Carnobacteriocin B2	*Carnobacterium piscicola* A9b	Proteases, Proteinase K, Trypsin	Pepsin, Lipase, Phospholipase C, α-amylase	~4.5 kDa	*Listeria*	Cold smoked salmon	Nilsson et al., 2002
Piscicocin CS526	*Carnobacterium piscicola* CS526	α-chymotrypsin, Papain, Proteinase K, Actinase, Trypsin	Catalase, RNase, Lipase	~4.4 kDa	*Listeria, Enterococcus, Pediococcus, Tetragenococcus, Leuconostoc*	Frozen surimi	Yamazaki et al., 2005; Suzuki et al., 2005
Piscicocin V1a	*Carnobacterium piscicola* V1	NA	NA	4.4 kDa	*Listeria, Enterococcus, Pediococcus, Leuconostoc, Lactobacillus, Carnobacterium*	Fish	Bhugaloo-Vial et al., 1996
Piscicocin V1b (or carnobacteriocin BM1)	*Carnobacterium piscicola* V1	NA	NA	4.5 kDa	*Listeria, Enterococcus, Pediococcus, Lactobacillus, Carnobacterium*	Fish	Bhugaloo-Vial et al., 1996
BLIS	*Enterococcus feacium* CHG 2-1 and Ch 1-2	NA	NA	NA	*Enterococcus*	Venus clams, Horse macker	Valenzuela et al., 2010
BLIS	*Enterococcus feacium* CHG 2-2, CHG 2-3, C 2-3, PE 1-2, PE 2-1, PE 3-2, M 2-1, and M 2-2	NA	NA	NA	*Enterococcus, Listeria*	Venus clams, Dogfish fillet, Swordfish fillet, Shark fillet	Valenzuela et al., 2010
BLIS	*Enterococcus feacium* C-K, C-S, M 2-1, and PEF 2-2	NA	NA	NA	*Listeria*	Anchovy, Shark fillet, Swordfish fillet	Valenzuela et al., 2010
Enterocin B-like BLIS	*Enterococcus feacium* ALP7	Trypsin, Proteinase K, Pronase E, Papain	Lipase, α−amylase	<6.5 kDa	*Listeria, Staphylococcus, Bacillus, Enterococcus, Lactobacillus, Lactococcus, Leukonostoc*	Non-fermented shellfish	Pinto et al., 2009
BLIS	*Enterococcus feacium* LHICA46	Proteinase K	Pepsin, Pancreatin	NA	*Bacillus, Carnobacterium, Listeria, Staphylococcus*	Refrigerated turbot	Hosseini et al., 2009
Enterocin A	*Enterococcus feacium* PE 2-2	NA	NA	NA	*Enterococcus, Listeria, Staphylococcus*	Swordfish fillet	Valenzuela et al., 2010

Table 3. Bacteriocin and BLIS activity characterized from marine bacteria (NA: Not available)

Bacteriocin	Producer Strain	Enzyme Sensitivity	Enzyme Resistance	Molecular Weight	Killing Breadth	Source of isolation	Reference
BLIS	Gram-positive marine bacterium ZM-81	Pronase, Trypsin	Pepsin, Lyzozyme, Lipase	>10 kDa	Gram-positive marine bacterium ZM-19	Water sample, Karachi coast, Pakistan	Pirzada & Ali, 2004
BLIS	*Lactobacillus lactis*	Proteinase K, Pepsin	α-amylase, DNase, RNAse, Lipase	94 kDa	*Bacillus, Staphylococcus, Enterococcus, Escherichia coli , Pseudomonas, Shigella*	Sediment sample from the Bay of Bengal, Indian Ocean	Rajaram et al., 2010
Pediocin PA1-like BLIS	*Pediococcus pentosaceus* ALP57	Trypsin, Proteinase K, Pronase E, Papain	Lipase, α−amylase	<6.5 kDa	*Listeria, Bacillus, Enterococcus, Lactobacillus, Lactabacillus, Lactococcus, Leukonostoc*	Non-fermented shellfish	Pinto et al., 2009
BLIS	*Pediococcus pentosaceus* Iz3.13	α-chymotrypsin, Proteinase K	Catalse, Lipase, , α-[illegible]	4.6 kDa	*Clostridium botulinum*	Fermented izushi	Bagenda et al., 2008
BLIS P-153	*Pseudoalteromonas* spp. X153	NA	NA	87 kDa	*Pseudomonas, Escherichia coli, Staphylococcus, Propionibacterium, Candida, Pityrosporum, Vibrio* spp., *Deleya, Halomanas, Cytofaga* spp., *Bacillus* spp., *Pseudomonas* spp.	A pebble collected at St. Anne du Portzic, France	Longeon et al., 2004
Phocaecin PI80	*Streptococcus phocae*	Trypsin, Protease, Pepsin, Chymotrypsin	Catalase, Peroxidase, Diastase	9.2 kDa	*Listeria, Vibrio*	Indian white shrimp	Kumar & Arul, 2009
BLIS	*Streptomyces* sp. BTL-7	NA	NA	NA	*Micrococcus, Pseudomonas, Escherichia coli, Bacillus, Salmonella, Staphylococcus, Vibrio, Klebsiella, Clostridium*	Marine sponge	Selvin et al., 2004
BLIS	*Vibrio* sp. Strain NM10	Trypsin, Proteinase K	α−chymotrypsin, Protease, Lysozyme, Achromopeptidise, α-amylase, Ribonuclease A,	< 5 kDa	*Bacillus* spp., *Coryneforms, Enterobacteriaceae, Flavobacterium* spp., *Pseudomonas* spp., *Vibrio* spp., *Pasteurella*	Intestine of a spotnape ponyfish	Sugita et al., 1997
BLIS	*Vibrio anguillarum* AVP10	Proteinase, Proteinase K, Trypsin	Lipase	NA	*Vibrio* spp.	Catfish	Zai et al., 2009
BLIS-BC2	*Vibrio cholerae*	Protease, α-amylase	Lipase	1.35 kDa	*Vibrio, E. coli*	Water samples, Wilmington, NC	Shehane & Sizemore, 2002
Harveyicin	*Vibrio harveyi* SY	Trypsin, Papain	Pepsin, Lipase	~24 kDa	*Vibrio*	Galveston Island, Texas	McCall & Sizemore, 1979; Hoyt & Sizemore, 1982
BLIS	*Vibrio harveyi* VIB 571	Proteinase K, Pepsin, Trypsin, Pronase E, Lipase	Lysozyme, α-amylase	~32 kDa	*Vibrio* spp.	Sea bass	Prasad et al., 2005
BLIS-400	*Vibrio mediterranei* 1	Proteinase K	Trypsin, α-chymotrypsin	63-65 kDa	*Aeromonas, Vibrio*	Seafood	Carraturo et al., 2006
BLIS-BC1	*Vibrio vulnificus*	Protease, α-amylase	Lipase	7.5 kDa	*Vibrio*	Water samples, Wilmington, NC	Shehane & Sizemore, 2002
BLIS-IW1	*Vibrio vulnificus*	Protease, α-amylase	Lipase	9.0 kDa	*Vibrio*	Water samples, Wilmington, NC	Shehane & Sizemore, 2002

Table 3. Continued

BLIS from the marine environment are a relatively uncharacterized group of inhibitory substances. While they may not be well characterized, they are abundant. Indeed, numerous BLIS have been identified from marine species belonging to *Vibrio, Aeromonas, Carnobacterium, Lactococcus, Streptomyces, Pseudoalteromonas, Enterococcus,* and *Pediococcus* genera (Table 3) (Bagenda et al., 2008; Carraturo et al., 2006; Longeon et al., 2004; Messi et al., 2003; Moro et al., 1997; Pinto et al., 2009; Pirzada et al., 2004; Prasad et al., 2005; Rajaram et al., 2010; Satish Kumar & Arul, 2009; Selvin et al., 2004; Shehane & Sizemore, 2002; Sugita et al., 1997; Valenzuela et al., 2010; Zai et al., 2009). These species are important in the seafood industry and in human health.

The bacteriocins and BLIS isolated from marine microorganisms are diverse. However, they do share common characteristics with bacteriocins from Bacteria and Archaea (Fig. 3). They can be small peptides (5-10 kDa) like microcins of Gram-negative bacteria, microhalocins of halobacteria, and class I and II bacteriocins of Gram-positive bacteria. They can also be larger in size (10-90 kDa) like colicins and colicin-like bacteriocins of Gram-negative bacteria. The majority of marine BLIS have been tested against a number of proteases including trypsin, proteinase K, and pronase A, which are commonly used to identify bacteriocin activity. Some have unique characteristics. For example, BLIS-IW1, BLIS-BC1, and BC2 from *Vibrio* species possess a carbohydrate moiety while BLIS VIB 571 from *V. harveyi* has a lipid moiety (Shehane & Sizemore, 2002). It is not clear if these moieties are

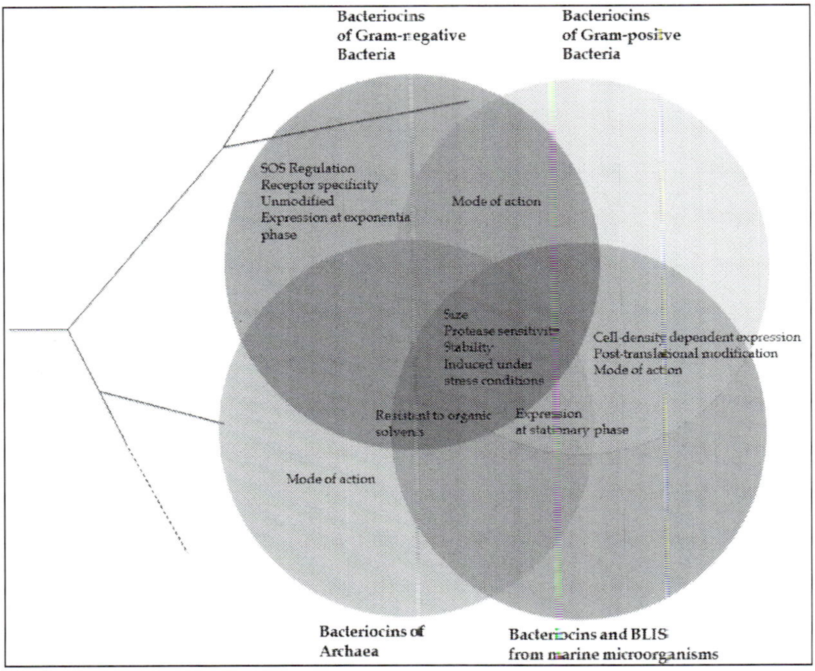

Fig. 3. Shared and unique characteristics of bacteriocins and BLIS from marine Bacteria with bacteriocins of terrestrial Bacteria and Archaea

involved in BLIS activity. The mode of actions of these BLIS is unknown and their killing ranges are variable; either narrow - like bacteriocins of Gram-negative bacteria - or quite broad - like that observed in bacteriocins of Gram-positive bacteria (Bhugaloo-Vial et al., 1996; Moro et al., 1997; Sugita et al., 1997; Suzuki et al., 2005; Yamazaki et al., 2005).

Further, the stability of BLIS activity varies. BLIS IW1 is inactive after cold storage at -70C; like phage-tail-like bacteriocins. Others, such as BLIS-BC1, BLIS-BC2, and BLIS-400, are still active after cold treatment, as is observed in colicin and colicin-like bacteriocins. Some BLIS are also resistant to treatment with organic solvents. For example, BLIS-400 is resistant to organic chemicals including ethanol, methanol, acetone, and chloroform (Carraturo et al., 2006). This feature is similar to some bacteriocins isolated from Gram-positive bacteria and Archaea (Carraturo et al., 2006). Finally, most BLIS from marine bacteria are produced during the stationary phase of growth, similar to bacteriocins from Gram-positive bacteria (Pinto et al., 2009; Tahiri et al., 2004).

4. Applications and implications of marine bacteriocins

The international seafood industry is one of the world's most profitable commodities, worth more than $75 billion per year (Food and Agriculture Organization of the United Nations [FAO], 2006). Fish and seafood are major proteins in some areas of the world. In 2006, of the 143 million tons of total fishery production (including fish, crustacean, and mollusks), 110 million tons was for direct human consumption (Pilet & Leroi, 2011). To meet this demand, we have seen a marked rise in aquaculture (the farming of aquatic plants and animals) in the last decade (FAO, 2006). Recently, there have also been dramatic changes in the seafood industry due to technological advances, consumer habits, and globalization of the food market (Galvez et al., 2008). In particular, there has been an increase in consumer preference for foods that are minimally processed or preserved (especially those that claim health-promoting benefits). Consumers are also demanding that these foods be fresh tasting and ready-to-eat (Galvez et al., 2008). The demands on the industry to provide fresh, minimally preserved products in the ever-growing globalized food market is requiring a longer and more complex food-chain and increasing the risk of microbial contamination and spoilage (García et al., 2010).

4.1 Challenges in the seafood Industry: Spoilage and disease

With the expansion of the seafood industry and the rise in seafood consumption, spoilage and disease in fish are the main challenges the industry faces (Gram & Dalgaard, 2002; Toranzo et al., 2005). Microorganisms are the major cause of spoilage and diseases in the seafood industry. It is estimated that nearly 25% of all the seafood produced is lost to microbial spoilage (Baird-Parker, 2000). Microbes cause changes in the sensory properties of the seafood (smell, taste, color), which make it less appealing and often dangerous to eat (Gram & Dalgaard, 2002). Disease severely affects the production and trade of farmed seafood, creating high economic impacts for many countries (Bondad-Reantaso et al., 2005).

4.2 Common microbial diseases in aquaculture

Fear of disease, as well as climate change, are acting as deterrents to aquaculture farming (Bondad-Reantaso et al., 2005). It has been shown that bacteria exhibit greater pathogenesis at higher temperatures, leading to greater and more virulent disease in aquaculture (Desriac et al., 2010). The bacterial microflora of fish can become pathogenic under stress conditions such as sudden temperature changes, overcrowding, and poor water quality conditions and thus can cause diseases in fish. Furunculosis, vibriosis, columnaris disease, streptococcosis, pasteurellosis, fish tuberculosis, and enteric septicemia are common microbial diseases observed in economically important aquaculture fish species. These diseases are predominantly caused by *Aeromonas, Vibrio, Cytophaga, Streptococcus, Pasteurella, Mycobacterium,* and *Edwardsiella* genera (Table 4).

Microorganism	Hosts	Diseases	Antibiotic Treatment	References
Aeromonas	Salmonids, catfish, carp, tilapia, trout, perch, goby	Frunculosis MAS Erythrodermatitis, Ulcer	Florfenicol Sulfadimethoxine and Ormetoprim Oxytetracycline dihydrate	Noga, 2010
Vibrio	Salmonids, cod, ayu, Japanese eels	Vibriosis Cold water vibriosis, Hemorrhagic septicmia Skin ulcers	Oxytetracycline Potentiated sulfonamides Oxolinic acid	Noga, 2010 Toranzo et al., 2005
Cytophaga	Salmonids, catfish, tilapia, stripped bass	Columnaris disease	Oxytetracycline dihydrate Florfenicol	Noga, 2010
Streptococcus	Stripped bass, tilapia, turbot, barramundi, Atlantic Salmon	Streptococcosis	Oxytetracycline Amoxicillin Erythromycin	Noga, 2010 Toranzo et al., 2005
Pasteurella	Yellow tail, seabass	Pasteurellosis	Oxytetracycline	Noga, 2010 Toranzo et al., 2005
Mycobacterium	Striped bass, seabass, Atlantic salmon	Fish tuberculosis	Ampicillin Erythromycin thiocyanate	Noga, 2010 Jacobs et al., 2009
Edwardsiella	Channel catfish	Enteric septicemia Fish gangrene	Florfenicol Sulfadimethoxine and Ormetoprim	Noga, 2010 Mohanty & Shaoo, 2007

Table 4. Fish pathogens, corresponding diseases, and antibiotic treatment regimens

Furunculosis (skin infection) and motile Aeromonas septicemia (MAS) are two major fish diseases caused by *Aeromonas* species (Noga, 2010). Nonmotile, psychrophilic *Aeromonas salmonicida* is the causative agent of furunculosis in salmonids (O'Brien et al., 1994). Besides furunculosis in salmonids, it causes erythrodermatitis in carp and ulcers in trout (Noga, 2010). Furthermore, motile mesophilic *Aeromonas hydrophila* and *Aeromonas veronii* are the causative agents of MAS in carp, tilapia, perch, catfish, and salmon. They also cause ulcer disease in catfish, cod, carp, and goby (Noga, 2010).

Vibrio species cause systemic infections in fish (Vibriosis). The common symptoms are loss of appetite and skin ulcers, which are associated with septicemia. *Vibrio anguillarum* and *Vibrio salmonicida* are the causative agents of vibriosis and cold water vibriosis in salmon and cod, respectively (Colwell & Grimes, 1984). Further, *Vibrio vulnificus* causes hemorrhagic septicemia in Japanese eels and *V. damsela* causes skin ulcers (Toranzo et al., 2005).

Columnaris disease affects the skin and gills of freshwater fish including salmonids, catfish, striped bass, and tilapia (Noga, 2010). Infection causes degradation of the cartilage tissue, the major cause of death. *Cytophaga columnaris* as well as *Bacillus columnaris, Flexibacter columnaris,* and *Flavobacterium columnare* are the causative agents of columnaris disease (Noga, 2010).

Streptococcosis is a fish disease caused by *Streptococcus* species. Some of the clinical symptoms of streptococcosis are anorexia, loss of orientation, erratic swimming, lethargy, and hemorrhaging (Noga, 2010). *Streptococcus iniae* is the causative agent of streptococcosis in at least 27 species of farmed finfish. The worldwide impact of streptococcosis was estimated globally to be around $100 million (USD) (Agnew &Barnes, 2007).

Pasteurellosis (pseudotuberculosis) is one of the most important diseases in Japanese aquaculture affecting commercial yellowtail, ayu, black sea bream, red sea bream, and hybrid striped seabass (Noga, 2010). *Pasteurella piscidium* is the causative agent of this disease (Romalde, 2002). Hemorrhaging around the gills and lesions in the skin, liver, and kidneys are common symptoms in acute and chronic forms of this disease (Noga, 2010).

Mycobacteriosis (or fish tuberculosis) can be observed in nearly 200 species including salmonids, seabass, turbo, cod, and halibut (Noga, 2010). *Mycobacterium marinum* is the most common *Mycobacterium* species causing tuberculosis. Greyish-white nodules in the spleen, liver, and kidney, and hemorrhagic lesions are among the symptoms of this disease (Jacobs et al., 2009).

Edwardsiellosis (emphysematous putrefactive disease) is commonly observed in carp, tilapia, eel, catfish, salmon, and trout (Noga, 2010). *Edwardsiella tarda* and *Edwardsiella ictaluri* are the main species causing Edwardsiellosis. Gas-filled lesions in the skeletal musculature are the major clinical sign of this disease (Mohanty & Sahoo, 2007). Further, *Edwardsiella ictaluri* is associated with enteric septicemia in channel catfish (Mohanty & Sahoo, 2007)

4.3 Current practices in the seafood industry to combat spoilage and disease

The seafood industry employs numerous techniques to eliminate microorganisms from their products. The oldest and still widely used form of seafood preservation is drying/salting. Besides keeping the seafood chilled in cold-water or on ice, this is the most low-tech preservation technique. There are many variations to this method such as wet-salting or additional acidification, but most achieve the same results. In this method, the fish is dried (with or without salt), which creates an environment devoid of vast nutrients (and possibly with high salinity). This prevents most bacterial growth, but spoilage can still occur due to filamentous fungi growth or insect infestation. Yeast is also able to grow in heavily wet-salted fish (Gram & Dalgaard, 2002).

Other preservation methods include washing with disinfectants, including chlorinated water, iodophores, salts, organic compounds, aldehydes, hydrogen peroxide, quaternary ammonium compounds, and antiseptic dyes (Calo-Mata et al., 2007; Shao, 2001). Disinfectants are mostly used to kill fungi and parasites. However, they may also select for antibiotic resistance in bacteria (Murray et al., 1984). Seafood can also be marinated in an acidic solution to prevent bacterial growth. Vacuum-packing and preservatives such as sorbate and benzoate have also been employed to prevent microbial growth (Einarsson & Lauzon, 1995; Gram & Dalgaard, 2002). Recently, complex solutions such as carbon-dioxide packing, spray-drying of antimicrobials, and radiofrequency heating have been applied to fight these ever-present problems of spoilage and contamination (Calo-Mata et al., 2007; Galvez et al., 2007; Gram & Dalgaard, 2002).

The use of vaccines and antibiotics in aquaculture is aimed at preventing the initial colonization by microorganisms. The means of administration of these prophylactics is simple, they are either added to the water or feed, or given by injection (Shao, 2001). A large body of research in the 1970s resulted in vaccines against numerous seafood pathogens, primarily species of *Vibrio* (Shao, 2001). Although vaccines are effective (and cost-effective), there are still no vaccines against shrimp and mollusc pathogens (Subasinghe, 2009).

An alternative to vaccination is the use of antibiotics to prevent bacterial infections. Tetracycline has become one of the most popular antibiotics in aquaculture due to its low-cost, low-toxicity, and high efficacy. Further, florfenicol, sulfadimethoxine/ormetoprim, oxytetracycline, and sulfonamides, are used to treat common bacterial infections (Table 4). However, the rampant use of antimicrobials in this industry has created massive selective pressure for bacteria to develop resistance (World Health Organization [WHO], Fact sheet 194). While these drugs are effective in killing bacteria, they also play a much more nefarious role in aquaculture. Antibiotics have varying half-lives, meaning they degrade at different rates. Some antibiotics degrade slowly and thus proliferate in the aquatic environment (Cabello, 2006). Worse still, the drugs flow into open waterways, sewage systems, sediments, and can even remain in the flesh of the farmed seafood (Benbrook, 2002; Cabello, 2006). These antibiotics continue to impose selective pressures leading to resistance until they are eventually degraded. This resistance is not only seen in the bacteria that inhabit the seafood produced in antibiotic-using aquaculture facilities, but in the animals neighboring the facilities as well (Benbrook, 2002). It has also been shown that these resistant bacteria are able to horizontally transfer their resistance-conferring genes to other human pathogens (Benbrook, 2002). For these reasons, governing agencies, such as the US Food and Drug Administration (FDA) and Environmental Protection Agency (EPA), strictly regulate the use of antibiotics in aquaculture.

Due to the problems associated with antibiotic use, the seafood industry is exploring the use of probiotics to promote the growth of healthy microflora in the seafood that can combat infecting pathogens. Probiotics are live microbial feed supplements that beneficially affect the host animal by improving its intestinal health (Fuller, 1989). While this definition needs a bit of tweaking to fit the seafood industry, it does still apply.

One area of active research in seafood aquaculture is the utilization of bacteriocins as antimicrobials. Bacteriocins have a long history of use in dairy or meat applications and there is an increasing number of studies on the effect of bacteriocins as antimicrobials in the seafood industry (Table 5) (Aasen et al., 2003; Al-Holy et al., 2004; Budu-Amoako et al., 1999; Einarsson & Lauzon, 1995; Elotmani & Assobhei, 2004; Luders et al., 2003; Neetoo et al., 2008; Nilsson et al., 1997; Nykanen et al., 2000; Szabo &Cahill, 1999; Tahiri et al., 2009; Tsironi & Taoukis, 2010; Zuckerman & Ben Avraham, 2002). These studies have focused largely on the effects of nisin, a Gram-positive bacteriocin that has been generally recognized as safe (GRAS) by the FDA. Early studies of nisin indicated that it delayed growth of *L. monocytogenes* in cold-smoked salmon. Later research revealed that the addition of CO_2 atmospheric packing significantly increased the effectiveness of nisin against *Listeria* (Nilsson et al., 1997).

There has also been encouraging research into nisin-coated packaging. Neetoo et al. (2008) investigated the effect of nisin-coated plastic films on the survival of *L. monocytogenes* on vacuum-packed cold smoked salmon. This study showed that nisin-coated plastic films

Bacteriocin	Target	Seafood product	Reference
Bavaricin A	Extended shelf-life	Shrimp	Einarsson et al., 1995
Carnocin U149	Extended shelf-life	Shrimp	Einarsson et al., 1995
Divergicin M35	*L. moncytogenes*	Salmon	Tahiri et al., 2009
Nisin	*L. moncytogenes*	Salmon	Nilsson et al., 1997
Nisin	*L. moncytogenes*	Salmon	Nilsson et al., 1997
Nisin	*L. moncytogenes*	Salmon	Szabo and Cahill, 1999
Nisin	*L. moncytogenes*	Salmon	Neetoo et al., 2008
Nisin	*L. moncytogenes*	Salmon	Zuckerman and Ben Avraham, 2002
Nisin	*L. moncytogenes*	Trout	Nykanen et al., 2000
Nisin	*L. moncytogenes*	Lobster	Budu-Amoako et al., 1999
Nisin	*L. innocua*	Caviar and ikura	Al-Holy et al., 2004
Nisin	Aerobic bacteria	Salmon	Zuckerman and Ben Avraham, 2002
Nisin	Bacterial flora	Sardines	Elotmani et al., 2004
Nisin	Extended shelf-life	Fish	Tsironi and Taoukis, 2010
Nisin Z	Extended shelf-life	Shrimp	Einarsson et al., 1995
Pediocin	*L. moncytogenes*	Salmon	Szabo and Cahill, 1999
Sakacin P	*L. moncytogenes*	Salmon	Aasen et al., 2003
Sakacin P	*E. coli*	Salmon	Luders et al., 2003

Table 5. Examples of bacteriocin trials in seafood products (Adapted from Galvez et al. 2008).

reduced the number of *L. monocytogenes* by 3.9 log CFU/cm^2 at 4 $^{\circ}$C and 10 $^{\circ}$C after 56 and 49 days of incubation, respectively. Further, this study also showed that nisin-coated plastic films suppressed the growth of other aerobic and anaerobic spoilage microorganisms in a concentration-dependent manner (Neetoo et al., 2008).

The combination of nisin with heat has also been shown as an effective method to prevent *L. monocytogenes* contamination. Budu-Amoako et al. (1999) applied nisin along with moderate heating on cold-packed lobster and showed a reduction of L. *monocytogenes* by 3-5 logs in comparison to nisin and/or heat treatment alone (Budu-Amoako et al., 1999). Further, Al-Holy et al. (2004) used a radio-frequency heating method to apply heat treatment in conjunction with nisin. The combination of nisin and radio-frequency heating caused reduction of *L. innocua* by 100% (Al-Holy et al., 2004). With an industry moving away from traditional preservation techniques, bacteriocins (such as nisin) offer a promising alternative as antimicrobials in the seafood industry.

4.4 Bacteriocin potential in the seafood industry

There are a number of factors that play a significant role in the potential to use bacteriocins as probiotics and/or bio-preservatives in the seafood industry. The natural microbiota of the seafood needs to continue to be surveyed for its sensitivity to bacteriocins. This information should be incorporated into the guidelines for bacteriocin use in order to use these proteins prudently against relevant pathogens. The environmental conditions, such as pH and temperature, during seafood growth and processing could also affect the activity of applied bacteriocins and requires further investigation (Galvez et al., 2007).

Despite these factors, research on aquatic microorganisms has shown that bacteriocin production and diversity in aquatic environment is abundant (Bagenda et al., 2008; Bhugaloo-Vial et al., 1996; Carraturo et al., 2006; Hosseini et al., 2009; Hoyt & Sizemore, 1982; Longeon et al., 2004; McCall & Sizemore, 1979; Messi et al., 2003; Metivier et al., 1998; Moro et al., 1997; Nilsson et al., 2002; Pinto et al., 2009; Pirzada et al., 2004; Prasad et al., 2005; Rajaram et al., 2010; Selvin et al., 2004; Shehane & Sizemore, 2002; Stoffels et al., 1992; Sugita et al., 1997; Suzuki et al., 2005; Tahiri et al., 2004; Valenzuela et al., 2010; Yamazaki et al., 2005; Zai et al., 2009). Bacteriocins have numerous qualities that make them attractive as alternatives to antibiotics. They have been shown to be non-toxic to eukaryotic cells and are GRAS, making them a safe alternative to traditional antimicrobials (Galvez et al., 2008). It has also been shown that purified bacteriocins do not effect the sensory qualities of seafood and that they are stable up to a salinity concentration of 10%. Additionally, the relatively narrow killing spectrum of bacteriocins compared to traditional antibiotics limits the selective pressure for bacteria to evolve resistance to these antimicrobials and thus reduces the incidence of drug-resistant pathogens.

Because of the above stated reasons, some have suggested that bacteriocins should be applied to foods by spray-drying as either dried bacteriocins or probiotic bacteriocinogenic strains (Calo-Mata et al., 2007; Galvez et al., 2007). For example, Brillet et al. (2005) has shown that bacteriocin producer *Carnobacterium divergens* V41 can be used as a biopreservative to inhibit the growth of *Listeria monocytogenes* in cold smoked salmon (Brillet et al., 2005). This study showed that spray application of *C. divergens* V41 on commercial smoked salmon did not affect the sensory qualities of the salmon (Brillet et al., 2005). Additionally, Schobitz et al. (1999) directly applied a BLIS from *Carnobacterium piscicola* into vacuum-packed meat, which inhibited the growth of *L. monocytogenes* in the vacuum-packed meat after 14 days of storage at 4 °C (Schobitz et al., 1999). These studies aid in the argument that bacteriocins should be used as a biopreservation technique in the seafood industry. This technology has already emerged in the terrestrial food industry as we see with nisin (an FDA approved food additive) and Microgard™ (a milk-based BLIS).

It has also been suggested that bacteriocins could be combined with current methods of antimicrobial treatment and preservation to produce synergistic effects, such as incorporating bacteriocins into bio-active packaging (Calo-Mata et al., 2007; Galvez et al., 2007; Pilet & Leroi, 2011). For example, bacteriocins can be impregnated into gel coatings and/or polyethylene films and can be applied to seafood during packaging (Neetoo et al., 2008). The application of bacteriocins on packaged seafood is steadily being seen as a very promising biopreservation method (Aasen et al., 2003; Al-Holy et al., 2004; Budu-Amoako et al., 1999; Einarsson & Lauzon, 1995; Elotmani & Assobhei, 2004; Luders et al., 2003; Neetoo et al., 2008; Nilsson et al., 1997; Nykanen et al., 2000; Szabo &Cahill, 1999; Tahiri et al., 2009; Tsironi & Taoukis, 2010; Zuckerman & Ben Avraham, 2002). In fact, immobilization of bacteriocins on coating materials for biopreservation may actually reduce the cost of packaging due to the reduced amount and cost of the antibacterial needed to attach to the film (Galvez et al., 2008). Creating combinations of bacteriocins and current methods used in the seafood industry has the potential to increase the guarantee of freshness by assuring the inhibition of spoilage causing microorganisms.

One trouble that the industry was having was the scale-up of these bacteriocins to levels that were high enough for use in pilot studies and/or on the industrial scale. However, there is technology in the pipeline that will make this an issue of the past (Galvez et al., 2008). As the exploration of the aquatic environments of our planet increases, we are sure to find new and exciting bacteriocins, which could play a vital role in antimicrobial effects and biopreservation in the seafood industry.

5. Conclusions and future research

Bacteriocins have been the focus of an extensive number of studies for the past sixty years due to their important role in nature and more recently, their potential for use as therapeutics and probiotics. Most studies of bacteriocins initially focused on phenotypic and molecular characterization of these toxins. However, due to their high potency and relatively narrow killing spectrum, they were quickly recognized as a natural alternative to antibiotics.

Knowledge regarding bacteriocins and their potential applications from terrestrial bacteria is vast. Gram-negative bacteriocins such as colicins and microcins of enteric bacteria and pyocins of *P. aeruginosa* have great promise in human and veterinary medicine. In addition, Gram-positive bacteriocins such as nisin, pediocin, and lacticin have been developed for use as food preservatives.

The focus of this chapter was to explore the bacteriocin and BLIS activity characterized from marine microorganisms and assess their potential applications in aquaculture. The bacteriocins and BLIS from marine microorganisms are under-studied relative to their terrestrial counterparts. Thus far, most studies are limited to the identification of BLIS activity and characterization of its killing breadth. Some studies have further characterized these proteinecous killing agents and classified them based on similarities to known bacteriocins. However, the abundance and diversity of bacteriocins in marine microorganisms remains to be fully explored. We predict that a wealth of interesting bacteriocin proteins can be easily identified as the screening efforts proceed.

Further, there is increased interest in the use of bacteriocins as alternatives to classical antibiotics in aquaculture. Bacteriocins are highly potent against marine pathogens and environmentally safe, due to the fact that they do not create intensive selection pressures for antibiotic resistance. Clearly, bacteriocins could prove extremely beneficial to the seafood industry and more research should be dedicated to exploring their potential applications as probiotics and therapeutics. Given the fact that all species of bacteria have the potential to produce bacteriocins, and only a handful have thus far been identified from marine microorganisms, we are confident that existing studies have exposed only the tip of the iceberg, in terms of bacteriocin diversity and potential use in the seafood industry.

6. References

Aasen, I.M., Markussen, S., Moretro, T., Katla, T., Axelsson, L., & Naterstad, K. (2003). Interactions of the bacteriocins sakacin P and nisin with food constituents. *International journal of food microbiology*, Vol.87, No.1-2, pp. 35-43, ISSN 0168-1605

Abee, T., Krockel, L., & Hill, C. (1995). Bacteriocins: modes of action and potentials in food preservation and control of food poisoning. *International journal of food microbiology*, Vol.28, No.2, pp. 169-185, ISSN 0168-1605

Agnew, W., & Barnes, A.C. (2007). Streptococcus iniae: an aquatic pathogen of global veterinary significance and a challenging candidate for reliable vaccination. *Veterinary microbiology*, Vol.122, No.1-2, pp. 1-15, ISSN 0378-1135

Al-Holy, M., Ruiter, J., Lin, M., Kang, D.H., & Rasco, B. (2004). Inactivation of Listeria innocua in nisin-treated salmon (Oncorhynchus keta) and sturgeon (Acipenser transmontanus) caviar heated by radio frequency. *Journal of food protection*, Vol.67, No.9, pp. 1848-1854, ISSN 0362-028X

Bagenda, D.K., Hayashi, K., Yamazaki, K., & Kawai, Y. (2008). Characterization of an antibacterial substance produced by Pediococcus pentosaceus Iz3.13 isolated from Japanese fermented marine food. *Fisheries Science*, Vol.74, No.2, pp. 439-448, ISSN 0919-9268

Baird-Parker, T., (2000). The production of microbiologically safe and stable foods, In:*The microbiological Safety and Qaulity of Food*, Lund, B.M., Baird-Parker, T., Gaithersburg, G.G.W. (Eds.), pp. 3-18, ISBN 0-8342-1323-0, Gaithersburg, Maryland

Benbrook, C.M., (2002). Antibiotic Drug Use in U.S. Aquaculture, The Northwest Science and Environmental Policy Center, Sandpoint, Idaho.

Bhugaloo-Vial, P., Dousset, X., Metivier, A., Sorokine, O., Anglade, P., Boyaval, P., & Marion, D. (1996). Purification and amino acid sequences of piscicocins V1a and V1b, two class IIa bacteriocins secreted by Carnobacterium piscicola V1 that display significantly different levels of specific inhibitory activity. *Applied and Environmental Microbiology*, Vol.62, No.12, pp. 4410-4416, ISSN 0099-2240

Bondad-Reantaso, M.G., Subasinghe, R.P., Arthur, J.R., Ogawa, K., Chinabut, S., Adlard, R., Tan, Z., & Shariff, M. (2005). Disease and health management in Asian aquaculture. *Veterinary parasitology*, Vol.132, No.3-4, pp. 249-272, ISSN 0304-4017

Bradley, D.E. (1967). Ultrastructure of bacteriophage and bacteriocins. *Bacteriological reviews*, Vol.31, No.4, pp. 230-314, ISSN 0005-3678

Brillet, A., Pilet, M.F., Prevost, H., Cardinal, M., & Leroi, F. (2005). Effect of inoculation of Carnobacterium divergens V41, a bio-preservative strain against Listeria monocytogenes risk, on the microbiological, chemical and sensory quality of cold-smoked salmon. *International journal of food microbiology*, Vol.104, No.3, pp. 309-324, ISSN 0168-1605

Buchanan, R.L., & Bagi, L.K. (1997). Microbial competition: Effect of culture conditions on the suppression of Listeria monocytogenes Scott A by Carnobacterium piscicola. *Journal of food protection*, Vol.60, No.3, pp. 254-261, ISSN 0362-028X

Budu-Amoako, E., Ablett, R.F., Harris, J., & Delves-Broughton, J. (1999). Combined effect of nisin and moderate heat on destruction of Listeria monocytogenes in cold-pack lobster meat. *Journal of food protection*, Vol.62, No.1, pp. 46-50, ISSN 0362-028X

Cabello, F.C. (2006). Heavy use of prophylactic antibiotics in aquaculture: a growing problem for human and animal health and for the environment. *Environmental microbiology*, Vol.8, No.7, pp. 1137-1144, ISSN 1462-2912

Calo-Mata, P., Arlindo, S., Boehme, K., Miguel. T., Pascoal, A., & Barros-Velazquez, J. (2007). Current Applications and Future Trends of Lactic Acid Bacteria and their Bacteriocins for the Biopreservation of Aquatic Food Products. *Food and Bioprocess Technology*, Vol.1, No.1, pp. 43-63, ISSN 1935-5130

Carraturo, A., Raieta, K., Ottaviani, D., & Russo, G.L. (2006). Inhibition of Vibrio parahaemolyticus by a bacteriocin-like inhibitory substance (BLIS) produced by

Vibrio mediterranei 1. *Journal of applied microbiology*, Vol.101, No.1, pp. 234-241, ISSN 1364-5072

Cascales, E., Buchanan, S.K., Duche, D., Kleanthous, C., Lloubes, R., Postle, K., Riley, M., Slatin, S., & Cavard, D. (2007). Colicin biology. *Microbiology and molecular biology reviews : MMBR*, Vol.71, No.1, pp. 158-229, ISSN 1092-2172

Chavan, M.A., & Riley, M.A., (2007). Molecular Evolution of Bacteriocins in Gram-Negative Bacteria, In:*Bacteriocins: Ecology and Evolution*, Riley, M.A., Chavan, M.A. (Eds.), pp. 5-18, ISBN 3-540-36603-2, Heidelberg, Germany

Cheung, J., Danna, K.J., O'Connor, E.M., Price, L.B., & Shand, R.F. (1997). Isolation, sequence, and expression of the gene encoding halocin H4, a bacteriocin from the halophilic archaeon Haloferax mediterranei R4. *Journal of bacteriology*, Vol.179, No.2, pp. 548-551, ISSN 0021-9193

Cleveland, J., Montville, T.J., Nes, I.F., & Chikindas, M.L. (2001). Bacteriocins: safe, natural antimicrobials for food preservation. *International journal of food microbiology*, Vol.71, No.1, pp. 1-20, ISSN 0168-1605

Colwell, R.R., & Grimes, D.J. (1984). Vibrio Diseases of Marine Fish Populations. *Helgolander Meeresuntersuchungen*, Vol.37, No.1-4, pp. 265-287, ISSN 0174-3597

Daw, M.A., & Falkiner, F.R. (1996). Bacteriocins: nature, function and structure. *Micron*, Vol.27, No.6, pp. 467-479, ISSN 0968-4328

De la Rosa-Garcia, S.C., Munoz-Garcia, A.A., Barahona-Perez, L.F., & Gamboa-Angulo, M.M. (2007). Antimicrobial properties of moderately halotolerant bacteria from cenotes of the Yucatan peninsula. *Letters in Applied Microbiology*, Vol.45, No.3, pp. 289-294, ISSN 0266-8254

Desriac, F., Defer, D., Bourgougnon, N., Brillet, B., Le Chevalier, P., & Fleury, Y. (2010). Bacteriocin as weapons in the marine animal-associated bacteria warfare: inventory and potential applications as an aquaculture probiotic. *Marine drugs*, Vol.8, No.4, pp. 1153-1177, ISSN 1660-3397

Duffes, F., Corre, C., Leroi, F., Dousset, X., & Boyaval, P. (1999). Inhibition of Listeria monocytogenes by in situ produced and semipurified bacteriocins of Carnobacterium spp. on vacuum-packed, refrigerated cold-smoked salmon. *Journal of food protection*, Vol.62, No.12, pp. 1394-1403, ISSN 0362-028X

Dufour, A., Rince, A., Uguen, P., & Le Pennec, J.P. (2000). IS1675, a novel lactococcal insertion element, forms a transposon-like structure including the lacticin 481 lantibiotic operon. *Journal of bacteriology*, Vol.182, No.19, pp. 5600-5605, ISSN 0021-9193

Duport, C., Baysse, C., & Michel-Briand, Y. (1995). Molecular characterization of pyocin S3, a novel S-type pyocin from Pseudomonas aeruginosa. *The Journal of biological chemistry*, Vol.270, No.15, pp. 8920-8927, ISSN:0021-9258

Drider, D., Fimland, G., Hechard, Y., McMullen, L.M., & Prevost, H. (2006). The continuing story of class IIa bacteriocins. *Microbiology and molecular biology reviews : MMBR*, Vol.70, No.2, pp. 564-582, ISSN:1092-2172

Eijsink, V.G., Axelsson, L., Diep, D.B., Havarstein, L.S., Holo, H., & Nes, I.F. (2002). Production of class II bacteriocins by lactic acid bacteria; an example of biological warfare and communication. *Antonie van Leeuwenhoek*, Vol.81, No.1-4, pp. 639-654, ISSN 0003-6072

Einarsson, H., & Lauzon, H.L. (1995). Biopreservation of Brined Shrimp (Pandalus borealis) by Bacteriocins from Lactic Acid Bacteria. *Applied and Environmental Microbiology*, Vol.61, No.2, pp. 669-676, ISSN 0099-2240

Ellen, A.F., Rohulya, O.V., Fusetti, F., Wagner, M., Albers, S.V., & Driessen, A.J. (2011). The sulfolobicin genes of Sulfolobus acidocaldarius encode novel antimicrobial proteins. *Journal of bacteriology*, pp. ISSN 1098-5530

Elotmani, F., & Assobhei, O. (2004). In vitro inhibition of microbial flora of fish by nisin and lactoperoxidase system. *Letters in Applied Microbiology*, Vol.38, No.1, pp. 60-65, ISSN 0266-8254

Field, D., Cotter, P., Hill, C., & Ross, R.P., (2007). Bacteriocin Biosynthesis, Structure, and Function, In:*Research and Applications in Bacteriocins*, Riley, M.A., Gillor, O. (Eds.), pp. 5-41, ISBN 1-904933-23-8, Norfolk, U.K.

Fredericq, P. (1957). Colicins. *Annual review of microbiology*, Vol.11, pp. 7-22, ISSN 0066-4227 Food and Agriculture Organization of the United Nations. http://www.fao.org/

Fuller, R. (1989). Probiotics in man and animals. *The Journal of applied bacteriology*, Vol.66, No.5, pp. 365-378, ISSN 0021-8847

Galvez, A., Abriouel, H., Lopez, R.L., & Ben Omar, N. (2007). Bacteriocin-based strategies for food biopreservation. *International journal of food microbiology*, Vol.120, No.1-2, pp. 51-70, ISSN 0168-1605

Galvez, A., Lopez, R.L., Abriouel, H., Valdivia, E., & Omar, N.B. (2008). Application of bacteriocins in the control of foodborne pathogenic and spoilage bacteria. *Critical reviews in biotechnology*, Vol.28, No.2, pp. 125-152, ISSN 1549-7801

García, P., Rodríguez, L., Rodríguez, A., & Martínez, B. (2010). Food biopreservation: promising strategies using bacteriocins, bacteriophages and endolysins. *Trends in Food Science & Technology*, Vol.21, No.8, pp. 373-382, ISSN 09242244

Gillor, O., & Ghazaryan, L. (2007). Recent advances in bacteriocin application as antimicrobials. *Recent patents on anti-infective drug discovery*, Vol.2, No.2, pp. 115-122, ISSN 1574-891X

Gillor, O., Kirkup, B.C., & Riley, M.A. (2004). Colicins and microcins: the next generation antimicrobials. *Advances in applied microbiology*, Vol.54, pp. 129-146, ISSN 0065-2164

Gillor, O., Nigro, L.M., & Riley, M.A., (2007). Potential application of bacteriocins as antimicrobials, In:*Research and Applications in Bacteriocins*, Riley, M.A., Gillor, O. (Eds.), pp. 67-79, ISBN 1-904933-23-8, Norfolk, U.K.

Gillor, O., Etzion, A., & Riley, M.A. (2008). The dual role of bacteriocins as anti- and probiotics. *Applied microbiology and biotechnology*, Vol.81, No.4, pp. 591-606, ISSN 1432-0614

Govan, J.R., & Harris, G. (1985). Typing of Pseudomonas cepacia by bacteriocin susceptibility and production. *J Clin Microbiol*, Vol.22, No.4, pp. 490-494, ISSN 0095-1137

Gram, L., & Dalgaard, P. (2002). Fish spoilage bacteria – problems and solutions. *Current Opinion in Biotechnology*, Vol.13, No.3, pp. 262-266, ISSN 09581669

Heng, N.C.K., Wescombe, P.A., Burton, J.P., Jack, R.W., & Tagg, J.R., (2007). The Diversity of Bacteriocins in Gram-Positive Bacteria, in:*Bacteriocins: Ecology and Evolution*, Riley, M.A., Chavan, M.A. (Eds.), pp. 45-92, ISBN 3-540-36603-2, Heidelberg, Germany

Hosseini, S.V., Arlindo, S., Bohme, K., Fernandez-No, C., Calo-Mata, P., & Barros-Velazquez, J. (2009). Molecular and probiotic characterization of bacteriocin-producing

Enterococcus faecium strains isolated from nonfermented animal foods. *Journal of applied microbiology*, Vol.107, No.4, pp. 1392-1403, ISSN:1365-2672

Hoyt, P.R., & Sizemore, R.K. (1982). Competitive Dominance by a Bacteriocin-Producing Vibrio harveyi Strain. *Applied and Environmental Microbiology*, Vol.44, No.3, pp. 653-658, ISSN 0099-2240

Jacobs, J.M., Stine, C.B., Baya, A.M., & Kent, M.L. (2009). A review of mycobacteriosis in marine fish. *Journal of Fish Diseases*, Vol.32, No.2, pp. 119-130, ISSN 1365-2761

Joerger, M.C., & Klaenhammer, T.R. (1986). Characterization and purification of helveticin J and evidence for a chromosomally determined bacteriocin produced by Lactobacillus helveticus 481. *Journal of bacteriology*, Vol.167, No.2, pp. 439-446, ISSN 0021-9193

Kerr, B., Riley, M.A., Feldman, M.W., & Bohannan, B.J. (2002). Local dispersal promotes biodiversity in a real-life game of rock-paper-scissors. *Nature*, Vol.418, No.6894, pp. 171-174, ISSN 0028-0836

Kuipers, O.P., Beerthuyzen, M.M., Siezen, R.J., & De Vos, W.M. (1993). Characterization of the nisin gene cluster nisABTCIPR of Lactococcus lactis. Requirement of expression of the nisA and nisI genes for development of immunity. *European journal of biochemistry / FEBS*, Vol.216, No.1, pp. 281-291, ISSN 0014-2956

Lee, H., & Kim, H.Y. (2011). Lantibiotics, class I bacteriocins from the genus Bacillus. *Journal of Microbiology and Biotechnology*, Vol.21, No.3, pp. 229-235, ISSN 1738-8872

Leisner, J.J., Laursen, B.G., Prevost, H., Drider, D., & Dalgaard, P. (2007). Carnobacterium: positive and negative effects in the environment and in foods. *FEMS microbiology reviews*, Vol.31, No.5, pp. 592-613, ISSN 0168-6445

Longeon, A., Peduzzi, J., Barthelemy, M., Corre, S., Nicolas, J.L., & Guyot, M. (2004). Purification and partial identification of novel antimicrobial protein from marine bacterium Pseudoalteromonas species strain X153. *Marine biotechnology*, Vol.6, No.6, pp. 633-641, ISSN 1436-2228

Luders, T., Birkemo, G.A., Fimland, G., Nissen-Meyer, J., & Nes, I.F. (2003). Strong synergy between a eukaryotic antimicrobial peptide and bacteriocins from lactic acid bacteria. *Applied and Environmental Microbiology*, Vol.69, No.3, pp. 1797-1799, ISSN 0099-2240

Maqueda, M., Galvez, A., Bueno, M.M., Sanchez-Barrena, M.J., Gonzalez, C., Albert, A., Rico, M., & Valdivia, E. (2004). Peptide AS-48: prototype of a new class of cyclic bacteriocins. *Current protein & peptide science*, Vol.5, No.5, pp. 399-416, ISSN 1389-2037

Matsui, H., Sano, Y., Ishihara, H., & Shinomiya, T. (1993). Regulation of pyocin genes in Pseudomonas aeruginosa by positive (prtN) and negative (prtR) regulatory genes. *Journal of bacteriology*, Vol.175, No.5, pp. 1257-1263, ISSN 0021-9193

McAuliffe, O., Ross, R.P., & Hill, C. (2001). Lantibiotics: structure, biosynthesis and mode of action. *FEMS microbiology reviews*, Vol.25, No.3, pp. 285-308, ISSN 0168-6445

McCall, J.O., & Sizemore, R.K. (1979). Description of a bacteriocinogenic plasmid in Beneckea harveyi. *Applied and Environmental Microbiology*, Vol.38, No.5, pp. 974-979, ISSN 0099-2240

Messi, P., Guerrieri, E., & Bondi, M. (2003). Bacteriocin-like substance (BLS) production inAeromonas hydrophilawater isolates. *FEMS Microbiology Letters*, Vol.220, No.1, pp. 121-125, ISSN 03781097

Metivier, A., Pilet, M.F., Dousset, X., Sorokine, O., Anglade, P., Zagorec, M., Piard, J.C., Marion, D., Cenatiempo, Y., & Fremaux, C. (1998). Divercin V41, a new bacteriocin with two disulphide bonds produced by Carnobacterium divergens V41: primary structure and genomic organization. *Microbiology*, Vol.144 (Pt 10), pp. 2837-2844, ISSN 1350-0872

Michel-Briand, Y., & Baysse, C. (2002). The pyocins of Pseudomonas aeruginosa. *Biochimie*, Vol.84, No.5-6, pp. 499-510, ISSN 0300-9084

Mierau, I., & Kleerebezem, M. (2005). 10 years of the nisin-controlled gene expression system (NICE) in Lactococcus lactis. *Applied microbiology and biotechnology*, Vol.68, No.6, pp. 705-717, ISSN 0175-7598

Mohanty, B.R., & Sahoo, P.K. (2007). Edwardsiellosis in fish: a brief review. *Journal of biosciences*, Vol.32, No.7, pp. 1331-1344, ISSN 0250-5991

Moro, E.M., Weiss, R.D., Friedrich, R.S., & Nunes, M.P. (1997). Bacteriocin-like substance of Aeromonas hydrophila. *Memorias do Instituto Oswaldo Cruz*, Vol.92, No.1, pp. 115-116, ISSN 0074-0276

Murray, G.E., Tobin, R.S., Junkins, B., & Kushner, D.J. (1984). Effect of chlorination on antibiotic resistance profiles of sewage-related bacteria. *Applied and Environmental Microbiology*, Vol.48, No.1, pp. 73-77, ISSN 0099-2240

Nakayama, K., Takashima, K., Ishihara, H., Shinomiya, T., Kageyama, M., Kanaya, S., Ohnishi, M., Murata, T., Mori, H., & Hayashi, T. (2000). The R-type pyocin of Pseudomonas aeruginosa is related to P2 phage, and the F-type is related to lambda phage. *Molecular microbiology*, Vol.38, No.2, pp. 213-231, ISSN 0950-382X

Neetoo, H., Ye, M., Chen, H., Joerger, R.D., Hicks, D.T., & Hoover, D.G. (2008). Use of nisin-coated plastic films to control Listeria monocytogenes on vacuum-packaged cold-smoked salmon. *International journal of food microbiology*, Vol.122, No.1-2, pp. 8-15, ISSN 0168-1605

Nilsson, L., Huss, H.H., & Gram, L. (1997). Inhibition of Listeria monocytogenes on cold-smoked salmon by nisin and carbon dioxide atmosphere. *International journal of food microbiology*, Vol.38, No.2-3, pp. 217-227, ISSN 0168-1605

Nissen-Meyer, J., Holo, H., Havarstein, L.S., Sletten, K., & Nes, I.F. (1992). A novel lactococcal bacteriocin whose activity depends on the complementary action of two peptides. *Journal of bacteriology*, Vol.174, No.17, pp. 5686-5692, ISSN 0021-9193

Noga, E.J. (2010). *Fish Disease: Diagnosis and Treatment,* Iowa State University Press, ISBN 978-0-8138-0697-6, Iowa, USA

Nykanen, A., Weckman, K., & Lapvetelainen, A. (2000). Synergistic inhibition of Listeria monocytogenes on cold-smoked rainbow trout by nisin and sodium lactate. *International journal of food microbiology*, Vol.61, No.1, pp. 63-72, ISSN 0168-1605

O'Brien, D., Mooney, J., Ryan, D., Powell, E., Hiney, M., Smith, P.R., & Powell, R. (1994). Detection of Aeromonas salmonicida, causal agent of furunculosis in salmonid fish, from the tank effluent of hatchery-reared Atlantic salmon smolts. *Applied and Environmental Microbiology*, Vol.60, No.10, pp. 3874-3877, ISSN 0099-2240

O'Connor, E.M., & Shand, R.F. (2002). Halocins and sulfolobicins: the emerging story of archaeal protein and peptide antibiotics. *Journal of industrial microbiology & biotechnology*, Vol.28, No.1, pp. 23-31, ISSN 1367-5435

Pilet, M.F., & Leroi, F., (2011). Applications of protective cultures, bacteriocins, and bacteriophages in fresh seafood and seafood products, In: *Protective cultures, antimicrobial metabolites and bacteriophages for food and bevarage biopreservation*, pp. 1-21, ISBN 978-1-84569-669-6, Zurich, Switzerland

Pinto, A.L., Fernandes, M., Pinto, C., Albano, H., Castilho, F., Teixeira, P., & Gibbs, P.A. (2009). Characterization of anti-Listeria bacteriocins isolated from shellfish: potential antimicrobials to control non-fermented seafood. *International journal of food microbiology*, Vol.129, No.1, pp. 50-58, ISSN 0168-1605

Pirzada, Z.A., Ali, S.A., Khan, B.M., & Rasool, S.A. (2004). Production and Physico-chemical Characterization of Bacteriocins-like. Inhibitory Substances from Marine Bacterium ZM81. *Pakistan Journal of Biological Sciences*, Vol.7, No.12, pp. 2026-2030,

Prangishvili, D., Holz, I., Stieger, E., Nickell, S., Kristjansson, J.K., & Zillig, W. (2000). Sulfolobicins, specific proteinaceous toxins produced by strains of the extremely thermophilic archaeal genus Sulfolobus. *Journal of bacteriology*, Vol.182, No.10, pp. 2985-2988, ISSN 0021-9193

Prasad, S., Morris, P.C., Hansen, R., Meaden, P.G., & Austin, B. (2005). A novel bacteriocin-like substance (BLIS) from a pathogenic strain of Vibrio harveyi. *Microbiology*, Vol.151, No.Pt 9, pp. 3051-3058, ISSN 1350-0872

Price, L.B., & Shand, R.F. (2000). Halocin S8: a 36-amino-acid microhalocin from the haloarchaeal strain S8a. *Journal of bacteriology*, Vol.182, No.17, pp. 4951-4958, ISSN 0021-9193

Pugsley, A.P., & Oudega, B., (1987). Methods for studying colicins and their plasmids, In:*Plasmids: a practical approach*, Hardy, K.G. (Ed.), pp. 105-161, ISBN 0-947946-81-0 , Oxford, England

Rajaram, G., Manivasagan, P., Thilagavathi, B., & Saravanakumar, A. (2010). Purification and Characterization of a Bacteriocin Produced by *Lactobacillus lactis* Isolated from Marine Environment. *Advance Journal of Food Science and Technology*, Vol.2, No.2, pp. 138-144

Reeves, P. (1965). The Bacteriocins. *Bacteriological reviews*, Vol.29, pp. 24-45, ISSN 0005-3678

Rihakova, J., Belguesmia, Y., Petit, V.W., Pilet, M.F., Prevost, H., Dousset, X., & Drider, D. (2009). Divercin V41 from gene characterization to food applications: 1998-2008, a decade of solved and unsolved questions. *Letters in Applied Microbiology*, Vol.48, No.1, pp. 1-7, ISSN 1472-765X

Riley, M.A., & Gordon, D.M. (1992). A survey of Col plasmids in natural isolates of Escherichia coli and an investigation into the stability of Col-plasmid lineages. *Journal of general microbiology*, Vol.138, No.7, pp. 1345-1352, ISSN 0022-1287

Riley, M.A., & Gordon, D.M. (1999). The ecological role of bacteriocins in bacterial competition. *Trends in microbiology*, Vol.7, No.3, pp. 129-133, ISSN 0966-842X

Riley, M.A., & Wertz, J.E. (2002a). Bacteriocin diversity: ecological and evolutionary perspectives. *Biochimie*, Vol.84, No.5-6, pp. 357-364, ISSN 0300-9084

Riley, M.A., & Wertz, J.E. (2002b). Bacteriocins: evolution, ecology, and application. *Annual review of microbiology*, Vol.56, pp. 117-137, ISSN 0066-4227

Riley, M.A., Goldstone, C.M., Wertz, J.E., & Gordon, D. (2003). A phylogenetic approach to assessing the targets of microbial warfare. *Journal of evolutionary biology*, Vol.16, No.4, pp. 690-697, ISSN 1010-061X

Romalde, J.L. (2002). Photobacterium damselae subsp. piscicida: an integrated view of a bacterial fish pathogen. *International microbiology : the official journal of the Spanish Society for Microbiology*, Vol.5, No.1, pp. 3-9, ISSN 1139-6709

Sano, Y., Matsui, H., Kobayashi, M., & Kageyama, M. (1993). Molecular structures and functions of pyocins S1 and S2 in Pseudomonas aeruginosa. *Journal of bacteriology*, Vol.175, No.10, pp. 2907-2916, ISSN 0021-9193

Satish Kumar, R., & Arul, V. (2009). Purification and characterization of phocaecin PI80: an anti-listerial bacteriocin produced by Streptococcus phocae PI80 Isolated from the gut of Peneaus indicus (Indian white shrimp). *Journal of Microbiology and Biotechnology*, Vol.19, No.11, pp. 1393-1400, ISSN 1017-7825

Schindler, C.A., & Schuhardt, V.T. (1964). Lysostaphin: A New Bacteriolytic Agent for the Staphylococcus. *Proceedings of the National Academy of Sciences of the United States of America*, Vol.51, pp. 414-421, ISSN 0027-8424

Schobitz, R., Zaror, T., Leon, O., & Costa, M. (1999). A bacteriocin from Carnobacterium piscicola for the control of Listeria monocytogenes in vacuum-packaged meat. *Food microbiology*, Vol.16, No.3, pp. 249-255, ISSN 0740-0020

Selvin, J., Joseph, S., Asha, K.R., Manjusha, W.A., Sangeetha V.S., Jayaseema, D.M., Antony, M.C., & Denslin Vinitha, A.J. (2004). Antibacterial potential of antagonistic Streptomyces sp. isolated from marine sponge Dendrilla nigra. *FEMS microbiology ecology*, Vol.50, No.2, pp. 117-122, ISSN 1574-6941

Shand, R.F., & Leyva, K.J., (2007). Peptide and Protein Antibiotics from the Domain Archaea: Halocins and Sulfolobicins, In:*Bacteriocins: Ecology and Evolution*, Riley, M.A., Chavan, M.A. (Eds.), pp. 93-109, ISBN 3-540-36603-02, Heidelberg, Germany

Shao, Z.J. (2001). Aquaculture pharmaceuticals and biologicals: current perspectives and future possibilities. *Advanced drug delivery reviews*, Vol.50, No 3, pp. 229-243, ISSN 0169-409X

Shehane, S.D., & Sizemore, R.K. (2002). Isolation and preliminary characterization of bacteriocins produced by Vibrio vulnificus. *Journal of applied microbiology*, Vol.92, No.2, pp. 322-328, ISSN 1364-5072

Stoffels, G., Nes, I.F., & Guthmundsdottir, A. (1992). Isolation and properties of a bacteriocin-producing Carnobacterium piscicola isolated from fish. *The Journal of applied bacteriology*, Vol.73, No.4, pp. 309-316, ISSN 0021-8847

Subasinghe, R. (2009). Disease control in aquaculture and the responsible use of veterinary drugs and vaccines: The issues, prospects, and challenges. *Options Mediterraneennes*, Vol.86, pp. 5-11

Sugita, H., Matsuo, N., Hirose, Y., Iwato, M., & Deguchi, Y. (1997). Vibrio sp. strain NM 10, isolated from the intestine of a Japanese coastal fish, has an inhibitory effect against Pasteurella piscicida. *Applied and Environmental Microbiology*, Vol.63, No.12, pp. 4986-4989, ISSN 0099-2240

Sun, C., Li, Y., Mei, S., Lu, Q., Zhou, L., & Xiang, H. (2005). A single gene directs both production and immunity of halocin C8 in a haloarchaeal strain AS7092. *Molecular microbiology*, Vol.57, No.2, pp. 537-549, ISSN 0950-382X

Suzuki, M., Yamamoto, T., Kawai, Y., Inoue, N., & Yamazaki, K. (2005). Mode of action of piscicocin CS526 produced by Carnobacterium piscicola CS526. *Journal of applied microbiology*, Vol.98, No.5, pp. 1146-1151, ISSN 1364-5072

Szabo, E.A., & Cahill, M.E. (1999). Nisin and ALTA 2341 inhibit the growth of Listeria monocytogenes on smoked salmon packaged under vacuum or 100% CO2. *Letters in Applied Microbiology*, Vol.28, No.5, pp. 373-377, ISSN 0266-8254

Tagg, J.R., Dajani, A.S., & Wannamaker, L.W. (1976). Bacteriocins of gram-positive bacteria. *Bacteriological reviews*, Vol.40, No.3, pp. 722-756, ISSN 0005-3678

Tahiri, I., Desbiens, M., Kheadr, E., Lacroix, C., & Fliss, I. (2009). Comparison of different application strategies of divergicin M35 for inactivation of Listeria monocytogenes in cold-smoked wild salmon. *Food microbiology*, Vol.26, No.8, pp. 783-793, ISSN 1095-9998

Tahiri, I., Desbiens, M., Benech, R., Kheadr, E., Lacroix, C., Thibault, S., Ouellet, D., & Fliss, I. (2004). Purification, characterization and amino acid sequencing of divergicin M35: a novel class IIa bacteriocin produced by Carnobacterium divergens M35. *International journal of food microbiology*, Vol.97, No.2, pp. 123-136, ISSN 0168-1605

Toranzo, A., Magarinos, B., & Romalde, J. (2005). A review of the main bacterial fish diseases in mariculture systems. *Aquaculture*, Vol.246, No.1-4, pp. 37-61, ISSN 00448486

Torreblanca, M., Meseguer, I., & Ventosa, A. (1994). Production of Halocin Is a Practically Universal Feature of Archaeal Halophilic Rods. *Letters in Applied Microbiology*, Vol.19, No.4, pp. 201-205, ISSN 0266-8254

Tsironi, T.N., & Taoukis, P.S. (2010). Modeling microbial spoilage and quality of gilthead seabream fillets: combined effect of osmotic pretreatment, modified atmosphere packaging, and nisin on shelf life. *Journal of food science*, Vol.75, No.4, pp. M243-251, ISSN 1750-3841

Valenzuela, A.S., Benomar, N., Abriouel, H., Canamero, M.M., & Galvez, A. (2010). Isolation and identification of Enterococcus faecium from seafoods: antimicrobial resistance and production of bacteriocin-like substances. *Food microbiology*, Vol.27, No.7, pp. 955-961, ISSN 1095-9998

Wilson, G.S., Raftos, D.A., Corrigan, S.L., & Nair, S.V. (2010). Diversity and antimicrobial activities of surface-attached marine bacteria from Sydney Harbour, Australia. *Microbiological research*, Vol.165, No.4, pp. 300-311, ISSN 1618-0623

Word Health Organization http://www.who.int/mediacentre/factsheets/fs194/en/

Yamazaki, K., Suzuki, M., Kawai, Y., Inoue, N., & Montville, T.J. (2005). Purification and characterization of a novel class IIa bacteriocin, piscicocin CS526, from surimi-associated Carnobacterium piscicola CS526. *Applied and Environmental Microbiology*, Vol.71, No.1, pp. 554-557, ISSN 0099-2240

Zai, A.S., Ahmad, S., & Rasool, S.A. (2009). Bacteriocin production by indigenous marine catfish associated Vibrio spp. *Pakistan journal of pharmaceutical sciences*, Vol.22, No.2, pp. 162-167, ISSN 1011-601X

Zuckerman, H., & Ben Avraham, R. (2002). Control of growth of L-monocytogenes in fresh salmon using Microgard (TM) and Nisin. *Lebensmittel-Wissenschaft Und-Technologie-Food Science and Technology*, Vol.35, No.6, pp. 543-548, ISSN 0023-6438

13

The Atlantic Salmon (*Salmo salar*) Vertebra and Cellular Pathways to Vertebral Deformities

Elisabeth Ytteborg[1], Jacob Torgersen[1],
Grete Baeverfjord[1] and Harald Takle[1,2]
[1]*Nofima AS, Norwegian Institute of Food, Fisheries and Aquaculture Research*
[2]*AVS Chile S.A.*
[1]*Norway*
[2]*Chile*

1. Introduction

Stagnating fish stocks and a growing population demanding for aquatic food products have been the major driving forces behind the rapid increase in aquaculture production. Federation of European Aquaculture Producers (FEAP) estimates that 650 000 tons of fish are farmed in the EU annually (compared to 60 000 tons in 1970). Within Europe as a whole, the total production is more than 1.6 million tons (FEAP, 2009). Norway is a major contributor to Europe's aquaculture sector with over 860 000 tons of Atlantic salmon (*Salmo salar*) and trout produced each year, a production that has been more than doubled the last ten years (Directorates of fisheries, Norway 2009). Forecasts predict that production will need to increase for decades to come if demands are to be met (Brugere & Ridler 2004). To keep up with the growing demand, the aquaculture industry is constantly searching for new strategies to improve the rearing conditions and reduce production time and cost. However, as a relatively new industry, and as a consequence of intensified production regimes, the aquaculture sector faces growth constraints.

Farmed salmon is bred for rapid growth, and the industry aim at obtaining the optimal growth rate by optimizing both diets and environmental factors accordingly. However, intensive rearing conditions are linked to increased occurrence of production related diseases and malformations. Elevated temperature during the fresh water period was commonly used in the 90`ies to speed up developmental rate. An increasing number of fish developing manufacturing defects, such as skeletal abnormalities (figure 1), heart failure and jaw deformities was observed. Recommendations limiting temperatures to safe levels, ≤8°C during egg rearing and ≤12°C after fist feeding, led to substantial reductions in skeletal malformations (Baeverfjord et al., 1999). However, in the last few years, the start feeding temperature has been increased again, due to the stakeholders demand for reduced production time. Further, the growing need of replacing fish meal in commercial fish feeds have come into focus and deformities related to feed ingredient replacements, malnutrition and mineral deficiency are investigated.

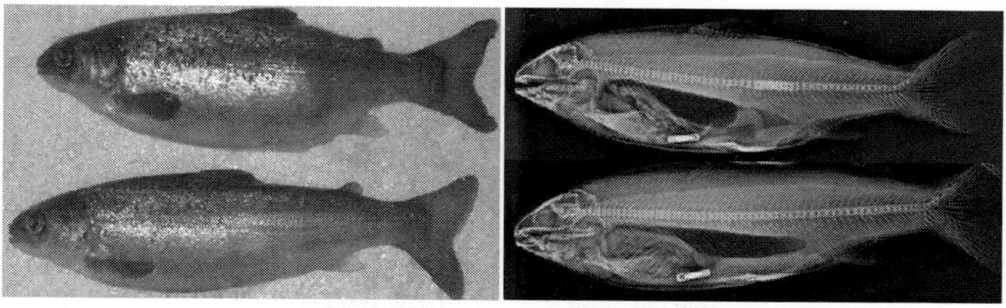

Fig. 1. Deformed (top) and non-deformed Atlantic salmon and corresponding radiographic pictures. Photo: Grete Baeverford, Nofima.

In the present situation, fast growth in combination with unpredictable and potentially low bioavailability of nutrients is considered the main challenge for adequate skeletal development. Suboptimal supply of minerals (phosphorous, magnesium, zinc) and nutritional imbalances of fatty acids, vitamins (A, C and D) and amino acids are considered the main challenges in regard to skeletal malformations. The challenges related to bioavailability are further amplified with the introduction of vegetable meals, some of which are rich in antinutrients (e.g. phytic acid) that may further impair absorption. It is therefore important to completely understand the molecular and cellular events in bone development in salmon in order to deal with upcoming questions

Most of the knowledge currently available on cellular mechanisms for bone development is adopted from studies using mammalian species. However, information from teleosts, like zebrafish (*Danio rerio*), sea bass (*Dicentrarchus labrax)*, sea bream (*Sparus aurata)* and Atlantic salmon and related *in vitro* studies, is emerging. Molecular tools like in situ hybridization, microarray, real time quantitative PCR, immunohistochemistry and cell culture systems have allowed researchers working with teleosts to do more comparative and functional studies. Also, transgenic model organisms, such as zebrafish, and *in vitro* transfection with reporter gene constructs are now being more common and will provide valuable information on processes involved in bone metabolism. Relevant examples include studies on gene expression of several bone and cartilage associated marker genes, such as bone morphogenetic protein 2 (BMP-2) (Rafael et al., 2006), osteocalcin (Wargelius et al., 2009; Ytteborg et al.,2010b), osteopontin (Fonseca et al., 2007), vitamin D receptor (Lock et al., 2009), parahyroid related hormone (Flanagan et al., 2000) and proteoglycans (Pedersen et al 2010;Conceição et al., 2008). The Sea bream vertebrae cell lines described by Braga et al. (2006) as well as two reports on zebrafish (DeLaurier et al., 2010; Kimmel et al., 2010) show the possibilities with fluorescent reporter gene constructs in bone research. The work developed during the last few years has provided clear evidence that fish can be adequate supplementary model systems to study bone and cartilage biology. Teleosts have been successfully used to analyze molecular and cellular mechanisms involved in different developmental pathways and revealed that the key genetic factors regulating lineage determination and differentiation of stem cells are conserved among vertebrates at the molecular level in both sequence and expression pattern (Kikuta et al., 2007; Shafizadeh et al., 2004; Nakashima et al., 2003; Aubin, 1998; Pinto et al., 2001; Renn et al., 2006; Wise et al., 2006; Ytteborg et al., 2010a). Due to the similar physiologic pathways and genetic background of fish and mammals, this alternative system is also an interesting model to unveil some of the molecular determinants of human bone related diseases and malformations, like osteogenesis imperfecta, degenerated disc disease, persistent notochordal canal and scoliosis (Gorman and Breden, 2007;Nissen et al., 2006;Fisher et al., 2003). A number of animal models have been used to explore the pathology of spinal deformities and revealed that vertebral pathology presents a complex but comparable cross species etiology. With regard to complex disorders in humans, multiple models are critical for the investigation and manipulation of etiological factors.

Fish systems could be of benefit to vertebral research because they exhibit a diverse range of deformities, are free from skeletal appendicles and substantial genomic resources have been developed for several species. Skeletal deformities in commercial salmon production have been recognized as a problem of obvious relevance to economy as well as animal welfare. Much effort has been put into understanding malformed development of Atlantic salmon vertebrae during the years due to the importance of this organism to the aquaculture industry. As a consequence, Atlantic salmon is emerging as an excellent model to study vertebral deformities and other relevant vertebral pathological states. In this review the current knowledge on the cellular and molecular mechanisms for skeletal homeostasis in the mature Atlantic salmon vertebrae is discussed. Further, the cellular mechanisms for differentiation and activation of osteoblasts and chondrocytes are described in relation to pathways for pathological development and discussed in the light of related pathological conditions in mammalian species.

2. Cellular and molecular mechanisms controlling bone formation

Bone formation basically occur via two mechanisms in both mammals and teleosts: mesenchymal stem cells (MSC) either differentiate directly into bone producing osteoblasts (intramembranous ossification) or by first forming a cartilaginous template secreted by chondrocytes which is later replaced by bone (endochondral ossification) (Erlebacher et al., 1995). However, similarities and differences in tissue structure between teleost and mammalian bone have been described (Witten et al., 2009; Huysseune et al., 2000). In general, fish possess rather few long bones with growth plate-like arrangements exhibiting typical endochondral bone formation as seen in mammals. In the Atlantic salmon vertebrae, compact bone of the amphicoel and trabeculae is formed directly through intramembranous ossification, whereas the arch centra are modelled through endochondral ossification. Both mechanisms lead to the formation of mineralized extracellular matrix (ECM), consisting of fibers, mainly collagen embedded in a matrix of proteoglycans (PGs) and proteins. An overview of the two different processes in the Atlantic salmon vertebra is shown in figure 2a and b.

2.1 Linage determination and cellular differentiation

The cellular lineage determination and differentiation of osteoblasts and chondrocytes from the MSC lineage are determined by a number of transcription factors, regulatory mechanisms, environmental conditions and mineral availability. The pathways are interconnected during vertebral formation and must be coordinated. In particular, the transcription factors Runx2, Osterix, Sox9, Twist and Mef2c have distinct functions both in the establishment of the vertebral bodies and later in the differentiation and maturation of

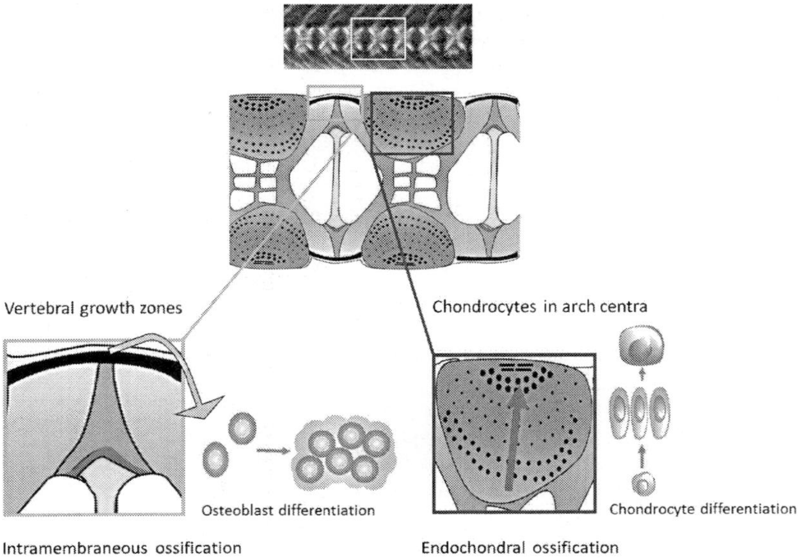

Fig. 2a. Overview of the intramembraneous and endochondral ossification in the Atlantic salmon vertebra. See the text below for details.

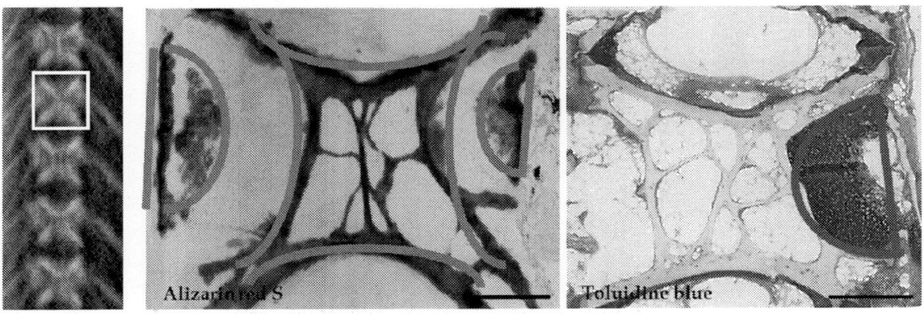

Fig. 2b. Bone and cartilage in the Atlantic salmon vertebra. Both endochondral and intramembraneous ossification leads to mineralized bone formation (red - Alizarin red S staining of bone in the centra and arches; blue - Toluidine blue staining of cartilage in the arches). See text below for further details on bone and cartilage formation. Vertebrae from 15g fish, scale bar= 200 µm

specific skeletal cell types (Karsenty et al., 2009). Similarly, signaling molecules like bone morphogenetic proteins (Bmp2 and Bmp4) and hedgehog proteins (Ihh and Shh) play different roles both during cell differentiation and skeletal tissue ontogeny (Karp et al., 2000; Hogan et al., 1996; Spinella-Jaegle et al., 2001). Important signalling pathways that induce transcription of matrix producing and mineralizing genes in osteoblasts and chondrocytes include the downstream targets of Bmps; Runt-related transcription factor 2 (Runx2) the zinc finger containing transcription factor Osterix and Sex determining region Y box 9 (Sox9). Whereas Runx2 and Osterix activates genes in the osteoblastic lineage (Karsenty et al., 1999; Otto et al., 1997; Nakashima et al., 2002), Sox9 regulates transcription of chondrocytic genes (Bell et al., 1997).

The differentiation of MSC into mature osteoblasts involves several phases, which may be divided into three subsequent stages; commitment, extracellular matrix production and mineralization. Estrogen and 1,25-dihydroxy vitamin D_3 are among the hormones shown to increase osteogenic differentiation via up-regulation of osteogenic growth factors, such as BMP2. Among the many transcription factors expressed early in osteogenesis, runx2 is noteworthy because it is required for bone formation and is an important early indicator of osteogenic capacity of cells. Downstream targets of Runx2 and Osterix include genes encoding both collagenous (e.g. Collagen 1α and 1ß) and non-collagenous (e.g. Osteopontin, Osteocalcin, Osteonectin, Bone sialoprotein and Alp) proteins, which make osteoblasts capable of producing and mineralizing bone matrix (osteoid). In both teleosts and mammalian MSCs, alkaline phosphatase (Alp), col1a and osteopontin serve as useful markers of early osteogenesis and the expression of these genes usually increases throughout maturation. Col1 is the major structural component of bone, whereas the non-collagenous proteins binds inorganic minerals and are involved in the mineralization process (Cowles et al., 1998; Ikeda et al., 1992; Bolander et al., 1988; Termine et al., 1981). Upon maturation, osteoblasts start secreting osteoid and mineralizing components, leading to direct formation of bone via the intramembraneous ossification pathway. The key markers involved in osteogenesis are shown in Figure 3.

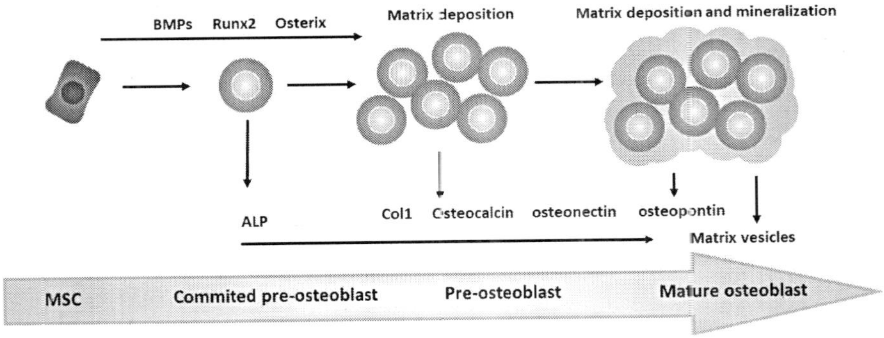

Fig. 3. Osteoblast differentiation, maturation and key factors involved. After commitment to the osteoblast lineage, matrix deposition starts. Mature osteoblasts are responsible for both osteoid production and mineralization. See text for details.

The chondrocytes undergo a more synchronized process of proliferation, differentiation and maturation so that three pronounced zones can be identified in the growing cartilage: resting, proliferating and hypertrophic zones (Hunziker et al., 1994). Chondrocytes in the resting zone are irregularly scattered in cartilage matrix, whereas chondrocytes in the proliferating and hypertrophic zones are arranged in columns. The chondrocytes in the resting zone serves as stem-like cells in the growth plate, stimulated by e.g. growth hormone (GH) and insulin like growth factor (IGF). The proliferating zone is the region for active cell replication and chondrocytes in this zone are mostly devoted to cell cycle processes. Chondrocyte hypertrophy is the final step of chondrocyte maturation, regulated by the transcription factors Myocyte enhancer factor 2c (Mef2c) and Runx2 (Arnold et al., 2007; Kim et al., 1999). Parathyroid hormone related protein (PTHrP) and Ihh appear to play important roles in proliferating chondrocytes by maintaining cells in a proliferative condition, hence preventing chondrocyte hypertrophy. After commitment to the hypertrophic state, chondrocytes start expressing Col10 (Ytteborg et al., 2010b; Arnold et al., 2007), a unique component of the matrix produced by hypertrophic cells and extensively used as a marker for chondrocyte hypertrophy (Iyama et al., 1991). Once hypertrophy is reached, endochondral ossification can be initiated (Mackie et al., 2008). Hypertrophic chondrocytes induce angiogenesis by secreting angiogenetic factors, such as the Matrix metalloproteinases (Mmps) and Vascular endothelia growth factor (VEGF) so that osteoblasts and osteoclasts may enter via newly formed blood vessels (Blavier et al., 1995). The key markers involved in chondrogenesis are shown in Figure 4.

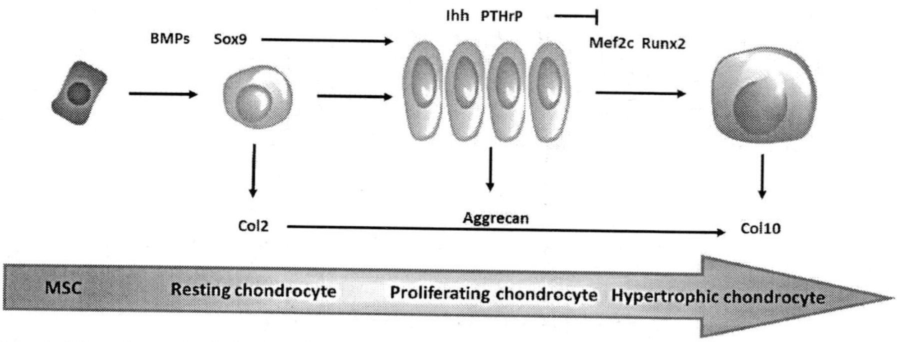

Fig. 4. Chondrocytic differentiation, maturation and key factors involved. Resting, proliferating and hypertrophic chondrocytes are clearly visible as zones in the growth plate. See text for details.

Osteoclasts are cells involved in removing damaged bone, repair mechanisms, mineral homeostasis and replacement of cartilage with bone, both in mammals (review in Boyle et al., 2003) and teleosts (reviewed in Witten et al., 2009). Osteoclasts provide an acidic environment where mineralized matrix may be dissolved through secretion of cathepsins, mmps and tartrate resistante acid phosphatase (TRAP) (Delaisse et al., 2003; Motyckova et al., 2001; Ortega et al., 2003; Engsig et al., 2000). As in mammals, osteoclasts in Atlantic salmon are multinucleated and the mechanisms involved in activation and differentiation of osteoclasts are conserved (review in Witten et al., 2009). Mononuclear cells respond to macrophage colony stimulating factor (M-CSF) produced by nearby stromal cells and osteoblasts, through activation of c-fms, the receptor for M-CSF (Wiktorjedrzejczak et al., 1990; Yoshida et al., 1990). The other signaling system essential for osteoclast differentiation is triggered when receptor activator of nuclear factor kappa (κ) B ligand (RANKL), a member of the tumor necrosis factor (TNF) family, activates its receptor RANK (reviewed in Collin-Osdoy et al., 2004). Among the downstream genes of RANKL are genes directly ivolved osteoclast function (e.g. TRAP and Cathepsin K). The key markers involved in osteoclastogenesis are shown in Figure 5. In addition, mononucleated osteoclasts are also found in both mammals and teleosts and are considered to participate in minor, fine tuning bone resorption (Witten et al., 2009). However, since teleost lack haemapoietic tissue in bone marrow, the question of the origin of these cells remains unknown. In the vertebrae of Atlantic salmon, multinucleated osteoclasts have been identified in the arch centra and trabeculae but not in the compact bone of the amphicoel (Witten et al., 2009).

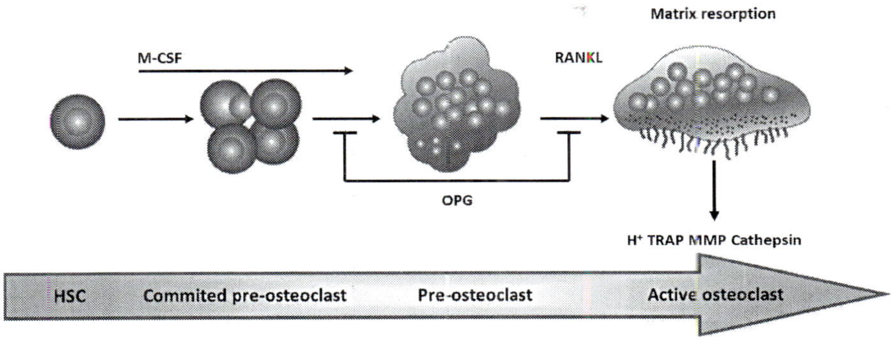

Fig. 5. Osteoclast differentiation, maturation and key factors involved. Fully mature osteoclasts are able to dissolve bone. See text for details.

2.2 Matrix mineralization

Skeletal formation and growth occurs as a result of mineralization of ECM. A time lag where collagen synthesis decreases and mineralization increase appears to be required for allowing modifications of the osteoid so that it is able to support mineralization and hydroxyapatite ($Ca_{10}[PO_4]^6[OH]^2$) formation (Hernandez et al., 2000). Mineralization of both bone and cartilage occurs by deposition of inorganic hydroxyapatite crystals inthe ECM. This process has not yet been described in teleosts. In mammals, the initiating step of hydroxyapatite formation occurs in ECM vesicles secreted from mature osteoblasts (Anderson et al., 2005, 1996, 1995). These vesicles create an environment where deposition of minerals (mainly Ca^{2+} and P_i) occurs and hydroxyapatite is produced, a process involving proteins like Annexins and Alp (Balcerzak et al., 2003; Kirsch et al., 2005). The attachment of the vesicles to bone is not well understood, but Alp and Annexin in the vesicle membrane are reported to anchor to collagen fibrils (Wu et al., 1991). Vesicle formation is followed by the linking of hydroxyapatite crystals to ECM components (Balcerzak et al., 2003) using the Ca^{2+} and hydroxyapatite binding properties of Osteonectin, Osteopontin, Osteocalcin and Bone sialoprotein (Hoffmann et al., 1996; Pinto et al., 2001; Furie et al., 1991). Hypertrophic chondrocytes are also capable of initiating calcification processes by releasing similar matrix vesicles as osteoblasts and it has been suggested that hypertrophic chondrocytes may participate actively in bone formation (Anderson et al., 1975; Kirsch et al., 1997). Moreover, hypertrophic chondrocytes from both mammals and teleosts express genes like osteocalcin, osteonectin and alp (Ytteborg et al., 2010b; Ishizeki et al., 1996; Lian et al., 1993). Cancedda et al. (1992) showed that hypertrophic chondrocytes from chicken can be induced to obtain a strictly osteoblastic phenotype *in vitro*. These findings are supported by Yasui et al. (1997)

who suggested that hypertrophic chondrocytes are able to trans-differentiate into osteoblasts and produce bone through a process called trans-chondroid ossification. More than 10 different forms of cartilage and several other tissues with histological characteristics between bone and cartilage have so far been identified in fish (Huysseune et al., 1986; Huysseune et al., 1990). This makes bone studies in Atlantic salmon more complicated, as strict lines between cell types and distinct borders between tissue structures are difficult to define. However, intermediate tissue is instructive due to the many molecular pathways and cellular adaptations during pathological development and normal growth.

3. Pipeline for studying vertebral development

Bone deformities in Atlantic salmon are a complex problem, which may have diverse causes, acting either one by one or in combination, hence, a number of different tools are important to establish in order to cover different mechanisms involved in their development. The pipeline for studying bone development in teleosts is shown in figure 6. In vertebrates, both bone and cartilaginous structures coexist during development of the vertebral column and both tissues are built up mainly of the organic ECM. Cartilage and bone cellular activity largely depends on the interaction with ECM components. ECM components regulate cell growth and differentiation by interacting with growth factors and enzymes, provide the tissue with mechanical strength and resilience and constitute the template for mineralization during development of the vertebral column. The composition and structure of molecules in the ECM are shown to play pivotal roles in bone formation and changes therein may result in deformities in the spine of both mammals and teleosts (Pedersen et al., 2010). Radiography, or the use of X-rays for analysis, is the preferred method for fish skeletal deformity diagnostics. X-rays have enough energy to penetrate soft tissues, but not bone and other hard substances. Moreover radiography thus allows the creation of a negative image of the skeletal structures of the fish, which allows the evaluation of calcification level and for identification of pathology in the bones, without cutting into or even killing the fish. However, fish radiography has its limits and it is difficult to diagnose fish before the deformity has developed. More sensitive techniques are therefore necessary. So far *in vivo* trials with Atlantic salmon using different temperatures and light regimes, water speed for studying the effect of training and feeding trials using custom-made feeds for studying mineral and vitamin components has been applied for deformity studies. In addition to radiography and measuring rate of development and growth, essential minerals have been followed from uptake and secretion in the intestines using quantitative real time PCR, to incorporation in the bone matrix using mineral analysis and Fourier Transform InfraRed (FT-IR), histological staining techniques and screening techniques such as microarray. Important pathways for cellular differentiation of bone and cartilage have been followed using gene expressional tools, like quantitative real time PCR, *in situ* hybridization and immunohistochemistry. In a recent puplication, (Ytteborg et al., 2010b), it was shown by using molecular markers and gene transcription techniques, that fish susceptible for developing vertebral fusions could be detected already at 2g size. Atlantic salmon *in vitro* based systems are also developed, where cellular differentiation and lineage determination can be studied in more controlled environments (Ytteborg et al, 2010a). Combining radiography, histological staining techniques and molecular tools has led to a more complete understanding of how normal and pathological bone formation in Atlantic salmon progress and opens up for prospective advanced functional studies in

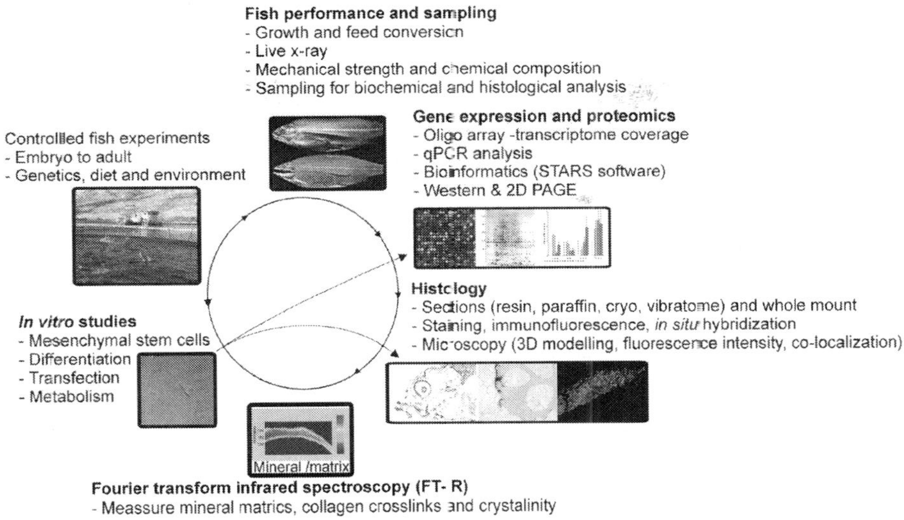

Fig. 6. Pipeline for studying vertebral development.

commercial teleosts species. Importantly, management control of deformities and health in general demands precise tools and knowledge to depict any problem as early as possible in the production line. The reliable correlation between defined skeletal markers and the risk of developing vertebral deformities has indicated that these genes can be developed as prognostic markers and further be used to investigate how the progression of skeletogenesis is modulated in response to other stimuli.

4. The teleost vertebra

The vertebral column is the defining feature of all vertebrates, composed of an alternating pattern of vertebral bodies (centra) and intervertebral regions. While centra give support and strength to the organism, intervertebral regions provide flexibility. The segmented pattern of the spine is established during embryogenesis when the precursors of the vertebrae, the somites, are formed (review in Brand-Saberi et al., 2000). The mature Atlantic salmon vertebra consists of approximately 58 vertebral bodies with neural and heamal arches protruding from the top and bottom of the centrum, respectively (Kacem et al., 1998). Grotmol and co-workers (Grotmol et al., 2006, 2005, 2003) have previously described the early development of the Atlantic salmon vertebrae in details. However, few studies have defined the nutritional needs or described the functions needed to keep continuous growth, remodelling and homeostasis in the mature vertebrae. An overview of the Atlantic salmon vertebra features is shown in figure 7.

4.1 The intervertebral regions

The notochord is found in embryos of all chordates, being well conserved between species as the forerunner of the spinal column. However, whereas only remnants of the notochord

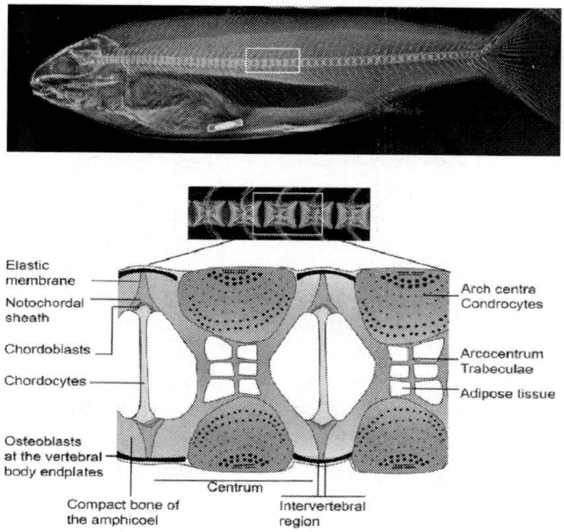

Fig. 7. Overview of the Atlantic salmon vertebra.

exist in the mammalian intervertebral disc (IVD) between adjacent vertebrae (Walmsley et al., 2009), the notochord persists throughout all life stages and throughout the entire length of the fully developed vertebral column in many teleosts, including Atlantic salmon. The morphology and function of the notochord of the mature vertebrae has not been thoroughly described. However, the layers and cell types found at early larval stages persist throughout all life stages in the salmonides. Hence, the mature notochord of Atlantic salmon consists of a core of chordocytes, a layer of chordoblasts, an acellular fibrous sheath and an outer elastic membrane (Grotmol et al., 2006). The chordoblasts continue to divide throughout life in accordance with sustained notochordal growth (Grotmol et al., 2006) and maturate into chordocytes, containing large fluid filled vacuoles (Adams et al., 1990; Glickman et al., 2003; Nordvik et al., 2005). The chordoblasts also produce the basal membrane and ECM components of the notochordal sheath, which in both mammals and teleosts like Atlantic salmon, has been shown to consist of mainly Col2 fibrils (Domowicz et al., 1995;Linsenma et al., 1973; Sandell et al., 1994). In mammals, the remnants of the notochord, the chordoblasts and their subsequent matrix, develop into the intervertebral discs (IVDs), which separate the vertebral bodies. The *annulus fibrosus* (AF) surrounding the *nucleus pulposus* (NP) of mammalian discs consists of overlapping collagen and elastin fibrils, forming transversing bands crossing the joint in opposite directions, hence, stabilizing and supporting the intervertebral regions. The NP consists of a fluid filled matrix which distributes the hydraulic pressure in all directions within each disc under compressive loads. Similarly in teleosts, the helical geometry shift between adjacent collagen lamella in the acellular notochordal sheath restricts expansion of the vacuolated chordocytes (Grotmol et al., 2005; Grotmol et al., 2006; Koehl et al., 2000). The elastic membrane surrounding the notochordal sheath has a thickened structure in the intervertebral regions, further contributing to increased strength in these regions. At more mature stages, the notochordal sheath consists of folded structures (Ytteborg et al., 2010d), which may be the consequence of compressions of the notochordal sheath upon formation and mineralization of the centra. As the cross-helical architecture of parallel Col2 fibrils probably is important for flexural stiffness of the larval body during development (Grotmol et al., 2006) the folded pattern may contribute to increased flexibility and normal functioning of the mature spinal column.

In addition to its structural role, the notochord secretes factors to surrounding tissues and contributes to vertebral patterning during embryogenesis (Cleaver et al., 2001; Fleming et al., 2004). The role of the notochord in patterning of the somites is known from several studies from chicken, mouse and zebrafish, in which secretion of Sonic hedgehog (Shh) from the notochord appears to be essential both for somite survival during the early somitogenesis and for induction of the sclerotome during later somitogenesis (review in Monsoro-Burq et al., 2005). In vertebrate species with limited growth, such as humans, the notochord ceases its regulating role for vertebral development as part of the normal ontogeny, followed by the transformation of notochordal tissue into cartilage (Hunter et al., 2003; Oegema et al., 2002). In Atlantic salmon, however, the notochord should fulfil its regulating role for vertebral body differentiation throughout life, since salmon and other fish species do not stop growing. Immunohistochemistry with the proteoglycan component Perlecan has revealed that this protein is abundantly present in the notochordal sheath of Atlantic salmon (Ytteborg et al., 2010d). Perlecan has structural roles in mammalian cartilage and IVD (Sivan et al., 2006) and is important for proper establishment of basement membranes in different vertebrates including teleosts (Parsons et al., 2002; Aviezer et al., 1994). An interesting aspect of Perlecan is its link to nutritional transportation over the notochordal sheath. Parsons et al. (2002) have previous suggested similarities between the structural role of the teleost notochordal sheath and the mammalian glomerular kidney membrane (GBM). GBM is an important part of the filtration machinery in the kidneys and involved in hydrostatic pressure maintenance (Timpl et al., 1996). The heparan sulfate chains of perlecan have further been shown to play important roles in glomerular filtration (Morita et al., 2005) and to be involved in diffusion of nutrients during tooth development in mice (Ida-Yonemochi et al., 2005). The mammalian IVD basically relies on diffusion for nutrient supplies and removal of waste products. As no evidences for vascularization of the Atlantic salmon notochord exists today, it seems likely that a similar transportation system must apply for the vacuolated chordocytes in the notochord core.

4.2 The centra

The Atlantic salmon spinal column is formed directly in bone, in contrasts to the formation of the vertebrae of avian and mammalian species, which are first formed in cartilage (Arratia et al., 2001; Smith et al., 2009). At early stages, the precursors for the osteoblasts are situated on the external elastic membrane only interrupted by the neural and haemal arch cartilages. The segmentation process leading to formation of vertebral and intervertebral regions starts with the formation of the chordacentra, where matrix in the outer half of the notochordal sheath becomes mineralized (Fleming et al., 2004; Arratia et al., 2001; Laerm et al., 1976; Grotmol et al., 2003). Osteoblast at the vertebral growth zones and osteoblasts lining the trabeculae are involved in intramembraneous ossification. Denser osteoblast populations are located along the cranial and caudal rims of each vertebral body, leading to the biconid hour-glass shaped vertebra. *In situ* hybridization has confirmed transcription of osteogenic marker genes like runx2, col1a, osteocalcin and osteonectin in these populations at mature stages in Atlantic salmon ontogeny, confirming their actively involvement in osteoid production throughout life (Krossøy et al., 2009; Ytteborg et al., 2010b and c), shown in figure 8. In the arch centra of Atlantic salmon, *in situ* hybridization have identified sub-populations of chondrocytes corresponding to the resting, proliferating and hypertrophic chondrocytes described in mammals (Ytteborg et al., 2010b; Hunziker et al., 1994). Chondrogenic marker genes, like *col2a, col10a, sox9* and *mef2c,* are characteristic for specific maturation zones and have been used to characterize the maturation process in the arches of Atlantic salmon (Ytteborg et al., 2010b). TRAP secreting osteoclasts has further been identified at the ossifying borders of the salmon arch centra, marking the ossification front during endochondral ossification (Witten et al., 2009, Helland et al., 2006; Ytteborg et al., 2010c). In the vertebrae of Atlantic salmon, multinucleated osteoclasts have also been identified in the trabeculae but not in the compact bone of the amphicoel (Witten et al., 2009). As the vertebra grow through the activity of osteoblasts located along the distal ridges, the trabeculae becomes more branched and filled with adipose tissue. After finishing shaping the scaffold for the vertebral bodies, the Atlantic salmon vertebrae continue to grow throughout life (Nordvik et al., 2005). Compared to mammals, where bone is constantly remodeled, the shape and constant growth of the salmon vertebrae have indicated that the need for bone remodeling is scarce. During stressful or unfavorable conditions or during periods of rapid growth, the mammalian skeleton is used as a mineral reservoir, where minerals are released through the activity from the osteoclasts. In Atlantic salmon however, such reservoirs are mostly found in the scales. Experiments have shown that long-term

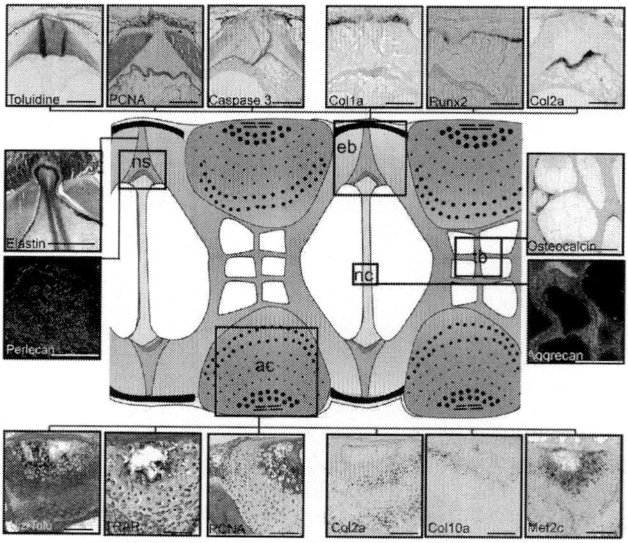

Fig. 8. Overview of histological, immunohistochemical and molecular findings in non-deformed vertebrae. Vertebral endbones (top): Toluidine, PCNA, Caspase 3, *col1a, runx2, col2a* Elastin in elastic membrane (left), Perlecan in notochordal sheath (left). *Osteocalcin* in trabeculae (right). Aggrecan in chordocytes (right) Arch centra (bottom): Alizarin red/Toluidine blue, TRAP, PCNA, *col2a, col10a, mef2c*. Trabeculae, tb; Notochordal sheath, ns; Notochord, nc; End bone, eb; Arch centra, ac. Scale bare = 100 µm.

stressful conditions rather manifests in salmon as overall improper bone formation and cellular disturbances rather than increased bone resorption. This has been shown through x-ray visualization of lower radiodensity (e.g. in "ghost" and "hyperdence" vertebrae), development of "soft" bone phenotype, transcriptional analysis (e.g. reduced transcription of genes involved in production and mineralization of ECM) and immunohistochemistry showing disturbed cell cycling (e.g. using PCNA antibodies) in vertebrae not yet possessing skeletal malformations. However, these disturbances might further develop into vertebral deformities at later stages.

5. Vertebral deformities

Deformities in the spinal column have been observed in a diverse array of vertebrates and a number of causatives have been suggested. Spinal disorders are a major concern for human health and often related to painful conditions (Freemont et al., 2009). Spinal lesions observed in wild animals, such as brown bear, sandtiger shark and smallmouth bass are occasionally found and often reflect environmental problems (Preziosi et al., 2006; Bengtsson et al., 1979; Vandenavyle et al., 1989; Wagner et al., 2005). Deformities in domesticated animals like chicken, broilers, pigs and farmed fish are recognized as a reoccurring problem in intensive production system and represent both ethical and economical challenges (Berg et al., 2006; Hammond et al., 2007; Julian et al., 1998; Reiland et al., 1978; Sullivan et al., 2007). Fish with spinal deformities, such as salmon, trout, cod, halibut, sea bass and sea bream, do not swim efficiently, are less capable of acquiring food, are at a greater risk of predation and are more susceptible to physiological imbalance, in addition to being down-graded at slaughter (Silverstone et al., 2002). Most deformity studies in teleosts have been largely descriptive and primarily performed to reveal factors contributing to increased occurrence of skeletal deformities, e.g. genetics, infections, fast growth, light regimes, vaccination, water current and quality, pollution, malnutrition and elevated temperatures (Berg et al., 2006; Berntssen et al., 2003; Cahu et al., 2003; Divanach et al., 1997; Gjerde et al., 2005; Koumoundouros et al., 2001; Lall et al., 2007; Madsen et al., 2000; Roy et al., 2002; Vagsholm et al., 1998). Spinal deformities in Atlantic salmon have been intensively studied during the past years due to the importance of this specie to the aquaculture industry. Bone deformities in Atlantic salmon are a complex problem, which may have diverse causes that may act alone or in combination. Among these causes of bone deformities, the effect of temperature stress during the early developmental stages is best documented (Ytteborg et al., 2010a,b; Wargelius et al,. 2005). Malformations later in life are often related to abnormal nutritional preferences, malnutrition or fast growth. Until recently, the molecular development of spinal deformities in fish has received relatively little attention and few deformities have been explored beyond the level of association with particular causative factors. However, accumulated studies on intensive production regimes and incidence of deformities have been followed by more and more advanced studies on vertebral development and bone biology. Below is the current state of knowledge on cellular mechanisms for pathological bone development. In figure 9, major causatives, radiography and histological staining of normal and deformed salmon is shown.

5.1 Cellular mechanisms behind weakened bone structures and development

Conditions that accompany fast growth in farmed animals, e.g. light and feeding regimes, elevated temperatures and breeding, are linked to increased numbers of spinal deformities

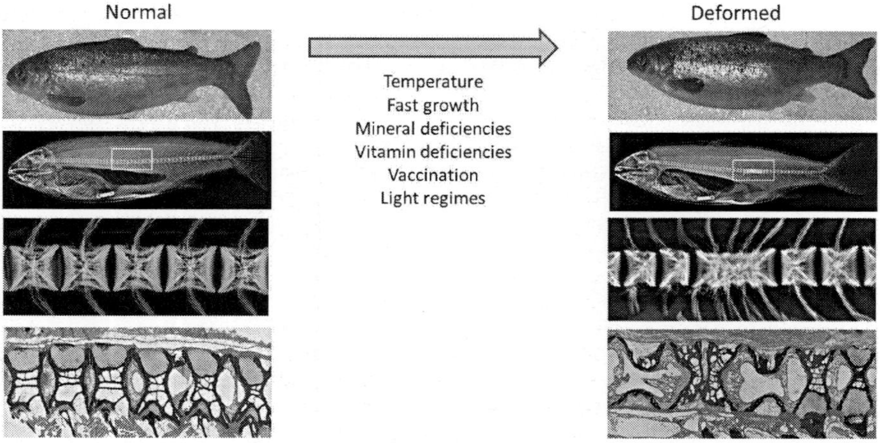

Fig. 9. Normal (left) and deformed (right) Atlantic salmon. From the top: photography of the fish, radiographic image, enlarged radiography, Alizarine red S and Tolouidine blue double staining.

(Julian et al., 1998; Reiland et al., 1978; Wargelius et al., 2009). Fast growing Atlantic salmon has been shown to develop soft, low mineralized, bone compared to fish with lower growth rates (Fjelldal et al., 2006) and to have an increased risk of developing vertebral deformities (Fjelldal et al., 2007; Fjelldal et al., 2005). In fast growing Atlantic salmon, elevated muscle mass exercise pressure on under-calcified bone that increases the mechanical pressure, which might trigger formation of intermediate tissues and malformations (Witten et al., 2005). Comparative studies have been performed in commercially farmed chicken, which are the product of long-term selective breeding for high growth rates (Leterrier et al., 1992). Fast growing chicken have weaker bone structures and increased rates of skeletal abnormalities than slower growing broilers, which reduces the bone's ability to adapt to the higher loads induced by the increasing body weight (Rawlinson et al., 2009). In Atlantic salmon, however, high genetic growth rates have not been correlated to increased rates of deformities (Gjerde et al., 2005). To fulfil the requirements for bone mineralization, fast growing animals needs to assimilate a higher proportion of the mineral intake (Hernandez et al., 2000). However, knowledge concerning mineral uptake and transportation in the fish intestines are lacking and needs to be studied further. The current change in fish feed production, switching to a vegetable based lipid diet, may further change the intestinal uptake of minerals, vitamins and amino acids (Jutfelt et al., 2007). Achieving predictable production of high-quality fish that perform well later in life therefore requires a high level of control of various factors influencing normal development and growth during early phases of life. Understanding the interactions between dietary mineral levels, n−6/n−3 fatty acid ratios, bioavailability, growth rate, temperature and intestinal uptake is imperative to be able to balance diet composition and use available feed ingredients adequately.

At the cellular level, a general trade-off between proliferation and differentiation has been suggested as a cause for delayed skeletal development in fast growing species of birds (Arendt et al., 2000; Rawlinson et al., 2009). It has further been suggested that during rapid growth the time required for bone matrix to be produced and mineralized may be reduced to a critical level (Hernandez et al, 2000); hence development of a soft bone phenotype. This causative relation has been suggested for fast growing under-yearling Atlantic salmon smolt that has a higher incidence of vertebral deformities than slower growing yearling smolt (Fjelldal et al., 2006). Temperature and light regimes are factors shown to speed up developmental rate in Atlantic salmon, but also to delay production of osteoid. It therefore seems that bone remodeling in Atlantic salmon is generally sensitive to elevated growth rates (Ytteborg et al., 2010a). Osteoblasts and chondrocytes are cell types producing large quantities of ECM and may therefore be particularly sensitive to stressful conditions, due to reduced normal protein synthesis (Tsang et al., 2007; Haynes et al., 2004). Quantification of mRNA in vertebrae from fast growing Atlantic salmon has revealed a reduced transcription of important genes encoding structural proteins taking part in the bone matrix and mineralization, e.g. *col1a1*, *osteocalcin* and *osteonectin* (Ytteborg et al., 2010b). Furthermore, generally weaker *in situ* hybridization signals were detected for probes targeting these ECM transcripts in areas where intramembranous ossification takes place. These findings further correlated to an impaired mineralization and supported the assumption that disturbances in bone formation constitute an important part of the mechanisms involved in soft bone formation. These observations are further consistent with an Atlantic salmon osteoblast *in vitro* experiment, where long-term 16°C heat exposed cells showed a decreased transcription of alp, col1a1 and osteocalcin. Based on *in vitro* and *in vivo* results it seems that Atlantic salmon osteoblasts may be particularly sensitive to elevated temperatures during the early stages of differentiation.

In mammals and teleosts like Atlantic salmon, elevated temperatures and fast growth may also interrupt the normal chondrocytic differentiation pattern and delay endochondral bone formation, further weakening the bony structures (Tsang et al., 2007). A number of studies have linked skeletal malformations to disturbances in chondrocytic maturation (Kieswetter et al., 1997; Farquharson et al., 2000; Julian et al., 1998). Recent results have suggested that fast growth caused by elevated temperatures leads to an arrest prior to the final maturation of chondrocytes in the Atlantic salmon vertebral arch centra (Ytteborg et al., 2010c). Morphological studies of the arch centra of juvenile Atlantic salmon reared under intensive temperatures have identified chondrocytes with a distorted maturation pattern and an increased zone of hypertrophic chondrocytes (Ytteborg et al., 2010c). In this study, an increased zone of hypertrophic chondrocytes correlated with increased transcription of hypertrophic marker genes such as col10a1 and mef2c. Fast growing chickens are also characterized by disturbed chondrocytic maturation where cartilage do not mature enough to ossify (Julian et al., 2005; Farquharson et al., 2000) and increased mechanical load is associated with an increased hypertrophic zone in the growth plate of rat ulnae along with a suppressed mineralization rate (Robling et al., 2001; Ohashi et al., 2002). Furthermore, mammalian osteoclasts are temperature sensitive and hypothermic conditions may stimulate their activity (Patel et al., 2009). Similar observations have been described in Atlantic salmon where no TRAP activity was observed in the arch centra of fish reared at intensive temperatures. Also transcription of osteoclast associated marker genes, like Mmps and Cathepsin K was reduced (Ytteborg et al., 2010c). Absence of Mmps may cause delays in endochondral ossification and *runx2* deficiency may inhibit *mmp* expression and lead to mild disturbances of chondrocyte differentiation (Inada et al., 1999, Kirsch et al., 1997; Pratap, 2005). Disturbances in chondrocytic maturation and endochondral ossification will overall weaken the vertebrae, and may be an explanation for wrinkled and shortened ribs observed in Atlantic salmon suffering from P deficiency (reviewed in Sugiura et al 2005).

Overall, both bone and cartilage formation seems disturbed during fast growth and may equally contribute to weakened skeletal structures. In Atlantic salmon, experiments have indicated that during rapid growth, both endochondral and intramembraneous ossification is affected. Moreover, fast growth leading to weakened bone and cartilage structures at juvenile stages increases the risk of developing severe deformities later in ontogeny. This might be a result of local cellular compensation and an effort to restore and strengthen a weakened area in the vertebrae, as described in the next chapter.

5.2 Cellular mechanisms behind vertebral deformities

Witten et al. (2009) recently published a survey on commonly observed vertebral malformations in Atlantic salmon which included different grades and combinations of platyspondyly (compressions), ankylosis (fusions), lordosis (V-shaped vertebral column), kyphosis (Λ-shaped vertebral column) and scoliosis (S-shaped vertebral column). Histological characterization of compressions and fusions have described shape alterations of vertebral body endplates, reduced intervertebral space, transformation of intervertebral notochord tissue into cartilage, mineralization of the intervertebral cartilage and replacement of intervertebral cartilage by bone (Witten et al., 2005; Kvellestad et al., 2000; Witten et al., 2006), independent of the factor inducing the malformation. Changes in transcriptional processes in osteoblasts and chondrocytes from both mammals and teleosts are involved in pathological vertebral formation (Hammond et al., 2007; Breen et al., 1999; Wargelius et al., 2005). The development of vertebral fusions is a dynamic process but recent publications have shown that the underlying cellular and molecular mechanisms may be summarized as four key events (Ytteborg et al., 2010a, b and c). These events are illustrated in figure 10 and described in the text below.

I: Disorganization

The initiation of the fusion process includes disorganization and proliferation of osteoblasts and chordoblasts. Osteoblasts at the growth zones of the vertebral body endplates have a markedly increased cell proliferation rate and the growth zones extend spatially along the rims of fusing vertebral bodies. As the intervertebral space narrows, proliferating chordoblasts and denser packet chordocytes appear. With a progressing pathology, proliferating chordoblasts occupy most of the intervertebral space and vacuolated chordocytes disappear.

II: Metaplastic shift

Proliferating cells at the border between the osteoblast growth zones and the arch centra show a transcriptional shift, where co-transcription of osteogenic (*col1a, runx2, osteocalcin* and *osteonectin*) and chondrogenic (*col2a, mef2c* and *col10a*) marker genes are prominent. The marked border between the osteoblast growth zones and the chondrocytic areas connected to the arches becomes less distinct, as proliferating cells and chondrocytes blend through an intermediate zone. A similar shift is found in the notochord where co-transcription of genes such as col2a, sox9, col1a and runx2 increase with proliferation of chordoblasts. In the central notochord of developing fusions, hyperdense regions of denser packet chordocytes lacking vacuoles appear as the number of proliferating cell increase.

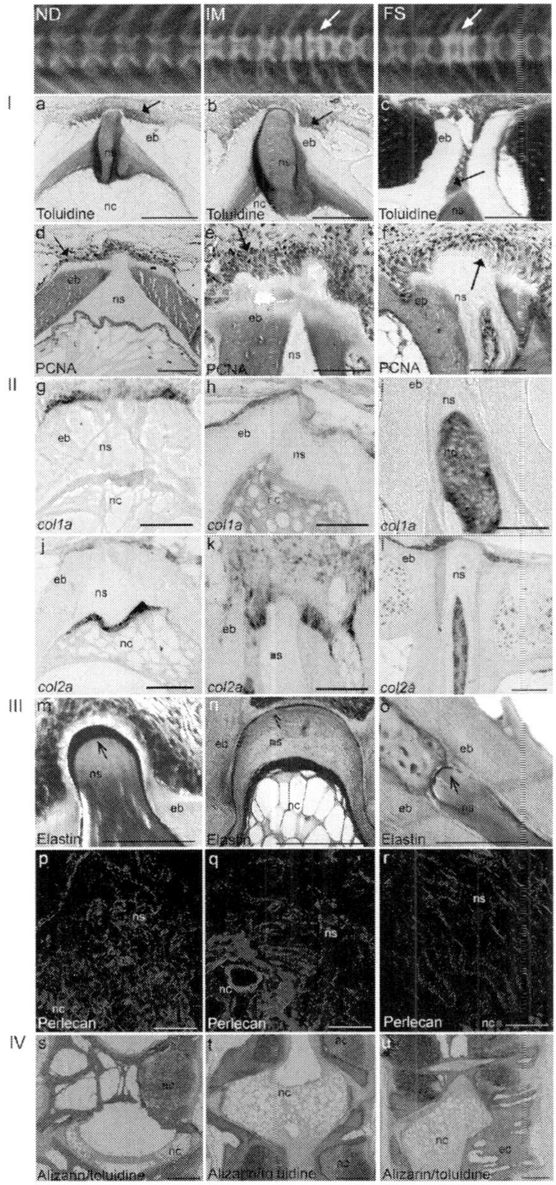

Fig. 10. Major findings during the development of vertebral fusions in Atlantic salmon. I Disorganization, II Metaplsatic shift, III Loss of notochordal sheath integrity, and IV Ectopic bone formation. See text for details. Scale bar = 100 µm (a-l), 50 µm (m-r), 200 µm (s-u).

III: Loss of notochordal sheath integrity

The elastic membrane surrounding the notochord becomes fragmented and the notochordal sheath loose its integrity. Verhoeff`s hematoxylin staining has visualized a thinner elastic membrane surrounding the notochordal sheath of developing vertebral fusions. In the most severe cases, the elastic membrane is fragmented. Furthermore, the highly folded structures in the notochordal sheath are lost during development of spinal fusions.

IV: Ectopic bone formation

Ectopic bone formation in the affected areas gives the vertebral bodies a squared morphology as the arch centra fuse and ossify. Ectopic mineralization of intervertebral regions and arch centra is formed, indicating that the proliferating and metaplastic cells not only differentiates towards osteoblast-like cells, but also complete the differentiation to cells that are capable of producing mineralized matrix. The intervertebral space narrows completely down and the notochord mineralizes.

The overall structural and molecular features of bone and cartilage development in vertebral fusions in Atlantic salmon have shown resemblance with similar pathological spinal conditions in mammals (Ytteborg et al., 2010c and d; Gorman et al., 2007; Witten et al., 2006). For example several mammalian studies have suggest that changes in the balance between cell death and cell proliferation is involved in bone and cartilage defects which may lead to malformations (Cockroft et al., 1978; Miura et al., 2004; Breen et al., 1999; Farquharson et al., 2000). Spinal fusions in Atlantic salmon are characterized by changes in ECM components and mineralization of the intervertebral regions (Ytteborg et al., 2010c; Witten et al., 2006). Similarly, intervertebral disc degeneration (IDD) in mammals involves breakdown of ECM components in the AF and calcification of the NP (Takaishi et al., 1997; Kanemoto et al., 1996; Antoniou et al., 1996). Fusion, compression and chondrogenic transformation of skeletal tissue have also been reported from lordosis and kyphosis in sea bass. Histological examinations of both lordosis and hyperdense vertebrae have further indicated cellular plasticity (like metaplastic shifts and trans-differentiation) and development of intermediate tissues as pathological events (Ytteborg et al., 2010c; Helland et al., 2006; Kranenbarg et al., 2006; Witten et al., 2006; Witten et al., 2005). It has previously been suggested that a metaplastic shift is involved in the development of spinal fusions, leading to the formation of chondroid bone which at later stages in the fusion process is replaced by bone. As previously discussed, chondrocytes associated with calcifying cartilage can acquire properties of osteoblasts (Cancedda et al., 1992) and are able to change their phenotype from a primarily cartilage synthesizing cell type to a bone synthesizing cell type (Lian et al., 1993). Co-transcription of chondrogenic and osteogenic marker genes in the arch centra and notochord supports the suggestion of an adaptation through metaplastic shifts during development of vertebral fusions, which may be induced to produce more robust cells that are able to withstand increased mechanical load. A pathway to bone formation through chondrocytes might be possible during development of vertebral fusions and fast growth, which could be similar to trans-chondroid ossification, as described by Yasui et al. (1997). Trans-differentiation and ectopic calcification has also been suggested as pathological pathways in lordotic sea bass where deformations stimulate ectopic bone formation in the intervertebral regions between two affected vertebral bodies and along the rims of the vertebral body endplates (Kranenbarg et al., 2006). Similarly, a shift in the mammalian IVD NP cell population coincides with spinal disorders like intervertebral disc degeneration and changes in the synthesis of matrix molecules differ with the degree of degeneration (Handa et al., 1997). The mammalian AF is further strengthened through cartilage formation upon elevated mechanical load (Lotz et al., 2002; Prescher et al., 1998). Moreover, breakdown of PG components, like Aggrecan and Perlecan, may lead to reduced hydrostatic pressure, invasion of nerves and blood vessels and loss of transportation of nutrients and waste products in degenerating IVD (Kauppila, 1995; Urban et al., 2003 and 2004; Melrose et al., 2002; Yasuma et al., 1993). Loss of Aggrecan resulting in tissue dehydration, reduces the ability of mammalian IVD to transmit and absorb compressive load (Kanemoto et al., 1996; Urban et al., 1985). Loss of Aggrecan and Perlecan has also been observed in the notochord of Atlantic salmon during development of vertebral fusions (Ytteborg et al. 2010d), which may possibly reduce the hydrostatic pressure and hence the transportation of nutrients and alteration of pH values. Another comparative pathological process to teleost vertebral fusions is the mammalian "Bamboo spine", describing a condition where vertebral bodies have fused and reshaped through ectopic bone formation (Bakay et al., 1970; Resnick et al., 1983). Witten et al.(2005) have described similar processes in Atlantic salmon. Fusing vertebral bodies may either stabilize as on large vertebral body or continue to develop through neighbouring vertebrae. What kind of cellular actions leading to a stabilized or aggravating fusion remains to be answered. However, it seems that different types of deformities have similar pathways of cellular pathological development, processes involving proliferation, metaplastic shifts, cellular instability and trans-differentiation.

6. Conclusion

During the last decade, fish have emerged as suitable animal models for studying bone and cartilage biology and have shown to be a suitable supplement to mammalian systems aiming to uncover the corresponding fundamental cellular and molecular mechanisms of action. In the light of metaplastic shifts during skeletal deformities in Atlantic salmon, a cell culture based system allowing for cellular differentiation and lineage determination studies have been developed. In this particular system, precursor cells are stimulated to myogenic, adipogenic and osteogenic differentiation, and opens up for studies where these cells can be manipulated upon different stimuli to undergo metaplastic shifts. Hence, functional studies can be performed to better characterize the pathology, define particular requirements and minimize the occurrence of bone disorders. Advanced methods and defined molecular markers should enable us to detect the risk of developing deformities early in ontogeny. Similar diagnostics and medications as those existing in the human medicine will not be applicable for farmed animals. However, treatments and diets have shown to be well suited also for teleosts. Exercise and addition of minerals in the feed have already shown positive effects in regards of bone quality and should be further addressed in future research.

7. Acknowledgments

The authors would like to thank Johanne Halvorsen (www.forestry.no) for help with graphic illustrations.

8. References

Adams DS, Keller R, Koehl MAR (1990) The Mechanics of Notochord Elongation, Straightening and Stiffening in the Embryo of Xenopus-Laevis. *Development*, 110: 115-130.

Anderson HC, Čecil R, Sajdera SW (1975) Calcification of Rachitic Rat Cartilage Invitro by Extracellular Matrix Vesicles. *American Journal of Pathology*, 79: 237-254.

Anderson HC, Garimella R, Tague SE (2005) The role of matrix vesicles in growth plate development and biomineralization. *Frontiers in Bioscience*, 10: 822-837.

Anderson HC (1996) Molecular-Biology of Matrix Vesicles. *Clinical Orthopaedics and Related Research*, 266-280.

Antoniou J, Steffen T, Nelson F, Winterbottom N, Hollander AP, Poole RA, Aebi M, Alini M (1996) The human lumbar intervertebral disc: evidence for changes in the biosynthesis and denaturation of the extracellular matrix with growth, maturation, ageing, and degeneration. *Journal of Clinical Investigation*, 98: 996-1003.

Arendt JD (2000) Allocation of cells to proliferation vs. differentiation and its consequences for growth and development. *Journal of Experimental Zoology*, 288: 219-234.

Arnold MA, Kim Y, Czubryt MP, Phan D, McAnally J, Qi X (2007) MEF2C transcription factor controls chondrocyte hypertrophy and bone development. *Developmental Cell*, 12: 377-389.

Arratia G, Schultze HP, Casciotta J (1998) Vertebral column and associated elements in dipnoans and comparison with other fishes: Development and homology. *Journal of Morphology* 2001, 250: 101-172.

Aubin JE (1998) Bone stem cells. *Journal of Cellular Biochemistry*, 73-82.

Aviezer D, Hecht D, Safran M, Eisinger M, David G, Yayon A (1994) Perlecan, basal lamina proteoglycan, promotes basic fibroblast growth factor-receptor binding, mitogenesis, and angiogenesis. *Cell*, Volume 79:6, 1005-1013.

Baeverfjord G, Åsgård T, Rye M, Storset A (1998) High temperatures during egg incubation may induce malformations in Atlantic salmon (*Salmo salar L.*). *Aguaculture and water*, EAS special publication no 26, pp. 24-25.

Baeverfjord G, Åsgård T, Shearer KD (1998) Development and detection of phosphorus deficiency in Atlantic salmon (*Salmo salar*). *Aquacquaculture Nutrition*, 4:1-11.

Baeverfjord G. (1999) Deformities induced by temperature stress in farmed Atlantic salmon and Rainbow trout. *Report to the Norwegian Research Council of Norway*, pp. 19.

Bakay L, Cares HL, Smith RJ (1970) Ossification in Region of Posterior Longitudinal Ligament As A Cause of Cervical Myelopathy. *Journal of Neurology Neurosurgery and Psychiatry*, 33: 263-371.

Balcerzak M, Hamade E, Zhang L, Pikula S, Azzar G, Radisson J (2003) The roles of annexins and alkaline phosphatase in mineralization process. *Acta Biochimica Polonica*, 50: 1019-1038.

Bell DM, Leung KKH, Wheatley SC, Ng LJ, Zhou S, Ling KW (1997) SOX9 directly regulates the type-II collagen gene. *Nature Genetics* 1997, 16: 174-178.

Bengtsson BE (1979) Biological Variables, Especially Skeletal Deformities in Fish, for Monitoring Marine Pollution. *Philosophical Transactions of the Royal Society of London Series B-Biological Sciences*, 286: 457-464.

Berg A, Rodseth OM, Tangeras A, Hansen T (2006) Time of vaccination influences development of adhesions, growth and spinal deformities in Atlantic salmon (*Salmo salar*). *Diseases of Aquatic Organisms*, 69: 239-248.

Berntssen MHG, Waagbo R, Toften H, Lundebye AK (2003) Effects of dietary cadmium on calcium homeostasis, Ca mobilization and bone deformities in Atlantic salmon (*Salmo salar L.*) parr. *Aquaculture Nutrition,* 9: 175-183.

Blavier L, Delaisse JM (1995) Matrix Metalloproteinase Activity Is Obligatory for the Recruitment of Osteoclasts to Future Bone-Resorption Sites *Journal of Bone and Mineral Research,* 10: S224.

Bolander ME, Young MF, Fisher LW, Yamada Y, Termine JD (1998) Osteonectin cDNA Sequence Reveals Potential Binding Regions for Calcium and Hydroxyapatite and Shows Homologies with Both A Basement-Membrane Protein (Sparc) and A Serine Proteinase-Inhibitor (Ovomucoid). Proceedings of the *National Academy of Sciences of the United States of America,* 85: 2919-2923.

Boyle WJ, Simonet WS, Lacey DL (2003) Osteoclast differentiation and activation. *Nature,* 423: 337-342.

Braga D, Laizé V, Tiago DM, Cancela ML (2006) Enhanced DNA transfer into fish bone cells using polyethylenimine. *Mol Biotechnol* 34:51–54.

Brand-Saberi B, Christ B (2000) Evolution and development of distinct cell lineages derived from somites. Current *Topics in Developmental Biology,* 48, 48: 1-42.

Breen JG, Claggett TW, Kimmel GL, Kimmel CA (1999) Heat shock during rat embryo development *in vitro* results in decreased mitosis and abundant cell death. *Reproductive Toxicology,* 13: 31-39.

Bruger, C, Ridler, N (2004) Global aquaculture outlook in the next decades: an analysis of national aquaculture production forecasts to 2030. *Fisheries Circular* No. 1001 (FAO ed.), pp. 47. FAO, Rome

Cahu C, Infante JZ, Takeuchi T (2003) Nutritional components affecting skeletal development in fish larvae. *Aquaculture,* 227: 245-258.

Cancedda FD, Gentili C, Manduca P, Cancedda R (1992) Hypertrophic Chondrocytes Undergo Further Differentiation in Culture. *Journal of Cell Biology* 1992, 117: 427-435.

Cleaver O, Krieg PA (2001) Notochord patterning of the endoderm. Developmental Biology 234: 1-12.

Cockroft DL, New DAT (1978) Abnormalities Induced in Cultured Rat Embryos by Hyperthermia. *Teratology,* 17: 277-283.

Collin-Osdoby P (2004) Regulation of vascular calcification by osteoclast regulatory factors RANKL and osteoprotegerin. Circulation Research 95: 1046-1057.

Conceição N, Laizé V, Simões B, Pombinho AR, Cancela ML (2008) Retinoic acid is a negative regulator of matrix Gla protein gene expression in teleost fish *Sparus aurata. Biochim. Biophys. Acta* 1779, 28-39.

Cowles EA, Derome ME, Pastizzo G, Brailey LL, Gronowicz GA (1998) Mineralization and the expression of matrix proteins during *in vivo* bone development. *Calcified Tissue International,* 62: 74-82.

Delaisse JM, Andersen TL, Engsig MT, Henriksen K, Troen T, Blavier L (2003) Matrix metalloproteinases (MMP) and cathepsin K contribute differently to osteoclastic activities. *Microscopy Research and Technique* 61: 504-513.

DeLaurier A, Eames BF, Blanco-Sánchez B, Peng G, He X, Swartz ME, Ullmann B, Westerfield M, Kimmel CB (2010) Zebrafish sp7:EGFP: a transgenic for studying otic vesicle formation, skeletogenesis, and bone regeneration. *Genesis.* 48(8):505-11.

Denstadli V, Witten PE, Hjelde K, Storebakken T, Skrede A (2006) Hyper dense vertebrae and mineral content in Atlantic salmon (*Salmo salar L.*) fed diets with graded levels of phytic acid. *Aquaculture* 2006, 261: 603-614.

Divanach P, Papandroulakis N, Anastasiadis P, Koumoundouros G, Kentouri M (1007) Effect of water currents on the development of skeletal deformities in sea bass (*Dicentrarchus labrax L.*) with functional swimbladder during postlarval and nursery phase. *Aquaculture,* 156: 145-155.

Domowicz M, Li H, Hennig A, Henry J, Vertel BM, Schwartz NB (1995) The Biochemically and Immunologically Distinct Cspg of Notochord Is A Product of the Aggrecan Gene. *Developmental Biology* 1995, 171: 655-664.

Engsig MT, Chen QJ, Vu TH, Pedersen AC, Therkidsen B, Lund LR (2000) Matrix metalloproteinase 9 and vascular endothelial growth factor are essential for osteoclast recruitment into developing long bones. *Journal of Cell Biology* 151: 879-889.

Erlebacher A, Filvaroff EH, Gitelman SE, Derynck R (1995) Toward A Molecular Understanding of Skeletal Development. *Cell* 80: 371-378.

Farquharson C, Jefferies D. (2000) Chondrocytes and longitudinal bone growth: The development of tibial dyschondroplasia. *Poultry Science* 79: 994-1004.

Fisher S, Jagadeeswaran P, Halpern ME (2003) Radiographic analysis of zebrafish skeletal defects. *Dev. Biol.* 264, 64-76.

Fjelldal G, Hansen TJ, Berg AE (2007) A radiological study on the development of vertebral deformities in cultured Atlantic salmon (*Salmo salar L.*). *Aquaculture,* 273: 721-728.

Fjelldal PG, Hansen T, Breck O, Sandvik R, Waagbo R, Berg A (2009)Supplementation of dietary minerals during the early seawater phase increase vertebral strength and reduce the prevalence of vertebral deformities in fast-growing under-yearling Atlantic salmon (*Salmo salar L.*) smolt. *Aquaculture Nutrition,* 15: 366-378.

Fjelldal PG, Lock EJ, Grotmol S, Totland GK, Nordgarden U, Flik G (2006)Impact of smolt production strategy on vertebral growth and mineralisation during smoltification and the early seawater phase in Atlantic salmon (*Salmo salar). Aquaculture,* 261: 715-728.

Fjelldal PG, Nordgarden U, Berg A, Grotmol S, Totland GK, Wargelius A (2005) Vertebrae of the trunk and tail display different growth rates in response to photoperiod in Atlantic salmon, *Salmo salar L.,* post-smolts. *Aquaculture,* 250: 516-524.

Flanagan JA, Power DM, Bendell LA, Guerreiro PM, Fuentes J, Clark MS, Canario AVM, Danks JA, Brown BL, Ingleton PM (2000). Cloning of the cDNA for Sea Bream (*Sparus aurata*) Parathyroid Hormone-Related Protein. *General and Comparative Endocrinology* 118:3, 373-382.

Fleming A, Keynes R, Tannahill D (2004) A central role for the notochord in vertebral patterning. *Development,* 131: 873-880.

Fonseca VG, Laizé V, Valente MS, Cancela ML (2007) Identification of an osteopontin-like protein in fish associated with mineral formation. *FEBS J.* 274, 4428-4439.

Freemont AJ (2009) The cellular pathobiology of the degenerate intervertebral disc and discogenic back pain. *Rheumatology* 48: 5-10.

Furie B, Furie BC (1991) Molecular-Basis of Gamma-Carboxylation, Role of the Propeptide in the Vit K-Dependent Proteins. *Annals of the New York Academy of Sciences* 614: 1-10.

Gjerde B, Pante MJR, Baeverfjord G (2005) Genetic variation for a vertebral deformity in Atlantic salmon (*Salmo salar*). *Aquaculture* 244: 77-87.

Glickman NS, Kimmel CB, Jones MA, Adams RJ (2003) Shaping the zebrafish notochord. *Development* 130: 873-887.

Gorman KF, Breden F (2007) Teleosts as models for human vertebral stability and deformity. *Comp. Biochem. Physiol. C Toxicol. Pharmacol.* 145, 28- 38.

Grotmol S, Kryvi H, Keynes R, Krossoy C, Nordvik K, Totland GK (2006) Stepwise enforcement of the notochord and its intersection with the myoseptum: an evolutionary path leading to development of the vertebra? *Journal of Anatomy* 209: 339-357.

Grotmol S, Kryvi H, Nordvik K, Totland GK (2003) Notochord segmentation may lay down the pathway for the development of the vertebral bodies in the Atlantic salmon. *Anatomical Embryology*, 207: 263-272.

Grotmol S, Nordvik K, Kryvi H, Totland GK (2005) A segmental pattern of alkaline phosphatase activity within the notochord coincides with the initial formation of the vertebral bodies. *Journal of Anatomy* 206: 427-436.

Hammond CL, Simbi BH, Stickland NC (2007) *In ovo* temperature manipulation influences embryonic motility and growth of limb tissues in the chick (*Gallus gallus*). *Journal of Experimental Biology*, 210: 2667-2675.

Handa T, Ishihara H, Ohshima H, Osada R, Tsuji H, Obata K (1997) Effects of hydrostatic pressure on matrix synthesis and matrix metalloproteinase production in the human lumbar intervertebral disc. *Spine* 22: 1085-1091.

Hayes AJ, Benjamin M, Ralphs JR (2001) Extracellular matrix in development of the intervertebral disc. *Matrix Biology*, 20: 107-121.

Haynes CM, Titus EA, Cooper AA (2004) Degradation of misfolded proteins prevents ER-derived oxidative stress and cell death. *Molecular Cell*, 15: 767-776.

Helland S, Denstadli V, Witten PE, Hjelde K, Storebakken T, Skrede A (2006) Hyper dense vertebrae and mineral content in Atlantic salmon (*Salmo salar L.*) fed diets with graded levels of phytic acid. *Aquaculture*, 261: 603-614.

Hernandez CJ, Beaupre GS, Carter DR (2000) A model of mechanobiologic and metabolic influences on bone adaptation. *Journal of Rehabilitation Research and Development*, 37: 235-244.

Hoffmann HM, Beumer TL, Rahman S, Mccabe LR, Banerjee C, Aslam F (1996) Bone tissue-specific transcription of the osteocalcin gene: Role of an activator osteoblast-specific complex and suppressor hox proteins that bind the OC box. *Journal of Cellular Biochemistry*, 61: 310-324.

Hogan BLM, Cooper AR, Kurkinen M (1980) Incorporation Into Reicherts Membrane of Laminin-Like Extracellular Proteins Synthesized by Parietal Endoderm Cells of the Mouse Embryo. *Developmental Biology*, 80: 289-300.

Hunter CJ, Matyas JR, Duncan NA(2003) The three-dimensional architecture of the notochordal nucleus pulposus: novel observations on cell structures in the canine intervertebral disc. *Journal of Anatomy*, 202: 279-291.

Hunziker EB (1994) Mechanism of Longitudinal Bone-Growth and Its Regulation by Growth-Plate Chondrocytes. *Microscopy Research and Technique*, 28: 505-519.

Huysseune A, Sire JY (1990) Ultrastructural Observations on Chondroid Bone in the Teleost Fish *Hemichromis bimaculatus*. *Tissue & Cell*, 22: 371-383.

Huysseune A, Verraes W (1986) Chondroid Bone on the Upper Pharyngeal Jaws and Neurocranial Base in the Adult Fish *Astatotilapia elegans*. *American Journal of Anatomy*, 177: 527-535.

Ida-Yonemochi H, Ohshiro K, Swelam W, Metwaly H, Saku T. (2005) Perlecan, a basement membrane-type heparin sulfate proteoglycan, in the enamel organ: its intraepithelial localization in the stellate reticulum. *J Histochem Cytochem* 53:763–772.

Ikeda T, Nomura S, Yamaguchi A, Suda T, Yoshiki S (1992) *In situ* Hybridization of Bone-Matrix Proteins in Undecalcified Adult-Rat Bone Sections. *Journal of Histochemistry & Cytochemistry*, 40: 1079-1088.

Inada M, Yasui T, Nomura S, Miyake S, Deguchi K, Himeno M (1999) Maturational disturbance of chondrocytes in Cbfa1-deficient mice. *Developmental Dynamics*, 214: 279-290.

Ishizeki K, Takigawa M, Harada Y, Suzuki F, Nawa T (1996) Meckel's cartilage chondrocytes in organ culture synthesize bone-type proteins accompanying osteocytic phenotype expression. *Anatomy and Embryology*, 193: 61-71.

Iyama K, Ninomiya Y, Olsen BR, Linsenmayer TF, Trelstad RL, Hayashi M (1991) Spatiotemporal Pattern of Type-X Collagen Gene-Expression and Collagen Deposition in Embryonic Chick Vertebrae Undergoing Endochondral Ossification. *Anatomical Record*, 229: 462-472.

Julian RJ (2005) Production and growth related disorders and other metabolic diseases of poultry - A review. *Veterinary Journal*, 169: 350-369.

Julian RJ (1998) Rapid growth problems: Ascites and skeletal deformities in broilers. *Poultry Science*, 77: 1773-1780.

Jutfelt F, Olsen RE, Björnsson BT, Sundell K (2007). Parr–smolt transformation and dietary vegetable lipids affect intestinal nutrient uptake, barrier function and plasma cortisol levels in Atlantic salmon. *Aquaculture* 273; 298-311.

Kacem A, Meunier FJ, Bagliniere JL (1998) A quantitative study of morphological and histological changes in the skeleton of *Salmo salar* during its anadromous migration. *Journal of Fish Biology*, 53: 1096-1109.

Kanemoto M, Hukuda S, Komiya Y, Katsuura A, Nishioka J (1996) Immunohistochemical study of matrix metalloproteinase-3 and tissue inhibitor of metalloproteinase-1 in human intervertebral discs. *Spine* 21:1-8.

Karp SJ, Schipani E, St-Jacques B, Hunzelman J, Kronenberg H, McMahon AP (2000) Indian hedgehog coordinates endochondral bone growth and morphogenesis via Parathyroid Hormone related-Protein-dependent and -independent pathways. *Development*, 127: 543-548.

Karsenty G, Ducy P, Starbuck M, Priemel M, Shen J, Geoffroy V (1999) Cbfa1 as a regulator of osteoblast differentiation and function. *Bone*, 25: 107-108.

Karsenty G, Kronenberg HM, Settembre C. (2009). Genetic Control of Bone Formation. *Annual Review of Cell and Developmental Biology* 25: 629-648.

Kauppila LI (1995). Ingrowth of blood vessels in disc degeneration. Angiographic and histological studies of cadaveric spines. *J Bone Joint Surg Am.* 77(1):26-31.

Kieswetter K, Schwartz Z, Alderete M, Dean DD, Boyan BD (1997) Platelet derived growth factor stimulates chondrocyte proliferation but prevents endochondral maturation. *Endocrine*, 6: 257-264.

Kikuta H, Laplante M, Navratilova P, Komisarczuk AZ, Engstrom PG, Fredman D (2007) Genomic regulatory blocks encompass multiple neighboring genes and maintain conserved synteny in vertebrates. *Genome Research*, 17: 545-555.

Kim IS, Otto F, Zabel B, Mundlos S (1999) Regulation of chondrocyte differentiation by Cbfa1. *Mechanisms of Development*, 80: 159-170.

Kimmel CB, DeLaurier A, Ullmann B, Dowd J, McFadden M (2010). Modes of developmental outgrowth and shaping of a craniofacial bone in zebrafish. *PLoS* One. 5;5(3):e9475.

Kirsch T, Nah HD, Shapiro IM, Pacifici M (1997) Regulated production of mineralization-competent matrix vesicles in hypertrophic chondrocytes. *Journal of Cell Biology*, 137: 1149-1160.

Kirsch T (2005) Annexins - Their role in cartilage mineralization. *Frontiers in Bioscience*, 10: 576-581.

Koehl MAR, Quillin KJ, Pell CA (2000) Mechanical design of fiber-wound hydraulic skeletons: The stiffening and straightening of embryonic notochords. *American Zoologist*, 40: 28-41.

Koumoundouros G, Divanach P, Kentouri M (2001) The effect of rearing conditions on development of saddleback syndrome and caudal fin deformities in *Dentex dentex (L.)*. *Aquaculture*, 200: 285-304.

Kranenbarg S, Schipper H, van Leeuwen J (2006) Local biophysical stimuli and chondroid bone in lordotic vertebrae of the sea bass. *Comparative Biochemistry and Physiology A-Molecular & Integrative Physiology*, 143: S95.

Kranenbarg S, van Cleynenbreugel T, Schipper H, van Leeuwen J (2005) Adaptive bone formation in acellular vertebrae of sea bass (*Dicentrarchus labrax L.*). *Journal of Experimental Biology*, 208: 3493-3502.

Krossøy C, Ørnsrud R, Wargelius A (2009) Differential gene expression of bgp and mgp in trabecular and compact bone of Atlantic salmon (*Salmo salar L.) Journal of Anatomy* 215:6, 663–672.

Laerm J (1976) Development, Function, and Design of Amphicoelous Vertebrae in Teleost Fishes. *Zoological Journal of the Linnean Society*, 58: 237-254.

Lall SP, Lewis-McCrea LM (2007) Role of nutrients in skeletal metabolism and pathology in fish - An overview. *Aquaculture*, 267: 3-19.

Leterrier C, Nys Y (1992) Clinical and Anatomical Differences in Varus and Valgus Deformities of Chick Limbs Suggest Different Etiopathogenesis. *Avian Pathology*, 21: 429-442.

Lian JB, Mckee MD, Todd AM, Gerstenfeld LC (1993) Induction of Bone-Related Proteins, Osteocalcin and Osteopontin, and Their Matrix Ultrastructural-Localization with Development of Chondrocyte Hypertrophy Invitro. *Journal of Cellular Biochemistry*, 52: 206-219.

Linsenma TF, Trelstad RL, Gross J (1973) Collagen of Chick Embryonic Notochord. Biochemical and Biophysical Research Communications, 53: 39-45.

Lock EJ, Ornsrud R, Aksnes L, Spanings FAT, Waagbo R, Flik G (2007) The vitamin D receptor and its ligand 1 alpha,25-dihydroxyvitamin D-3 in Atlantic salmon (*Salmo salar*). *Journal of Endocrinology*, 193: 459-471.

Lotz JC, Hsieh AH, Walsh AL, Palmer EI, Chin JR (2002) Mechanobiology of the intervertebral disc. *Biochemical Society Transactions*, 30: 853-858.

Mackie EJ, Ahmed YA, Tatarczuch L, Chen KS, Mirams M (2008) Endochondral ossification: How cartilage is converted into bone in the developing skeleton. *International Journal of Biochemistry & Cell Biology*, 40: 46-62.

Madsen L, Arnbjerg J, Dalsgaard I (2000) Spinal deformities in triploid all-female rainbow trout. *Bulletin of the European Association of Fish Pathologists*, 20: 206-208.

Melrose J, Roberts S, Smith S, Menage J, Ghosh P (2002) Increased nerve and blood vessel ingrowth associated with proteoglycan depletion in an ovine anular lesion model of experimental disc degeneration. *Spine* 27:1278-1285.

Miura M, Chen XD, Allen MR, Bi YM, Gronthos S, Seo BM (2004): A crucial role of caspase-3 in osteogenic differentiation of bone marrow stromal stem cells. *Journal of Clinical Investigation*, 114: 1704-1713.

Morita H, Yoshimura A, Inui K, Lodeura T, Watanabe H, Wang L (2005) Heparan sulfate of perlecan is involved in glomerular filtration. *Journal of the American Society of Nephrology*, 16: 1703-1710.

Motyckova G, Weilbaecher KN, Horstmann M, Rieman DJ, Fisher DZ, Fisher DE (2001) Linking osteopetrosis end pycnodysostosis: Regulation of cathepsin K expression by the microphthalmia transcription factor family. *Proceedings of the National Academy of Sciences of the United States of* America, 98: 5798-5803.

Nakashima K, de Crombrugghe B (2003) Transcriptional mechanisms in osteoblast differentiation and bone formation. *Trends Genet*, 19: 458-466.

Nakashima K, Zhou X, Kunkel G, Zhang ZP, Deng JM, Behringer RR (2002) The novel zinc finger-containing transcription factor Osterix is required for osteoblast differentiation and bone formation. *Cell*, 108: 17-29.

Nissen RM, Amsterdam A, Hopkins N (2006) A zebrafish screen for craniofacial mutants identifies *wdr68* as a highly conserved gene required for endothelin-1 expression. *BMC Dev. Biol.* 6, 28.

Nordvik K, Kryvi H, Totland GK, Grotmol S (2005) The salmon vertebral body develops through mineralization of two preformed tissues that are encompassed by two layers of bone. *Journal of Anatomy*, 206: 103-114.

Oegema TR (2002) The role of disc cell heterogeneity in determining disc biochemistry: a speculation. *Biochemical Society Transactions*, 30: 839-844.

Ohashi N, Robling AG, Burr DB, Turner CH (2002) The effects of dynamic axial loading on the rat growth plate. *Journal of Bone and Mineral Research*, 17: 284-292.

Ortega N, Behonick D, Stickens D, Werb Z (2003) How proteases regulate bone morphogenesis. *Tissue Remodeling*, 995: 109-116.

Otto F, Thornell AP, Crompton T, Denzel A, Gilmour KC, Rosewell IR (1997) Cbfa1, a candidate gene for cleidocranial dysplasia syndrome, is essential for osteoblast differentiation and bone development. *Cell*, 89: 765-771.

Parsons MJ, Pollard SM, Saude L, Feldman B, Coutinho P, Hirst EMA (2002) Zebrafish mutants identify an essential role for laminins in notochord formation. *Development*, 129: 3137-3146.

Patel JJ, Orriss IR, Key ML, Taylor SEB, Karnik K, Arnett R (2009) Hypothermia stimulates osteoclastogenesis but inhibits osteoblast differentiation and bone formation. *Bone* 2009, 44: S305.

Pedersen ME, Takle H, Ytteborg E, Veiseth E, Enersen G, Færgestad E, Bæverfjord G, Hannesson KO (2010) Matrilin-1 in the vertebral column of Atlantic Salmon (*Salmo Salar* L) with skeletal deformities. *Fish Phys Biochem* PMID 21452016.

Pinto JP, Ohresser MCP, Cancela ML (2001) Cloning of the bone Gla protein gene from the teleost fish *Sparus aurata*. Evidence for overall conservation in gene organization and bone-specific expression from fish to man. *Gene*, 270: 77-91.

Pratap J, Javed A, Languino LR, van Wijnen AJ, Stein JL, Stein GS (2005) The Runx2 osteogenic transcription factor regulates matrix metalloproteinase 9 in bone metastatic cancer cells and controls cell invasion. *Mol Cell Biol*, 25: 8581-8591.

Prescher A (1998) Anatomy and pathology of the aging spine. *European Journal of Radiology* 1998, 27: 181-195.

Preziosi R, Gridelli S, Borghetti P, Diana A, Parmeggiani A, Fioravanti ML (2006) Spinal deformity in a sandtiger shark, *Carcharias taurus Rafinesque*: a clinical-pathological study. *Journal of Fish Diseases*, 29: 49-60.

Rafael MS, Laizé V, Cancela ML (2006) Identification of *Sparus aurata* bone morphogenetic protein 2: molecular cloning, gene expression and *in silico* analysis of protein conserved features in vertebrates. *Bone* 39, 1373-1381.

Rawlinson SCF, Murray DH, Mosley JR, Wright CDP, Bredl JC, Saxon LK (2009) Genetic selection for fast growth generates bone architecture characterised by enhanced periosteal expansion and limited consolidation of the cortices but a diminution in the early responses to mechanical loading. *Bone*, 45: 357-366.

Reiland S (1978) Effect of Decreased Growth-Rate on Frequency and Severity of Osteochondrosis in Pigs. *Acta Radiologica-Diagnosis*, 107-122.

Renn J, Winkler C, Schartl M, Fischer R, Goerlich R (2006) Zebralish and medaka as models for bone research including implications regarding space-related issues. *Protoplasma*, 229: 209-214.

Resnick D (1983) Case Report-240 - Fibrodysplasia Ossificans Progessiva (Fop) - Radiological and Gross Pathological Abnormalities in A Macerated Cadaver. *Skeletal Radiology* 10: 131-136.

Robling AG, Duijvelaar KM, Geevers JV, Ohashi N, Turner CH (2001) Modulation of appositional and longitudinal bone growth in the rat ulna by applied static and dynamic force. *Bone*, 29: 105-113.

Roy PK, Witten PE, Hall BK, Lall SP (2002) Effects of dietary phosphorus on bone growth and mineralisation of vertebrae in haddock (Melanogrammus aeglefinus L.). *Fish Physiology and Biochemistry*, 27: 35-48.

Sandell LJ (1994) *In-Situ* Expression of Collagen and Proteoglycan Genes in Notochord and During Skeletal Development and Growth. *Microscopy Research and Technique*, 28: 470-482.

Shafizadeh E, Paw BH (2004) Zebrafish as a model of human hematologic disorders. *Current Opinion in Hematology*, 11: 255-261.

Silverstone AM, Hammell L (2002) Spinal deformities in farmed Atlantic salmon. *Canadian Veterinary Journal-Revue Veterinaire Canadienne*, 43: 782-784.

Sivan SS, Tsitron E, Wachtel E, Roughley PJ, Sakkee N, Ham F van der, DeGroot J, Roberts S, Maroudas A (2006) Aggrecan turnover in human intervertebral disc as determined by the racemization of aspartic acid. *J Biol Chem* 281:13009–13014.

Smith SM, Whitelock JM, Iozzo RV, Little CB, Melrose J (2009) Topographical variation in the distributions of versican, aggrecan and perlecan in the foetal human spine reflects their diverse functional roles in spinal development. *Histochemistry and Cell Biology*, 132: 491-503.

Spinella-Jaegle S, Rawadi G, Kawai S, Gallea S, Faucheu C, Mollat P, Courtois B, Bergaud B, Ramez V, Blanchet AM, Adelmant G, Baron R, Roman-Roman S (2001) Sonic hedgehog increases the commitment of pluripotent mesenchymal cells into the osteoblastic lineage and abolishes adipocytic differentiation. *Journal of cell science* 114(Pt 11):2085-94.

Sugiura SH, Hardy RW, Roberts RJ (2004) The pathology of phosphorus deficiency in fish – a review, *Journal of fish diseases*, 27:5, 255–265.

Sullivan M, Hammond G, Roberts RJ, Manchester NJ (2007) Spinal deformation in commercially cultured Atlantic salmon, *Salmo salar L.*: a clinical and radiological study. *Journal of Fish Diseases*, 30: 745-752.

Takaishi H, Nemoto O, Shiota M, Kikuchi T, Yamada H, Yamagishi M, Yabe Y (1997) Type-II collagen gene expression is transiently upregulated in experimentally induced degeneration of rabbit intervertebral disc *Jourrial of Orthopaedic Research* 15:4 528–538.

Termine JD, Kleinman HK, Whitson SW, Conn KM, Mcgarvey ML, Martin GR (1981) Osteonectin, A Bone-Specific Protein Linking Mineral to Collagen. *Cell*, 26: 99-105.

Timpl R (1996) Macromolecular organization of basement membranes. *Current Opinion in Cell Biology*, 8: 618-624.

Tsang KY, Chan D, Cheslett D, Chan WCW, So CL, Melhado IG (2007) Surviving endoplasmic reticulum stress is coupled to altered chondrocyte differentiation and function. PLOS Biology, 5: 568-585.

Urban JPG, Mcmullin JF (1985) Swelling Pressure of the Intervertebral Disk - Influence of Proteoglycan and Collagen Contents. *Biorheology* 22:145-157.

Urban JPG, Roberts S (2003) Degeneration of the intervertebral disc. Arthritis Research & Therapy 5:120-130

Urban JPG, Smith S, Fairbank JCT (2004) Nutrition of the intervertebral disc. *Spine* 29:2700-2709.

Vagsholm I, Djupvik HO (1998) Risk factors for spinal deformities in Atlantic salmon, *Salmo salar L. Journal of Fish Diseases*, 21: 47-53.

Vandenavyle MJ, Garvick SJ, Blazer VS, Hamilton SJ, Brumbaugh WG (1989) Skeletal Deformities in Smallmouth Bass, *Micropterus-Dolomieui*, from Southern Appalachian Reservoirs. *Archives of Environmental Contamination and Toxicology*, 18: 688-696.

Wagner WM, Hartley MP, Duncan NM, Barrows MG (2005) Spinal spondylosis and acute intervertebral disc prolapse in a European brown bear (*Ursus arctos arctos*). *Journal of the South African Veterinary Association*, 76: 120-122.

Walmsley R (1953) The development and growth of the intervertebral disc. *Edinburgh Medical.Journal* 60, 341-364.

Wargelius A, Fjelldal PG, Hansen T (2005) Heat shock during early somitogenesis induces caudal vertebral column defects in Atlantic salmon (*Salmo salar*). *Dev Genes Evol*, 215: 350-357.

Wargelius A, Fjelldal PG, Nordgarden U, Hansen T (2009) Continuous light affects mineralization and delays osteoid incorporation in vertebral bone of Atlantic salmon (*Salmo salar L.*). *Journal of Experimental Biology*, 212: 656-661.

Wiktorjedrzejczak W, Bartocci A, Ferrante AW, Ahmedansari A, Sell KW, Pollard JW (1990) Total Absence of Colony-Stimulating Factor 1 in the Macrophage-Deficient Osteopetrotic (Op Op) Mouse. Proceedings of the National Academy of Sciences of the United States of America, 87: 4828-4832.

Wise SB, Stock DW (2006) Conservation and divergence of Bmp2a, Bmp2b, and Bmp4 expression patterns within and between dentitions of teleost fishes. *Evolution & Development*, 8: 511-523.

Witten PE, Gil-Martens L, Hall BK, Huysseune A, Obach A (2005) Compressed vertebrae in Atlantic salmon *Salmo salar*: evidence for metaplastic chondrogenesis as a skeletogenic response late in ontogeny. *Diseases of Aquatic Organisms*, 64: 237-246.

Witten PE, Gil-Martens L, Huysseune A, Takle H, Hjelde K (2009) Towards a classification and an understanding of developmental relationships of vertebral body malformations in Atlantic salmon (*Salmo salar L.*). *Aquaculture*, 295: 6-14.

Witten PE, Hall BK (2002) Differentiation and growth of kype skeletal tissues in anadromous male Atlantic Salmon (*Salmo salar*). *International Journal of Developmental Biology*, 46: 719-730.

Witten PE, Huysseune A (2009) A comparative view on mechanisms and functions of skeletal remodelling in teleost fish, with special emphasis on osteoclasts and their function. *Biological Reviews*, 84: 315-346.

Witten PE, Obach A, Huysseune A, Baeverfjord G (2006) Vertebrae fusion in Atlantic salmon (*Salmo salar*): Development, aggravation and pathways of containment. *Aquaculture*, 258: 164-172.

Wu LNY, Genge BR, Wuthier RE (1991) Association Between Proteoglycans and Matrix Vesicles in the Extracellular-Matrix of Growth Plate Cartilage. *Journal of Biological.*

Yasui N, Sato M, Ochi T, Kimura T, Kawahata H, Kitamura Y (1997) Three modes of ossification during distraction osteogenesis in the rat. *Journal of Bone and Joint Surgery-British Volume*, 79B: 824-830.

Yasuma T, Arai K, Yamauchi Y (1993) The Histology of Lumbar Intervertebral Disc Herniation - the Significance of Small Blood-Vessels in the Extruded Tissue. *Spine* 18:1761-1765.

Yoshida H, Hayashi SI, Kunisada T, Ogawa M, Nishikawa S, Okamura H (1990) The Murine Mutation Osteopetrosis Is in the Coding Region of the Macrophage Colony Stimulating Factor Gene. Nature, 345: 442-444.

Ytteborg E, Baeverfjord G, Hjelde K, Torgersen J, Takle H (2010b) Molecular pathology of vertebral deformities in hyperthermic Atlantic salmon (*Salmo salar*). *BMC Physiology*, 10:12.

Ytteborg E, Torgersen J, Baeverfjord G, Takle H (2010c) Morphological and molecular characterization of developing vertebral fusions using a teleost model. *BMC Physiology*, 10:13.

Ytteborg E, Torgersen J, Pedersen M, Baeverfjord G, Hannesson K, Takle H (2010d) Remodeling of the notochord during development of vertebral fusions in Atlantic salmon (*Salmo salar*). *Cell and tissue research.*

Ytteborg E, Vegusdal A, Witten PE, Berge G, Takle H, Østbye TK, Ruyter B (2010a) Atlantic salmon precursor cells from white muscle differentiate into osteoblasts in vitro: Polyunsaturated fatty acids and temperature influence gene expression and PGE2 production. *Biochemica et Biophysica Acta Molecular Cell Research*, 1801; 127-137.

Part 5

Ecological Impacts of Fish Farming

14

Ecological Features of Large Neotropical Reservoirs and Its Relation to Health of Cage Reared Fish

Edmir Daniel Carvalho, Reinaldo José da Silva, Igor Paiva Ramos, Jaciara Vanessa Krüger Paes, Augusto Seawright Zanatta, Heleno Brandão, Érica de Oliveira Penha Zica, André Batista Nobile, Aline Angelina Acosta and Gianmarco Silva David

São Paulo State University (UNESP)
Brazil

1. Introduction

Environmental Brazilian legislation allow the implementation of cage culture fish farm enterprises in large public reservoirs (Ayroza et. al., 2006), aiming a qualitative and quantitative increase of inland aquaculture, regarding concerns with environmental, economic and social sustainability (Costa-Pierce, 2002; Valente, 2000). Fish farming is an important activity for animal protein production, and if well planned, is benefic to the economic development of the country. However, it is necessary continuous assistance of appropriate expertise and scientific support in order to organize and improve fishery and aquaculture (Agostinho et al., 2007). Due to low productivity of inland native fish stocks, Brazilian government aquaculture programs initially focused on change artisanal fisher to fish farmers, assuming that this could improve their economic situation. This is a mistaken philosophy, since the extrativist way of life contradicts the planned life of a modern aquaculturist. This is especially applicable for Brazilian Southeastern region, where this activity is basically maintained by intensive culture of tilapia (Furlaneto et al., 2006) by capitalized stakeholders. Food and Agriculture Organization of the United Nations [FAO] (2010) and Rojas & Wadsworth (2007) highlight that aquaculture annual growth rate surpasses other zootechnical activities. Currently, annual fish yield in Brazil, by means of aquaculture and extractive fishing, has accomplished a 1,240,000 metric tons/year baseline, of which approximately 10% are related to tilapias (*Ministério da Pesca* [MPA], 2010), predominantly *Oreochromis niloticus* and derived hibrids Lovshin, 1982). Within this context, Brazil has been showing vigorous growth (over 25% per year) in this decade, however this growth is still modest considering the prodigious potential of water resources and suitable weather that Brazil offers (Godinho, 2007; MPA, 2010). It is evident that Brazilian aquaculture production is behind its potential comparing to Chinese fish yield, that produces approximately 47.5 millions of metric tons/year (FAO, 2010). Fitzsimmons (2006) highlights Brazil as a prominent country that could compete with China as biggest fish producer in the world. The Paraná River is the second largest catchment in South America, with 3,780 km of extension, and is the main River of La Plata

River basin originating at the confluence of Paranaíba and Grande Rivers and has a watershed area of 2,800,000 km^2, which consists chiefly of sedimentary and volcanic rocks. The Paraná River stretches are divided into an upper course, from its source to Itaipu reservoir; a middle course along Paraguay-Argentina border; and a lower course from Paraguay River confluence to La Plata River estuary. The Upper Paraná River basin, with an extension of 809 km and area 820,000 km^2, has about 250 km without impoundments, resulting in a deeply altered hydrological and limnological regime (Stevaux, 1994). Currently, Brazilian inland net cage aquaculture is integrated to these large reservoirs. Within this context, in the last five decades it is noticed that Brazilian large Rivers have been impounded to build dams and power-plants, aiming hidroelectricity as a priority (Tundisi, 1993; Zocchi, 2002), to meet the increasing demand for energy in the country. This way of producing hidroelectricity energy represents 14.8% of all Brazilian energy matrix (Ministério das Minas e Energia [MME], 2006), with São Paulo state responsible for over 22% of this type of energy. These impoundments were built as a cascade system in large rivers (Grande, Tietê, Paranapanema and Paraná Rivers) (Agência Nacional de Energia Elétrica [ANEEL], 2009; Agostinho et al., 2007). Under an ecological perspective and environmental legislation, a good water quality and aquatic ecosystem integrity are fundamental to allocate the multiple uses of these large reservoirs, especially to effective organization by policy makers for aquaculture and fishing activities. In limnological terms, the determination of trophic state index (TSI) *sensu* Carlson (1977), based upon phosphorus and *a* chlorophyll contents to a specific water body, is a satisfactory and practical tool as environmental indicator, considering different human interventions that induce artificial eutrophication process. Various studies in hydrographic sub-basins of Tietê and Paranapanema Rivers show that this index usually varies. As an example, the index varies between oligotrophic to mesotrophic state for upper and middle Paranapanema stretches (Nogueira et al. 2006), and also between oligotrophic to eutrophic for Tietê River (Barbosa et al. 1999; Moretto et al., 2008; Tundisi & Straškraba, 1999). These variations are mainly due to anthropogenic actions, such as occupation of lands for agriculture, livestock, increasing urbanization due to growth of human population, and emissions of organic wastes (Tundisi, 2005). Brazilian native ichthyofauna of large rivers has been subjected to negative impacts, such as these impoundments (Agostinho et al., 2007), introduction of non-native species (Brandão et al., 2009; Latini & Petrere, 2004; Orsi & Agostinho, 1999; Santos & Formagio, 2000; Souto et al., 2011), environmental contamination, loss of riparian vegetation, sedimentation, and erosion (Agostinho et al., 2007). Currently, a new form of impact in Brazil is the increasing development of fish farming in floating cages (Ramos et al., 2008). In cage systems, the input of organic matter and nutrients is done by artificial feeds, and output is done through the removal of fish produced, similar to what occurs in fish ponds (Beveridge, 2004). However, Beveridge (2004); Munday et al. (1992); Persson (1988); and Pillay (2004) report that in fish cage farming systems up to 30% of feed is lost into the aquatic environment, in the form of unconsumed feed and wastes. These feed losses can cause problems related to eutrophication (Beveridge, 1984) and/or be used as a food resource by local biota (Beveridge, 2004; Håkanson, 2005; Ramos et al., 2008; Vita et al., 2004), resulting in ecological changes around these systems (Beveridge, 2004; Håkanson, 2005; Ramos et al., 2008). Besides these impacts, several authors as Agostinho et al. (2007); Beveridge (1984, 1996 e 2004); Dempster et al. (2002); Håkanson (2005); Karakassis et al.

(2000, 2002 e 2005); Machias et al. (2004, 2005 e 2006); Pitta et al. (2005); Ramos et al. (2008); Yucel-Gier et al. (2007); and Zanatta et al. (2010) discuss the problems of this activity in coastal and inland waters. These authors cite impacts upon water quality and sediment which have implications on the structure of benthic communities, plankton and fish, and furthermore, the inherent scapes of caged fish. Thus, it is evident the necessity of developing new technologies aiming the enhancement of fish yield, associated to decrease of environmental impacts caused by this zootechnical activity. This is a big challenge for Brazilian aquaculture that needs to guarantee its economic and social sustainability with preservation of water resources and multiple uses of public reservoirs.

2. Situation of the aquaculture in cage farms in Brazilian Southeastern reservoirs: An overview

The effects of cage aquaculture enterprises in Brazilian inland waters upon the biota and water quality have not been satisfactory elucidated yet, thus these effects still require studies aiming a full comprehension to better ordinate these activities. However, despite divergent opinions of some sectors of Brazilian society, the Brazilian government has been sponsoring studies focusing on taking advantage of the potential for aquaculture of large public reservoirs (Ayroza et al., 2006). Thus, Brazilian government has been selecting, delimitating, and controlling areas to install Aquaculture Parks in different reservoirs, based upon premises that promote sustainable development, enhancement of Brazilian fish yield, social inclusion, and food safety. Effective success of production in aquaculture founds in limnological studies and environmental carrying capacity, regarding concerns with water quality and maintenance of water resources (Costa-Pierce, 2002; Tundisi, 2005), as well as its interference upon aquatic biota, through escapes of non-native species, and pathogen dispersion (Agostinho et al., 2007; Orsi & Agostinho, 1999). According to Dillon & Rigler (1975) and Beveridge (2004), the modeling of environmental carrying capacity is done based upon two fundamental equations, which define phosphorus emission to aquatic ecosystem without inducing eutrophication, and how much P is emitted per product unit. The lack of suitable information to define application values of this modeling has been repeatedly emphasized as the highest difficulty to appraise environmental impacts caused by aquaculture (Pillay, 2004). In Brazil, cage fish farming systems has gained impulse in the mid-90's, especially in Brazilian Southeastern (Medeiros, 2002; Ono, 1998; Rojas & Wadsworth, 2007). Nowadays, this activity is in accelerated expansion in Brazilian reservoirs, and at least 40 freshwater fish species are used in Brazil, considering all kinds of pisciculture (Godinho, 2007). However, despite this rich fish diversity, the most used species for inland aquaculture in Brazil is the non-native Nile tilapia (*O. niloticus*) due to its favorable zootechnical features (Castagnolli et al., 2000; David et al, 2006; Rojas & Wadsworth, 2007). The Middle Paranapanema River area has about 800 net-cages designated to Tilapia culture along its reservoirs and ponds. It is estimated in 200 kg of fish/m^3/cycle in small net-cages (up to 6m^3); and 100 kg of fish/m^3/cycle in net-cages of great capacity (over 10m^3) (Furlaneto et al., 2006). In this sense, three issues are eminent within this approach: 1) artificial eutrophication, which is interconnected with environmental carrying capacity; 2) dispersion of parasites and pathogens and; 3) escapes, relating to depletion of fish biodiversity *latu sensu* by ecological processes of competition and predation. Below, it is followed the results of our main researches aiming to identify and quantify the ecological interferences of tilapia farming.

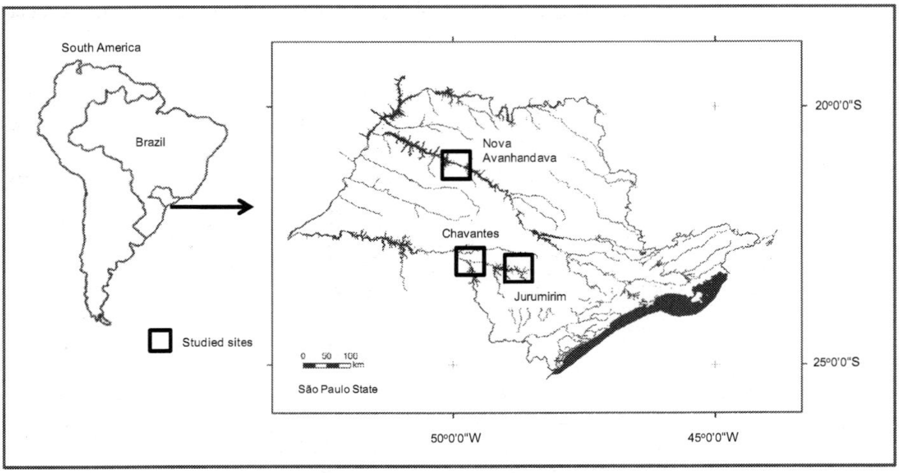

Fig. 1. Reservoirs studied in the Upper Paraná River Basin, São Paulo State, Brazil.

2.1 Limnological conditions and trophic state index

A comparative approach was used to evaluate the impact of three cage farms upon limnological conditions of the stretches were the cages are installed. These farms are located in three different reservoirs in the upper Paraná River Basin, São Paulo state, Brazil (fig. 1). The farms are small and medium-sized enterprises (ranging from 30 to 200 cages with 6 m^3 each), mainly for the culture of Nile tilapia (*O. niloticus*) using an intensive model, with high densities and fed with pelleted compound feed. An example of such farms is shown in fig. 2. Monthly surveys were performed in the farms in different years from 2003 to 2009, sampling limnological data and water samples inside the Farm Sites (referred as FS) and in similar Control areas sited upstream (CT). Differences among farm sites and control areas were regarded as effect of the nutrient loads from the farms. Limnological sampling included Secchi depth (Z_{DS}), water temperature, pH, dissolved oxygen (DO), electric conductivity (K_{25}) measured *in situ* using a water quality multiprobe Horiba model U-22; water samples were collected for nutrients (total Nitrogen - TN and total Phosphorus - TP following Valderrama, 1981; Strickland & Parsons, 1968), total suspended solids (Teixeira et al., 1978) and chlorophyll *a* (CHL, Golterman et al., 1978) analyses. TSI was determined based on CHL and TP by formulae 1 and 2, combined using formulae 3 (Carlson, 1996, as cited in Lin, 2001). The first reservoir studied by our research group was Nova Avanhandava, which is the fifth of a cascade of reservoirs on the Middle Tietê River, sampled from 2003 to 2004 (Paes, 2006). It is a run-of-river plant, that has been operated since 1982, located at 358 m above sea level, with surface area of 210 km^2, total water volume of 2,720 x 10^6 m^3, mean discharge rate of 688 m^3/s, maximum depth 30 m, level oscillation of less than 1 m throughout the year, and water residence time of 46 days (Torloni et al., 1993; Rodgher et al., 2002). The second reservoir was Jurumirim, where the surveys were done from 2004 to 2005; it is a storage plant at Upper Paranapanema River, operated since 1962, located at 568 m above sea level, total area 484 km^2, total volume of 7,900 x 10^6 m^3 mean discharge rate of 210

m^3 s^{-1}, mean depth 12.9 m, residence time 334 days (Henry et al., 2006) and significant level oscillation of 2.2 m along the year.

$$TSI\ (CHL) = 9.81*\ln(CHL)+30.60 \quad (1)$$

$$TSI\ (TP) = 14.42*\ln(TP)+4.15 \quad (2)$$

$$TSI = (TSI\ (CHL)+ TSI\ (TP))/2 \quad (3)$$

Fig. 2. Cage farm at Chavantes reservoir, Middle Paranapanema River Basin, São Paulo State, Brazil.

Lastly, it was studied Chavantes reservoir, that is also a storage plant in Paranapanema River, operating since 1970 in its middle stretch at 474 m above sea level. This reservoir area is 400 km^2, total volume of 8,800 x 10^6 m^3, maximum depth 80 m, residence time 281 days, mean outflow 322 m^3 s^{-1}, and seasonal water level oscillation is more than 3 m (Nogueira et al., 2006). The three reservoirs are in a region with annual precipitation above 1,500 mm, and with a rainy season from September to February and dry season from March to August. Limnological characterization of the three sites studied are shown in Table 1.

Limnological features of Tietê River basin are quite different from Paranapanema River basin (table 1). Ph values are higher in Tietê, reaching 9.6, while Paranapanema values were close to neutrality. Conductivity was three fold higher in Tietê, and CHL were almost ten fold higher. TP values were slightly higher at Tietê, and TN was two times higher in Nova Avanhandava than in Jurumirim, and five fold higher than Chavantes. OD mean values

Reservoirs	Nova Avanhandava[1]		Jurumirim[2]		Chavantes[3]	
Parameters/Sites	FS	CT	FS	CT	FS	CT
Temperature (°C)	20.3-31.3 (26.3±3.3)	20.0-31.2 (26.2±3.5)	18.9-27.4 (23.6±3.0)	18.4-28.2 (23.2±3.0)	19.3-27.4 (23.4±2.5)	19.6-28.7 (23.8±2.5)
pH	5.8-8.9 (8.0±0.7)	7.2-9.6 (8.2±0.8)	6.0-8.3 (6.8±0.6)	6.0-8.2 (6.7±0.8)	6.0-7.6 (7.0±0.4)	6.6-7.8 (7.1±0.3)
Dissolved Oxygen (mg/L)	6.4-9.9 (8.5±1.1)	7.0-10.0 (8.7±1.0))	6.5-8.9 (8.0±0.8)	6.8-9.0 (8.0±0.7)	6.5-11.4 (8.1±1.3)	2.2-11.5 (8.2±1.7)
Conductivity (µS/cm)	143.7-208.0 (184.8±18.5)	148.0-207.0 (183.9±19.0)	50.0-83.0 (59.3±9.4)	50.0-86.0 (59.8±9.1)	36.8-60.0 (40.5±6.0)	36.0-60.0 (40.3±6.2)
Secchi depth (m)	0.7-2.7 (1.4±0.6)	0.8-2.5 (1.3±0.5)	0.6-1.5 (0.9±0.2)	0.4-1.5 (1.0±0.3)	1.5-5.0 (3.1±0.7)	1.5-5.0 (3.1±0.8)
Chlorophyll a (µg/L)	2.2-30.5 (17.7±9.3)	5.1-30.1 (18.0±8.7)	0.5-8.0 (3.3±2.8)	0.9-3.8 (2.4±0.8)	0.9-2.3 (1.5±0.5)	0.7-5.4 (1.9±1.3)
Total dissolved solids (mg/L)	0.0-11.1 (3.8±2.8)	0.5-4.0 (2.7±1.1)	0.1-5.0 (2.0±1.5)	0.1-4.4 (1.3±1.4)	-	-
Total P (µg/L)	7.7-23.7 (14.1±3,7)	5.9-21.3 (12.7±4.3)	3.4-9.8 (7.0±2.0)	4.9-15.0 (8.2±3.6)	4.9-23.5 (11.1±5.2)	3.1-23.1 (9.7±4.1)
Total N (µg/L)	763.7-1950.2 (1282.6± 283.9)	871.0-1973.4 (1213.2± 342.5)	600.0-825.0 (669.4±72.0)	520.0-780.0 (621.3±83.5)	108.2-404.6 (244.9±74.0)	120.0-376.9 (236.0±78.7)

Table 1. Limnological characterization of surface waters of the three reservoirs studied in the Paranapanema and Tietê Rivers, Upper Paraná Watershed. Modified from [1]Paes (2006), [2]Zanatta (2007) and [3]David et al. (2011).

were above 6 mg/L in all sites, considered adequate for tilapia culture, and average Z_{DS} reached 3.1 m at Chavantes, while in the other sites were about 1 m. A concerning issue about the improvement of cage aquaculture in Upper and Middle Paranapanema reservoirs is that winter temperatures were far below 23°C (fig. 3) and in a less degree for Nova Avanhandava, assumed as a lower limit for efficient tilapia grow-out commercial systems (Shelton & Popma, 2006; Suresh, 2002). This fact can hinder aquaculture sustainability and profitability in this watershed, and suitable native species would be desirable. Paes (2006) studied a medium-sized farm, with 80 fish cages of six m³ volume, in Nova Avanhandava reservoir and concluded that the limnological variables measured showed no statistically significant differences between FS and CT (Table 1). Some of the variables were considered high, especially electrical conductivity and total nitrogen, indicating risks of water quality deterioration, a typical feature of Tietê River Watershed (Barbosa et al., 1999). The Trophic state index (TSI) was mesotrophic for both areas (fig. 4A). These conditions can be seriously aggravated by nutrient loads from cage farming activities, which can induce autopollution and loss of water quality for aquaculture purposes. Alves & Baccarin (2006) found similar values for these limnological variables, in a cage farm that uses more than 2 tons of feed per day to produce 1,500 to 1,800 kg of fish per cage per cycle. Water quality depletion is object of concerns in this basin because of widespread blooms of cyanobacteria (Fracácio et al., 2002, Tundisi, 2005). In Jurumirim reservoir, Zanatta (2007) studied a small farm with only 30 cages of six m³ volume, and no significant differences between FS and CT was found for all limnological parameters measured. The very limited scale of farming operations probably prevented impacts on water quality in this large reservoir. Thus, only natural

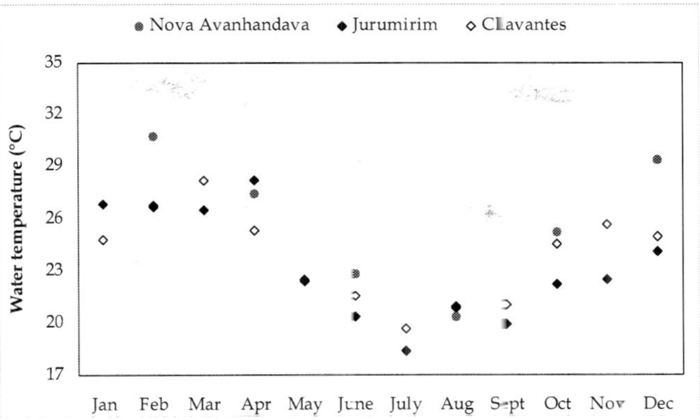

Fig. 3. Seasonal variation of water surface temperature in the studied sites.

and seasonal differences were reported. Jurumirim reservoir was oligotrophic in both areas throughout the year, except for FS (mesotrophic in August), (fig. 4B).

David et al. (2011) using data from a cage farm in Chavantes reservoir with 200 cages, found significant seasonal variation for all limnological parameters measured, but no significant differences among areas at monthly scale. In this same study, comparison of mean values (pooled data per year) of phosphorus in the water between FS and CT indicates that fish farming may be related to hipernutrification of FS in the euphotic layer ($\Delta TP=1.65mg/m^3$, $p<0.05$), although no significant increase in chlorophyll *a* was detected. However, in the euphotic layer other limnological variables measured showed no significant differences between areas, even though depletion of dissolved oxygen in the bottom layer in FS was verified. Most of the year, oligotrophic conditions were also found in Chavantes, but switched to mesotrophic in January for CT, and in December/January for FS (fig. 4C). Simulations on the carrying capacity, due to availability of more detailed data on hydrodynamics and field farming practices in Chavantes reservoir, was also performed by 18 David et al. (2011). Data used for carrying capacity modelling were: Feed Conversion Rate (FCR): 1.5; feed total phosphorus content: 1.5%; whole fish phosphorus content: 0.9% (Dantas & Athayde, 2007), resulting on of 13.5 kg P/ton of fish produced. It is worth emphasizing that phosphorus loads can be greatly influenced by FCR (fig. 5), which itself is related to the welfare of fish and their physiological condition. In addition to unprofitability, the cultivation of tilapia at temperatures below the optimal nutrient emissions increases, which brings higher risks of eutrophication. Hydrological data of FS used in modelling were: original total phosphorus water content of 10.85 mg/m^3, maximum allowable TP water content of 30.0 mg/m^3, and sedimentation rate of 0.083 (Larsen & Mercier, 1976). Retention time used was 1 day for precautionary purposes, according to estimates of Persson et al. (1994) and direct ADCP measurements was less than 1 day, with measured water current velocity from 0.1 to 0.3 m/s.

Maximum allowable fish production in the area was estimated to be 1666 tons/year, while the theoretical fish production that would result on the observed increase of TP was 144 tons/year while total tilapia production of approximately 55 tons/year. These results

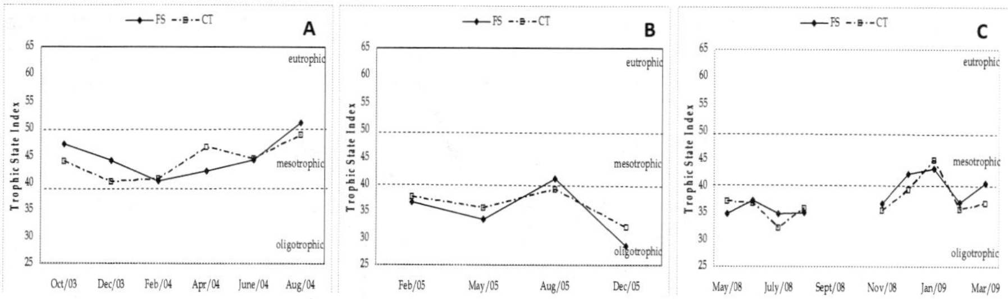

Fig. 4. Seasonal variation of Trophic State Index (TSI) for FS and CT for Nova Avanhandava (A), Jurumirim (B) and Chavantes (C) reservoirs.

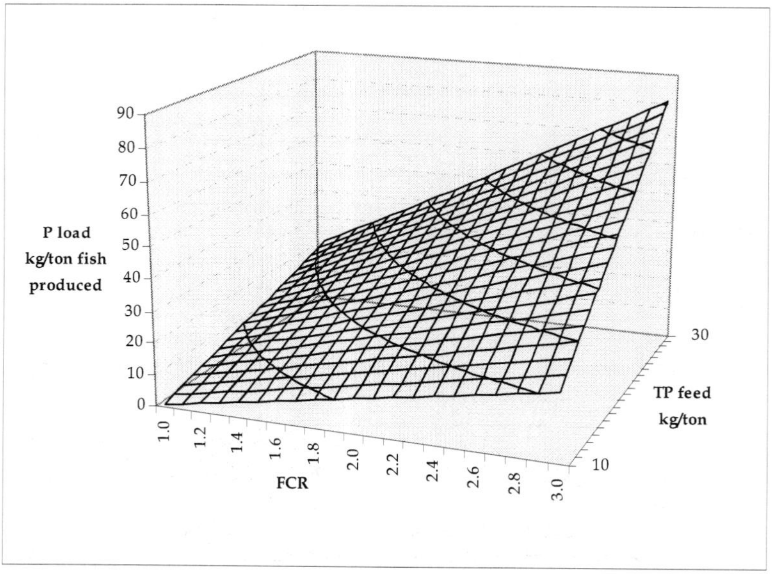

Fig. 5. Effect of FCR and feed composition on Phosphorus loads by fish cage culture

indicate that fish production at FS is compatible with the local carrying capacity for the assimilation and recycling of nutrients derived from aquaculture.

Regarding water quality, conditions may be considered adequate, with no signals of surpassing eutrophication thresholds. Dillon & Rigler mass balance model overestimated the fish production needed to cause the observed FS increase, indicating that specific models are needed for the management of aquaculture in the ecosystems studied here.

2.2 Main fish parasites in cage farms

The interest in fish parasites has increased in recent decades due to economic implications, particularly for intensive fish farms. The high rates of parasitic infestations or infections can cause considerable mortality in several species of farmed fish and its treatment in some cases is very difficult, while in others, there are no currently effective treatments. The economic losses caused by parasites can often be verified in an indirect way, either by reducing the rates of assimilation and growth of infected animals, or by decreasing the value of marketable final product (Eiras, 1993). Many organisms have been associated with fish diseases. Although parasites can occur in natural environments, these organisms become more abundant in conditions of intensive cultivation and depending on the conditions of culture, they can have a deleterious effect to fishes. In eutrophic environments this situation could worsen due to the presence of many intermediate host species, favouring the life cycle of many parasites (Pillay, 2004). Our studies performed with *O. niloticus* in cage farms have shown that the main parasites affecting fish health in intensive cage systems are monogenoids, *Trichodina* spp., *Ichthyophthirius multifiliis* and *Henneguya* sp. Monogeneans (fig. 6A) are ectoparasites responsible for the most important parasitic disease of fish farming in Brazil (Martins, 1998). They are characterized by the presence of a fixation structure generally located in the back of the body, called haptor, which contain hooks, bars and anchors, in different numbers and sizes according to the species and its function is the fixation of parasite in the hosts (Gerasev, 1990). The adult parasites are elongated, oval or round and measure one to 75 millimeters. The most important monogeneans in fish farms belong to the families Gyrodactylidae and Dactylogyridae. Gyrodactilids are viviparous species and they are mostly parasites of the body surface of fish. Dactylogirids are oviparous and are found in the gills, but can also become placed in the nasal cavities and, more rarely, other parts of the body (Kubtiza & Kubtiza, 1999). The studies performed with *O. niloticus* from fish farms from Chavantes reservoir showed that the dactylogyrids *Cichlidogyrus halli* (fig. 6B-C) and *Scutogyrus longicornis* were the most abundant monogeneans found in the gills of this fish species. Dactylogyrid monogeneans are hermaphrodites and have direct life cycle, which facilitates the parasitic re-infestation. Generally in monogeneans the eggs (dactylogyrids) or larvae (gyrodactylids) leave the uterus by the genital pore, attach to the host and develops in the same host (Cheng, 1986). According to Eiras (1993), pathogenesis caused by monogeneans varies with the species and the fixation site. Monogeneans which parasitize the gills frequently cause cell hyperplasia and mucus hypersecretion. The lesions are much more serious as the parasites are most abundant, which can reach high densities. When attached to the tegument, necrosis of cells, destruction of scales and mucus hypersecretion can be observed. In cases of high intensity of infection, monogeneans can cause mortalities especially in small fish (Noga, 1996), which has been recorded for numerous fish species (Cone et al., 1983; Ergens, 1983; Lester & Adams, 1974; Mackenzie, 1970). Considering the seriousness of this disease and also the difficulty of eradicating this disease in fish after installed, it is suggested that all new fish bought undergo to quarantine and prophylactic baths using commercial formalin diluted 1:4,000 for an hour or sodium chloride in 1 to 3% from 30 minutes to 3 hours (Pavanelli et al., 1999). Another important group of organisms that can affect fish health, especially those in breeding system, includes the protozoans of the phylum Ciliophora, highlighting, *Trichodina* spp. (fig. 6D) and *I. multifiliis* (fig. 6E). Apparently, these ciliates live as ectocomensal in the tegument and gills of the fish without causing major damage, except in cases of heavy infestations, which is particularly evident in species that multiply rapidly by successive binary divisions, especially in environments with excess of organic material and low amounts of dissolved oxygen in water (Eiras, 1993). The rearing of *O. niloticus* in cage farms in Chavantes reservoir have high infection rates by *Trichodina* spp. (fig. 6D), in some cases and periods, the prevalence is 100%. Trichodinids are ciliated protozoa commonly found in both freshwater and saltwater, and show no host specificity, which favors their widespread distribution. Their morphology is characteristic, with a circular shape and the presence of an adhesive disc with a series of denticles that help fix the parasites in the host. They are usually considered ectoparasites of skin and gills of the host and can rapidly proliferate in the presence of decaying material (Heckmann, 1996). The life cycle of *Trichodina* spp. occurs by binary fission, in which the parasite divides and fixes in the host's skin (Cheng, 1986). Its pathogenesis is related to the rotatory movements of these ciliates on the gills and tegument of the host, leading to an abrasive action of the skeletal structures and denticles present in the adhesive disk, which damage the epithelial cells. Trichodinidiasis signs include loss of appetite, lethargy, excessive mucus production in the gill epithelium and skin, erythema, and sometimes bleeding skin (Heckmann, 1996). This is more evident in cases of intense parasitism, which is observed when environmental conditions favour the reproduction of the parasite and weakens the host, which happens when there is a decline in water quality. These parasites are easily transmitted through infected fish, water, plants or utensils used on fish farms. The treatment can be accomplished by bath with malachite green 2-3 g/10 m^3, when the fish are not used for consumption, or a bath for two hours in commercial formalin 1:4.000-6.000 (Pavanelli et al., 1999). *Ichthyophthirius multifiliis* (fig. 6E) however, has not been frequently found in *O. niloticus* from Chavantes reservoir. It causes a disease commonly known as freshwater white spot disease. This protozoa is a ciliate parasite, ovoid, measuring 100-1000 µm, the cytoplasm is granular and contains numerous vacuoles and contractile structures. These protozoans are characterized by the presence of a horseshoe-shaped macronucleus and a micronucleus barely visible (Cheng, 1986). Although often cited as ectoparasites, it is located under the epidermis, presenting the appearance of small white spots on skin and gills of fish (Eiras, 1993). The life cycle of *I. multifiliis* is completed in 3-4 days at 25.5 °C, but can also occur in up to 5 weeks at 18 °C and in lower temperatures the parasite will remain dormant. The adult parasite called trophont reach the maturity, leaves the host as tomont forms and fixes in the substrate of fish cage forming a cyst. Inside it, multiple cell divisions form many tomites which are the infective forms to new fishes when the cyst breaks (Ewing & Kocani, 1987). Probably this protozoan is responsible for major economic losses in fish farms in the world. Young fish are usually more susceptible, and high infestations are usually associated with sudden drops in temperature in the water fish farm (Eiras, 1993). In Brazil, *I. multifiliis* was already reported infecting *Colossoma macropomum; Prochilodus cearensis, Cichla ocellaris, Piaractus mesopotamicus; Leporinus macrocephalus;* tambacu hybrid (*P. mesopotamicus* x *C. macropomum*); *Tilapia rendalli; O.niloticus* and *Cyprinus carpio* (Békési, 2002; Tavares-Dias et al., 2001).

The most appropriate way to avoid freshwater white spot disease is maintaining good water quality, adequate food, and avoiding stress, caused mainly by changes in temperature. In this case, fish are more sensitive to attack by parasites. It should be considered that freshwater white spot disease is difficult to treat, especially in large cages. Treatment should be done in special cages for therapeutic baths. One product that can be used is sodium chloride 0.3%, where the fish stay sunk for about 24 hours. More concentrated doses, 5% for example, may be used in severe cases, with the fish remaining in the solution for 30 minutes (Pavanelli et al., 1999). Myxozoa includes parasites frequently found in marine fish and freshwater. Currently the phylum Myxozoa includes the classes Myxosporea and Malacosporea, and almost all fish parasites belong to Myxosporea (Eiras et al., 2006). The life

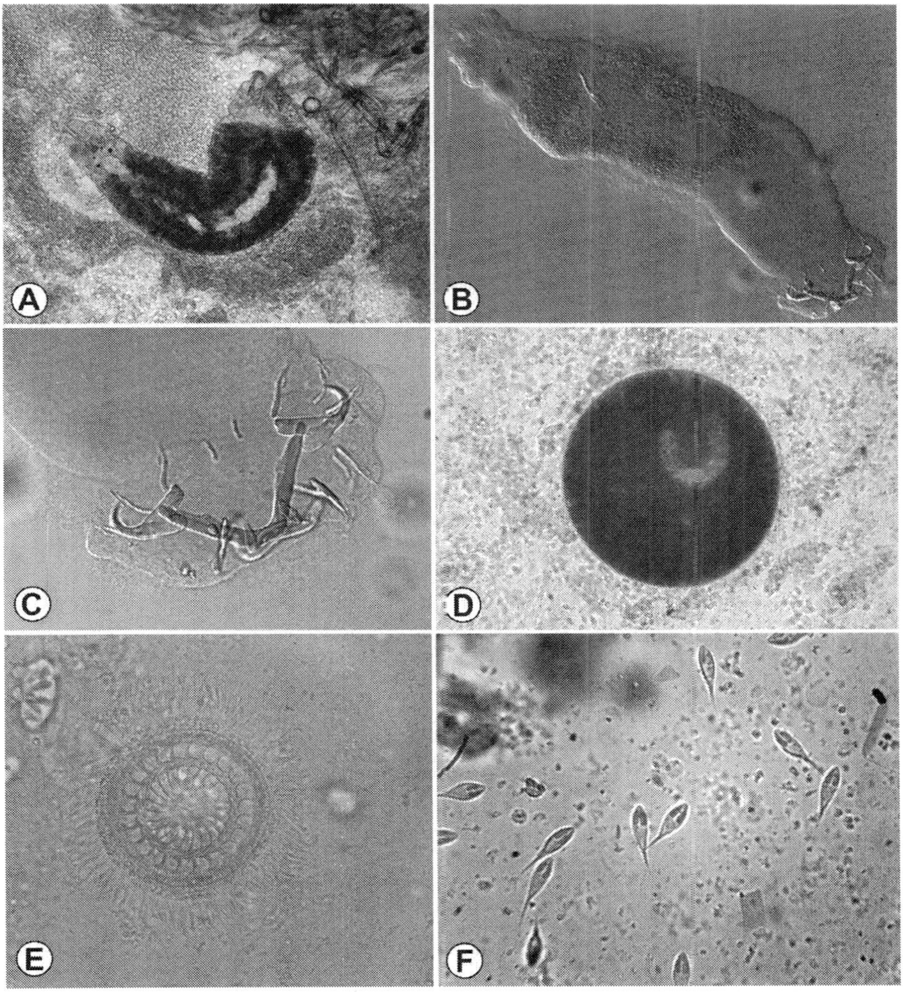

Fig. 6. A) Monogenoid in the gill of *Oreochromis niloticus* (10x); B) *Cichlidogyrus halli* specimen (10x); C) detail of the haptor and sclerotized structures of *Cichlidogyrus halli* (40x); D) *Ichthyophthirius multifiliis* specimen (40x); E) *Trichodina* sp. specimen (40x); F) *Henneguya* sp. spores in fish tegument (100x).

cycle of mixosporeans occurs in two hosts: a vertebrate (fish) and invertebrates (oligochaete). The spore is released when infected fish dies and decomposes, or through contaminated faeces of a predator fish that have eaten infected fish. The oligochaetes are infected when they ingest the sporoplasms that fix to the intestinal wall using polar filaments; these proliferate and remain incubated for 3 months. After this period the Actinospore are released to infect the fish in the water column and deposit the sporoplasm within the epidermis which will migrate to the organs and gills (Stevens et al., 2001). The main genera of Myxozoa infecting fishes are *Myxobolus* and *Henneguya* (fig. 6F). In Brazil there are only reports on the occurrence of *Henneguya* sp. infecting *O. niloticus* (Ranzani-Paiva et al., 2005). Histopathological analysis of the gills of fish infected *Henneguya* sp. reveal the presence of severe bleeding and inflammatory foci in the gill epithelium, where the cysts are located. Lesions such as compression of capillaries causing edema in superficial lamellae more frequently in primary and secondary lamellae occasionally can be observed. In later stages, cysts dilate the respiratory lamellae decreasing respiratory efficiency of infected fish (Martins et al., 1999). The treatment of this disease has been carried out using 10 ml of formalin/m^3, which has been quite effective (Martins et al., 1999). Other studies have been conducted evaluating the oral administration of chemotherapeutics (quinine, salinomycin) that had significant effect on the treatment of infections in the gills of fish mixosporeans (Dohle et al., 2002). However, studies on the treatment of infections *Henneguya* spp. are still scarce. Our studies with *O. niloticus* in Chavantes reservoir have also demonstrated that the occurrence of parasites in cage farms is associated with environmental variables. Therefore, we have observed a higher prevalence, abundance and intensity of infection of parasites in the summer. No difference was noted in the parasitism levels among different fish tanks, considering the groups established by the zootechnical methods in fish farm. Depending on the parasite species, fish size also influences the parasitism. For example, a negative correlation was observed between monogenoids and fish size and a positive correlation between *Trichodina* sp. and fish size. This is important because demonstrates that some parasites are more important in the initial breeding phase while others are problems in the final stages. In conclusion, we can note that despite the wide distribution and incidence of parasites in fish rearing, there are few studies on the prevalence, pathogenesis, and potential biological cycle of transmission of these parasites in Brazilian cage fish farms. It is important to consider that in cage farming there must be a balance between the health of the host, the proliferation of pathogens and conditions of the aquatic environment. Thus, poor water quality, reduction of dissolved oxygen, changes in temperature, high fish density, inadequate management or unbalanced nutrition are factors able to induce stress to the animals, predisposing them to various types of infections, including parasites. Water offers an extremely favorable environment for the proliferation of these agents and the parasites are responsible for major losses in fish farms worldwide, with more relevance in the neotropics, due to the climatic characteristics of these regions (Martins, 1998; Thatcher, 1994).

2.3 Impacts on the resident ichthyofauna and fish feeding ecology

Brazilian freshwater fish fauna has been managing to adjust to continuous environmental impacts, such as damming and deforestation, due to its great diversity of species, with different tactics of life cycle (Agostinho et al., 2007). Currently, a new expanding form of impact is fish farming in floating cages. However, knowledge of the impacts from this activity on the ichthyofauna is still precarious. This activity provides food resources and shelter to resident fish fauna, attracting a large number of fish (Brandão, 2010; Nobile, 2010; Paes, 2006; Zanatta, 2007, 2011), which is also observed for marine fish farms (Boyra et al., 2004; Machias et al., 2004, 2005, 2006). This attraction is due to availability of food resources, such as feed losses, fish scales, and fish faeces. According to Beveridge (2004) and Pillay (2004), these losses can reach 30% of all feed used in aquaculture enterprises. Thus, the contribution of allochthonous energy in aquatic ecosystems can cause changes in the food chain, especially upon plankton community, benthic community, and fish fauna, interfering with the dynamics of the aquatic ecosystem. Our studies in the Parana River basin (Middle Tietê River and Upper/Middle Paranapanema River) reported impacts on the ichthyofauna (Nobile, 2010; Paes, 2006; Zanatta, 2007, 2011) from this activity, such as changes in population structure of fish fauna (Ramos et al., 2008), changes in diet (Brandão, 2010; Ramos et al., 2008, 2009), and changes in bromatological composition of some fish species (Queiroz, 2010). A research made by Paes (2006) in 2003 and 2004 at Nova Avanhandava reservoir, evaluated some impacts of fish farms upon ichthyofauna. While in FS was recorded 18 species, a greater number of fish species (N=20) was recorded in CT. However, a greater number of individuals was captured in FS (n=684) than in CT (n=518). For species, the Shannon-Wiener diversity index H' (Krebs, 1999) showed no significant differences between these two sites ($p > 0.05$). For Simpson dominance 1/D (Krebs, 1999), similar values were recorded for FS (1/D = 5701) and CT (1/D = 5555). Even though this difference was small, the dominant species in the FS was *Metynnis maculatus*, an omnivorous species with tendency to herbivory, while in CT the dominant species was *Plagioscion squamosissimus*, a strict carnivore species. A study conducted by Ramos et al. (2008) aimed to compare the diet of the species *M. maculatus*, *Astyanax altiparanae* and *P. squamosissimus* between FS and CT. The results show that the omnivorous species *M. maculatus* changed its diet in FS. This species used a new alimentary source with high availability and low energy cost, and its diet consists almost exclusively of feed from fish farming (85%). Such a change in diet led to a change in population structure of this species, showing higher values of weight and length in FS ($p < 0.05$). Similar results were found at Jurumirim by Zanatta (2007). A greater number of species (N=24) was captured in CT than in FS (N=21), and a greater number of individuals were captured in CT (n=1,601) than in FS (n=1,470). There was no significant differences for H ' ($p > 0.05$) between the two sites, and Pielou evenness E (Krebs, 1999) values were similar for FS (E = 0.73) and CT (E = 0.76). At Chavantes reservoir, Nobile (2010) showed differences only between the abundance of individuals in both sites, recording the 78% of the capture (n=3096) in FS. Furthermore, it was recorded introduction of the non-native fish *Ictalurus punctatus* by aquaculture activities, whose impacts upon native fish fauna of Brazilian reservoirs are still unknown (Zanatta et al., 2010). There was also the capture of *O. niloticus* juveniles, which we believe that they are recruited from reproductive processes occurring in the reservoir rather than escaped, due to its great abundance, and size smaller than juveniles stocked in the cages (fig. 7). If these recruits are born from adult scaped fish from fish farms, than efficiency of the sexual reversion process in use is not enough to avoid breeding in this aquatic ecosystem, which may lead to disruptions of ecological interactions with native fish assemblages. In Chavantes reservoir, Brandão (2010) and Ramos (2009) observed that the most abundant species showed differences in diet composition between the FS and CT. Remains of feed was major component of the diet of *Pimelodus maculatus* (98%) and *A. altiparanae* (99%) in FS. For CT, the diet of *P. maculatus* was composed of detritus (27%), aquatic insects (22%), vegetables (21%) and molluscs (21%); and *A. altiparanae* by terrestrial insects (61%) and vegetables (30%). Another species analyzed was *Apareiodon affinis*, which diet was composed of detritus (80%) and feed remains (20%) in FS, and only by detritus in CT. Moreover, two carnivorous species were analyzed (*Galeocharax knerii* and *P. squamosissimus*), and there was no differences in their diet between the sites studied. Comparative analysis of weight and length show that omnivorous species grows faster at FS due to use of remains of feed as main food source. In addition,

Fig. 7. *Oreochromis niloticus* juveniles captured around FS at Chavantes reservoir.

Queiroz (2010) observed change in bromatological composition of muscle tissue of *P. maculatus*.

Significant differences for total lipids and gross protein were observed between fish caught around FS and CT ($p < 0.05$). It is evident that this type of fish farm attracts native ichthyofauna in these reservoirs. Attractiveness is directly related to availability of food resources, especially remains of feed. These are similar to the findings for marine aquaculture in coastal ecosystems, which also found the effect of attractiveness of these systems (Boyra et al., 2004; Dempster et al., 2002; Håkanson, 2005). The results presented allow the conclusion that remains of feed are used by omnivorous species with large feeding plasticity, which certainly justifies its dominance in sites used for cage aquaculture. Furthermore, the changes in the diet of these species have direct effects on their population structure, and the effects upon fish fauna still need to be clarified. Håkanson (2005) and Ramos et al. (2008) report that such changes in the diet of abundant species in aquatic ecosystems may affect the food chain in long term, and the effects upon ichthyofauna are still unknown. We conclude that where fish farms in cages are deployed, there is a change upon the population structure of fish communities, due to attractiveness and changes in the diet of omnivorous species in nearby sites, which may interfere in the local ecological dynamics, especially in the food web, altering ecological relationships between the components of local biota.

3. Conclusion

The results showed that cage aquaculture fish production in oligotrophic ecosystems, as Jurumirim and Chavantes reservoirs (Paranapanema River) is compatible with environmental carrying capacity, and wild fishes play a relevant role recycling nutrients derived from aquaculture. The main constraint to cultivation of tilapia in these reservoirs is the low temperatures during the winter months, with recent records of mass mortality of fish due to parasite infestation in this season. These aquaculture enterprises are an important source of dispersion of non-native fish species, as *O. niloticus* and *I. punctatus*. In Tietê river (Nova Avanhandava reservoir), the situation is more complex because of low water quality due to high nutrient loadings of sewage and agriculture runoff, related to the presence of cyanobacteria blooms, causing severe consequences for the environmental and economic sustainability of aquaculture. After some years operating cage farms in middle and lower Tietê River, events of fish mass mortality associated with eutrophication motivated fish farmers to move to sites with better water quality. It is also pertinent to note that fish farming in cages still need the development of scientific and technological knowledge, which besides basic technical support, could also allow the improvement of this activity. The lack of expertise and the inappropriate implementation of best management practices can lead to failure of these enterprises, making them economically unviable in the short and medium term. Measures for the effective planning for farming public waters require further discussion and guidance at governmental levels, in order to reach a truly sustainable aquaculture, in all its socioeconomic and environmental interfaces.

4. Acknowledgments

The authors are grateful to the Conselho Nacional de Desenvolvimento Científico e Tecnológico - CNPq, Financiadora de Estudos e Projetos - FINEP, Fundação de Amparo à Pesquisa do Estado de São Paulo - FAPESP, Fundação para o Desenvolvimento da Unesp - FUNDUNESP for grants and financial support. Authors also thank Botucatu Bioscience Institute (IBB) and Aquaculture Center - São Paulo State University (CAUNESP) for the use of laboratory facilities and logistics. This work complies with the current Brazilian laws.

5. References

Agência Nacional de Energia Elétrica - Aneel. 2009. Banco de Informações de geração. In: *ANEEL, Brasília, DF, Brasil*. Available from (http://www.aneel.gov.br)

Agostinho, A.A.; Gomes, L.C. & Pelicice, F.M. (2007). *Ecologia e Manejo de Recursos Pesqueiros em Reservatórios do Brasil*, EDUEM, ISBN 978-857-6280-95-8, Maringá, Brazil

Alves, R.C.P. & Baccarin, A.E. (2005). Efeito da produção de peixes em tanques-rede sobre a sedimentação de material em suspensão e de nutrientes no córrego da Arribada (UHE Nova Avanhandava, baixo rio Tietê, SP), In: *Ecologia de reservatórios: impactos potenciais, ações de manejo e sistemas em cascata*, Nogueira, M.G.; Henry, R. & Jorcin, A. (Eds.), 329-347, RiMa, ISBN 85-7656-092-5, São Carlos, Brasil

Ayrosa, D.M.M.R.; Furlaneto, F.P.B. & Ayrosa, L.M.S. (2005). Regularização dos projetos de tanques-rede em águas públicas continentais de domínio da União no Estado de São Paulo. *Boletim técnico do Instituto de Pesca*, No.36, (December 2006), pp.1-32, ISSN 0103-1767

Barbosa, F.A.R.; Padisák, J.; Espíndola, E.L.G.; Borics, G. & Rocha, C. (1999). *The cascading reservoir continuum concept and its application to the river Tietê-Basin, São Paulo State, Brazil*, In: *Theoretical reservoirs ecology and its applications*. J.G. Tundisi & M. Straskraba (Eds.), 425-437, International Institute of Ecology, ISBN 8587418025, São Carlos, Brazil

Békési, L.; Székel, C. & Molnár, K. (2002). Atuais Conhecimentos Sobre Myxosporea (Myxozoa), Parasitas de Peixes. Um Estágio Alternativo dos Parasitas no Brasil. *Brazilian Journal of Veterinary Research and Animal Science*, Vol.39, No.5, (July 2002), pp. 271-276, ISSN 1413-9596

Beveridge, M.C.M. (1984). *Cage and pen fish farming: carrying capacity models and environmental impact*, FAO, ISBN 92-5-102163-5, Rome, Italy

Beveridge, M.C.M. (1996). *Cage Aquaculture* (2ª edition), Fishing News Book, ISBN 0-85238-235-9, Oxford, United Kingdon

Beveridge, M.C.M. (2004). *Cage Aquaculture* (3ª edition), Blackwell Publishing, ISBN 978-140-5108-42-3, Oxford, United Kingdon

Boyra, A.; Sanchez-Jerez, P.; Tuya, F.; Espino, F. & Haroun, R. (2004). Attraction of Wild Coastal Fishes to an Atlantic Subtropical Cage Fish Farms, Gran Canária, Canary Islands. *Environmental Biology of Fishes*, Vol.70, (November 2003), pp. 393-401, ISSN 1573-5133

Brandão, H. (2010). Biologia Populacional com Ênfase na Dieta das Principais Espécies de Peixes Agregados a um Sistema de Piscicultura em Tanques-Rede na Pepresa de Chavantes (Médio Rio Paranapanema, SP/PR). In: *Pesquisas de teses e dissertações - Instituto de Biociências, Universidade Estadual Paulista, Botucatu*. Available from (http://www.ibb.unesp.br/posgrad/teses/zoologia_do_2010_heleno_brandao.pdf)

Brandão, H.; Vidotto-Magnoni, A.P.; Ramos, I.P. & Carvalho, E. D. (2009). Assessment of the ichthyofauna in stretches under the influence of Salto Grande Reservoir (Middle Paranapanema River, SP/PR, Brazil). *Acta Limnologica Brasiliensia*, Vol.21, No.4, (Novembro 2009), pp. 451-463, ISSN 0102-6712

Carlson, R.E. (1977). A Trophic State Index for Lakes. *Limnology and Oceanography*, Vol.22, No.2, (March 1977), pp. 361-369.

Castagnolli, N. (2000). Piscicultura intensiva e sustentável, In: *Aqüicultura no Brasil:* bases para um desenvolvimento sustentável, W.C. Valenti, (Ed.), 181-195, CNPq/Ministério da Ciência e Tecnologia, ISBN 85-87632-09-4, Brasília, Brasil

Cheng, T.C. (1986). *General Parasitology* (2ª edition), Academic Press, ISBN 978-012-1707-55-2, Orlando, United States

Cone, D.K.; Beverly-Burton, M.; Wiles, M. & MC Donald,T.E. (1983). The Taxonomy of *Gyrodactylus* (Monogenea) Parasitizing Certain Salmonid Fishes of North America, with Description of *G. nerkae* n. sp. *Canadian Journal of Zoology*, Vol.61, No.11, (January 1983), pp. 2587-2597, ISSN 1480-3283

Costa-Pierce, B.A. (2002). Sustainability of cage culture ecosystems for large scale resettlement from hydropower dams: an Indonesian case study, In: *Ecological aquaculture*: the evolution of the blue revolution, B.A. Costa-Pierce (Ed.), 286-313, Blackwell Science, ISBN 0632049618, Oxford, United Kingdon

Dantas, M.C. & Attayde, J.L. (2007). Nitrogen and phosphorus content of some temperate and tropical freshwater fish. *Journal of Fish Biology*, Vol.70, No.1, (January 2007), pp. 100–108, ISSN 0022-1112

David, G.S.; Carvalho, E.D.; Novaes, J.L.C.; Biondi, G.F. (2006). A tilápia do Tietê: desafios e contradições da pesca artesanal de tilápias nos reservatórios hipereutróficos do Médio Rio Tietê. *Panorama da Aqüicultura*, Vol.16, No.5, (September/October 2006), ISSN 1519-1141

David, G.S.; Carvalho, E.D.; Ramos, I.P.; Silva, R.J.; Ayroza, D.M.M.R.; Ayroza, L.M.S. & Lemos, D. (2011). Evaluation of mass balance models to predict impact of cage fish farming in large freshwater reservoirs, In: *Proceedings World Aquaculture WA 2011* (Conference Aquaculture for a changing World), pp. 322, abstracts, Natal, Brazil, June 2011

Dempster, T.; Sanchez-Jerez, P.; Bayle-Sempere, J.T.; Giménez-Casalduero, F. & Valle, C. (2002). Attraction of Wild Fish to Sea-Cage Fish Farms in the South-Western Mediterranean Sea: Spatial and Short-Term Temporal Variability. *Marine Ecology Progress Series*, Vol.242, (October 2002), pp. 237-252. ISSN 0171-8630

Dillon, P.J. & Rigler, F.H. (1975). A simple method for predicting the carrying capacity of lake for development, based on lake trophic status. *Jounal of Fisheries Research Board of Canada*, Vol.32, No.9, pp. 1519-1531, doi: 10.1139/f75-178

Dohle, A.; Schmahl, G.; Raether, W.; Schmidt, H. & Ritter, G. (2002). Effects of Orally Administered Chemotherapeutics (Quinine, Aalinomycin) Against *Henneguya* sp. Thelohán,1892 (Myxozoa: Myxobolidae), a Gill Parasite in the Tapir Fish *Gnasthonemus peterssii* Günther,1862 (Teleostei). *Parasitology Research*, Vol.88, No.9, (June 2002), pp. 861-867, ISSN 1432-1955

Eiras, J.C. (1993). *Elementos de Ictioparasitologia*. Fundação Engenheiro Antônio de Almeida, ISBN 972-9194-76-9, Porto, Portugal

Eiras, J.C.; Takemoto, R.M. & Pavanelli, G.C. (2006). *Métodos de estudo e técnicas laboratoriais em parasitologia de peixes*, EDUEM, ISBN 85-7628-032-9 Maringá, Brazil

Ergens, R. (1983). *Gyrodactylus* from Eurasian Freshwater Salmonidae and Thymallidae. *Folia Parasitologica*, Vol.30, No.1, (January 1983), pp. 15-26, ISSN 1803-6465

Ewing, M.S. & Kocan, K.M. (1987). *Ichthyophthirius multifilis* (Ciliophora) Exit from Gill Epithelium. *The Journal of Protozoology*, Vol.34, No.3, (August 1987), pp. 309-312, ISSN 0022-3921

Fitzsimmons, K. (2006). Prospect and potential for Global production, In: *Tilapia:* biology, culture and nutrition, C.E. Lim & C.D. Webster (Eds.), 51-72, Haworth Press, ISBN 9781560223184, New York, United State

Food And Agriculture Organization Of The United Nations - FAO. (2010). The State Of World Fisheries And Aquaculture, *FAO*, ISSN 1020-5489, Rome

Fracácio, R., Espíndola, E.L.G.; Rodgher, S.; Pereira, R.H.G.; Rocha, O. & Verani, N.F. (2002). Limnologia dos reservatórios em cascata do médio e baixo rio Tietê: uma análise espacial e temporal, In: *Recursos Hidroenergéticos* usos, impactos e planejamento integrado, E.L.G. Espíndola; F.F. Mauad; V. Schalch; O. Rocha; N. Felicidade & A.C. Rietzler, 145-163, RiMa, ISBN 8586552259, São Carlos, Brazil

Furlaneto, F.P.B.; Ayrosa, D.M.M.R. & Ayroza, L.M.S. (2006). Custo e rentabilidade da produção de tilápia (*Oreochromis* spp.) em tanque-rede no médio Paranapanema, Estado de São Paulo, safra 2004/05. *Informações Econômicas*, Vol.36, No.3, (March 2006) pp. 63-69, ISSN 0100-4909

Gerasev, P.I. (1990). Principles for Revision of the Genus *Dactylogyrus* (Monogenea). *Journal of Ichthyology*, Vol.30, No.5, (January 1990), pp. 110-119, ISSN 0032-9452

Godinho, H.P. (2007). Estratégias reprodutivas de peixes aplicadas à aqüicultura: bases para o desenvolvimento de tecnologias de produção. *Revista Brasileira de Reprodução Animal*, Vol.31, No.3, (July 2007), pp. 351-360. (http://www.cbra.org.br).

Golterman, H.L.; Clymo, R.S. & Ohnstad, M.A.M. (1978). *Methods for physical and chemical analysis of freshwater*, Blackwell (IBP Handbook, 8), ISBN 0632994592, 9780632994595, Oxford, United Kingdon

Håkanson, L. (2005). Changes to Lake Ecosystem Structure Resulting from Fish Cage Farm Emissions. *Lakes & Reservoirs: Research and Management*, Vol.10, (February 2005), pp. 71-80, ISSN 1440-1770

Heckmann, R. (1996). Protozoan Parasites of Fish, Part II. *Aquaculture Magazine*, Vol. 21, pp. 56-59, ISSN 0199-1388.

Henry, R.; Nogueira, M.G.; Pompeo, M.L.M. & Moschini-Carlos, V. (2006). Annual and short-term variability in primary productivity by phytoplankton and correlated abiotic factors in the Jurumirim reservoir (São Paulo, Brazil). *Brazilian Journal of Biology*, Vol.66, No.1B, (February 2006), pp. 239-261, ISSN 1519-6984

Karakassis, I.; Manolis, T.; Christopher, J. S.; & Heye, R. (2002). Fish farming impacts in the Mediterranean studied through sediment profiling imagery. *Marine Ecology*, Vol.27, No.4, (December 2006), pp. 310-319, ISSN 0173-9565

Karakassis, I.; Pitta, P. & Krom, M.D. (2005). Contribution of fish farming to nutrient loading of the Mediterranean. *Scientia Marina*, Vol.69, No.2, (November 2004), pp. 313-321, ISSN 0214-8358

Karakassis, I.; Tsapakis, M.; Hatzyanni, E.; Papadopoulou, K.N. & Plaiti, W. (2000). Impact of cage farming of fish on the seabed in three Mediterranean coastal áreas. *ICES J. Marine Science*, Vol.57, No.5, (October 2000), pp. 1462-1471, ISSN 1095-9289

Krebs, C.J. (1999). *Ecological methodology*. Benjamim/Cummings, Menlo Park, CA, United States, ISBN 0-321-02173-8

Kubitza, F.; Kubitza, L.M.M. (1999). *Principais Parasitoses e Doenças dos Peixes Cultivados* (3ª edition), F. Kubitza (1999), ISBN 85-98545-03-1 , Jundiaí, Brazil

Larsen, D.P. & Mercier, H.T. (1976). Phosphorus retention capacity of lakes. *Journal Fisheries Research Board of Canada*, Vol.33, No.8, pp. 1742-1750, doi: 10.1139/f76-221

Latini, A.O. & Petrere-Junior, M. (2004). Reduction of Native Fish Fauna by Alien Species: an Example from Brazilian Freshwater Tropical Lakes. *Fisheries Management and Ecology*, (April 2004), Vol.11, pp. 71-79, ISSN 1365-2400

Lester, R.J.G. & Adams, J.R. (1974). *Gyrodactylus alexanderi*: Reproduction, Mortality and Effect on it´s Host *Gasterosteus aculeatus*. *Canadian Journal of Zoology*, Vol.52, No.7, (July 1974), pp. 827-833, ISSN 1480-3283

Lin, S.D. (2001). *Water and wastewater calculations manual*, McGraw-Hill Professional, ISBN 0-07-137195-8, New York, United States

Lovshin, L.L. (1982). Tilapia Hybridization, In: *The Biology and Culture of Tilapias*, R.S.V. Pullin & R.H. Lowe-McConnel (Eds.), pp. 279-308, WorldFish Center/International Center for Living Aquatics resources Management , ISBN 9710400045/ 0009799710400041, Manila, Phillipines

Machias, A.; Giannoulaki, M.; Somarakis, S.; Maravelias, C.D.; Nofitou, C.; Koutsoubas, D.; Papadopoulou, K.N. & Karakassis, I. (2006). Fish Farming Effects on Local Fisheries Landings in Oligotrophic Seas. *Aquaculture*, (July 2006), Vol.261, pp .809-816, ISSN 0044-8486

Machias, A.; Karakassis, I.; Labropoulou, M.; Somarakis, S.; Papadopoulou, K.N. & Papaconstantinou, C. (2004). Changes in Wild Fish Assemblages After the Establishment of a Fish Farming Zone in an Oligotrophic Marine Ecosystem. *Estuarine, Coastal and Shelf Science*, Vol.60, (March 2004), pp. 771-779, ISSN 0272-7714

Machias, A.; Karakassis, I.; Somarakis, S.; Giannoulaki, M.; Papadopoulou, K.N. & Smith, C. (2005). The Response of Demersal Fish Communities to the Presence of Fish Farms. *Marine Ecology Progress Series*, Vol.288, (March 2005), pp. 241-250, ISSN 0171-8630

Mackenzie, K. (1970). *Gyrodactylus unicopola* Gluxhova, 1955, from Yong Plaice *Pleurinectes platessa* L., with Notes on the Ecology of the Parasite. *Journal Fish Biology*, Vol.2, No.1, (January 1970), pp. 23-34, ISSN 1095-8649

Martins, M.L. (1998). *Doenças Infecciosas e Parasitárias de Peixes* (2ª edition), Funep, Jaboticabal, Brazil

Martins, M.L.; Souza, V.N.; Moraes, J.R.E. & Moraes, F.R. (1999). Gill Infection of *Leporinus macrocephalus* Garavello & Britski, 1988 (Osteichthyes: Anostomidae) by *Henneguya leporinicola* n. sp. (Myxozoa: Myxobolidae). Description, Histopathology and Treatment. *Revista Brasileira de Biologia*, Vol.59, No.3, (August 1999), pp. 527-534, ISSN 034-7108

Medeiros, F.C. (2002). *Tanque-rede: mais tecnologia e lucro na piscicultura* (1ª. edition), F.C. Medeiros, Cuiabá, Brazil

Ministério da Pesca e Aquicultura - MPA (2010). Produção Pesqueira e Aquícola: Estatística - 2008 e 2009. In: MPA, Brasília, DF, Brasil Available from (http://www.mpa.gov.br/mpa/seap/Jonathan/mpa3/dados/2010/Docs/Caderno%20Consolida%C3%A7%C3%A3o%20dos%20dados%20estatiscos%20final%20curvas%20-%20completo.pdf)

Ministério de Minas e Energia - MME (2006). Matriz energética brasileira. In: *MME, Brasília, DF, Brasil*. Available from (http:// www.mme.gov.br)

Moretto, E.M.; Marciano, F.T.; Velludo, M.R.; Fenerich-Verani, N.; Espíndola, E.L.G. & Rocha, O. (2008). The recent occurrence, establishment and potential impacto f *Geophagus proximus* (Cichlidae, Perciformes) in the Tietê River reservoirs: an Amazonian fish species introduced in the Paraná Basin (Brazil). Biodiversity Conservations, Vol.17, No.12, (November 2008), pp. 3013-3025, ISSN 0960-3115

Munday, B.W.; A. Eleftheriou, M. Kentouri; & P. Divanach. (1992). The interactions of aquaculture and the environment: a bibliographical review. In: Commission of the European Communities, Directorate General for Fisheries, Brussels

Nobile, A.B. (2010). A Ictiofauna Agregada a um Sistema de Piscicultura em Tanques-Rede na Represa Oligotrófica de Chavantes (Médio Rio Paranapanema, SP/PR): Composição de Espécies e Atributos Ecológicos. In: *Pesquisas de teses e dissertações - Instituto de Biociências, Universidade Estadual Paulista, Botucatu*. Available from (http://www.ibb.unesp.br/posgrad/teses/zoologia_me_2010_andre_nobile.pdf)

Noga, E.J. (1996). *Fish Disease* (1ª edition.), Mosby-Year Book, ISBN 0-8138-2558-X, Iowa University Press, United States

Nogueira, M.G.; Henry, R.; Jorcin, A. (2006). *Ecologia de reservatórios: impactos potenciais, ações de manejo e sistemas em cascata* (2ª. edition), RiMa, ISBN 85-7656-092-5, São Carlos, Brazil

Ono, E.A. (1998). *Cultivos de peixes em tanques-rede*. Fundação Biblioteca Nacional, Rio de Janeiro, Brazil

Orsi, M.L. & Agostinho, A.A. (1999). Introdução de Espécies de Peixes por Escapes Acidentais de Tanque de Cultivo em Rios da Bacia do Rio Paraná, Brasil. *Revista Brasileira de Zoologia*, (March 1999), Vol.6, No.2, pp. 557-560, ISSN 0101-8175

Paes, J.V.K. (2006). A Ictiofauna Associada e as Condições Limnológicas numa Área de Influência da Criação de Tilápias em Tanques-Rede no Reservatório de Nova Avanhandava. In: *Catalogo Athena - Universidade Estadual Paulista*. Available from

(http://www.athena.biblioteca.unesp.br/exlibris/bd/bbo/33004064012P8/2006/paes_jvk_me_botib.pdf)

Pavanelli, G.C.; Eiras, J.C. & Takemoto, R.M. (1999). Doenças de Peixes: Profilaxia, Diagnóstico e Tratamento. EDUEM, ISBN 978-857-6281-17-7, Maringá, Brazil

Persson, G. (1988). Environmental impact band nutrient emissions from salmonid fish culture. In: French-Swedish Limnological Symposium, Thonon-les-Bains, Francia. Proceedings of French-Swedish Limnological Symposium, Thonon-les-Bains: INRA: 215-226

Persson, J.; Hakanson, L. & Pilesjo, P. (1994). Prediction of theoretical surface water turnover time in coastal waters using digital bathymetric information. *Environmetrics*, Vol.5, No.4, (December 1994), pp. 433 - 499, ISSN 1180-4009

Pillay, T.V.R. (2004). *Aquaculture and the Environment* (2ª edition). Blackwell Publishing, ISBN 978-140-5108-42-3, Oxford, United Kingdon

Pitta, P.; Apostolaki, E.T.; Giannoulaki, M. & Karakassis, I. (2005). Mesoscale Changes in the Water Column in Response to Fish Farming Zones in Three Coastal Areas in the Eastern Mediterranean Sea. *Estuarine, Coastal and Shelf Science*, (August 2005), Vol.65, pp. 501-512, ISSN 0272-7714

Queiroz, J.V. (2010). Mudança na composição bromatológica natural de *Pimelodus maculatus* (Siluriformes, Pimelodidae) sob influência de pisciculturas em tanques-rede. *Botucatu: Instituto de Biociências*. Universidade Estadual Paulista. Relatory (Scientific initiation)

Ramos, I.P. (2009). Aspectos da Biologia Populacional de *Pimelodus Maculatus* (Teleostei: Siluriformes), Sob Influência de Sistemas de Pisciculturas em Tanques-Rede. In: *Pesquisas de teses e dissertações - Instituto de Biociências, Universidade Estadual Paulista, Botucatu*. Available from (http://www.ibb.unesp.br/posgrad/teses/zoologia_me_2009_igor_ramos.pdf)

Ramos, I.P.; Vidotto-Magnoni, A.P. & Carvalho, E.D. (2008). Influence of Cage Fish Farming on the Diet of Dominant Fish Species of a Brazilian Reservoir (Tietê River, High Paraná River Basin). *Acta Limnolica Brasiliensia*, (October 2008), Vol.20, No.3, pp. 245-252, ISSN 0102-6712

Ranzani-Paiva, M. J. T; Felizardo, N. N.; Luque, J. L. (2005). Análise Parasitológica e Hematológica em Tilápia do Nilo, *Oreochromis niloticus* Linnaeus, 1757, da Represa de Guarapiranga, Estado de São Paulo, Brasil. *Acta Scientiarum - Biological Sciences*, Vol.27, No.3, (July 2005), pp. 231-237, ISSN 1807-863X

Rodgher, S.; Espíndola, E.L.G.; Fracácio, R.; Rodrigues, M.H.; Pereira, R.H.G. & ROCHA, O. (2002). Estudos ecotoxicológicos nos reservatórios em cascata do médio e baixo rio Tietê: uma avaliação dos impactos ambientais, In: *Recursos Hidroenergéticos*: usos, impactos e planejamento integrado, E.L.G. Espíndola; F.F. Mauad; V. Schalch; O. Rocha; N. Felicidade & A.C. Rietzler (Eds.), 131-144, RiMa, ISBN 8586552259, São Carlos, Brazil

Rojas, A. & Wadsworth, S. (2007). A review of cage aquaculture: Latin America and the Caribbean, In: *Cage aquaculture*: Regional reviews and global overview, M. Halwarth,; D. Soto & J.R. Arthur, pp. 73-104, FAO Fisheries technical paper Nº 498, ISBN 9251058016, 9789251058015, Rome

Santos, G.B. & Formagio, P.S. (2000). Estrutura da Ictiofauna dos Reservatórios do Rio Grande, com Ênfase no Estabelecimento de Peixes Piscívoros Exóticos. *Informe Agropecuário*, (January 2000), Vol.21, No.203, pp. 98-106, ISSN 0100-3364

Shelton, W.L. & Popma, T.J. (2006). Biology, In: *Tilapia*: Biology, Culture and Nutrition, C.E. Lim & C.D. Webster (Eds.), 1-27, Haworth Press, ISBN 1560223189, 9781560223184, Birmington, NY, United States of America

Souto, A. C.; Vidotto-Magnoni, A.P.; Brandão, H.; Ramos, I.P. & Carvalho, E.D. (2011). Actinopterygii, Siluriformes, Loricariidae, Loricariichthys platymetopon Isbrücker and Nijssen, 1979: First record in Reservoir of Canoas II, Middle Paranapanema River, border of the states of São Paulo and Paraná, Brazil. *Check List, Vol.* (São Paulo, online edition). , Vol.7, pp. 279-281, ISBN 1809-127X

Stevaux, J.C. (1994). The upper Paraná River (Brazil): geomorphology, sedimentology and paleoclimatology. *Quaternary International*, Vol.22, pp. 143-161, ISSN 1040-6182

Stevens, R; Kerans, B.L.; Lemon, J.C. & Charlotte, R. (2001). The Effects of *Myxobolus cerebralis* Myxospore Dose on Triactinomyxon Production and Biology of *Tubifex tubifex* from Two Geographic Regions. *Journal of Parasitology*, Vol.87, No.2, (April 2001), pp. 315-321, ISSN 1937-2345

Strickland, D.D. & Parsons, T.R. (1968). *A practical manual of seawater analysis.* Bulletin of Fisheries and Resources Aquatics of Canada, pp.1-311.

Suresh, V. (2003). Tilapias, In: *Aquaculture: farming aquatic animals and plants*, J.S. Lucas & P.C. Southgate (Eds.), 321-345, Blackwell Publishing, ISBN 0-85238-222-7

Tavares-Dias, M.; Martins, M.L. & Moraes, F.R. (2001). Fauna Parasitária de Peixes Oriundos de "Pesque-Pague" do Município de Franca, São Paulo, Brasil. I - Protozoários. *Revista Brasileira de Zoologia*, Vol.18, No.1, (June 2001), pp. 67-79, ISSN 0101-8175

Teixeira, C.; Tundisi, J.G. & Kutner, M.B. (1978). Plankton studies in a mangrove environment. II. The standing stock ando some ecological factors. *Boletim do Instituto Oceanográfico*, Vol.24, pp. 23-41, ISSN 0373-5524

Thatcher, V.E. & Brites-Neto, J. (1994). Diagnóstico, Prevenção e Tratamento das Enfermidades de Peixes Neotropicais de Água Doce. *Revista Brasileira de Medicina Veterinária*, Vol.16, No.3, (September 1994), pp. 111-128, ISSN 0100-2430

Torloni, C.E.C.; Corrêa, A.R.A.; Carvalho Jr., A.A.; Santos, J.J.; Gonçalves, J.L.; Gereto, E.J.; Cruz, J.A.; Moreira, J.A.; Silva, D.C.; Deus, E.F. & Ferreira, A.S. (1993). Produção pesqueira e composição das capturas em reservatórios sob concessão da CESP nos rios Tietê, Paraná e Grande, no período de 1986 a 1991. *Pesquisa e Desenvolvimento*, CESP, São Paulo, Brazil

Tundisi, J.G. (2005). *Água no século XXI*: enfrentando a escassez. RiMa, ISBN 85-86552-51-8, São Carlos, Brazil

Tundisi, J.G. & Straškraba, M. (1999). *Theoretical reservoir ecology and its applications.* International Institute of Ecology. Brazilian Academy of Sciences and Backhuis Publishers, Netherlands. pp. 99-124, ISBN. 85-87418-02-5/90-5782-034X, São Carlos, Brazil

Valderrama, J.C. (1981). The simultaneous analysis of total nitrogen and total phosphorus in natural waters. *Marine Chemistry*, Vol.10, No.2 , (January 1981), pp. 109-122, ISSN 0304-4203

Valente, W.C. (2000). Aquaculture for sustainable development, In: Aqüicultura no Brasil: bases para um desenvolvimento sustentável, W.C. Valenti; C.R. Poli; J.A. Pereira &

Borghetti, J.R., pp. 17-24, CNPq/Ministério da Ciência e Tecnologia, ISBN 85-87632-09-4, Brasília, Brasil

Vita, R.; Marín, A.; Madrid, J.A.; Jiménez-Brinquis, B.; Cesar, A. & Marín-Guirao, L. (2004). Effects of Wild Fishes on Waste Exportation from a Mediterranean Fish Farm. *Marine Ecology Progress Series,* (August 2004), Vol.277, pp. 253-261, ISSN 1616-1599

Yucel-Gier, G.; Kucuksezgin, F. & Kocak, F. (2007). Effects of Fish Farming on Nutrients and Benthic Community Structure in the Eastern Aegean (Turkey). *Aquaculture Research,* (January 2008), Vol.38, pp. 256-267, ISSN 1355-557X

Zanatta, A.S. (2007). Tilapicultura em Ecossistemas Aquáticos: Desenvolvimento Sustentável ou Degradação Ambiental? Estudo de Caso em Represa Oligotrófica. In: *Pesquisas de teses e dissertações - Instituto de Biociências, Universidade Estadual Paulista, Botucatu.* Available from (http://www.ibb.unesp.br/posgrad/teses/zoologia_me_2007_augusto_zanatta.pdf)

Zanatta, A.S. (2011). As Interferências das Pisciculturas em Tanques-Rede Sobre as Assembleias de Peixes em Grandes Reservatórios do Rio Paranapanema (Bacia do Alto Paraná). In: *Pesquisas de teses e dissertações - Instituto de Biociências, Universidade Estadual Paulista, Botucatu.* Available from (http://www.ibb.unesp.br/posgrad/teses/zoologia_do_2011_augusto_zanatta.pdf)

Zanatta, A.S.; Ramos, I.P.; Silva, R.J.; Langeani, F. & Carvalho, E.D. (2010). Pisces, Siluriformes, Ictaluridae, *Ictalurus punctatus* (Rafinesque, 1818): First Record in Middle Paranapanema River Reservoir, Aquaculture and Exotic Species Dispersion. *Check List – Journal of Species List and Distribution,* (November 2010), Vol.6, No.4, pp. 589-591, ISSN 1809-127X

Zocchi, P. (2002). *Paranapanema: da nascente à foz.* Editora Horizonte Geográfico/ Audichromo, São Paulo, ISBN 85-88031-13-2

Part 6

Work-Related Hazards – Prevention and Mitigation

15

Aquacultural Safety and Health

Melvin L. Myers[1] and Robert M. Durborow[2]
[1]*University of Kentucky, Lexington, Kentucky*
[2]*Kentucky State University, Frankfort, Kentucky*
USA

1. Introduction

Worldwide, 11,289,000 people worked in aquaculture in 2004—up nearly three-fold from 3,832,000 workers in 1990 (Watterson et al., 2008). Aquaculture, including mariculture, is a fast growing sector of worldwide agriculture but has unaddressed occupational safety and health issues. Many fish farming tasks are dangerous; working around water poses a particular danger, and Working at night and alone compounds the danger. A safety or health hazard is any work design or property (physiological, physical, chemical, biological, or psychological) that may cause harm to workers or bystanders.

As the cultivation of aquatic organisms, aquaculture may include the complete value chain of production including feed production and fish processing, but for our purposes, we delimit the scope of our writing to aquatic organism production that includes the hatchery, nursery, and grow-out phases of production. Feeding, controlling predators, applying chemicals, harvesting, and refurbishing or constructing structures or ponds are examples of typical fish farm operations. While onshore aquaculture is associated with many of the same hazards that are present in agriculture generally, offshore aquaculture is more closely akin to a combination of hazards associated with shallow water commercial fishing and offshore drilling.

Mariculture has many hazards. SINTEF, an independent research organization headquartered in Norway, has presented data for Norwegian aquaculture showing that the fatality rate (9.13 deaths/100,000 work yearsIs "work years" the correct unit here?) is 17 times the average rate for other industries (0.53 deaths/100,000 work years) and equivalent to that of its fishing fleet (Clausen, 2000). In another study with more detail, 16 fatalities occurred in the Norwegian salmon farming sector between 1980 and 1999. Ten of the deaths were associated with using a boat: five in small boats, three occurring: in one type of incident, boats capsized with overloading and shifting loads along with bad weather. Three deaths occurred in workboats when two workers were either stuck by a crane or loads from a crane, and another incident on a well boat, a worker was struck by an anchor that was propelled into his face from a hang-up under the boat by the recoil of the anchor line. Three additional workers died while diving, all of whom lacked a professional diving certificate (Norwegian Labor Inspection Authority, 2001).

Figure 1 shows the percentage and type of non-fatal mariculture-related injuries. Most of these injuries were associated with machinery followed by slips trips. Knife cuts and fish

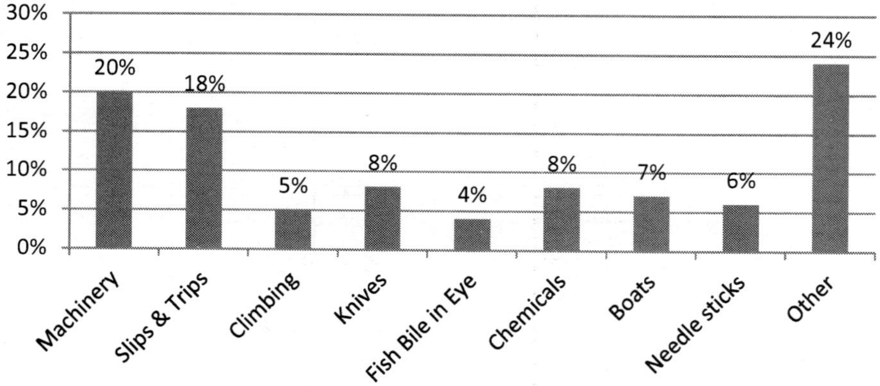

Fig. 1. Percentage of nonfatal occupational injuries associated with aquaculture in Norway, 1980-1999. Source: Norwegian Labor Inspection Authority, 2001

bile in the eye were likely processing-related injuries, but boat-related injuries and needle sticks were associated with fish production, as most probably were the climbing-related injuries.

Figure 2 shows the frequency of mariculture-related illnesses in Norway during the 1980-1999 period. The highest number of reported cases was for musculoskeletal disorders followed by skin allergies and hearing loss.

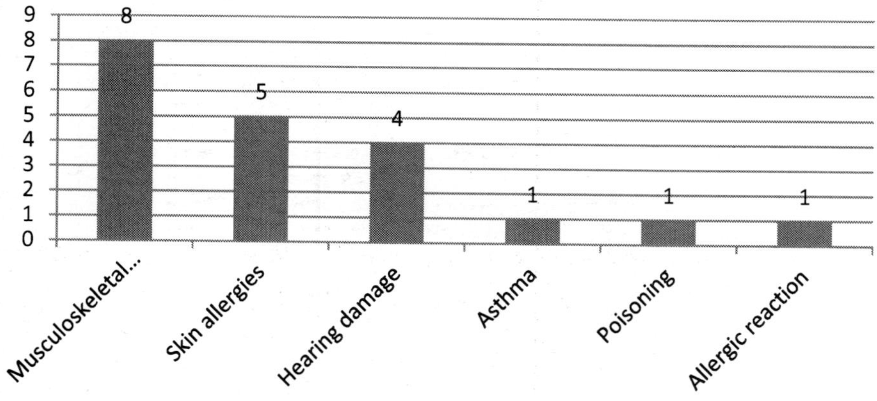

Fig. 2. Number and type of aquaculture-related occupational illness cases reported in Norway, 1980-1999. Source: Norwegian Labor Inspection Authority, 2001

In the United States, the non-fatal occupational injury rate in 2006 for onshore aquaculture was 6.8 injuries per 100 full-time employees according to the U.S. Bureau of Labor Statistics.

In comparison, rates were 5.3 and 7.8 injuries per 100 full-time employees for terrestrial crop and animal production, respectively, and 4.6 injuries per 100 full-time employees across all occupational sectors (Myers & Durborow, 2011; Cole et al., 2009).

Potential occupational hazards in aquaculture have been associated with fatalities that include drownings, electrocutions, crushing-related injuries, hydrogen sulfide poisonings, and fatal head injuries. Non-fatal injuries have been associated with slips, trips, falls, machine operation and repair, strains and sprains, chemicals, and fires. Risk factors include cranes (tip over and power line contact), aerators (entanglement and trauma), tractors and sprayer-equipped all-terrain vehicles (overturn), heavy loads (lifting), boat propellers, high pressure sprayers, slippery surfaces, rotting waste (hydrogen sulfide production), eroding levees (overturn hazard), storm-related rushing water, diving conditions (bends and drowning), night-time conditions, working alone, lack of training, no personal flotation devices (PFD), and all-terrain vehicle use (ATV, also known as a quad-bike). Other hazards include punctures or cuts from fish teeth or spines, needle-sticks, exposure to low temperatures, and bacterial and parasitic infections (Myers, 2010).A fatality of an aquaculture farm manager occurred in Kentucky when he was entangled in a tractor power-take-off during fence post installation, and a diver drowned offshore in Hawaii while working with submerged fish cages (Shikina, 2011).

In this chapter we follow an order that has been developed in industrial hygiene for protection against hazards: identify, evaluate, and control the hazards. In the following sections, we describe approaches for recognizing hazards, including descriptions of the known occupational hazards involved with aquacultural work, and for evaluation, a job hazard analysis approach is described. A risk matrix is used for priority setting so as to deal with the "worst" hazards "first" and risk assessment is described. We describe precedence models—a safety hierarchy—for valuing the effectiveness of hazard controls, and finally, we provide an outline that can be used for developing a safety manual for the individual enterprise.

2. Identifying occupational hazards in aquaculture

This section describes the known hazards of aquacultural work and identifies and discusses the hazards associated with different species and rearing technologies. This information will aid fish farmers in recognizing hazards associated with their operations. Recognized hazards involves employees and includes observed close calls and occupational injury or illness history. The endpoint for the recognition of hazards is an inventory that lists the hazards associated with all tasks, equipment, and substances:

- Injury and illness information and data regarding the industry and related industries, e.g., farming or fishing.
- Information from past incidents and workplace injuries.
- Information from your workers as well as family members and neighbors.
- Product literature and information from suppliers.
- Best industry practices.
- Examine areas or activities where children or visitors may be present.

To identify and better understand hazards before product use, employers need to obtain and read the manuals and safety sheets that are provided by equipment, machinery, and chemical manufacturers. Employers should also develop and implement communication and emergency plans to allow for a timely response in the event of an incident.

Five categories of hazards are 1) physiological (work design), 2) physical, 3) chemical, 4) biological, and 5) psychological (Moreau & Neis, 2009). These categories are described in Table 1. Onshore and offshore exposures to hazards differ greatly, and most species raised offshore involve onshore tasks for the hatchery and nursery phases.

Categories	Exposures	Potential Consequences
Physiological (work design)	Heavy lifting, prolonged standing, awkward postures, repetitive motion, overexertion, lack of visibility	Low back pain, neck and shoulder pain, bursitis, tendonitis, tenosynovitis, carpal tunnel syndrome
Physical	Slips and trips, falls from height, falls overboard, transport and trucking, machinery, electricity, fire, heat and cold, diving, noise, vibration, confined spaces, entanglement, underwater entrapment, solar radiation	Injuries, cuts, burns, broken bones, amputation, hypothermia, hyperthermia, drowning, electrocution, injury-related death, asphyxiation, decompression illness, sprains and strains
Chemical (toxic, flammable, corrosive, explosive)	Disinfectants, parasiticides, piscicides, fungicides, antifoulants, anesthetics, antibiotics, radon gas from water sources, hydrogen sulfide, carbon monoxide, sulfites, dusts, fumes, styrene, needlesticks, flammabilities, battery explosion	Respiratory illness, burns, cancer, central nervous system effects, birth defects, reproductive effects, poisoning, hematopoietic effects, and lung, eye, or skin irritations
Biological	Sharp teeth, spines, aerosolized proteins, bacteria, parasites, skin contact with shellfish and finfish tissues and fluids, enzymes, airborne proteins and endotoxins, fish feed dust	Bites, cuts, punctures and related infections; allergy, asthma, eczema, urticaria (hives), chapped skin, itching.
Psychological	High demand and low control situations, remote locations away from family, potential for large fish kills, abusive social environment	Work-related stress

Sources: Moreau & Neis, 2009; Myers, 2010; Durborow 1997; Erondu & Anyanwu, 2005.

Table 1. Occupational Hazards Associated with Aquaculture

Table 2 provides a summary of common occupational hazards and associated consequences that have been related with work in aquaculture, focusing on the hazards particular to rearing different fish and plant species. Hazards by species vary, but occupational hazards also vary by the phase of production. While hazards associated with the nursery and grow-

Species	Potential Fatal Incident	Potential Non-fatal Incident
Catfish	Injuries from traffic collision, unused seatbelt, or tractor overturns and electrocution from electrical contact.	Disease or illness from chemical exposure: formalin, potassium permanganate, hydrogen peroxide, fertilizers, lime, oxidants, disinfectants, algaecides, herbicides; burns from exposure to sulfuric acid or fires; injury related to fatigue or slips and falls; as well as hand infections, back injury, spine envenomation, and venomous injuries.
Trout	Injury related to falls from raceways or live tank trucks, roadway collisions or, crane overturn; drowning in raceways.	Injury related to high pressure water jet penetration, no guard rails, falls, and slips and trips; hearing loss from noise exposure; leptospirosis infection; toxic exposure to formaldehyde.
Shellfish	Drowning when scuba diving.	Electrical shock from broken electrical conduits; injuries related to loose hand rails, missing safety latches on electric hoists, and working on dredges; shrimp meal allergy; crab scrape-related infections.
Crawfish, snail (rice field flooding)	Crushing injury from tractor overturns.	Injury from boat propeller entanglement; wrist/hand and back strains
Tilapia		Electrical shock from extension cords in water; injury related to poor flooring, missing hand rails, working alone, forklift, algae growing on walking surfaces, trip hazards, or no fall protection; toxic effects or burns related to chlorine, hydrogen peroxide, or hydrochloric acid exposure; musculoskeletal disorders; and spine prick infection (finger amputation).
Ornamental fish		Salt and coral dermatitis, fish tank granuloma, skin infection, cercarial dermatitis, skin granulomas after abrasions, *Mycobacterium marinum* infection.
Salmon	Respiratory failure from H2S exposure; drowning related to fall overboard, boat capsizing, or diving; injury when struck by crane or crane loads or gear entanglement	Decompression illness; toxic effects from styrene and acetone exposure during fiberglass tank construction; anaphylactic shock or infection from self-injection of vaccine; and pesticide poisonings.
Tuna	Diving-related drowning.	Decompression illness.
Sturgeon		Toxic effects or burns from exposure to hydrogen peroxide, lime, or muriatic acid; fork lift-related injuries; fire danger from oxygen.
Plant production (taro, cress, water spinach, mimosa, dropwort, water lilies)	Drowning related to fall overboard or no PFD.	Dermatitis, wastewater-related dermatitis, leptospirosis.

Source: Myers 2010.

Table 2. Recognized Hazards Associated with Specific Species or Phases of Production out phases may differ between species and rearing technology, hatchery operations have much in common between species production. Deaths have occurred in hatcheries associated with hydrogen sulfide exposure and slips and falls. Other hazards associated with hatcheries include exposures to aerators, pumps, heaters, and other types of machinery; fuels, solvents, hypochlorite, formaldehyde, formalin, confined spaces, water jets, unguarded saws, ozone, and hair entanglement in hatching trough paddles.

Erondu & Anyanwu (2005) identified hazards in African aquaculture including noise, cuts, sprains, fractures, asthma, rhinitis, snake and fish bites, bronchitis, chemical burns, pesticides and disinfectant poisoning, parasites, and pathogens. In another review, Conway and RaLonde (1998) identified hazards associated with worldwide aquaculture including machine entanglements, hearing loss, slips and falls, drowning, lacerations, infections, electric shock, hypothermia, repetitive strains, sleep deprivation, decompression illness, organophosphate poisoning, respiratory illness, sunburn, keratotic injury, leptospirosis, and dermatitis. In another two reviews, Durborow (1997; 1999) identified several hazards associated with aquaculture including exposures to bacterial infection, hydrogen sulfide, sodium metasulfide, sodium bisulfite, anesthetics, antibiotics, vaccines, tractor overturns, electricity, overhead power lines, power-take-off or aerator entanglements, falls, muscle strains, drowning, hypothermia, and decompression illness.

Tables 1 and 2 can be helpful in listing known or potential hazards into a hazard inventory. This inventory should include the following (Myers, 2011):

- Process descriptions,
- Controls related to the hazards,
- Location organizationally and physically of each process,
- Supervisor name and contact information,
- Number of employees who work in the process,
- Medical information related to the hazards, and
- Historical information about the process and related hazards

3. Job hazard analysis

The evaluation step aids in decision-making about the nature and control of hazards given the circumstances of the work and exposure to the hazards (Myers, 2011). Priorities for focusing on the "worst first" are effected by using a risk matrix. The matrix maps the likelihood of an incident occurring versus the severity of the consequence, which informs the decision maker of the risk and the urgency for taking preventive action. A risk assessment tool summarizes and evaluates the workers' exposure, job redesign, and actual or potential control measures by the processes of the enterprise.

The job hazard analysis, which is currently used at some fish farms, identifies a job such as climbing a ladder and lists the steps involved in the job and the associated hazards with possible countermeasures to reduce exposure to the hazard. The job hazard analysis tool is presented in a table format in Table 3.

A likely hazardous exposure to workers or bystanders must be corrected immediately. When a hazard has been identified, the risk can be assessed by examining the likelihood of the hazard resulting in injury to workers or other persons (is it likely or unlikely to

Job: Lifting and Carrying Feed Bags | Date: 02.11.2003
Title of Person doing the job: | Analysis by:

Organization: Location: | Reviewed by:
Required or recommended personal protective equipment: | Approved by:

Sequence of job steps	Potential hazards	Recommended control
1. Planning ahead.	1. Holding the bag may cause strain, or it may be dropped.	1. Know where the bag is to go; Have a place to put the bag. Consider lighter loads with more trips.
2. Lifting bag.	2. Back injury.	2. Use arm and leg muscles, not your back; When in doubt make it a 2-person task.
3. Carrying the bag.	3. The bag may slip and fall resulting in strain from overexertion.	3. Grasp bag firmly and secure your footing; Keep your back straight and the bag close to your body. Avoid twisting your back while carrying the load; the back muscles may be strained or the foot could stick to a high traction surface while the rest of the body is twisting, causing knee sprain.
4. Seeing where you are walking.	4. Carrying the load in front of you can cause injury to you or another person.	4. Make sure the load does not block your view while walking; Use a hand truck or move bags stacked on pallets with a forklift.
5. Setting the load down or dumping the load.	5. Back injury.	5. Use your arm and leg muscles. Avoid bending back.

Table 3. Sample Job Hazard Analysis Form

occur) and if the incident occurs, what is its potential severity, i.e., death, serious injury, or minor injury?

Ask questions like: How many people come in contact with the hazard? How often? How seriously could someone be harmed? How quickly could a dangerous situation occur if something goes wrong (PEI, 2005)? Other factors to consider include abilities of the individual, the weather and terrain, and how equipment is used.

The UN Food and Agriculture Organization provides a typical risk matrix as shown in Table 4 that can be used to set priorities for controlling hazards. In this table "catastrophic" refers to death or a lethal disease, "critical" relates to severe injury or occupational disease, "serious" relates to injury or disease requiring medical care but is not critical or catastrophic, "minor"

Likelihood	Consequence				
	Insignificant	Minor	Serious	Critical	Catastrophic
Rare	N	L	L	M	M
Very low	N	L	M	H	H
Low	N	L	H	H	E
Moderate	N	M	H	E	E
High	N	M	E	E	E

Risk level denoted by: N = negligible, L = low, M = moderate, H = high, E = extreme
Source: Arthur et al. 2009.

Table 4. A Risk Matrix

refers to a minor injury or disease, and "insignificant" refers to property damage only (Myers, 2011).

In Norway, job hazard analyses are summarized by job location. This summary is a risk assessment that is made easy by asking three questions (Norwegian Labor Inspection Authority, 2001):

1. What can go wrong (hazard)?
2. What can we do to prevent this (recommended control)?
3. What can we do to reduce the consequences if something occurs (recommended control)?

The emphasis should be aimed at the greatest risk. The format for this assessment is to place answers to these questions as shown in Table 5. The examples developed in Norway suggest different tables for each location in salmon farming: (1) fish hatchery, (2) dock, (3) fish fry boat, (4) feed boat/work boat, (5) plant base, floating or on land, (6) plastic net pens, (7) steel installation, and (8) feeding station. One example is shown in Table 5, which is consistent with the Job Hazard Analysis and uses recommended control information from the US Occupational Safety and Health Administration and the US National Institute for Occupational Safety and Health.

4. Hierarchy of controls

This section introduces the reader to the hierarchy of controls as an extension of the identified countermeasures listed in the job hazard analysis. It distinguishes passive from active controls. Passive controls involve no human action for protection and include the elimination of the hazard at the top of the hierarchy, followed by substitution of a less hazardous technology or an engineered guard against the hazard. Active controls include awareness through warnings or training and the use of personal protective equipment. Farm operators are encouraged to adopt or develop inherently safety technologies by first (**eliminating**), then (**guarding against**), and finally (**warning about**) the hazard. Warnings are not always reliable in preventing contact with hazards. Examples of this hierarchy, which has evolved for safety engineering, are presented in Table 6 (Wogalter, 2006).

The hierarchy of controls is an approach for evaluating the inherently safer technologies with an emphasis on moving from active to passive controls. This simple two-step hierarchy was used in highway safety with the highest precedence based upon (1) passive control that

Work task	What can go wrong? (Potential hazards)	What can we do? (Recommended control)
Walking on the dock..	Slipping or tripping on dock.	Eliminate, to the extent possible, conditions causing slippery working and walking surfaces in immediate work areas, e.g., brush poured cement before it dries to provide traction to the walking surface. Active work areas shall be kept free of equipment and materials not in use, and clear of debris and other objects not necessary for the work in progress. Cargo and material shall not obstruct access to or egress from boats, cranes, vehicles, or buildings. PFDs must be available.
Climbing aboard and ashore from boats.	Falling into sea.	An adequate gangway must be provided. When a gangway is not practical, then a ladder or floating bridge must be used. The gangway must have clear access, have hand rails, and be properly trimmed and illuminated. When boarding, leaving, or working from small boats or floats, workers shall be protected by personal flotation devices.
Loading or unloading with cranes.	Crushed by load or crane contact.	Cranes are to be operated only by qualified and trained personnel. Inspect all rigging prior to use. Do not exceed the load chart capacity while making lifts. Suspended loads must not pass over the gangway or above workers. Shut down the operation during extreme wind or during storms.
Forklift loading or unloading.	Crushed by runover.	Do not operate a forklift unless you are trained and licensed. Do not handle loads that are heavier than the weight capacity of the forklift. Operate the forklift at a speed that will permit it to be stopped safely.

Source: Norwegian Labor Inspection Authority, 2001. Adapted to recommendations by US agencies: the Occupational Safety and Health Administration and the National Institute for Occupational Safety and Health.

Table 5. An Example of a Job Hazard Analysis Related to a Work Location: The Dock

requires no human intervention at the work interface, whereas the less safe approach was (2) the active control that depends upon human behavior at the work interface (Haddon, 1974). The passive control emphasized roadway and vehicle design features while active controls focused on the driver.

PASSIVE CONTROLS—protection does not depend upon the worker's actions (Haddon, 1974).

1. Eliminate hazards posed by equipment, animals, and the environment if at all possible or substitute something safer by using a different machine, material or work practice that poses less risk to perform the same task. For example, replace a faulty machine or use a safer chemical instead of a more dangerous chemical.
2. Guard against the hazard when it is not possible to eliminate hazards. Engineered controls include machinery guards and PTO shields. Design controls, such as locked fences, isolate the worker from the hazard.

ACTIVE CONTROLS—protection depends upon the worker's actions (Haddon, 1974).

3. Warn against the hazard if other controls are inadequate. Protect workers through training, supervision, and personal protective equipment (PPE). For example, supervise new workers until they are competent to deal with hazardous situations.Use and provide proper clothes and respirator protection for handling dangerous chemicals or biohazards.

With more than 50 fish farm visits, investigators were able to identify hazards on fish farms as well as different levels of hazard control on different farms (Durborow, et al., 2011; Ogunsanya et al.,2011). Farmers were generally aware of the hazards but were less aware of controls that different farmers had used to prevent injury from the hazards. Twelve hazards and a range of controls for each hazard are summarized below in Table 6, which classifies the interventions against the precedence hierarchy of controls (Myers & Durborow, 2011; Myers & Cole, 2009).

5. Model safety manual

The Global Aquaculture Alliance for Best Aquaculture Practices (BAP) is a standards-based certification system that combines site inspections and records review to help program participants meet the global demands for wholesome seafood produced in an environmentally and socially responsible manner. BAP has developed standards to certify shrimp hatcheries and shrimp, tilapia, channel catfish, pangasius, and salmon farms. One section of 13 sections in the certifications process includes Worker Safety and Employee Relations (BAPa, 2011). Other sections have an effect on worker safety and health: Storage and Disposal of Farm Supplies, Drug and Chemical Management, Microbial Sanitation, and Harvest and Transport.

In the certification process, the following standards regard worker safety: Living quarters provided by the employer shall be well ventilated and have an adequate shower and toilet and potable water; and wholesome meals should be available for workers. National labor laws or criteria in the International Labor Organization Conventions for minimum age and child labor shall be followed. Minimum standards for occupational safety and health include (1) medical care access, (2) an emergency response plan regarding serious illnesses or injuries, (3) workers trained in first aid and the emergency response plan, (4) available first aid kits, and (5) personal protective devices and clothing should be

Hazard	Warning	Guarding	Elimination
Falling lid on live tanks	Post sign urging caution when working near raised lid	Place wooden wedge under open lid	Install locking or pneumatic hinges
Impalement on electric fence rods (to deter otters)	Be careful and don't fall	Use top insulators as impalement caps	Place rods horizontally on raceway walls
Fall from feed bin roof	Hang on tight to ladder	Install a ladder guard or use a harness attached to a cable	Install hatch handles at ground level
Needle stick while vaccinating fish	Keep fingers away from injection site.	Use corrugated table top to immobilize the fish in the corrugated groove	Install automatic fish vaccination machine
Overhead electric power line contact	Flag areas under power lines	Raise power lines, e.g., 30′ to 45′	Bury power lines
Lifting fish with a dip net	Keep good posture while lifting	Lift smaller loads of fish making more frequent trips	Install a pulley to raise fish nets from tank and a track to slide loaded net to the weighing point
Tractor overturn	Stay off of slopes that could cause rollover	Install rollover protection and use the seatbelt	
Net entanglement & drowning while diving	Don't panic	Place regulator shrouds on O_2 tanks	
Traffic collision hazard	Don't drive sleepy	Maintain distance from other drivers	
Aerator PTO entanglement	Keep away from rotating PTO shafts	Place guards on power-take offs	Use electric-powered aerators
Hatchery paddle entanglement	Use panic wire to stop shaft when hair entangles	Replace bolted metal paddles with plastic paddles that slip upon contact	Mount motor that drives the paddles on a movable platform that completely disengages the drive belts if disturbed
Solar radiation	Don't expose bare skin	Wear sun block	Work in covered areas

Table 6. Examples of Prevention Effectiveness Related to the Hierarchy of Controls

provided as needed. Best Available Practices are established for the following categories of fish farms and hatcheries:

Tilapia Farms (BAPb, 2011):

- The above standards apply.

Shrimp Farms (BAPc, 2011):

- The above standards apply.

Shrimp Hatcheries (BAPd, 2011):

- The above standards apply, and
- Train and assure appropriate licensing of machinery operators, drivers, and repair personnel in machine safety.

Channel Catfish Farms (BAPe, 2011):

- The above standards apply.
- Electrical pumps and aerators must be wired according to standard and safe procedures.

Pangasius Farms (BAPf, 2011):

- The above standards apply, and
- Comply with laws that govern diving on fish farms.
- Dive safety plans that include diver training, maintenance of diving logs, and equipment maintenance.
- Written procedures and trained staff to handle diving emergencies, and regular audit of records and procedures.

Salmon Farms (BAPg, 2011):

- The above standards apply, and
- Initial training of workers for their assigned tasks and safety procedures and use of boats and related equipment.
- Familiarize workers with emergency response plans and train them in first aid, one of whom shall be present among untrained personnel.
- The employer and diving contractors shall comply with laws that govern diving on fish farms or implement a dive safety plan requiring diver training and certification.
    - Minimize the frequency of ascents during the diving day.
    - Maintain dive logs that document procedures and safety-related incidents.
    - Require records on equipment maintenance.
    - Dive safety equipment shall include the availability of bottled oxygen.
- Written safety policies for contractors.

A possible safety manual is provided by the Prince Edward Island Workers' Compensation Board in its Aquaculture Safety Code of Practice. It can be accessed online at http://www.wcb.pe.ca/DocumentManagement/Document/pub_aquaculturesafetycodeofpractice.pdf. An outline for this manual is shown in Table 7. It does not address confined spaces, but a US Occupational Safety and Health Administration publication can be used to develop policies regarding confined spaces (OSHA, 2004).

Responsibilities Under the Law	Hand & Power Tools
Aquaculture Safety Planning	Hoisting and Conveyor Systems
Boat, Deck & Navigational Safety	Hydraulic Safety
Chainsaw Safety	New & Young Workers
Chemical, Fuel & Lubricant Safety	Personal Protective Equipment
Diving Safety	Rescue Procedures
Electrical Safety	Sharp Objects Safety
Equipment & Machinery Safety	Slip, Trip and Fall Prevention
Ergonomics	Transportation Safety
Finfish Safety	Weather Hazards
Fire Prevention	Welding, Cutting or Soldering Safety
First Aid and Emergencies	Winter Harvesting Safety
Confined Spaces*	Workplace Housekeeping

Sources: PEI, 2005.
*OSHA 2004.

Table 7. Possible Sections That Can be Chosen for an Aquacultural Safety Manual

6. Conclusion

This chapter aims to provide information for establishing programs for protecting aquaculturalists from occupational hazards. It presents many occupational hazards associated with aquaculture with some regarding specific species and rearing technologies. These recognized hazards can help the aquaculture production enterprise identify potential hazards in its operation. Next, approaches are described for evaluating these hazards including the job hazard analysis and risk assessment approaches with a description of the risk matrix that can aid in setting priorities for controlling hazards. Section 4 addresses the use of the hierarchy of controls to implement the most effective protection by emphasizing passive controls (protection independent of the worker) over active controls (protection dependent on the actions of the worker). Finally, a possible table of contents for developing a safety manual for an operation is presented (Section 5).

We began with the model developed by industrial hygienists to protect against occupational hazards: the identification, evaluation, and control of hazards. More recently, industrial hygienists have added another purpose for their profession, the anticipation of hazards (Myers, 2005). The anticipation of hazards is of high importance to aquaculture, which is developing rapidly worldwide. One approach is to use known hazards and controls from other sectors. Procedures in the fish processing sector can be expanded into the fish production sector, and the unique procedures of the fishing sector can be adapted to offshore aquaculture. In addition, the traditional regulatory sector regimes for onshore and offshore operations need to come together to protect aquacultural workers (Claussen, 2000), many of whom work both onshore and offshore.

Other sources may aid in more specific approaches. As an example, regarding channel catfish, the Catfish Farmers of America, USDA Southern Regional Aquaculture Center, and National Aquaculture Association developed Safety for Fish Farm Workers program guidelines, which can be accessed at http://www.cdc.gov/nasd/docs/d001701-d001800/d001756/d001756.html. A manual for aquaculture safety in cold waters *Spawn,*

Spat, and Sprains is available at http://seagrant.uaf.edu/lib/an/17/AN-17.pdf, which deals not only with safety and ergonomics, but also with survival in the event of a vessel capsizing. A *Guide to Drug, Vaccine, and Pesticide Use in Aquaculture* produced by the Federal Joint Subcommittee on Aquaculture revised in 2007 can be accessed at http://www.aquanic.org/jsa/wgqaap/drugguide/drugguide.htm.

7. Acknowledgments

The project that informed this chapter was supported by the Centers for Disease Control and Prevention (NIOSH) Cooperative Agreement number 2 U50 OH007547-07, and human subjects review was provided by the University of Kentucky Office of Research Integrity under IRB protocol number 06-0564-P2G. We are grateful to Teresa Donovan of the University of Kentucky College of Public Health for her help in editing this paper.

8. References

Arthur, J.R., Bondad-Reantaso, M.G., Campbell, M.L., Hewitt, C.L., Phillips, M.J., & Subasinghe, R.P. (2009).*Understanding and applying risk analysis in aquaculture: A manual for decision-makers*. Rome: Food and Agriculture Organization of the United Nations, 2009, p. 34. Accessed on 07.27.2011 at http://www.fao.org/docrep/012/i1136e/i1136e.pdf.

BAPa (Best Aquaculture Practices).(2011). *Aquaculture Facility Certification*. Crystal River, FL: Aquaculture Certification Council, Inc. Updated on 03.16.2011. Accessed at http://www.aquaculturecertification.org/index.php?option=com_content&task=view&id=118&Itemid=89 on 07.25.2011.

BAPb (Best Aquaculture Practices).(2011). *Tilapia Farms: Best Aquacultural Practices Certification Standards, Guidelines, Sample Application/Audits*. Crystal River, FL: Aquaculture Certification Council, Inc. Updated on 03.16.2011. Accessed at http://www.aquaculturecertification.org/ACC-PDFS/BAP-TilapiaFG-609.pdf on 07.25.2011.

BAPc (Best Aquaculture Practices).(2011). *Shrimp Farms: Best Aquacultural Standards Certification Standards, Guidelines, Sample Application/Audits*. Crystal River, FL: Aquaculture Certification Council, Inc. Updated on 03.16.2011. Accessed at http://www.gaalliance.org/cmsAdmin/uploads/BAP-ShrimpF-909.pdf on 07.25.2011.

BAPd (Best Aquaculture Practices).(2011). Shrimp Hatcheries: Guidelines for BAP Standards. Crystal River, FL: Aquaculture Certification Council, Inc. Updated on 03.16.2011. Accessed at http://www.aquaculturecertification.org/images/PDF_files/hgud0909.pdf on 07.25.2011.

BAPe (Best Aquaculture Practices).(2011). *Channel Catfish Farms: Best Aquacultural Practices Certification Standards, Guidelines, Sample Application/Audits*. Crystal River, FL: Aquaculture Certification Council, Inc. Updated on 03.16.2011.Accessed http://www.gaalliance.org/cmsAdmin/uploads/BAP-CatfishF-909.pdf on 07.25.2011.

BAPf (Best Aquaculture Practices).(2011). *Pangasius Farms: Best Aquacultural Practices Certification Standards, Guidelines, Sample Application/Audits*. Crystal River, FL:

Aquaculture Certification Council, Inc. Updated on 03.16.2011.Accessed http://www.gaalliance.org/cmsAdmin/uploads/BAP-PangasiusF-810.pdf on 07.25.2011.

BAPg (Best Aquaculture Practices). (2011).*Salmon Farms: Best Aquacultural Practices Certification Standards, Guidelines, Sample Application/Audits*. Crystal River, FL: Aquaculture Certification Council, Inc. Updated on 03.16.2011. Accessed on http://www.gaalliance.org/cmsAdmin/uploads/BAP-SalmonF-611.pdf on 07.25.2011.

Claussen, C.B. (2011) High accident rates in aquaculture. *Gemini*. Accessed on 07.26.2000 at http://www.ntnu.no/gemini/2000-06e/10.htm .

Cole, D.W., Cole, R., Gaydos, S.J., Gray, J., Hyland, G., Jacques, M.L., Powell-Dunford, N., Sawhney, C.,& Au, W.W. (2009). Aquaculture: Environmental, toxicological, and health issues. *Int J Hygiene Environ Health. Vol. 212, 368-377.*

Conway, G,A.& RaLonde, R. (1998). Fish farming and aquaculture.*ILO Encyclopedia of occupational health and safety*. 4th ed. Geneva: International Labour Office (ILO). Vol. 3(70):33-34.

Durborow, R.M. (1997). Aquacultural hazards. In: Langley RL, editor. *Safety and health in agriculture, forestry, and fisheries*. Lanham, MD: Government Institutes, pp. 659-679.

Durborow, R.M. (1999). Health and safety concerns in fisheries and aquaculture. *Occup Med.* Vol. 14, 373-406.

Durborow, R.M., Myers, M.L., Cole, H.P., Ogunsanya, T.O., & Thompson, S.L. (2011).Presented at the National Symposium on Agriculture, Forestry, & Fishing Health and Safety in Boise, Idaho, June 28, 2011.

Erondu, E.S.&Anyanwu, P.E. (2005). Potential hazards and risks associated with the aquaculture industry. *Afr J Biotechnol.* Vol. 4,1622-1627.

Haddon, W. (1974). Strategies in preventive medicine: passive vs. active approaches to reducing human wastage. *J Trauma.* Vol. 14, 353-354.

Moreau, D.T.R.& Neis, B. (2009). Occupational health and safety hazards in Atlantic Canadian aquaculture: Laying the groundwork for prevention. *Marine Policy.* Vol. 33, 401-411.

Myers, M.L. (2005). Addressing risks and benefits: emerging technologies assessed for safety and health of farming people. *Resource.* Vol. 12, 13-14.

Myers, M.L. (2010). Review of occupational hazards associated with aquaculture. *J Agromedicine.* Vol. 15, 412-426.

Myers, M.L. & Cole, H.P. (2009). Simple solutions for reduced fish farm hazards. *J Agromedicine.* Vol. 14, 150-156.

Myers, M.L. (2011). Reducing hazards in the work environment. In *Praeger Handbook of Environmental Health*, R Friis, C Friis (Eds.) Vol. 4, Chapter 3, Santa Barbara, CA : Praeger (In press).

Norwegian Labor Inspection Authority (2001) Work Environment and Safety in Aquaculture: Tips for Risk Assessment and Precautionary Measures. A booklet (Translated from Norwegian). Trondheim, Norway: Norwegian Labor Inspection Authority.

Ogunsanya, T.O., Durborow, R.M., Myers, M.L., Cole, H.P., & Thompson, S.L. (2011).Safety on North Carolina and Kentucky Trout Farms. *J Agric Safety Health.* Vol. 17, 33-61.

OSHA.(2004). Permit Required Confined Spaces. Washington, DC: Occupational Safety and Health Administration, US Department of Labor, OSHA 3138-01 2004. http://www.osha.gov/Publications/osha3138.pdf

Prince Edward Island (PEI) Aquaculture Safety Code of Practice.(2005). Charlottetown, PEI: Workers Compensation Board of Prince Edward Island. Accessed at http://www.wcb.pe.ca/DocumentManagement/Document/pub_aquaculturesafetycodeofpractice.pdf. on 07.25.2011.

Shikina, R. (2011). Honolulu firefighter died in diving accident. Star Advertiser. May 24, 2011. Honolulu, Hawaii: Accessed at http://www.staradvertiser.com/news/breaking/122557309.html on 07.28.2011.

Watterson, A., Little, D., Young, J.A., Boyd, K., Azim, E.,& Murray, F. (2008).Towards Integration of Environmental and Health Impact Assessments for Wild Capture Fishing and Farmed Fish with Particular Reference to Public Health and Occupational Health Dimensions. *Int. J. Environ. Res. Public Health.* Vol. 5, 258-277.

Wogalter, M.S. (2006). Purposes and scope of warnings. In *Handbook of Warnings,* M.S. Wogalter (Ed). Mahwah, New Jersey: Lawrence Erlbaum Associates, pp. 3-10.

Part 7

Spread of Pathogens from Marine Cage

16

Spread of Pathogens from Marine Cage Aquaculture – A Potential Threat for Wild Fish Assemblages Under Protection Regimes?

Antonio Terlizzi*, Perla Tedesco and Pierpaolo Patarnello
Department of Biological and Environmental Sciences and Technologies, University of Salento, CoNISMa, Lecce
Italy

1. Introduction

The spread and the control of pathogens represent a serious problem in aquaculture, implying important environmental and economic issues (Huntington, 2006; Sapkota et al., 2008).

With the rapid development of aquaculture in recent decades (Food and Agriculture Organization [FAO], 2008), an increasing number of pathogens involved in fish disease, and the range of susceptible fish species have been identified and described. The increasing use of molecular tools has helped improve our knowledge in this area.

Few studies have been performed to evaluate the interactions host/pathogen in the natural environment (Hedrick, 1998), or to investigate the mechanisms of infection and interaction between farmed and wild specimens, nor the ecological role assumed by infected wild hosts has been explored yet.

In view of the current decline of fish stocks, which has recently been attributed not only to habitat loss and overfishing, but also even to the spread of pathogens (Daszak, 2000; Gozlan et al., 2005), understanding these mechanisms is essential for a proper conservation and management of marine and fisheries resources.

The effect of parasites and other pathogens on fish populations under protection regimes has received virtually no attention (McCallum et al., 2005). Marine Protected Areas (MPAs) can provide unique protection for critical areas and spatial escape for overexploited species and are regarded as essential tools for the conservation of marine environments (Lubchenco et al., 2003; Pauly, 2005; Roberts et al., 2005). However, the effectiveness of MPAs is greatly limited by several anthropogenic and natural impacts acting at scales much larger than that encompassing by the reserve boundaries. Examples of such impacts toward which MPAs cannot offer any direct protection are represented by climate change (Graham et al., 2008; McClanahan, 2000), with all the problems related to it (Keller et al., 2009), the spread of pollutants (Jarman et al., 1992; Loganathan & Kannan, 1994; Terlizzi et al., 2004), invasive

* Corresponding Author

species (Carlton et al., 1990; Katsanevakis et al. 2010; Trowbridge, 1995), and pathogens (Lessios, 1988; Littler & Littler, 1995; Rasmussen, 1977; Steinbeck et al., 1992).

In this chapter we briefly review the potential threats to wild population represented by the spread of pathogens from marine cage fish aquaculture. As an emblematic example we focus particularly on the effects of a serious disease, namely the *Viral Nervous Necrosis* (VNN), affecting several reared fish species worldwide.

In order to warn about how the spread of pathogens could represent a risk for natural populations, even under protection regimes, we report on the results of a preliminary survey for the detection of viral particles responsible for VNN in the gonads and brain homogenates in some wild specimens collected in two no take-no access Marine Protected Areas in the Southern Italy (South Adriatic and North Ionian Sea).

2. The spread of pathogens from marine cage aquaculture

There are several ways by which new pathogens can attain natural populations.

For example, a benign organism can undergo a genetic change that may increase its degree of pathogenicity (Bull, 1994; Cunningham, 2002). A pathogen may also be introduced through the expansion of the range of wild species in new regions as a result of global warming, the removal of barriers, or the transport of animals for commercial and recreational purposes (Gozlan et al., 2006).

It is now known that the activities of aquaculture farms, particularly those of offshore floating cages, are a source of various types of environmental (Pusceddu et al., 2007) and biological (Terlizzi et al., 2010) impact, including the facilitation of the spread of potential pathogens (Krkošek et al., 2006). The high density to which organisms are reared and the unnatural and stressful conditions to which they are subjected may in fact facilitate the emergence and spread of diseases, which can potentially extend to wild organisms (Naylor et al., 2005).

The use of net pen or cages systems involves a higher risk for disease transmission because there is no impermeable barrier between the farm and the aquatic environment (Huntington et al., 2006). There is therefore a strong potential for contamination between reared organisms and the wildlife of the surrounding environment.

Many species of wild fish are attracted by the presence of the cages, which provide them food and shelter (Diamant et al., 2000), often succeeding to get into cages, passing through the mesh. Farmed species may be able to escape in the environment too. This phenomenon may have important biological implications (e.g. in terms of genetic contamination) on wild populations. In addition, escaped individuals can carry infections they acquired at the culture systems (Conseil International pour L'Exploration Scientifique de la Mer Médetirranée [CIESM], 2007).

Farmed fish stocks may prove receptive to pathogens from the surrounding environment too. The exchange of pathogens between farmed and wild fish has been repeatedly observed in different geographical regions (Diamant et al., 2000; Kent, 2000; Mc Vicar, 1997; Nowak et al., 2004; Paperna, 1998; Sepulveda et al., 2004). Bacterial infections by *Mycobacterium marinum* and *Streptococcus iniae*, which are common among farmed fish stocks, have been described to cause mortality even in wild fish (Colorni et al., 2002).

Pathological events in populations of wild abalone (Alstatt et al., 1996) suggest a potential susceptibility that makes them able to influence or be influenced by, reared individuals. Such pathogen exchange relationship seems to occur frequently between farmed and wild fish populations, as in the case of a disease known as *Infectious Hematopoietic Necrosis*, which affects mainly salmons, and which appears to be transmitted between wild and farmed specimens in both directions (Naylor et al., 2003).

Transmission of pathogens from aquaculture to wild fish can occur at different stages of the life cycle: infection can start from the earliest stages of reproduction and larval rearing (hatchery) and spread in the later stages (growing) contaminating wild stocks and causing dangerous epidemics (Naylor et al., 2005).

Krkošek et al. (2011) studied the effects of sea lice (*Lepeophtheirus salmonis*) from salmon farming facilities on populations of wild salmon (*Onchorhynchus gorbusha* and *O. keta*), noting that juveniles of these species, when parasitized, show a reduced ability to escape predators that consume selectively parasitized salmon. Therefore, the mortality of juvenile wild salmon may be higher than previously believed and this has important implications for the development of appropriate conservation policies.

The importation of feed for aquaculture can also be considered vehicle of sources of pollutants and of pathogens between fish stocks very distant from each other (Dalton, 2004).

Parasites and pathogens can therefore play an important role for fish stocks, reducing both the number and yield, increasing mortality, altering the reproductive potential, affecting the structure of the population size and/or reducing the commercial value of harvested stocks (Dobson & May, 1987; Kuris & Lafferty, 1992).

Being able to influence the survival of the host and his reproductive success, pathogens should be taken into great consideration in the design and management of MPAs (McCallum et al., 2005). Also, even if the survival or reproduction are not affected, as in the case of infections by *Anisakis simplex* and *Pseudoterranova decipiens*, pathogens may pose a problem for fisheries management-oriented reserves, because the infected hosts have a lower market value and can represent a health risk when consumed as food by humans.

3. The Viral Nervous Necrosis (VNN)

The genus *Betanodavirus,* family Nodaviridae is a widespread and well-known viral pathogen in the field of fish farming. It is the etiologic agent responsible for a severe disease known as *Viral Encephalopathy and Retinopathy* (*VER*), or *Viral Nervous Necrosis* (*VNN*). It affects more than 40 species of marine fish worldwide (Panzarin et al., 2010). VNN disease is regarded as one of the most devastating viral infections among marine fish.

Affected fish, particularly juvenile forms, show spiralling swimming, darkened colour and hyper inflated swim bladders. Internal disease signs include pale livers, empty digestive tracts and intestines filled with brownish fluid. In affected fish the virus propagates in the eye, brain and distal spinal cord causing marked vacuolations leading to encephalopathy and retinopathy (Fig. 1). Such lesions in the Central Nervous System and in the retina (nervous tissue of the eye) have always been typical of larval and juvenile stages in the sea bass (Bovo et al., 1996; Breuil et al., 1991) and in other species sensitive to nodavirosis (Glazebrook et al., 1990; Grotmol et al., 1997; Mori et al., 1991; Nguyen et al., 1996). For this

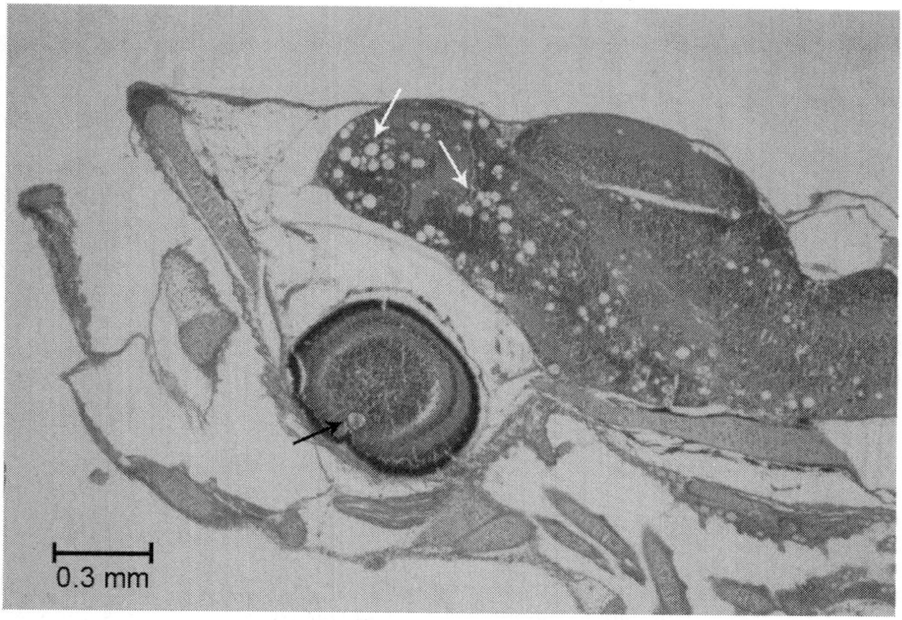

Fig. 1. Histological preparation of a sea bass larva (age 45 days) with the typical vacuolar necrosis of the brain tissue and in the retina (indicated by white and black arrows, respectively). After the collection, the sample has been readily fixed in 10% neutral buffered formalin for not more than 7 days. Thereafter it has been included in paraffin to be cut in to sections of 5 μm and then colored for histological analysis. The dye used in this histological section is hematoxylin-eosin.

reason histological analyses have often been performed in monitoring programs to evaluate the spread of the disease in the hatcheries (Sweetman et al 1996).

The virus also multiplies in the gonad, livers, kidney, stomach and intestine (Grotmol et al., 2000; Munday et al., 2002).

Experiments with *Pseudocaranx dentex* (Arimoto et al., 1992; Mushiake et al., 1994; Nishizawa et al., 1996) have shown that the virus transmission can occur either vertically, from broodstock to larvae through the eggs or sperm, either horizontally, through the contact between healthy fish and infected larvae (Arimoto et al., 1993). Experimental contamination with infected tissue homogenates (Arimoto et al., 1993; Glazebrook et al., 1990; Grotmol et al., 1999; Tanaka et al., 1998) or addition of purified virus from infected individuals (Nguyen et al., 1996) or produced in SSN-1 cells (Péducasse et al., 1999) is also able to transmit the virus. The ability of the betanodavirus to resist to a wide pH range, varying from 2 to 9 (Frerichs et al., 1996), and its resistance in sea water at 15° C for more than one year (Frerichs et al., 2000) increases the probability of horizontal transmission (Munday et al., 2002). The transmission of the betanodavirus from infected, asymptomatic specimens is also possible. Some Authors (e.g. Castric et al., 2001) have shown how specimens of *Sparus aurata* experimentally infected but apparently healthy, are able to transmit the virus to individuals of *Dicentrarchus labrax* placed in the same cage.

The finding, in Norway (Aspehaug et al., 1999), of adults of Atlantic halibut (*Hippoglossus hippoglossus*) positive for betanodavirus suggests a possible danger of using wild animals as breeding stock.

There is obviously also the possible risk of transmission occurring in the opposite direction, namely that the virus originating from aquaculture facilities can infect wild healthy individuals. Supports to this hypothesis, in addition to the above mentioned evidences, which show that betanodavirus can spread both horizontally and vertically within a system, are provided by the detection of the virus in several wild marine species, although the infection in the examined subjects was asymptomatic (Barker et al., 2002; Gagné et al., 2004; Gomez et al., 2004; Gomez et al., 2008; Thiéry et al., 2004).

Recent studies carried out along the Italian coast have identified the etiologic agent responsible for the disease in wild organisms belonging to a wide panel of species. From a virological examination conducted by Maltese et al. (2005) on samples from the northern Adriatic and the Strait of Sicily, the presence of viruses has been detected in the sea bass (*Dicentrarchus labrax*), gilthead seabream (*Sparus aurata*), white grouper (*Epinephelus aeneus*), dusky grouper (*Epinephelus marginatus*), red mullet (*Mullus barbatus*), axillary seabream (*Pagellus acarne*), poor cod (*Trisopterus minutus*) and in the black goby (*Gobius niger*). The study highlighted a possible widespread of the virus in nature, the role of *S. aurata* as an asymptomatic carrier and the high sensitivity of the genus *Epinephelus*, for which episodes of mortality in other geographical areas have been reported. The study also confirmed the considerable sensitivity of *D. labrax* to the betanodavirus, which had already been proven in the context of intensive farming.

4. VNN and protected populations: A limited effectiveness of MPAs?

The two areas considered for sampling are the MPA of Porto Cesareo, located along the southwestern coast of Apulia (Ionian Sea, 40°15' N - 17°52' E) and the MPA of Torre Guaceto, located along the east coast of Apulia (Adriatic Sea, 40°43' N - 17°48' E). Both areas are characterized by the past presence of extensive aquaculture activities within or close to the reserve boundaries. The survey focused mainly on specimens of *D. labrax* but, whenever possible, considered, as a preliminary screening, some other species of commercial interest for which positive serology for the betanodavirus has been already reported in literature. For all specimens, samples of brain and gonadal tissue, the main target organs of the virus were taken in double aliquot for DNA extraction and biomolecular analysis. Virological analyses were performed by the laboratories of the Istituto Zooprofilattico Sperimentale delle Venezie (Padua, Italy). Samples were analysed through real-time TaqMan PCR. This technique proved to be highly sensitive, is suitable for a wide range of organic matrices and is able to detect the four *betanodavirus* genotypes currently recognized. The experimental protocol followed the procedure described in Panzarin et al. (2010). More particularly, samples were processed for RNA extraction using the NucleoSpin® RNA II (Macherey-Nagel GmbH & C., Düren, Germany). Real-time PCR was performed using the LightCycler 2.0 system and carried out in 20 µl with "LightCycler® TaqMan® Master" (Roche Diagnostics GmbH, Mannheim, Germany), 0.9 µM of each primer, 0.75 µM of probe and 5 µl of cDNA template. The thermal profile consisted of a 10-min incubation at 95°C followed by 45 cycles of 10 s denaturation at 95°C, 35 s annealing at 58°C and 1 s elongation at 72°C, followed by an additional 30-s cooling step at 40°C.

The analysis involved the collection of 28 specimens, 9 of which coming from the MPA of Torre Guaceto and 19 from the MPA of Porto Cesareo. Results are summarised in Table 1.

Several specimens resulted positive to the VNN test. The individuals showing positivity to the molecular analyses were mostly represented by the sea bass (*Dicentrarchus labrax*) but the presence of the virus was also detected in the scorpion fish (*Scorpaena scrofa*), the gray mullet (*Mugil* sp.), the goby (*Gobius* sp.) and the comber (*Serranus cabrilla*). All positive individuals appeared asymptomatic.

Fish species	No. of examined fish	No. of positive samples
	Torre Guaceto MPA	
Dicentrarchus labrax	4	3
Diplodus sargus	5	0
	Porto Cesareo MPA	
Dicentrarchus labrax	3	3
Scorpaena scrofa	3	1
Mugil sp.	3	1
Gobius sp.	2	1
Serranus cabrilla	2	1
Synodus saurus	2	0
Serranus scriba	1	0
Sarpa salpa	2	0
Oblada melanura	1	0

Table 1. Summary of VNN tests performed in the two Marine Protected Areas (MPAs)

5. Discussion

The lack of clinical signs in infected fish specimens observed in this study is consistent with the results obtained in other recent studies of wild marine fauna (Baeck et al., 2007; Gomez et al., 2004), thus confirming how rare are the observations of clinical forms in wild fish populations (but see Gomez et al., 2009).

Some authors argue that most infections in wildlife are presumably latent and they do not evolve into a pathological condition until the fish becomes stressed (Barker et al., 2002).

Barker et al. (2002) firstly reported piscine nodavirus in wild winter flounder *Pleuronectes americanus* in Passamaquoddy Bay (New Brunswick, Canada) considering the possibility that the virus could be transferred between wild winter flounder and sea-caged halibut or cod.

Baeck et al. (2007) detected nodavirus in 21 different apparently healthy wild marine fish collected in coastal areas of Korea, close to aquaculture facilities and Gomez et al. (2004) reported high prevalence for betanodavirus in apparently healthy cultured and wild marine fish near mariculture areas in Japan. In both studies, sampling areas were located near aquaculture facilities and the authors support the hypothesis of an horizontal transmission of the virus.

The results here reported, although preliminary, make us suppose that the distribution of the betanodavirus is currently extending and not confined to marine cage aquaculture. The genetic characterization of isolated viral strains is currently under investigation and will likely help determining more precisely the range of dispersion and residence time of these possible sources of contamination, thus improving our understanding about the epidemiological mechanisms of the disease and consequently our ability to control it.

Our findings also confirm that MPAs are not able to cope with this particular form of impact, which clearly acts at a scale larger than that covered by the measures of protection (Allison et al., 1998).

The specimens considered, although asymptomatic, might still be reservoirs of infection, by acting as virus carriers, and further favoring the spread of the pathogen. The ecological role of these asymptomatic carriers in nature remains however unclear and the possible consequences on natural population still remain to be clarified. This is not an easy task, however, as disease disorders are difficult to observe in the wild because affected individuals are rapidly removed from the population by predators (Gozlan et al., 2006). These consequences are nevertheless conceivable since in some reared species such as the sea bass, the virus causes, in addition to the above-described damages to the nervous system, failure in reproduction (often sterilization), reduced hatching rate and high larval mortality.

In the view of the impact caused by the over-exploitation of marine resources and given the current trend of decline of commercially important fish stocks, increased mortalities and failure in reproduction due to diseases should be carefully considered and adequately addressed, especially in sound quantification of the effectiveness of MPAs.

Attempts to protect biodiversity through MPAs and/or restore ecosystem functioning (e.g., through the protection of over-exploited species that generate positive community-wide effect) might be frustrated and/or biased in their quantification by external pressures.

Given the wide spread of this virus and its possible consequences in asymptomatic forms it should be important to include the issue of pathogens in disease surveillance program and monitoring of biological health of each MPA as well as any area that has or has had a significant number of intensive aquaculture plants.

6. Acknowledgments

The research leading to these results has received funding from the European Community's seventh Framework programme (FP7 / 2007 – 2013) under Grant Agreement No. 266445 for the project Vectors of Change in the Oceans and Seas Marine Life, Impact on Economic Sectors (VECTORS).

7. References

Allison, G. W.; Lubchenco, J. & Carr, M. H. (1998). Marine reserves are necessary but not sufficient for marine conservation. *Ecological applications*, Vol.8, pp. S79–S92.

Alstatt, J. M.; Ambrose, R. F.; Engle, J. M.; Haaker, P. L.; Lafferty, K. D. & Raimondi, P. T. (1996). Recent declines of black abalone *Haliotis cracherodii* on the mainland coast of central California. *Marine Ecology Progress Series,* Vol.142, pp 185-192.

Arimoto, M.; Mori, K.; Nakai, T.; Muroga, K. & Furusawa, I. (1993). Pathogenicity of the causative agent of viral nervous necrosis disease in striped jack, *Pseudocaranx dentex* (Block & Schneider). *Journal of Fish Diseases,* Vol.16, pp. 461–469.

Arimoto, M.; Mushiake, K.; Mizuta, Y.; Nakai, T.; Muroga, K. & Furusawa, I. (1992). Detection of Striped Jack Nervous Necrosis Virus (JSNNV) by enzyme-linked immunosorbent assay (ELISA). *Gyobyo Kenkyu,* Vol.27, pp. 191–195.

Aspehaug, V.; Devold, M. & Nylund, A. (1999). The phylogenetic relationship of nervous necrosis virus from halibut (*Hippoglossus hippoglossus*). *Bulletin of European Association of Fish Pathologists,* Vol.19, pp. 196-202.

Baeck, G.W.; Gomez, D.K.; Oh, K.S.; Kim, J. H.; Choresca, C. H. Jr & Park, S. C. (2007). Detection of piscine nodaviruses from apparently healthy wild marine fish in Korea. *Bulletin of the European Association of Fish Pathologists,* Vol.27, pp. 116–122.

Barker, D. E.; MacKinnon, A. M.; Boston, L.; Burt, M. D. B.; Cone, D. K.; Speare, D. J.; Griffiths, S.; Cook, M.; Ritchie, R. & Olivier, G. (2002). First report of piscine nodavirus infecting wild winter flounder *Pleuronectes americanus* in Passamaquoddy Bay, New Brunswick, Canada. *Diseases of Aquatic Organisms,* Vol.49, pp. 99-105.

Bovo, G.; Borghesan, F.; Mutinelli, F.; Montesi, F. & Comuzzi, M. (1996). Viral Encephalo-retinopathy of reared sea bass: first detection in Italy. *Bollettino della Società Italiana di Patologia Ittica,* Vol.19, pp. 52-64.

Breuil, G.; Bonami, J. R.; Pepin, J. F. & Pichot, Y. (1991). Viral infection (picorna-like) associated with mass mortalities in hatchery-reared sea bass *Dicentrarchus labrax* larvae and juveniles. *Aquaculture* Vol.97, pp. 109–116.

Bull, J. J. (1994). Perspective e virulence. *Evolution,* Vol.48, pp. 1423-1437.

Carlton, J. T.; Thompson, J. K.; Schemel, L. E. & Nichols, F. H. (1990). Remarkable invasion of San Francisco Bay (California, USA) by the Asian clam *Potamocorbula amurensis.* I. Introduction and dispersal. *Marine Ecology Progress Series,* Vol.66, pp. 81-94.

Castric, J.; Thiery, R.; Jeffroy, J.; de Kinkelin, P. & Raymond, J. C. (2001). Sea bream *Sparus aurata,* an asymptomatic contagious fish host for nodavirus. *Diseases of Aquatic Organisms,* Vol.47, pp. 33-38.

Colorni, A.; Diamant, A.; Eldar, A.; Kvitt, H. & Zlotkin, A. (2002). *Streptococcus iniae* infections in Red Sea cage-cultured and wild fishes. *Diseases of Aquatic Organisms,* Vol.49, pp. 165-170.

Conseil International pour L'Exploration Scientifique de la Mer Médetirranée [CIESM] (2007). Impact of mariculture on coastal ecosystems. Ciesm Workshop Monographs n°32, Monaco www.ciesm.org/online/monographs/lisboa07.pdf

Cunningham, C. O. (2002). Molecular diagnosis of fish and shellfish diseases: present status and potential use in disease control. *Aquaculture,* Vol.206, pp. 19-55.

Dalton, R. (2004). Fishing for trouble. *Nature,* Vol.431, pp. 502–504.

Daszak, P. (2000). Emerging infectious diseases of wildlife - threats to biodiversity and human health. *Science,* Vol.287, pp. 443-449.

Diamant, A.; Banet, A.; Ucko, M.; Colorni, A.; Knibb, W. & Kvitt, H. (2000). Mycobacteriosis in wild rabbitfish *Siganus rivulatus* associated with cage farming in the Gulf of Eilat, Red Sea. *Diseases of Aquatic Organisms,* Vol.39, pp. 211–219.

Dobson, A. P. & May, R. M. (1987). The effects of parasites on fish populations: theoretical aspects. *International Journal for Parasitology,* Vol.17, pp. 363–370.

Food and Agriculture Organization [FAO] (2008). State of World Fisheries and Aquaculture (SOFIA), Food and Agricultural Organization of the United Nations, Rome.

Frerichs, G. N.; Rodger, H. D. & Peric, Z. (1996). Cell culture isolation of piscine neuropathy nodavirus from juvenile sea bass, *Dicentrarchus labrax. Journal of General Virology,* Vol.77, pp. 2067–2071.

Frerichs, G. N.; Tweedie, A.; Starkey, W. G. & Richards, R. H. (2000). Temperature, pH and electrolyte sensitivity, and heat, UV and disinfection inactivation of sea bass (*Dicentrarchus labrax*) neuropathy virus. *Aquaculture,* Vol.185, pp. 13–24.

Gagné, N.; Johnson, S. C.; Cook-Versloot, M.; MacKinnon, A. M. & Olivier, G. (2004). Molecular detection and characterization of nodavirus in several marine fish species from the northeastern Atlantic. *Diseases of Aquatic Organisms,* Vol.62, pp. 181-189.

Glazebrook, J. S.; Heasman, M. P. & de Beer, S. W. (1990). Picorna-like viral particles associated with mass mortalities in larval barramundi, *Lates calcarifer* Bloch. *Journal of Fish Diseases,* Vol.13, pp. 245–249.

Gomez, D. K.; Baeck, G. W.; Kim, J. H.; Choresca, C. H. Jr & Park, S. C. (2008). Molecular detection of betanodaviruses from apparently healthy wild marine invertebrates. *Journal of Invertebrate Pathology,* Vol.97, pp. 197–202.

Gomez, D. K.; Sato, J.; Mushiake, K.; Isshiki, T.; Okinaka, Y. & Nakai, T. (2004). PCR-based detection of betanodaviruses from cultured and wild marine fish with no clinical signs. *Journal of Fish Diseases,* Vol.27, pp. 603-608.

Gomez, D.K.; Matsuoka, S.; Mori, K.; Okinaka, Y.; Park, S.C. & Nakai, T. (2009). Genetic analysis and pathogenicity of betanodavirus isolated from wild redspotted grouper *Epinephelus akaara* with clinical signs. *Archives of Virology,* Vol.154, pp. 343-346.

Gozlan, R. E.; St-Hilaire, S.; Feist, S. W.; Longshaw, M. & Peeler, E. J. (2006). The effect of microbial pathogens on the diversity of aquatic populations, notably in Europe. *Microbes and Infection,* Vol.8, pp. 1358–1364.

Gozlan, R. E.; St-Hilaire, S.; Feist, S. W.; Martin, P. & Kent, M. L. (2005). Biodiversity: disease threat to European fish. *Nature,* Vol.435, p. 1046.

Graham, N. A. J.; McClanahan, T. R.; MacNeil, M. A.; Wilson, S. K.; Polunin, N. V. C.; Jennings, S.; Chabanet, P.; Clark, S.; Spalding, M. D.; Letourneur, Y.; Bigot, L.; Galzin, R.; Ohman, M. C.; Garpe, K. C.; Edwards, A. J. & Sheppard, C. R. C. (2008). Climate Warming, Marine Protected Areas and the Ocean-Scale Integrity of Coral Reef Ecosystems. *PLoS ONE,* Vol.3, pp. e3039. doi:10.1371/journal.pone.0003039

Grotmol, S.; Bergh, O. & Totland, G. K. (1999). Transmission of viral encephalopathy and retinopathy (VER) to yolk-sac larvae of the Atlantic halibut *Hippoglossus hippoglossus*: occurrence of nodavirus in various organs and a possible route of infection. *Diseases of Aquatic Organisms,* Vol.36, pp. 95–106.

Grotmol, S.; Nerland, A. H.; Biering, E.; Totland, G. K. & Nishizawa, T. (2000). Characterization of the capsid protein gene from a nodavirus strain affecting the Atlantic halibut *Hippoglossus hippoglossus* and design of an optimal reverse-transcriptase polymerase chain reaction (RT-PCR) detection assay. *Diseases of Aquatic Organisms,* Vol.39, pp. 79–88.

Grotmol, S.; Totland G. K.; Thorud, K. & Hjeltnes, B. K. (1997). Vacuolating encephalopathy and retinopathy associated with a nodavirus-like agent: a probable cause of mass mortality of cultured larval and juvenile Atlantic halibut *Hippoglossus hippoglossus*. *Diseases of Aquatic Organisms*, Vol.29, pp. 85-97.

Hedrick, R. P. (1998). Relationships of the host, pathogen and Environment: implications for disease of cultured and wild fish populations. *Journal of Aquatic Animal Health*, Vol.10, pp. 107-111.

Huntington, T. C.; Roberts, H.; Cousins, N.; Pitta, V.; Marchesi, N.; Sanmamed, A.; Hunter-Rowe, T.; Fernandes, T. F.; Tett, P.; McCue, J. & Brockie, N. (2006). Some Aspects of the Environmental Impact of Aquaculture in Sensitive Areas. Report to the DG Fish and Maritime Affairs of the European Commission. Rep. No. 221-EC/R/02/B. Poseidon Aquatic Resource Management Ltd, Windrush, Warborne Lane, Portmore, Lymington, Hampshire SO41 5RJ, UK.

Jarman, W. M.; Simon, M.; Norstrom, R. J.; Burns, S. A.; Bacon, C. A.; Slmonelt, B. R. T. & Risebrough, R. W. (1992). Global distribution of tris(4-chlorophenyl) methanol in high trophic level birds and mammals. *Environmental Science and Technology*, Vol.26, pp. 1770–1774.

Katsanevakis, S.; Issaris, Y.; Poursanidis, D. & Thessalou-Legaki, M. (2010). Vulnerability of marine habitats to the invasive green alga *Caulerpa racemosa* var. *cylindracea* within a marine protected area. *Marine Environmental Research*, Vol.70, pp. 210-218.

Keller, B. D.; Gleason, D. F.; McLeod, E.; Woodley, C. M.; Airamé, S.; Causey, B. D.; Friedlander, A. M.; Grober-Dunsmore, R.; Johnson, J. E.; Miller, S. L. & Steneck, R. S. (2009). Climate Change, Coral Reef Ecosystems, and Management Options for Marine Protected Areas. *Environmental Management*, Vol.44, pp. 1069-1088.

Kent, M. L. (2000). Marine net-pen farming leads to infections with some unusual parasites. *International Journal for Parasitology, Vol.*30, pp. 321-326.

Krkošek, M.; Connors, B. M.; Ford, H.; Peacock, S.; Mages, P.; Ford, J. S.; Morton, A.; Volpe, J. P.; Hilborn, R.; Dill, L. M. & Lewis, M. A. (2011). Fish farms, parasites, and predators: implications for salmon population dynamics. *Ecological Applications*, Vol.21, pp. 897–914.

Krkošek, M.; Lewis, M. A.; Morton, A.; Neil Frazer, L. & Volpe, J. P. (2006). Epizootics of wild fish induced by farm fish. *Proceedings of the National Academy of Sciences*, Vol.103, pp. 15506–15510.

Kuris, A. M. & Lafferty, K. D. (1992). Modeling crustacean fisheries – effects of parasites on management strategies. *Canadian Journal of Fisheries and Aquatic Sciences*, Vol.49, pp. 327-336.

Lessios, H. A. (1988). Mass mortality of *Diadema antillarum* in the Caribbean: what have we learned? *Annual Review of Ecology and Systematics*, Vol.19, pp. 371-393.

Littler, M. M. & Littler, D. S. (1995). Impact of CLOD pathogen on Pacific coral reefs. *Science*, Vol.267, pp. 1356-1360.

Loganathan, B. G. & Kannan, K. (1994). Global organochlorine contamination trends: an overview. *Ambio*, Vol.23, 187-191.

Lubchenco, J.; Palumbi, S. R.; Gaines, S. D. & Andelman, S. (2003). Plugging a hole in the ocean: The emerging science of marine reserves. *Ecological Applications*, Vol.13, pp. S3–S7.

Maltese, C.; Antonetti, P.; Quartesan, R.; Ormelli, S.; Borghesan, F.; Manfrin, A.; Selli, L.; Castiglione, F.; Ferrantelli, V.; Guercio, A. & Bovo, G. (2005). Isolation of viral encephalopathy and retinopathy virus (VERV) from wild marine fish species in the Mediterranean sea. In: 12th International Conference "Diseases of Fish and

Shellfish", European Association of Fish Pathologists, Copenhagen (Denmark), 11-16 September 2005: Poster 8.19.

McCallum, H.; Gerber, L. & Jani, A. (2005). Does Infectious Disease Influence the Efficacy of Marine Protected Areas? A Theoretical Framework. *Journal of Applied Ecology,* Vol.42, pp. 688-698.

McClanahan, T. R. (2000). Bleaching Damage and Recovery Potential of Maldivian Coral Reefs. *Marine Pollution Bulletin,* Vol.40, pp. 587–597.

McVicar, A. H. (1997). Interaction of pathogens in aquaculture with wild fish populations. *Bulletin of European Association of Fish Pathologists,* Vol.17, pp. 197-200.

Mori, K.; Nakai, T.; Nagahara, M.; Muroga, K.; Mekuchi, T. & Kanno, T. (1991). A viral disease in hatchery-reared larvae and juveniles of redspotted grouper. *Fish Pathology,* Vol.26, pp. 209-210.

Munday, B. L.; Kwang, J. & Moody, N. (2002). Betanodavirus infections of teleost fish: a review. *Journal of Fish Diseases,* Vol.25, pp.127–142.

Mushiake, K.; Nishizawa, T.; Nakai, T.; Furusawa, I.; & Muroga, K. (1994). Control of VNN in striped jack: selection of spawners based on the detection of SJNNV gene by polymerase chain reaction (PCR). *Fish Pathology,* Vol.29, pp. 177–182.

Naylor, R.; Eagle, J. & Smith, W. (2003). Salmon aquaculture in the Pacific North-west: a global industry with local impacts. *Environment,* Vol.45, pp. 18–39.

Naylor, R.; Hindar, K.; Fleming, I. A.; Goldburg, R.; Williams, S.; Volpe, J.; Whoriskey, F.; Eagle, J.; Kelso, D. & Mangel, M. (2005). Fugitive salmon: assessing the risks of escaped fish from net-pen aquaculture. *BioScience,* Vol.55, pp. 427–37.

Nguyen, H. D.; Nakai, T. & Muroga, K. (1996). Progression of striped jack nervous necrosis virus (SJNNV) infection in naturally and experimentally infected striped jack *Pseudocaranx dentex* larvae. *Diseases of Aquatic Organisms,* Vol.24, pp. 99-105.

Nishizawa, T.; Muroga, K. & Arimoto, M. (1996). Failure of the polymerase chain reaction (PCR) method to detect striped jack nervous necrosis virus (SJNNV) in striped jack *Pseudocaranx dentex* selected as spawners. *Journal of Aquatic Animal Health,* Vol.8, pp. 332–334.

Nowak, B. F.; Dawson, D.; Basson, L.; Deveney, M. & Powell, M. D. (2004). Gill histopathology of wild marine fish in Tasmania: potential interactions with gill health of cultured Atlantic salmon, *Salmo salar* L. *Journal of Fish Diseases,* Vol.27, pp. 709-717.

Panzarin, V.; Patarnello, P.; Mori, A.; Rampazzo, E.; Cappellozza, E.; Bovo, G. & Cattoli, G. (2010). Development and validation of a real-time TaqMan PCR assay for the detection of betanodavirus in clinical specimens. *Archives of Virology,* Vol.155, pp. 1193-1203.

Paperna, I. (1998). Mariculture versus environment: risks and impacts. *Biologia Marina Mediterranea,* Vol.5, pp. 381-389.

Pauly, D. (2005). Global trends in world fisheries: Impacts on marine ecosystems and food security. *Philosophical Transactions of the Royal Society* B, Vol.360, pp. 5–12.

Péducasse, S.; Castric, J.; Thiéry, R.; Jeffroy, J.; Le Ven, A. & Baudin Laurencin, F. (1999). Comparative study of viral encephalopathy and retinopathy in juvenile seabass *Dicentrarchus labrax* infected in different ways. *Diseases of Aquatic Organisms,* Vol.36, pp. 11–20.

Pusceddu, A.; Fraschetti, S.; Holmer, M.; Mirto, S. & Danovaro, R. (2007). Effects of intensive mariculture on sediment biochemistry. *Ecological Applications,* Vol.17, pp. 1366–1378.

Rasmussen, E. (1997). The wasting disease of eelgrass (*Zostera marina*) and its effects on environmental factors and fauna, In: *Seagrass ecosystems: a scientific perspective*, McRoy, C. P. & Helfferich, C., (Ed), pp. 1-51, Marcel Dekker, New York, USA.

Roberts, C. M.; Hawkins, J. P. & Gell, F. (2005). The role of marine reserves in achieving sustainable fisheries. *Philosophical Transactions of the Royal Society* B, Vol.360, pp. 123–132.

Sapkota, A.; Sapkota, A. R.; Kucharski, M.; Burke, J.; McKenzie, S.; Walker, P. & Lawrence, R. (2008). Aquaculture practices and potential human health risks: Current knowledge and future priorities. *Environment International*, Vol.34, pp. 1215–1226.

Sepulveda, F.; Marin, S. L. & Carvajal, J. (2004). Metazoan parasites in wild fish and farmed salmon from aquaculture sites in southern Chile. *Aquaculture*, Vol.235, pp. 89-100.

Steinbeck, J. R.; Groff, J. M.; Friedman, C. S.; McDowell, T. & Hedrick, R. P. (1992). Investigations into mortality among populations of the California black abalone, *Haliotis cracherodii*, on the central coast of California, USA, In: *Abalone of the World*, Shephard, S. A.; Tegner, M. J.; Guzmán del Próo, S. A., (Ed), pp. 203–213, Blackwell Scientific, Oxford.

Sweetman, E.; Sweetman, J.; Le Breton, A. & Grisez, L. (1996). Nodavirus: a review of the findings of the XIV/NODA/95 investigation. In: "Seabass and seabream culture: problems and prospects" European Aquaculture Society (ed.). Verona, Italy, 16-18 October 1996, pp. 87-101.

Tanaka, S.; Aoki, H. & Nakai, T. (1998). Pathogenicity of the nodavirus detected from diseased sevenband grouper *Epinephelus septemfasciatus*. *Fish Pathology*, Vol.33, pp. 31–36.

Terlizzi, A.; De Falco, G.; Felline, S.; Fiorentino, D.; Gambi, M. C. & Cancemi, G. (2010). Effects of marine cage aquaculture on macrofauna assemblages associated to *Posidonia oceanica* meadows. *Italian Journal of Zoology*, Vol.77, pp. 362–371.

Terlizzi, A.; Delos, A. L.; Garaventa, F.; Faimali, M. & Geraci, S. (2004). Limited effectiveness of marine protected areas: imposex in *Hexaplex trunculus* (Gastropoda, Muricidae) populations from Italian marine reserves. *Marine Pollution Bulletin*, Vol.48, pp. 188–192.

Thiéry, R.; Cozien, J.; De Boisséson, C.; Kerbart-Boscher, S. & Névarez, L. (2004). Genomic classification of new betanodavirus isolates by phylogenetic analysis of the coat protein gene suggests a low host-fish species specificity. *Journal of General Virology*, Vol.85, pp. 3079-3087.

Trowbridge, C. D. (1995). Establishment of the green alga *Codium fragile* spp. *Tomentosoides* on New Zealand rocky shores: current distribution and invertebrate grazers. *Journal of Ecology*, Vol.83, pp. 949-965.